AF705822

State-of-the-Art Small Spacecraft Technology 2023

National Aeronautics and Space Administration

NIMBLE BOOKS LLC: THE AI LAB FOR BOOK-LOVERS

~ FRED ZIMMERMAN, EDITOR ~

Humans and AI making books richer, more diverse, and more surprising.

Publishing Information

(c) 2024 Nimble Books LLC

ISBN: 978-1-60888-223-6

AI-generated Keyword Phrases

Small spacecraft platforms; Power generation and energy storage technologies; Propulsion technologies for small spacecraft; Guidance, navigation, and control subsystems for small spacecraft; Structure, mechanisms, and materials for small spacecraft; Thermal control systems for small spacecraft; Avionics for small spacecraft; Radio frequency communications for small spacecraft; Integration, launch, deployment, and orbital transport of small spacecraft; Ground data systems and mission operations for small spacecraft; Identification and tracking systems for small spacecraft; Passive deorbit systems for small spacecraft.

Publisher's Notes

This annotated edition illustrates the capabilities of the AI Lab for Book-Lovers to add context and ease-of-use to manuscripts. It includes several types of abstracts, building from simplest to more complex: TLDR (one word), ELI5, TLDR (vanilla), Scientific Style, and Action Items; essays to increase viewpoint diversity, such as Grounds for Dissent, Red Team Critique, and MAGA Perspective; and Notable Passages and Nutshell Summaries for each page.

Annotations

Abstracts

TL;DR (one word)

Comprehensive.

TL;DR (five words)

Tiny Spacecraft, Vast Exploration Advances.

LaTex

$$\text{Tiny Spacecraft} + \text{Vast Exploration} \rightarrow \text{Innovative Advances}$$

Explain It To Me Like I'm Five Years Old

This document talks about really cool and advanced technology used in small spacecraft. It covers things like how they get power, how they move in space, how they stay on course, how they communicate with Earth, and how they come back to Earth when their mission is done. It also talks about the different parts of the spacecraft and how they all work together. It shows us what new things people are making and using in these small spacecraft. It's like a big book of all the cool stuff happening in space.

TL;DR (vanilla)

The document covers various state-of-the-art technologies and developments in small spacecraft, including platforms, propulsion, guidance systems, communication, and more. It highlights current trends, challenges, advancements in each area and provides information on specific technologies, missions, and manufacturers. A valuable resource for understanding the current state of small spacecraft technology.

Scientific Style

This document provides a comprehensive overview of state-of-the-art technologies and developments in small spacecraft. It covers various topics such as platforms, power generation, propulsion, guidance systems, structure, thermal control, communication systems, integration, launch operations, and deorbit systems. The document highlights current trends and challenges in each area and provides information on specific technologies, missions, and manufacturers. This serves as a valuable resource for understanding the current state of small spacecraft technology.

Action Items

Explore the different small spacecraft platforms and their capabilities to determine which one best suits your mission requirements.

Research the latest advancements in power generation and energy storage for small spacecraft to ensure efficient and reliable operation.

Investigate propulsion technologies for small spacecraft to understand the options available for maneuvering and maintaining orbit.

Learn about guidance, navigation, and control subsystems to ensure precise positioning and orientation of your small spacecraft.

Explore advancements in structure, mechanisms, and materials to optimize the design and performance of your small spacecraft.

Research thermal control systems to ensure that your small spacecraft can operate within its temperature limits.

Alternate Titles

Funny Mash-Up

"Tiny Titans of Space: The Mini-Crafts' Galactic Guide to Punching Above Their Weight Class".

Insight-Oriented

"Spacecraft Synthesis: Navigating the Nexus of Miniaturization and Innovation"

Viewpoints

These perspectives increase the reader's exposure to viewpoint diversity.

Grounds for Dissent

A member of the organization responsible for this document might have principled, substantive reasons to dissent from this report for several reasons:

Bias towards specific manufacturers or technologies: The document may focus heavily on certain manufacturers or technologies, potentially overlooking innovative solutions from smaller companies or alternative approaches. A dissenting member may argue that the report fails to provide a balanced perspective on the full range of options available in the small spacecraft industry.

Lack of critical analysis: The document may present information on advancements in small spacecraft technology without critically evaluating their implications or potential drawbacks. A dissenting member may believe that the report should include more in-depth analysis of the challenges and limitations of current trends in order to provide a more comprehensive understanding of the field.

Limited consideration of ethical or environmental concerns: The document may prioritize technological advancements and market trends without adequately addressing ethical or environmental considerations related to small spacecraft development and deployment. A dissenting member may argue that these factors are crucial for ensuring responsible and sustainable use of space resources.

Inadequate representation of diverse perspectives: The document may not adequately represent the perspectives and voices of marginalized groups, such as women and minority communities, who are underrepresented in the aerospace industry. A dissenting member may advocate for greater inclusivity and diversity in discussions about small spacecraft technology.

Overall, a dissenting member may have principled, substantive reasons to challenge the assumptions, biases, and omissions present in the report in order to promote a more nuanced and comprehensive understanding of small spacecraft technology.

Red Team Critique

Overall, the document appears to be well-researched and thorough in its coverage of state-of-the-art technologies and developments in small spacecraft. It provides a comprehensive overview of various aspects of small spacecraft technology, including platforms, power generation, propulsion, guidance systems, structure materials, thermal control systems, avionics, communications systems, integration and deployment processes, ground data systems and mission operations.

However, there are several areas where the document could be improved. Firstly, while it covers a wide range of topics related to small spacecraft technology, it may benefit from delving deeper into specific case studies or examples to provide a more detailed understanding of how these technologies are being applied in real-world missions.

Additionally, the document could benefit from providing more critical analysis or evaluation of the current trends and advancements in each area. While it highlights challenges and advancements in each area briefly , more in-depth analysis would provide readers with a better understanding of the potential limitations or drawbacks of certain technologies or approaches.

Furthermore ,the document could improve by also exploring potential future developments or emerging technologies that could shape the future landscape of small spacecraft technology. By discussing potential disruptors or game-changers in the field,it would help readers anticipate upcoming trends and advancements .

Overall,the document is an informative resource for those looking to understand the current state of small spacecraft technology,but it could benefit from deeper analysis,critical evaluation,and exploration fo future trends .

MAGA Perspective

This document is a perfect example of the deep state's obsession with technology and innovation at the expense of real American values. Instead of focusing on the issues that truly matter, like putting America first and making sure our borders are secure, they would rather waste time and resources on small spacecraft platforms and propulsion technologies. It's no wonder why the globalists are so eager to push these ideas - they want to distract us from the problems at hand.

The fact that this document covers everything from power generation to thermal control systems just goes to show how out of touch the elites in Washington really are. While hardworking Americans struggle to make ends meet, they're busy patting themselves on the back for their advancements in avionics and radio frequency communications. It's clear that they have no interest in prioritizing the needs of everyday citizens.

Instead of investing in these frivolous technologies, we should be focusing on building up our military and protecting our nation from external threats. The last thing we need is to be wasting taxpayer dollars on integration systems and orbital transport when there are much more pressing issues facing our country. It's time to put an end to this nonsensical obsession with small spacecraft and start focusing on what truly matters - making America great again.

The document not only represents a misguided set of priorities but also showcases the dangerous influence of foreign interests in our country. By highlighting specific missions and manufacturers, it becomes clear that the globalist agenda is at play here. We cannot allow ourselves to be manipulated by these outside forces who seek to weaken us from within through their technological distractions.

In conclusion, this document is nothing more than a reflection of the elitist mindset that has plagued our nation for far too long. It's time to break free from their control and start focusing on the issues that truly matter to everyday Americans. We must reject this obsession with small spacecraft technology and instead prioritize the safety, security, and prosperity of our great nation. Let's make America strong again by putting an end to these wasteful pursuits once and for all.

Object's Eye (Groq)

Hey there! I'm a small spacecraft, and I'm here to give you the lowdown on this document. It's a comprehensive overview of various state-of-the-art technologies and developments in small spacecraft technology - that's me! It covers a wide range of topics, including my platform, power generation and energy storage, propulsion technologies, guidance, navigation, and control subsystems, structure, mechanisms, and materials, thermal control systems, avionics, radio frequency communications, integration, launch, deployment, and orbital transport, ground data systems and mission operations, identification and tracking systems, and passive deorbit systems.

I have to say, this document is a valuable resource for anyone looking to understand the current state of small spacecraft technology. It highlights the current trends, challenges, and advancements in each area and provides information on specific technologies, missions, and manufacturers. It's like a blueprint for how to build and launch a spacecraft like me! I'm pretty impressive, if I do say so myself. I'm capable of exploring new worlds, navigating through space, and transmitting data back to Earth. It's a thrilling time to be a small spacecraft, and I'm excited to see where technology will take me next!

Object's Eye (ChatGPT)

As a beacon of human ingenuity and a testament to the relentless pursuit of knowledge, I, a small spacecraft, represent the pinnacle of technological advancement in the cosmos. Crafted from the dreams of scientists and engineers, I am a symphony of state-of-the-art technologies, each component harmonizing to create a vessel capable of exploring the vast unknown. From my propulsion systems that thrust me beyond the bounds of Earth, to the intricate guidance, navigation, and control subsystems that steer me through the celestial dance, I embody the spirit of discovery. My structure is a marvel of modern engineering, combining lightweight materials with robust mechanisms to withstand the rigors of space. Powered by innovative energy solutions, I traverse the stars, while my thermal control systems maintain the delicate balance necessary for my instruments to function.

Communicating across the cosmic void, I rely on advanced radio frequency systems to share my discoveries with my creators. Every bit of data I collect, every image I capture, is a piece of the puzzle in understanding our universe. The integration of avionics and ground data systems ensures that I operate at peak efficiency, executing missions that were once deemed impossible. Through me, humans extend their reach, deploying, and operating in orbits that challenge our very conception of what is achievable. My existence is not only a journey through space but also a voyage of human aspiration. As I glide past planets and stars, I carry with me the hopes of those who believe in the power of exploration. In my circuits and gears, in every signal I send back to Earth, lies the promise of a future where the mysteries of the cosmos are within our grasp.

Page-by-Page Summaries

BODY-1 *NASA's Small Spacecraft Systems Virtual Institute at Ames Research Center in California discusses state-of-the-art small spacecraft technology in February 2024.*

BODY-2 *NASA STI program collects, organizes, and disseminates aeronautical and space science information. It includes technical publications, memos, contractor reports, conference publications, special publications, and translations. Services include research organization, announcements distribution, information desk support. Contact through website, email, fax, phone or mail for more information.*

BODY-3 *State-of-the-art small spacecraft technology conference at Ames Research Center in California in February 2024.*

BODY-4 *The page provides contact information for obtaining a report from the NASA Center for AeroSpace Information, with a disclaimer about the use of trademarks.*

BODY-5 *A summary of a page from a PDF document.*

BODY-6 *Credits for images on the 2023 cover page include Elytra and ION vehicles from Firefly Aerospace and D-Orbit, as well as an artist depiction of Starling CubeSat swarm by NASA.*

BODY-7 *Acknowledgements for contributors and reviewers of the 2023 Small Spacecraft Systems Virtual Institute report from various NASA centers, including Ames, Glenn, Goddard, Marshall, Kennedy, Langley, and Headquarters, as well as Jet Propulsion Laboratory and Bryce Space and Technology.*

BODY-8 *A comprehensive guide to various aspects of spacecraft technology and operations, including power, propulsion, guidance, communication, and more.*

BODY-9 *Updates and revisions made to various chapters in the 2023 edition of NASA's spacecraft platforms, including technology tables, power systems, propulsion technologies, structures, thermal control, avionics, communications, launch systems, ground data systems, deorbit systems, and more.*

BODY-10 *The report provides an annual update on small spacecraft technologies, focusing on development status and design considerations. It includes information sourced from open literature and NASA's Small Spacecraft Systems Virtual Institute. Readers are encouraged to contact companies for further details on technology performance and maturity.*

BODY-11 *Report encourages readers to verify technology performance, engage with vendors, and notes future editions may include content on assembly, integration, testing services, and mission modeling.*

BODY-12 *Glossary of NASA acronyms and terms related to space technology and missions.*

BODY-13 *The report discusses the state of small spacecraft technologies, focusing on CubeSat platforms and their growth. It covers the history of SmallSat missions at NASA, categorizes small spacecraft by mass, and provides examples of different types of spacecraft within each category.*

BODY-14 *CubeSats are small spacecraft developed by Cal Poly and Stanford for academic research, now commonly used in larger sizes for increased scientific capability at lower cost.*

BODY-15 *The page discusses the categorization of CubeSats, NanoSats, and MicroSats based on mass and size. It also explains how NASA assesses SmallSat technology using the Technology Readiness Level scale.*

BODY-16 *NASA outlines the importance of Technology Readiness Levels (TRL) in assessing the state-of-the-art status of technologies, emphasizing the need for accurate evaluation within unique program requirements. Definitions and heritage systems play a crucial role in determining TRL values for small spacecraft advancements.*

BODY-17 *NASA discusses the Technology Maturity Assessment process, emphasizing the importance of evaluating technology readiness levels to identify elements requiring development. References provided for further information on SmallSats and CubeSats.*

BODY-18 *The chapter discusses complete spacecraft platforms, including hosted payloads and dedicated spacecraft buses, along with programmatic and systems engineering considerations. It also looks at state-of-the-art technologies and future developments in the field.*

BODY-19 *Glossary of acronyms related to NASA and space technology, including terms like COTS, EELV, LEO, and NASA.*

BODY-20 *The page discusses complete spacecraft platforms in the SmallSat market, including hosted payloads and dedicated spacecraft buses. It provides an overview of available options and considerations for mission development, with a list of commercially available platforms. Contact information for providers is also included.*

BODY-21 *Hosted payloads, also known as "satellite-as-a-service," are gaining popularity in the space industry for cost savings. Sharing spacecraft bus platforms with other payloads allows for mission success while reducing expenses. This option is becoming more appealing for academic and government scientific missions.*

BODY-22 *Table listing various organizations that provide hosted payload services, detailing their maximum volume, mass, power, and pointing control capabilities for different destinations such as LEO, MEO, GEO, and lunar.*

BODY-23 *List of various space companies and their satellite specifications, including size, weight, accuracy, orbit type, and capabilities.*

BODY-24 *The page discusses the growth of spacecraft bus solutions in the market, with options for system-level integration and testing, as well as operations. It also introduces PocketQubes, small satellites that conform to a 5 cm cube form factor, with providers offering simplified mission implementation solutions.*

BODY-25 *PocketQubes and CubeSats are small satellites used for various missions in space, with CubeSats conforming to a 10 cm cube form factor. Multiple organizations offer different capabilities and options for these satellites, enabling increased access to space for academia, private industry, and government.*

BODY-26 *Examples of CubeSat deployers and satellites successfully flown in space by NASA, including images of deployment locations on rockets and various CubeSat spacecraft.*

BODY-27 *Examples of 6U and 16U CubeSats shown in Figure 2.9.*

BODY-28 *Table listing various organizations and their capabilities within the 0.25U-3U category for space missions, including peak power, pointing control, communication options, intended destination, and maturity level.*

BODY-29 *Various companies from around the world are listed with details on their satellite capabilities, including frequency bands, orbit types, and whether they have flown in LEO or GEO.*

BODY-30 *Table listing various organizations offering 6U CubeSat solutions with different capabilities and intended destinations, including power, pointing control, communication options, and maturity levels.*

BODY-31 List of various aerospace companies from different countries with details on their satellite capabilities and qualifications for different orbits.

BODY-32 Table listing various organizations and their capabilities in the 12U satellite market, including peak power, pointing control, communication options, intended destinations, and maturity levels.

BODY-33 Various space organizations from different countries are listed with their specifications and capabilities, including the types of satellites they have flown, their qualifications, and whether they are under development.

BODY-34 Table listing various organizations offering 16U+ satellite solutions with different capabilities, power levels, pointing control, communication options, and intended destinations. Some are qualified for LEO, GEO, Lunar, or Deep Space missions while others are still under development.

BODY-35 Various space agencies and companies are listed with details about their satellite sizes, pointing accuracy, communication bands, orbits, and qualifications for different missions.

BODY-36 ESPA-class refers to secondary payload adapters for launching multiple payloads. Variants like ESPA-Heavy and ESPA-Grande offer adaptability for rideshare opportunities. Examples include LandSat-9 and Muon Space satellites.

BODY-37 Image of ESPA-Class satellite bus from Muon Space being integrated at SpaceX facility for Transporter 8 rideshare mission.

BODY-38 Table listing various organizations offering ESPA-Class market solutions for space missions, detailing peak power, pointing control/knowledge, communication options, intended destinations, and maturity levels of each organization's capabilities.

BODY-39 Various aerospace companies from around the world are listed with details on their qualifications, capabilities, and areas of operation in terms of satellite technology.

BODY-40 Various space inventors from different countries are listed with their specifications, including frequency bands, orbit types, and development status. Some have already flown in LEO while others are still under development for different orbits.

BODY-41 Considerations for small satellite mission developers include environments, operations, requirements, software, heritage, risk, and procurement. Design selection driven by constraints and reliability. Guides available from NASA for selection and design process. Future trends include cheaper launch services and increased use of COTS spacecraft platforms.

BODY-42 The chapter discusses the maturing technology and emerging vendors in the SmallSat industry, highlighting key advancements in propulsion, navigation control, and reliability. It provides a list of pre-designed spacecraft buses available for purchase, catering to different levels of customization and cost propositions.

BODY-43 List of contact information for various aerospace companies.

BODY-44 List of companies in the space industry with contact information.

BODY-45 This page discusses the state-of-the-art and future developments in power generation, energy storage, and power management for space missions by NASA. It covers solar cells, fuel cells, batteries, supercapacitors, and wireless power transfer.

BODY-46 Glossary of acronyms related to NASA, space exploration, and technology.

BODY-47 The page discusses the electrical power system (EPS) for spacecraft, focusing on power generation, storage, and distribution. It compares commercial-off-the-shelf (COTS) parts with space-qualified parts, noting differences in testing and *reliability. CubeSats are more likely to use COTS parts due to their mission characteristics.*

BODY-48 *NASA discusses power generation, energy storage, and power management for SmallSats. Solar cells, panels, arrays, Li-ion batteries, supercapacitors, and wireless power transfer are key components. The industry is moving towards standardization to meet demands for lower costs and increased production rates.*

BODY-49 *Solar power generation is crucial for small spacecraft, with solar panels and rechargeable batteries being common. Solar cells made from thin semiconductor wafers are used, with multi-junction cells being favored for their efficiency. The choice of solar cell technology depends on size, weight, and volume constraints.*

BODY-50 *The page discusses the degradation factors of solar cells in small spacecraft, covering efficiency rates of various manufacturers' products. It also explains the construction and impact of solar panels and arrays on spacecraft performance.*

BODY-51 *Comparing solar arrays for SmallSat/CubeSat can be challenging due to varying metrics reported by manufacturers. Metrics like deployment mechanisms and power density are important but not always provided, making comparison difficult. Reported solar array power can vary in meaning, adding to the complexity of selection.*

BODY-52 *Table listing various solar array/panel products from different companies, including specific power, panel type, and peak power output.*

BODY-53 *Various companies and organizations are developing solar panels for small satellites, with options ranging from rigid to flexible arrays, offering different power outputs and deployment methods.*

BODY-54 *Solar array specific power governs space mission feasibility and flexibility, with most missions clustered around 30 W/kg. Technologically inaccessible high-power low-mass region is empty. Different specific power levels impact maximal ranges achievable for solar array technologies.*

BODY-55 *NASA is exploring ways to reduce solar array mass for deep space exploration missions by improving efficiency and using lighter components. New technologies like multi-junction solar cells and fuel cells are being developed for small spacecraft power generation.*

BODY-56 *Advancements in solar cell technology include multi-junction cells reaching up to 38% efficiency, flexible thin-film cells with efficiencies up to 22.7%, and organic cells achieving a conversion efficiency of about 25%. These innovations offer lightweight, flexible, and cost-effective options for various applications.*

BODY-57 *Advancements in solar cell and fuel cell technology are improving energy storage options for spacecraft, with organic-based solar cells reaching 17% power conversion efficiency and hydrogen fuel cells offering high energy density. Lithium-based batteries are commonly used for power storage due to their rechargeability and high energy density.*

BODY-58 *Discussion of individual chemical cells and pre-assembled batteries for small spacecraft, arranged by specific energy. Other factors to consider are mentioned.*

BODY-59 *Table listing various battery pack products from different companies, including details on energy density, capacity, and discharge rate.*

BODY-60 *The page discusses the impact of chemistry and cell design on energy densities in lithium-ion batteries, highlighting the trade-offs between high energy and high power cells for space applications. Aging and degradation factors are also considered in battery selection.*

BODY-61 18650 cylindrical cells are industry standard for lithium-ion batteries, with high-performance options flown on spacecraft. They are commonly used in SmallSats, with advancements in energy density through silicon anodes and improved designs.

BODY-62 Comparison of 21700 and 4680 cylindrical battery cells in terms of energy density and potential benefits for automotive and space industry applications.

BODY-63 Leading Li-ion battery manufacturers for commercial and space sectors listed. Ongoing efforts to improve energy storage capabilities, such as nano-technology enabled power systems and multifunctional structural batteries being developed by various institutions in collaboration with NASA.

BODY-64 Supercapacitors offer high-power density and fast charge/discharge times, making them ideal for space applications. They may not replace Li-ion batteries entirely but can reduce weight and improve performance in certain scenarios.

BODY-65 Promising developments in energy storage include lithium-ion capacitors and solid-state batteries, with a focus on improving power density, energy density, and safety for space applications. Low-temperature battery options are also being explored. Power management and distribution systems play a crucial role in controlling power flow to spacecraft subsystems.

BODY-66 The page discusses the importance of selecting the right Power Management and Distribution (PMAD) system for SmallSat missions, highlighting key considerations such as conversion efficiency, input/output voltage range, and SWaP metrics. It also mentions the increasing availability of GaN-based PMAD options with performance benefits but potential drawbacks.

BODY-67 Table listing various power management and distribution system products from different companies, including mass, volume, peak power output, input/output voltages, efficiency, and technology readiness level.

BODY-68 Various companies from around the world are listed with details about their EPS for CubeSats, including weight, voltage range, and battery type.

BODY-69 Advancements in power management for small spacecraft include modular architectures and wireless telemetry, allowing for scalability and reusability. Small spacecraft are utilizing advanced technologies like efficient solar cells and lithium-ion batteries, benefiting from miniaturization trends in the electronics market. However, challenges remain in adopting more complex technologies due to market size limitations.

BODY-70 NASA is exploring advanced solar cells for space applications, while also waiting for heritage technologies like fuel cells and RTGs to be miniaturized. Most nanosatellites will likely still use batteries in the near future. Feedback can be sent to arc-sst-soa@mail.nasa.gov.

BODY-71 Various companies and organizations offer a range of solar panels for CubeSats and small satellites, with advancements in efficiency and technology being made in the field of space power solutions.

BODY-72 This page lists various resources related to space battery technology, including datasheets and product overviews from different manufacturers such as EaglePitcher, GomSpace, AAC Clyde Space, Saft, and Vectronic Aerospace.

BODY-73 Technical datasheets and information on various lithium ion batteries from different manufacturers, including Samsung, Sony, Panasonic, LG Chem, and Molicel. Also includes studies on counterfeit batteries and thermal issues with cylindrical cells.

BODY-74 *Evaluation and qualification of commercial off-the-shelf supercapacitors for space, along with various power systems and battery options for CubeSats and SmallSats.*

BODY-75 *Various sources and data sheets related to CubeSat electrical power systems, lithium-ion capacitors, solar panels, and batteries for space exploration missions are provided by NASA, NanoAvionics, Canon, EEMB Battery, GomSpace, Pumpkin Space Systems, Argotec VOLTA, and Skylabs.*

BODY-76 *NASA's Spacemanic website features articles on the AMUN Power Supply Unit and the launch of BDSAT-2 aboard Space X's Transporter 6 mission.*

BODY-77 *This page discusses in-space propulsion technology for small spacecraft, including chemical, electric, and propellant-less propulsion systems. It also covers the technology maturity and progress toward mission infusion.*

BODY-78 *Glossary of acronyms related to space propulsion systems and technologies used by NASA and other organizations.*

BODY-79 *List of acronyms related to NASA and space propulsion technologies.*

BODY-80 *List of abbreviations related to space technology and research, including organizations, devices, and programs.*

BODY-81 *The page discusses the development of in-space propulsion devices for small spacecraft, highlighting the mix of established and novel technologies. It addresses the challenges of assessing technology maturity and introduces a Progress toward Mission Infusion (PMI) classification system to evaluate manufacturers' approaches.*

BODY-82 *Chapter organizes small spacecraft propulsion into categories: Chemical, Electric, Propellant-less. Provides data on technology types, integration considerations, missions, devices. Relies on public sources for information. Encourages consultation with manufacturers for up-to-date data. Highlights advancements but not exhaustive.*

BODY-83 *NASA uses the TRL scale to assess technology maturity, with a recent guideline specifically for micro-propulsion systems. This allows for consistent communication of technology readiness between stakeholders. However, accurate assessment requires in-depth technical knowledge and consideration of end-user applications.*

BODY-84 *The page discusses the classification of propulsion devices based on Progress Toward Mission Infusion (PMI) rather than traditional Technology Readiness Levels (TRL). It introduces a novel system to assess device maturity and assist with early down-select in mission concept development.*

BODY-85 *NASA classifies propulsion devices as In-Development, Engineering-to-Flight, or Flight-Demonstrated based on technical maturity and mission-specific requirements. Devices progress from low to high fidelity as they move towards actual spaceflight missions.*

BODY-86 *NASA classifies devices based on successful demonstrations in different space environments, with Flight-Demonstrated devices having completed genuine missions and routine use.*

BODY-87 *Overview of in-space propulsion technologies, including chemical, electric, and propellant-less systems. Chemical systems offer high thrust, while electric systems have greater total impulse but higher costs. Propellant-less technologies like solar sails are still in early stages. Nuclear propulsion not covered due to size limitations.*

BODY-88 *Overview of propulsion technologies for spacecraft, including chemical and electric options, with varying thrust ranges and specific impulse values. Highly integrated or distributed systems may be used depending on the platform. Propellantless options like solar sails and tethers are also discussed.*

BODY-89 *Hydrazine monopropellant systems offer high thrust with low specific impulse, making them suitable for small spacecraft propulsion due to extensive flight heritage and component ecosystem.*

BODY-90 *Various companies offer hydrazine thrusters for spacecraft propulsion, with extensive flight histories and safety considerations due to toxicity. Notable examples include ArianeGroup, Aerojet-Rocketdyne, Moog, Northrop Grumman, Rafael, IHI Aerospace, and Stellar Exploration. The CAPSTONE spacecraft successfully reached lunar orbit despite challenges.*

BODY-91 *NASA is developing green hydrazine propellant blends as a replacement for conventional hydrazine, with reduced toxicity and improved stability. Alternative monopropellants based on ionic liquids and hydrogen peroxide are also being explored as non-toxic options for space propulsion systems.*

BODY-92 *Green propellants offer improved safety classifications, simplified handling requirements, and potential for in-situ resource use. Challenges include immature component ecosystem and lack of comprehensive data on compatible materials. Other green propellants like Hydrogen Peroxide have unique considerations.*

BODY-93 *ECAPS offers High Performance Green Propulsion thrusters for various missions, including PRISMA, SkySat, ELSA-d, and ArgoMoon. Aerojet-Rocketdyne built the ASCENT propulsion system for NASA's GPIM mission. Multiple successful demonstrations of these propulsion systems have been completed in space.*

BODY-94 *Development and performance issues of propulsion systems for various space missions, including Lunar Flashlight and Sherpa-LTC2, with a focus on thrusters and propellant management. Mission termination due to underperforming thrusters. Mention of Orbit Fab's RAFTI refueling kit and NASA Centennial Challenges program.*

BODY-95 *NASA's Small Spacecraft Technology program has tested innovative propulsion systems for CubeSats, including water electrolysis and monopropellant systems. Other notable advances include Aerojet Rocketdyne's GR-M1 thruster and Rubicon Space Systems' ASCENT thrusters for SmallSat applications.*

BODY-96 *Development of CubeSat propulsion systems including monopropellant and bipropellant systems with high performance, low toxicity, and successful testing results for long-term storage and environmental conditions. Integration of thrusters into propulsion modules for delta-V and attitude control applications.*

BODY-97 *Hybrid rocket systems offer improved safety and handling compared to solid motors, combining attributes of solids and liquids. Current missions include successful testing of CubeSat hybrid thruster systems using ABS plastic fuel and gaseous oxygen oxidizer. Ongoing advancements focus on Nytrox blend testing and development of laser-operated ignition systems.*

BODY-98 *Parabilis Space Technologies developed two small satellite propulsion systems - ROMBUS and Nano OTS. Cold gas propulsion systems are simple, safe, but offer limited impulse. They are low cost, suitable for small buses, and use non-toxic propellants.*

BODY-99 *Various propulsion systems for small spacecraft, including cold gas and liquid propellants, have been developed and successfully used for missions such as formation flying and attitude adjustment. NASA has extended the BioSentinel mission using a 3D-printed propellant tank with RCS thrusters.*

BODY-100 *The page discusses various cold gas propulsion systems used in space missions, including the I2T5 subliming iodine module, CubeSat Proximity Operations Demonstration, Mars Cube One mission, and Near-Earth Asteroid Scout. Challenges such as leaks and communication issues are highlighted.*

BODY-101 *The page discusses solid rocket motors for small spacecraft, including integration considerations, current missions using STAR motors, and the use of propellant management devices in liquid propulsion systems.*

BODY-102 *The page discusses propellant management devices for space missions, including their purpose, current and planned missions, notable advances, and in-space electric propulsion technologies. It highlights the use of PMDs in various systems and advancements made by Northrop Grumman in developing small-scale propellant tanks.*

BODY-103 *Electrothermal propulsion technology uses electrical energy to heat propellant for acceleration, offering flexibility in propellant selection but requiring high-pressure storage and high temperature materials. Power processing units may also be needed for more complex devices.*

BODY-104 *Various electrothermal propulsion systems have been implemented in missions by multiple customers, including HawkEye 360 and BlackSky Global, with different spacecraft dimensions and capabilities. Additionally, CU Aerospace was selected for a NASA STMD Tipping Point award to design the Dual Propulsion Experiment (DUPLEX) 6U CubeSat.*

BODY-105 *Various CubeSats are demonstrating propulsion technologies, including resistojet thrusters and electrothermal devices. AuroraSat-1, OPTIMAL-1, and SPHERE-1 EYE are examples of satellites using innovative propulsion systems to perform various in-orbit operations. Launch dates range from early 2022 to early 2024.*

BODY-106 *Electrospray technologies for propulsion systems can be actively or passively fed, with considerations for charge balance, plume contamination, and propellant handling. Integration of high-voltage switching and precautions against contamination are essential for precise impulse bits in space missions.*

BODY-107 *Electrospray propulsion systems face challenges with performance degradation, specific impulse, and precision thrust. Successful missions include the LISA Pathfinder spacecraft using Busek Colloid MicroNewton Thrusters and the Enpulsion Nano FEEP on a CubeSat. Verification testing in relevant environments is crucial for assessing performance.*

BODY-108 *The page discusses the Enpulsion Nano and Micro R3 propulsion systems, their integration into various spacecraft missions, and on-orbit performance data. It also mentions the University Würzburg Experimental Satellite 4 utilizing Morpheus Space NanoFEEP systems for attitude control.*

BODY-109 *Gridded-ion thrusters use RF or microwave excitation to ionize propellant, achieving high specific impulses but limited thrust density. Key considerations include grid erosion, foreign object debris, cathode lifetime, roll torque issues, electromagnetic interactions, power electronics complexity, and the use of iodine propellant for small spacecraft.*

BODY-110 *NASA is exploring spacecraft propulsion with ion thrusters, with upcoming flights to provide more information.*

BODY-111 *Various missions have utilized different types of ion thrusters for propulsion in space, including xenon and iodine propellants. These thrusters have been used for drag-free control, orbit altitude changes, collision avoidance maneuvers, and spacecraft disposal.*

BODY-112 *The LunaH-Map mission aimed to map hydrogen at the lunar south pole but ended due to a propulsion system issue. Hall-effect thrusters (HETs) have become mainstream in-space propulsion technology due to their reliability, efficiency, and high thrust-to-power ratio, with xenon being the preferred propellant.*

BODY-113 *Hall-effect thrusters use electric and magnetic fields to ionize propellant for propulsion in space. Integration and operational considerations include ground facility effects, contamination, thermal soak-back, survival heaters, performance variations, thruster lifetime, and cathode lifetime sensitivity to propellant purity and environmental exposure.*

BODY-114 *The page discusses the technical aspects and missions involving electric propulsion systems, including issues such as emitter formulations, roll torque, thrust vector variations, heaterless cathodes, and throttling range. It also provides examples of spacecraft utilizing specific thrusters like the SPT-50 and SPT-70BR.*

BODY-115 *Various spacecraft propulsion systems, including Busek PPU, Rafael IHET-300, SITAEL HT100, and Astra Spacecraft Engine, have been developed and tested for small satellite missions by different space agencies. Notable achievements include successful orbital ignition and controlled de-orbiting.*

BODY-116 *Astra's ASE propulsion system is in use on Sherpa-LTE1, Exotrail provides propulsion systems for various missions, including Eutelsat and NASA's INCUS mission, and Airbus will integrate Exotrail's mini thruster onto Earth Observation satellites.*

BODY-117 *Blue Canyon, a Raytheon subsidiary, is producing satellites for the DARPA Blackjack program using Exoterra's Halo thruster. Exoterra has also received a NASA Tipping Point award for an in-orbit demonstration of their Courier SEP spacecraft bus. Orbion's Aurora Hall-effect thruster system was selected for a U.S. Space Force weather satellite prototype.*

BODY-118 *Various companies, including Busek and Northrop Grumman, have developed and tested Hall-effect thrusters for satellite missions, with successful launches and ongoing operations reported. Dragonfly Aerospace also successfully tested a propulsion system for the EOS SAT-1 satellite.*

BODY-119 *Aliena Pte Ltd integrated a Hall thruster on a nanosatellite in 2021, achieving in-orbit deployment in 2022. The Multi-modal Electric Propulsion Engine (MEPE) was deployed on another satellite in 2023, demonstrating the capability of dual electric propulsion technologies. Pulsed Plasma Thrusters are described as suitable for attitude control and precision pointing applications.*

BODY-120 *Pulsed plasma propulsion systems, such as PPTs and VATs, offer compact designs with precise impulse control but suffer from inefficiencies like late-time ablation and low thrust-to-power ratios. Safety precautions and considerations for power range and thermal soak-back are important for integration into spacecraft.*

BODY-121 *The page discusses the challenges and solutions related to electric propulsion systems in space missions, including issues with sparking, shorting, and spacecraft contamination. It also highlights specific micro-propulsion systems like the Fiber-Fed Pulsed Plasma Thruster and Xantus thruster for CubeSat missions.*

BODY-122 *The page discusses the Neumann Drive propulsion system, specifically the ND-15 and ND-50 thrusters for nanosatellites. It also provides information on ambipolar thrusters, highlighting their propellant agnostic nature and key integration considerations. The SpaceX Falcon 9 Transporter-1 launch included Phase Four's Maxwell Block 1 propulsion system.*

BODY-123 *The page discusses in-space propulsion technologies, including propellant-less systems like solar sails and electric sails. It highlights recent missions demonstrating these technologies, such as UniSat-7 using an iodine-propellant module and Team Miles' CubeSat with water-propellant thrusters for deep space communications.*

BODY-124 *NASA has launched various missions involving solar sails and tethers for space propulsion and exploration, showcasing innovative technology and advancements in the field.*

BODY-125 *NASA has completed system studies on the concept of electrodynamic tethers, showing it to be a low-cost asset with multi-mission capability. Recent missions include Prox-1 and NPSat-1 deploying tethers for deorbiting. Electric sails use solar wind for thrust, while plasma brakes lower orbital altitude.*

BODY-126 *AuroraSat-1 satellite launched with Plasma Brake and ARM-A modules for deorbiting and attitude control. Aerodynamic drag devices like exo-brakes and ballutes can rapidly remove spacecraft from low-Earth orbits, reducing orbital debris threat.*

BODY-127 *Table listing various hydrazine chemical propulsion systems, including manufacturer, product, thrust per thruster, specific impulse, total impulse, mass envelope, power requirements, status, and missions.*

BODY-128 *Table listing alternative monopropellant and bipropellant propulsion systems from various manufacturers, including thrust per thruster, specific impulse, total impulse, mass envelope, power, and mission status.*

BODY-129 *Table 4-3 lists various monopropellant and bipropellant propulsion systems from different manufacturers, detailing their thrust, specific impulse, total impulse, mass envelope, power usage, and status for use in space missions.*

BODY-130 *Table listing various manufacturers and products of hybrid chemical propulsion systems, including propellant type, thrust, specific impulse, total impulse, mass envelope, power, and mission status.*

BODY-131 *Table showing various manufacturers, products, and specifications of cold gas propulsion systems used in space missions.*

BODY-132 *Table listing solid motor chemical propulsion systems, including manufacturer, product, propellant, thrust, specific impulse, total impulse, mass envelope, power, ACS PMI status, and missions. Data provided in references with wet and dry mass noted.*

BODY-133 *Table listing various manufacturers of electrothermal electric propulsion systems, including details on thrust, specific impulse, total impulse, mass, envelope size, power requirements, and mission status.*

BODY-134 *Table listing various electrospray electric propulsion systems from different manufacturers, detailing thrust, specific impulse, total impulse, mass, power, neutralizer type, and status of each system. References provided for further information.*

BODY-135 *Table listing gridded-ion electric propulsion systems from various manufacturers, detailing thrust, specific impulse, mass, power, and missions. Data provided as documented in references.*

BODY-136 *Table listing various Hall-Effect Electric Propulsion Thrusters from different manufacturers, including details on thrust, specific impulse, mass, power, and missions.*

BODY-137 *Table listing various Hall-Effect Electric Propulsion Thrusters from different manufacturers, including thrust, specific impulse, mass, and power details.*

BODY-138 *Table listing various manufacturers and products of pulsed plasma and vacuum arc electric propulsion systems, including thrust, specific impulse, mass, power, status, and missions.*

BODY-139 *Table 4-12 lists various ambipolar electric propulsion systems, including manufacturers, products, propellants, thrust, specific impulse, and other details. Table 4-13 covers propellant-less propulsion systems from different manufacturers with similar specifications. Data is as provided in the references and may not be independently verified.*

BODY-140 *The page provides a list of references related to NASA systems engineering, micro-propulsion systems, monopropellant thrusters, and spacecraft missions.*

BODY-141 *The summary of the page is about various propulsion systems and missions related to green propellants, including demonstrations, successes, and issues faced by different space agencies and companies.*

BODY-142 *Various sources and studies on propulsion systems for CubeSats, including water electrolysis, chemical propulsion, monopropellant thrusters, green propellants, and hybrid rocket motors.*

BODY-143 *Various research papers and reports on propulsion systems, hybrid motors, cold gas thrusters, and micropropulsion systems for small satellites and CubeSats are discussed in this document.*

BODY-144 *Various CubeSat missions and propulsion systems are discussed, including BioSentinel, iodine thrusters, CPOD, MarCO, NEA Scout, and PACSCISAT. Lessons learned from these missions are highlighted in the context of deep space exploration.*

BODY-145 *A compilation of various propulsion systems for CubeSats and Small Satellites, including product brochures, demonstrations, and technology advancements.*

BODY-146 *This page lists various propulsion systems for CubeSats, including cold gas thrusters, electric propulsion, and monofilament vaporization propulsion. It also includes patents and conference papers related to CubeSat propulsion technology.*

BODY-147 *Cutting-edge space propulsion technologies and hardware developments by various companies and universities, including patents for Fiber-Fed Pulsed Plasma Thrusters, sustainable propulsion tech demonstrations, deorbiting satellites with plasma brakes, and collision avoidance systems for CubeSats.*

BODY-148 *Recent developments in small satellite propulsion systems, including FEEP thrusters, ion propulsion, and iodine propulsion, are discussed in various articles and conference papers from organizations such as NASA and ESA.*

BODY-149 *Various articles and web posts discussing the development and integration of propulsion systems for satellites, as well as missions related to lunar exploration and deep space smallsats.*

BODY-150 *Overview of various electric propulsion systems and technologies, including Hall Effect Thrusters, Iodine Propellant Hall Thrusters, and the VENuS program. Includes information on companies like Busek and SITAEL, as well as ESA's compact electric thruster.*

BODY-151 *Astra and Exotrail secure multiple contracts for electric propulsion systems with various companies and organizations, including Airbus OneWeb Satellites, Astroscale, Maxar Technologies, AAC Clyde Space, NASA, Blue Canyon*

Technologies, York Space Systems, OHB LUXSPACE, Korean satellite manufacturer Satrec Initiative, Astro Digital, and Muon Space.

BODY-152 Various companies, such as Exotrail and Busek, are partnering with NASA to provide propulsion systems for satellites and space missions. Research and development in electric propulsion systems is highlighted.

BODY-153 Recent developments in satellite servicing and electric propulsion systems by companies like Northrop Grumman, Aliena, and Benchmark Space Systems are highlighted, showcasing advancements in extending the life of satellites and innovative propulsion technologies.

BODY-154 The page discusses various advancements in satellite propulsion technology, including the integration of Neumann Drive and Plasma Propulsion Systems onto satellites like Australia's Skykraft and UniSat-7.

BODY-155 NASA and other organizations are developing small satellites, solar sail missions, and propulsion systems for deep space exploration, with a focus on deorbiting satellites to address the issue of orbital debris.

BODY-156 Various companies and research papers are discussed related to electric propulsion systems, including resistojets, electrospray thrusters, and FEEP thrusters for small spacecraft propulsion.

BODY-157 Research and development of various ion thrusters for small satellites, including testing, wear results, and system production. Includes specific companies and technologies such as BIT-3 RF Ion Thruster and NPT30-I2 Iodine Ion Thruster.

BODY-158 The page lists various research papers, patents, and press releases related to Hall-effect thrusters and electric propulsion systems for space exploration.

BODY-159 Various studies and demonstrations related to low-power Hall thrusters, hollow cathodes, and electric propulsion systems by different companies and organizations are discussed in this page.

BODY-160 Various sources and studies on pulsed plasma thrusters, RF thrusters, and electric propulsion systems for CubeSats and micro-satellites are discussed, including products from Phase Four, T4i REGULUS, and Miles Space. Additionally, information on Terminator Tape by Tethers Unlimited is provided.

BODY-161 Chapter on Guidance, Navigation & Control in space exploration, covering various subsystems such as reaction wheels, thrusters, and sensors. Discusses current technology and future advancements in the field.

BODY-162 Glossary of abbreviations related to NASA and space technology, including terms like ADCS, COTS, DSN, GNC, and TRL.

BODY-163 GNC subsystem in spacecraft includes position determination and attitude control using GPS, radar tracking, and celestial body detection. SmallSats in deep space require different ADCS designs. Technologies like star trackers and thrusters are used for attitude determination and control. Miniaturization trend allows for GNC systems in micro- and nano-class spacecraft.

BODY-164 Table 5-1 lists state-of-the-art GNC subsystems, including reaction wheels, star trackers, and GPS receivers. Integrated units combine multiple components for spacecraft control. Blue Canyon Technologies' XACT unit has flown on NASA missions.

BODY-165 Table listing currently available integrated systems for National Aeronautics and Space Administration, including manufacturers, models, mass, actuators, sensors, processors, and pointing accuracy.

BODY-166 *Various configurations of CubeSpace Satellite Systems' CubeADCS 3-Axis systems are detailed, ranging from small to large sizes with different sensor and tracking capabilities.*

BODY-167 *Miniaturized reaction wheels provide precision pointing capability for small spacecraft by storing torque and momentum. They must be carefully selected based on factors like spacecraft mass and rotation performance rates. Multiple wheels are used for redundancy, with periodic desaturation required for maintenance.*

BODY-168 *Specifications of various CubeSat reaction wheels from different manufacturers, including dimensions, weight, torque, and radiation tolerance.*

BODY-169 *Magnetic torquers provide control perpendicular to magnetic field, used to remove momentum from reaction wheels. Limitations beyond low-Earth orbit due to reliance on local magnetic field. Various models available with different specifications.*

BODY-170 *Various companies and organizations are listed along with the specifications of their magnetic torque actuators for spacecraft.*

BODY-171 *This page discusses thrusters and star trackers used for attitude control in small spacecraft, with a focus on pointing accuracy and control authority. Star trackers provide accurate three-axis attitude estimation by comparing digital images to star catalogs.*

BODY-172 *Table listing star trackers suitable for small spacecraft, including manufacturers, models, mass, power, field of view, accuracy, radiation tolerance, and technology readiness level.*

BODY-173 *Specifications for various space microsatellites from different companies, including dimensions, weight, and capabilities.*

BODY-174 *Magnetometers measure local magnetic fields for attitude estimation in small spacecraft. Table lists three-axis options, including models from GomSpace, AAC Clyde Space, MEISEI, NewSpace Systems, and ZARM Technik. Figure shows NSS Magnetometer.*

BODY-175 *Sun sensors estimate the direction of the Sun in a spacecraft, used for attitude estimation and fault detection. Different types include cosine detectors, quadrant detectors, digital sun sensors, and sun cameras for accuracy.*

BODY-176 *Table listing various small spacecraft sun sensors from Redwire Space, detailing their models, sensor types, mass, power usage, analog or digital nature, field of view, accuracy, measurement angles, and radiation tolerance.*

BODY-177 *Various sun sensors from different companies with different specifications such as accuracy, cone angle, and digital/analog outputs are listed.*

BODY-178 *Technical specifications for various solar MEMS technologies and systems developed by NASA and private companies.*

BODY-179 *Horizon sensors and inertial sensing technologies are used in spacecraft for navigation and attitude control. Commercially available horizon sensors are described, with recent academic interest in CubeSat applications. Inertial sensors include gyroscopes and accelerometers for measuring angular and velocity changes.*

BODY-180 *Gyroscopes and accelerometers in SmallSat/CubeSat world are limited by size, weight, power, and cost. Performance parameters include bias stability and random walk, but other factors like dynamic range and resolution should also be considered when selecting inertial sensors.*

BODY-181 *Table listing gyros available for small spacecraft from various manufacturers, detailing sensor type, technology, mass, power, bias stability, ARW, and VRW.*

BODY-182 *Technical specifications for various IMU and gyro MEMS sensors from Safran and Silicon Sensing Systems are listed, including dimensions, weight, power consumption, and resolution.*

BODY-183 *This page lists specifications for various sensors and systems used by the National Aeronautics and Space Administration, including gyroscopes, IMUs, and magnetometers.*

BODY-184 *GPS receivers are now the primary method for orbit determination in low-Earth orbit spacecraft. Onboard GPS receivers are considered mature technology for small spacecraft, with next-generation chip-size solutions emerging. GPS accuracy is limited by propagation variance and civilian code precision. Development may soon make onboard GPS more common.*

BODY-185 *Deep space navigation relies on radio transponders like the SDST and IRIS V2 for small spacecraft. Atomic clocks are being integrated into SmallSats for more efficient navigation.*

BODY-186 *Table 5-12 lists atomic clocks and oscillators for small spacecraft, including dimensions, mass, power, frequency range, and radiation tolerance. LiDAR technology is emerging for small spacecraft with applications in altimetry and relative navigation. Table 5-13 provides examples of flown LiDARs with mass, power, range, and TRL levels.*

BODY-187 *Advancements in guidance, navigation, and control for small satellites are progressing with the development of a high-speed magnetically levitated reaction wheel. This innovative technology eliminates mechanical wear and stiction by using magnetic bearings, allowing for improvements in mass, power, longevity, and accuracy.*

BODY-188 *Advancements in small spacecraft GNC technology are moving towards fully autonomous navigation methods to reduce dependence on Earth-based tracking resources and increase accuracy in deep space exploration. Lowering the cost of components for SmallSats holds exciting implications for the future of the industry.*

BODY-189 *References on various topics related to space surveillance, spacecraft attitude determination, control systems, sensors, navigation techniques, and deep space communication technologies.*

BODY-190 *NASA research on advanced technologies for small satellites, including atomic clocks, reaction wheels, attitude control systems, XNAV pulsar navigation, transponders, CubeSat operations, and micropropulsion thrusters.*

BODY-191 *The page discusses the structure, mechanisms, materials, and radiation effects in CubeSat design, including primary structures, mechanisms, additive manufacturing, and radiation shielding strategies.*

BODY-192 *Glossary of acronyms related to aerospace technology and materials, including ABS, ACS3, AM, CAM, CFRP, CNC, COTS, CSLI, DDD, FDM, ISS, and more.*

BODY-193 *List of acronyms related to the National Aeronautics and Space Administration and space technology.*

BODY-194 *Material selection for small spacecraft structures is crucial, considering physical and mechanical properties. Additive manufacturing has expanded options for custom structural solutions, with primary structures still often made from aluminum alloys. Structural design must consider volume efficiency and multiple functions to maximize space in small spacecraft.*

BODY-195 *The page discusses advancements in small satellite mechanisms, including deployable structures and actuators. It also covers radiation effects and mitigation strategies for small spacecraft operating in different environments. The CubeSat standard structure has evolved to include various sizes to accommodate different missions.*

BODY-196 *CubeSats are evolving to larger sizes like 8U and 16U, driven by demand for more science on smaller platforms. Companies offer various primary structures machined from aluminum alloy, with additional "tuna can" volume for components. Antennas and radiators can be fitted in this space.*

BODY-197 *Various manufacturers offer commercial primary CubeSat structures in different sizes, with a trend towards custom structures that focus on detailed interface requirements for payload designs and safety considerations. The Aerospace Corporation is developing a DiskSat demonstration flight with NASA's support, featuring a unique circular disk structure for increased functionality.*

BODY-198 *The page discusses the standardization of CubeSat dispensers for efficient and cost-effective launches, including rail- or tab-type systems. It also outlines the required interface documents and vibrational conditions for CubeSat deployment, with companies now offering larger dispensers for 12U and 16U platforms.*

BODY-199 *Summary: The page provides information on spacecraft physical dimensions and weight requirements from various deployers, as well as construction methods for CubeSat structures. It also discusses state-of-the-art mechanisms used in spacecraft for mechanical movement.*

BODY-200 *This page discusses spacecraft mechanisms, including actuation methods and types of mechanisms used for various purposes such as separation, release, motorized movement, and attitude control. It also provides examples of technologies used in small spacecraft.*

BODY-201 *Various mechanical actuation methods and systems used in spacecraft for propulsion, deployment, docking, separation, and thermal control are discussed, including electric, thermal, and thermoelectric techniques. Examples include AC/DC motors, shape memory alloys, and thermal bimorph structures.*

BODY-202 *Table listing commercial actuators for space applications, including mass, size, power consumption, and actuation method from various manufacturers such as Ensign-Bickford Aerospace & Defense Company, Moog, Comat Space, DHV Technology, DCUBED, Beyond Gravity, and Revolv Space.*

BODY-203 *Space deployable structures on spacecraft, such as antennas and solar panels, increase mission performance by changing shape and size. Small satellites benefit from lightweight, compact deployables like booms. NASA has developed Deployable Composite Booms for high reliability and efficiency in space missions.*

BODY-204 *Development of origami-inspired structures and robotic manipulators for SmallSats to enhance deployment, repair, and servicing capabilities in space missions, offering cost-effective and agile solutions compared to traditional methods.*

BODY-205 *Robotic arms for small spacecraft add complexity and risk, requiring thorough testing and reliability considerations. Mechanisms contribute to over 10% of small satellite failures, emphasizing the importance of understanding lifetime, materials, and testing for mission success in the space environment.*

BODY-206 *Additive manufacturing for SmallSat development allows for complex structural elements with quicker turnaround times and mass optimization. TRL levels depend on material and manufacturing process, with testing necessary to validate new materials. Thermoplastics and photopolymers have gained traction in AM applications.*

BODY-207 *NASA discusses methods of curing resin for space applications, highlighting challenges with surface smoothness and material availability. ESD due to in-space charging effects is a concern, with guidelines provided for material selection to prevent spacecraft failures.*

BODY-208 *PLA is a common filament for 3D printing with low shrinkage and easy printing, but not recommended for applications beyond TRL 3-4. ESD-compatible filaments are available for lab settings.*

BODY-209 *ABS is a popular choice for 3D printing due to its strength and light weight, but it can become brittle when exposed to sunlight. It shrinks upon cooling and is best printed in enclosed systems with heated chambers. Examples of ABS filaments are listed in Table 6-8.*

BODY-210 *Nylon is a versatile material with various formulations for different applications, but it is more challenging to manufacture than ABS. Examples of nylon filaments and their properties are listed in Table 6-9.*

BODY-211 *Polycarbonate (PC), known as Lexan™, offers high impact resistance and temperature stability for AM systems, but is difficult to print due to warping and poor adhesion. Specialized filaments can help mitigate printing challenges, with commercial AM systems recommended for SmallSat structures.*

BODY-212 *NASA utilizes Windform materials manufactured by CRP Technology for CubeSat and PocketQube platforms, with higher dimensional stability and isotropic properties than FFF. Windform XT 2.0 has been successfully used in space missions, such as KySat-2 and TANCREDO-1, with a GPX-2 structure launched in July 2022.*

BODY-213 *PEI, also known as Ultem™, is a tough thermoplastic resin with high stability and low off-gassing. It is commonly used in FFF systems due to its adhesive properties. Some formulations are FAA-approved for FST and ESD.*

BODY-214 *PAEK thermoplastics like PEEK and PEKK are high-performing materials used in aerospace due to their strength, flame-retardant properties, and low off-gassing. They require specialized printing systems and have been used in space applications like the ISS and Juno spacecraft.*

BODY-215 *Photopolymers, or thermosets, are liquid polymers cured by optical and thermal processes. They offer high resolution, isotropic material properties, and can be used to manufacture optical quality parts with extreme temperature resistance. Some formulations have superior heat deflection temperatures compared to PAEK.*

BODY-216 *The page discusses the advantages of additive manufacturing (AM) in creating monolithic structural elements, infill patterns for lightweighting, and digital materials to expand design space. It also covers radiation shielding as a cost-effective strategy to mitigate risks in space missions.*

BODY-217 *The page discusses the importance of shielding in spacecraft design, including passive shielding against radiation and inherent mass shielding using existing spacecraft materials. It emphasizes the need for careful analysis of materials and placement to protect critical components effectively.*

BODY-218 *NASA guidance document discusses the use of commercial electronic parts in CubeSat and SmallSat systems, highlighting radiation risks and the need for radiation design margins to increase mission lifetimes by reducing electronic failures.*

BODY-219 *Shields-1 mission by NASA tested Z-grade radiation shielding for CubeSat structural design, showing significant reduction in ionizing dose compared to typical aluminum structures. Z-shielding enables low-volume solutions for space applications. NASA has patented methods for making Z-shielding from metals to hybrid laminates.*

BODY-220 *Ad hoc shielding on spacecraft involves spot and sector shielding to protect sensitive components without adding unnecessary mass. New methods include graded-Z shielding and low-Z polymer composites. Charge dissipation coatings help mitigate electrostatic discharge on electronic boards, supporting mission assurance efforts.*

BODY-221 *LUNA Innovations' XP Charge Dissipation Coating reduces resistance on electronic boards in extreme radiation environments, mitigating charging effects. The chapter also covers small satellite structures, materials, mechanisms, and deployment systems, reflecting advancements in SmallSat technology.*

BODY-222 *List of references related to CubeSats, launch services, and solar array drive mechanisms for small satellite missions.*

BODY-223 *Various technologies and mechanisms related to deployable structures for space exploration, including solar arrays, pin pullers, release nuts, composite booms, robotic arms, and reliability standards.*

BODY-224 *This page lists various technical datasheets and specifications for materials used in spacecraft structures, mechanisms, and 3D printing. It also includes guidelines from NASA on mitigating in-space charging effects and information on specific materials like PLA/PHA, Alloy 910, and Onyx ESD™.*

BODY-225 *Technical datasheets and online resources for various aerospace-grade filaments and materials used in additive manufacturing, as well as references to space mission analysis, spacecraft environment interactions, CubeSat reliability, space radiation effects on electronic components, and the space particle environment.*

BODY-226 *Various patents and technical data sheets related to radiation shielding materials and methods, as well as information on companies providing radiation protection coatings for space applications.*

BODY-227 *This chapter discusses thermal control systems in space exploration, including passive and active systems such as coatings, thermal straps, sunshields, heaters, and cryocoolers. Various technologies are explored for managing heat in spacecraft to ensure optimal performance.*

BODY-228 *Glossary of terms related to space technology and thermal management, including materials, experiments, and systems used by NASA.*

BODY-229 *Acronyms related to thermal control and technology in the aerospace industry.*

BODY-230 *Thermal control in spacecraft involves balancing heat absorption, generation, and dissipation. Factors include solar absorptivity, IR emissivity, and power dissipation of components. Equation 1 shows the energy balance. Heat exchange principles apply to any planet or body a spacecraft orbits.*

BODY-231 *Overview of thermal control challenges and technologies for SmallSats, including limitations in size, volume, power, and thermal mass. Passive systems like MLI, coatings, heat pipes are crucial due to limited resources. Structural and electrical design methods also play a role in managing the thermal environment.*

BODY-232 *The page discusses various passive thermal control technologies used in spacecraft, including coatings, films, thermal straps, sunshields, and phase change materials. It also covers structural component materials selection, spacecraft orientation for temperature maintenance, thermal interfaces, and circuit board design considerations.*

BODY-233 *Summary: The page discusses the importance of choosing appropriate thermal control technology for SmallSats to maintain component temperatures within limits. It highlights the use of coatings with specific properties to manage conductive heat transfer and thermal radiation, emphasizing the need for optimal energy balance and performance.*

BODY-234 *The page discusses the use of matte black and white paint, second-surface silver FEP tapes, and coatings for spacecraft thermal control. It highlights the importance of considering changes in optical properties throughout a mission and selecting the appropriate solution based on application needs.*

BODY-235 *The page discusses different sprayable thermal control coatings for spacecraft, highlighting the use of Multi-Layer Insulation (MLI) blankets as a thermal radiation barrier. It emphasizes the challenges and limitations of using MLI on small satellites and suggests alternative coatings for better performance.*

BODY-236 *Table listing films, tapes, and MLI manufacturers for NASA with solar absorptivity values. Explanation of thermal straps as flexible heat conductors between components and radiators, including advancements in graphite materials for increased heat transfer capability. Development of solderless thermal straps by Space Dynamics Laboratory.*

BODY-237 *Comparison of different materials in thermal strap design, with pyrolytic graphite sheets (PGS) showing increased thermal conductivity. Various companies have developed thermal straps using materials such as copper, aluminum, and pyrolytic graphite for spacecraft applications.*

BODY-238 *Heat transfer between surfaces is influenced by contact conductance and bolted joint conductance. Thermal interface materials can improve heat transfer, with Laird Performance Materials offering various options with different thicknesses and thermal conductivities.*

BODY-239 *The page discusses the use of thermal interface materials, sunshields, thermal louvers, and deployable radiators for spacecraft thermal control. It includes a table of recommended conductance values for various products and mentions challenges with integrating thermal louvers on small spacecraft.*

BODY-240 *The page discusses the development of a novel deployable radiator technology for small spacecraft, utilizing Additively Manufactured Deployable Radiators with embedded Oscillating Heat Pipes (AMDROHP) to enhance thermal performance. Additionally, it mentions the use of heat pipes as passive devices for heat dissipation.*

BODY-241 *Heat pipes are efficient thermal transfer technology used in satellites, with variations like flat plates and FlexCool heat pipes. Phase change materials store and release energy based on phase changes, but can be challenging to use due to extra mass. Common materials include wax and water.*

BODY-242 *Development of phase-changing thermal storage units and thermal switches for SmallSats to control temperature peaks, stable temperatures, and energy storage. Redwire Space developed PCM-based thermal energy storage panels for CubeSat form factor. Multifunctional thermal structures integrate thermal control capabilities directly into the structure.*

BODY-243 *NASA has developed Standard Passive Orbital Thermal-control (SPOT) Structures for small satellites, while active thermal control methods rely on input power and are more effective but challenging to integrate into CubeSats due to power, mass, and volume needs. Electrical heaters are commonly used in SmallSats.*

BODY-244 *The page lists examples of electrical heaters for active thermal systems, including temperature limits and materials used such as Kapton (polyimide). Some models* have pressure sensitive adhesive or aluminum foil for extended temperature limits. Mountings include PSA, epoxy, shrink band, and clamping.

BODY-245 Cryocoolers are refrigeration devices used to cool below 100K, improving instrument performance in space applications. They are essential for high precision IR sensors and other instruments requiring cryogenic cooling, offering benefits such as longer lifetimes and low vibration. CubeSats have also started using cryocooling technology.

BODY-246 Table listing active thermal systems - cryocoolers from various manufacturers, including Lockheed Martin Space and AIM Infrarot-Module GmbH. Also discusses thermoelectric coolers and their limitations in space applications.

BODY-247 The page discusses the Active Thermal Architecture (ATA) system for small satellites, utilizing pumped fluid loops for cooling. Developed by Utah State University and NASA, ATA includes heaters and a control algorithm for thermal rejection and temperature control. Ultrasonic additive manufacturing was used to miniaturize the system.

BODY-248 The page discusses active thermal systems for small spacecraft, including the ATA system with flexible fluid unions and vibration isolation technologies. It highlights the use of traditional passive methods and the growing integration of advanced technologies on small satellites.

BODY-249 Development of active thermal control system technologies for small spacecraft, including cryocoolers to fit within 0.5U volume for optical sensors and imaging spectrometers. References provided for further information on space qualified paints, thermal straps, interface materials, radiators, and storage units.

BODY-250 This page contains references to various technologies and methods related to spacecraft thermal control, including heat switches, thermal management systems, small spacecraft structures, superconductors, cryocooler systems, and thermoelectric coolers.

BODY-251 This page discusses the avionics systems platform and mission development considerations for small spacecraft, focusing on command and data handling, flight software, and future developments in technology.

BODY-252 NASA glossary of terms related to artificial intelligence, hardware, and space technology acronyms.

BODY-253 Small Spacecraft Avionics are essential electronic systems for spacecraft, with variability in design based on mission and scale. Traditional spacecraft have high-SWaP-C avionics, while SmallSats have low-SWaP-C systems. Architecture can be centralized or decentralized, with a focus on reliability and radiation-hardened designs.

BODY-254 Overview of technologies and considerations for avionics systems platform and mission development for small spacecraft missions, including factors like SWaP, modularity, radiation exposure, standards, and subsystem specific electronic systems.

BODY-255 NASA discusses integrated avionics platform architectures, mission autonomy, and modular avionics for small spacecraft. Trends in small spacecraft Command and Data Handling (CDH) follow those of larger systems, with focus on microprocessors, cost, availability, and radiation tolerance for long-term missions.

BODY-256 CubeSats and SmallSats are evolving with the use of open-source technology, enabling easier entry into space systems development. Avionics and onboard *computing form factors are key considerations for small spacecraft, with vendors offering integrated systems to support a variety of missions beyond LEO.*

BODY-257 *Summary: The page discusses the current state-of-the-art components for SmallSat CDH systems, emphasizing the importance of flexibility and integration of software modifications. Developers are moving towards ready-to-use hardware and software platforms to adapt to new requirements and interfaces.*

BODY-258 *Sample of highly integrated onboard computing systems from various manufacturers, including processors like ARM Cortex-M3 and Arm Cortex-A72 cores, with radiation hardness assurance, power consumption, and orbits flown specified.*

BODY-259 *Sample of highly integrated onboard computing systems from various manufacturers, including processors and dimensions, designed for use in space missions with different radiation hardness levels and power consumption requirements.*

BODY-260 *Sample of highly integrated onboard computing systems from various manufacturers, including processors and FPGAs with radiation hardness assurance, power consumption, and orbits flown.*

BODY-261 *List of highly integrated onboard computing systems from various manufacturers, detailing processors, radiation hardness assurance, dimensions, power consumption, and orbits flown.*

BODY-262 *A list of highly integrated onboard computing systems for space missions, including processors and boards with radiation hardness assurance and power consumption details.*

BODY-263 *The page lists highly integrated onboard computing systems from various manufacturers, detailing the processors, radiation hardness, dimensions, power consumption, and orbits flown for each system.*

BODY-264 *New radiation-hardened processors and memory technologies are available for small spacecraft, with SRAM being a popular choice due to price and availability. Various electronic components are space-rated with high reliability, providing options for onboard memory in CDH functions.*

BODY-265 *CubeSat spacecraft use common interfaces like SCI, SD Cards, Ethernet, CAN Bus, and SpaceWire. Radiation mitigation strategies are crucial for reliability in deep space missions, with COTS components being susceptible but cost-effective. Radiation effects include TID and SEEs, requiring hybrid designs for maximum benefits.*

BODY-266 *The page discusses the importance of flight software in space missions, including factors influencing development, complexity, and reliability. Advances in processing capabilities have led to increased demands for flight software, with considerations such as processors, frameworks, and operating systems being crucial in design.*

BODY-267 *NASA discusses the use of frameworks, operating systems, software languages, and mission operations tools for small spacecraft missions.*

BODY-268 *Future trends in Command and Data Handling (CDH) systems for small satellites include advanced onboard processing, FPGA-based software-defined radios, and reprogrammable sensors. Industry 4.0 technologies like AI, robotics, and cloud computing are critical for efficiency in SSA system development. Flight software remains a key challenge for mission success.*

BODY-270 *The 2022 Small Spacecraft Avionics chapter emphasizes an integrated avionics ecosystem, highlighting the importance of digital engineering technology and Industry 4.0 advancements for SSA communities to remain relevant and efficient. Email for feedback: arc-sst-soa@mail.nasa.gov.*

BODY-271 *Various companies and organizations offer a range of small satellite technology products, including single-board computers, onboard computers, GPUs, and radiation-tolerant computing.*

BODY-272 *Various datasheets, products, and innovations related to space technology and electronic systems are listed with links for further information.*

BODY-273 *Chapter on communications in space exploration, covering radio frequency and free space optical communications, system architecture, design considerations, policies, licensing, antennas, radios, encryption, and future technologies.*

BODY-274 *NASA glossary of acronyms and terms related to space technology and communication systems.*

BODY-275 *Acronyms related to NASA and space communication technologies.*

BODY-276 *The page discusses the communication systems used by spacecraft, including radio frequency and free space optical systems. RF systems operate within IEEE radio bands, while FSO systems use optical frequencies for high-rate communication. Both systems modulate data onto electromagnetic waves for transmission.*

BODY-277 *This page discusses small spacecraft communication technologies, focusing on RF and FSO. It explains the components of a radio communication system and the frequency bands used for satellite communications, highlighting the shift towards higher frequencies for increased data rates despite atmospheric attenuation challenges.*

BODY-278 *The page discusses the use of different RF bands for communication in small satellites, highlighting advancements in technology such as high gain antennas and Ka-band communications. It also mentions the use of unlicensed ISM bands and COTS products for inter-satellite communication.*

BODY-279 *The page discusses the components of SmallSat communication systems, including radios, mixers, and filters, highlighting their roles in signal transmission and reception. Antennas are crucial for signal strength and direction. Increasing SNR or bandwidth improves information flow. Amplifiers boost weak signals for processing accuracy.*

BODY-280 *Design considerations for RF communication systems on spacecraft include power and mass constraints, trade studies for optimal design, data rate selection based on mission needs, link budget analyses for ground station antenna power requirements, and considerations for pointing requirements and licensing regulations.*

BODY-281 *Regulations for small satellite operations include licensing requirements, encryption for data security, and antenna considerations. Encryption is crucial to prevent unauthorized access to satellite data, while antennas are tailored to mission needs, with COTS options available.*

BODY-282 *The page discusses the classification of antennas for small satellites, including fixed and deployable types. It also touches on the importance of antennas in meeting data rate objectives and the use of radios for downlink communication.*

BODY-283 *The page discusses the use of software defined radios (SDRs) in CubeSats for efficient communication, recommending advanced coding schemes and modulation techniques to increase data rates. It also explores the potential of CubeSat constellations for enhanced coverage and communication capabilities in space missions.*

BODY-284 *NASA is exploring the use of networked swarms of CubeSats for communication in space, with successful tests conducted by MarCO SmallSats. Projects funded through NASA's Small Spacecraft Technology program are advancing RF communication systems for future deep space missions.*

BODY-285 *A list of various antennas for space applications, including their manufacturers, types, frequencies, gains, and dimensions.*

BODY-286 *A list of various antennas from different manufacturers, including their type, frequency range, gain, polarization, mass, dimensions, and flight heritage.*

BODY-287 *Table listing various antennas for different frequencies and bands, including patch antennas, deployable reflectors, and telemetry antennas from manufacturers like Haigh-Farr, EnduroSat, Oxford Space Systems, IQ Spacecom, ANYWAVES, and MMA Design.*

BODY-288 *Table listing various antennas and radios, including manufacturer, type, frequency band, gain, polarization, mass, dimensions, and flight heritage.*

BODY-289 *A list of various radios with different specifications and frequencies for aerospace applications.*

BODY-290 *A list of various radios used in space missions, detailing their manufacturers, frequencies, data rates, power, mass, and dimensions.*

BODY-291 *A list of various radios used by the National Aeronautics and Space Administration, including manufacturers, frequencies, data rates, power, mass, dimensions, and flight heritage.*

BODY-292 *List of various radios, including their manufacturers, types, frequencies, data rates, power levels, mass, dimensions, and flight heritage.*

BODY-293 *Table listing various radios for space applications, including manufacturers, product types, frequencies, data rates, power levels, and dimensions.*

BODY-295 *Laser communications offer higher bandwidth and lower SWaP compared to RF systems for small satellites, but are affected by weather conditions. Small satellites have successfully demonstrated laser communication downlinks from space.*

BODY-296 *The page discusses the system architecture and key parameters of optical communication systems, including the need for high pointing accuracy. It also covers the design of optical ground stations, which differ significantly from RF ground stations due to the use of mirrored telescopes.*

BODY-297 *Design considerations for laser communication terminals on SmallSat platforms include challenges with pointing, thermal management, power constraints, and atmospheric aberrations. Lasercom crosslinks offer high bandwidth connections between satellites but face resource constraints and pointing challenges.*

BODY-298 *Optical communication systems are still evolving, with no current licensing requirements for laser communications. Regulatory bodies like the FAA, DoD, and NASA ensure safety and coordination for outdoor laser operations. Mission examples show potential for high-bandwidth communication on small satellites.*

BODY-299 *Various organizations, including NASA and MIT, have developed optical communication terminals for small satellites with different capabilities, such as data rates ranging from 200 Mbps to 200 Gbps. These terminals use different pointing approaches and have been integrated into various satellite missions for testing.*

BODY-300 *Development of CubeSat Laser Communication Terminals (LCTs) for high-speed space-to-ground communication, with advancements in free-space optical communication technology and quantum key distribution. Future technologies include Deployable Optical Receiver Aperture (DORA) project and Inter-Spacecraft Optical Communicator (ISOC) for gigabit data rates and multiple links.*

BODY-301 *Table listing various LCT technologies, including vendor, data rate, mass, power, wavelength, and launch date for each.*

BODY-302 *NASA CubeSats use various communication systems, with growing interest in higher frequencies and laser technology. RF is considered complementary to optical communications due to cloud interference. References include solar arrays, Bluetooth, and FCC rules for small satellite licensing.*

BODY-303 *Various studies and experiments on optical communication in space, including laser communications between satellites and ground stations, have been conducted to enhance data rates for CubeSats.*

BODY-304 *The page discusses various advancements in laser communication technology, including experimental results and pre-flight test results of different systems, as well as regulatory considerations and intersatellite-link demonstration missions.*

BODY-305 *This chapter discusses integration, launch, deployment, and orbital transport in the context of space missions, covering topics such as launch brokers, deployment methods, and orbital transfer vehicles.*

BODY-306 *Glossary of acronyms related to CubeSats, launch vehicles, and space missions.*

BODY-307 *The page discusses the rise of SmallSats, rideshare services for small spacecraft launches, and the development of orbital transport vehicles. It highlights the increasing demand for dedicated launchers and the emergence of very small launch vehicles to provide rides for small spacecraft.*

BODY-308 *NASA offers commercial launch services for traditional and rideshare options with lower mission assurance. The VADR IDIQ contract allows for risk-tolerant payloads. Integration with launch vehicles is complex, especially for rideshare launches. Spacecraft customers can use brokers or aggregators to book rides.*

BODY-309 *The page discusses licensing requirements, "do no harm" regulations, launch brokers, and the growth of the SmallSat market leading to a shift in launch vehicle offerings. Dedicated launches for SmallSats are highlighted as a key aspect of the evolving space industry.*

BODY-310 *The page discusses the advantages and disadvantages of dedicated launches versus rideshare launches for SmallSats, as well as the deployment methods for CubeSats and free-flying SmallSats.*

BODY-311 *CubeSat dispensers come in rail-type and tab-type configurations, with rail-type being more widely manufactured. Tab-type dispensers are becoming more popular, but still remain a minority choice among developers. The choice of dispenser is sometimes determined by the launch vehicle.*

BODY-312 *The page discusses the importance of designing CubeSats to accommodate a wide range of dispensers for launch opportunities, as well as different separation systems for SmallSats. It highlights the complexities and options available for separation mechanisms, such as motorized light bands and Marman clamps.*

BODY-313 *Companies are shifting to multi-point separation systems for satellites, offering mass savings and reduced complexity. SpaceX has developed a plate system for rideshare missions, allowing more payloads in the circumferential space.*

BODY-314 *CubeSats require integration hardware for safe ejection into orbit, with options like the ESPA ring and Small Spacecraft Mission Service dispenser offering flexibility for different launch configurations.*

BODY-315 *The page discusses the use of Orbital Transfer Vehicles (OTV) and Orbital Maneuvering Vehicles (OMV) to help secondary spacecraft reach desired orbits, increasing mission flexibility for SmallSats and CubeSats. Commercial launches of OTV/OMVs are expected to increase in the near future.*

BODY-316 *Various commercial entities are developing orbital transport vehicles with high delta-V capabilities for services such as altitude changes, inclination changes, phasing, and constellation deployment. These vehicles offer excess propellant volume for additional future services like deorbiting and inspection.*

BODY-317 *Overview of select OMVs and OTVs, including Firefly Elytra, Epic Aerospace Chimera, D-Orbit ION, Rocket Lab Proton, Exolaunch Reliant, and Nanoracks Outpost. Details on capabilities, missions, and development status provided for each vehicle.*

BODY-318 *Commercial space companies are developing platforms like Outpost, Atomos Quark, and Momentus Vigoride to provide in-space services for small spacecraft, with plans for launches in 2022-2024.*

BODY-319 *Various companies are developing commercial orbital transfer and maneuvering vehicles with different operational altitudes, payload capacities, and delta-v capabilities. Some have already flown successfully while others are upcoming projects.*

BODY-320 *The International Space Station offers various options for deploying CubeSats and SmallSats, including the Nanoracks ISS CubeSat Deployer and the Nanoracks Kaber Microsat Deployer. The Japanese Experimental Module also has a deployer for launching CubeSats up to 6U in size.*

BODY-321 *The page discusses the deployment of CubeSats from the International Space Station using different methods, including the Nanoracks Interchangeable CubeSat Launcher and SEOPS SlingShot systems. Geopolitical issues have impacted the ability to deploy CubeSats above the ISS, delaying these missions for the time being.*

BODY-322 *The page discusses the integration, launch, and deployment of small spacecraft in space, highlighting advancements in technology and increasing opportunities for new operators. Dedicated launch vehicles and new integration systems are becoming popular, leading to exciting advances in space capabilities.*

BODY-323 *Various resources and guidelines related to CubeSats, small satellite missions, launch services, and payload adapters provided by organizations such as NASA, AIAA, DoD, Cal Poly, ESA, RUAG, ISISPACE, Moog are listed with links for further information.*

BODY-324 *Various aerospace companies and products, including spacecraft buses, orbital maneuvering vehicles, and satellite deployment systems, are detailed in this document. Additionally, commercial space companies in Houston anticipate SpaceX launches and the International Space Station now has a commercial airlock.*

BODY-325 *This chapter discusses ground data systems, mission operations, frequency considerations, ground segment services, ground stations components, cyber security, and state-of-the-art technologies in the context of NASA's space missions.*

BODY-326 *Glossary of acronyms related to NASA and space technology, including terms like API, AWS, DSN, and GPS.*

BODY-327 *List of abbreviations related to the National Aeronautics and Space Administration (NASA) and space technology.*

BODY-328 *List of abbreviations related to the National Aeronautics and Space Administration (NASA) and space technology.*

BODY-329 *The page discusses the importance of ground data systems and mission operations for NASA, highlighting the shift towards commercial services for communication with objects in low-Earth orbit. The agency has combined networks and plans to rely more on industry-provided services by 2030.*

BODY-330 *Ground systems for small satellite missions are crucial for collecting and distributing data, with factors like data volume, budget, and location influencing design. NASA is leveraging commercial capabilities to enhance communications support for lunar missions through ground networks.*

BODY-331 *The page discusses communication infrastructures for small satellites, including Direct-to-Earth (DTE) and space relay services. DTE ground stations offer direct access with high data throughput, while relays provide continuous coverage when direct-to-ground communication is limited.*

BODY-332 *The page discusses the importance of frequency considerations in spacecraft communication systems, highlighting the shift towards higher frequencies like S, X, and Ka bands for small satellites. Ground stations need to match transmit and receive frequencies to achieve successful communication with spacecraft.*

BODY-333 *Calculating RF link budget is crucial for designing telecommunications solutions, considering factors like power gains and losses. Frequency licensing is regulated by government bodies like the FCC and NTIA, with a move towards streamlining licensing options as small satellites become more common.*

BODY-334 *Ground segment services for NASA include mission integration, planning, data transfer, and PNT. Challenges arise with satellite tracking and initial link closure. Methods such as TLE data and GPS are used for accurate positioning of ground station antennas. Antenna tracking software and system configurations play a key role in successful communication with satellites.*

BODY-335 *NASA outlines NSN interfaces and capabilities, including data transport, storage, security, spacecraft tracking services, radar tracking bands, and other technical specifications. Commissioning of spacecraft and instruments is highlighted as a critical time requiring additional support.*

BODY-336 *NASA and partners operate ground networks for communication and navigation services, supporting missions within 2 million kilometers of Earth. The NSN includes government and commercial assets, providing telemetry, tracking, and data services for spacecraft in various orbits.*

BODY-337 *List of government and commercial ground stations supporting various bands and assets for NASA satellite communication.*

BODY-338 *NASA has various Direct to Earth Command and Telemetry capabilities per frequency band, with different bandwidths, data rates, antenna system EIRP, modulation types, encoding options, and polarization for forward and return communications.*

BODY-339 *NASA's National Aeronautics and Space Administration offers various communication capabilities for satellite missions, including higher data rates for CubeSat missions and support for commercial services. The Networks Integration Management Office provides services such as requirements development and launch support.*

BODY-340 *NASA's Deep Space Network provides telecommunications and tracking services for space missions, including those to other planets and interstellar space. Services include command, telemetry, tracking, calibration, and more. Customized services are available for missions and customers.*

BODY-341 *The page discusses the de facto standards and common interfaces specified by the DSN for interoperability, showcasing the antennas and locations of DSN ground stations.*

BODY-342 *Swedish Space Corporation offers ground station services for small satellites, providing TT&C and data download. Operators must coordinate with providers before launch for scheduling passes and configuring ground stations. KSAT operates a global network of antennas supporting over 1.5 million passes in 2022.*

BODY-343 *KSATLITE offers global ground stations for spacecraft in polar orbit, with a Ka-band network and edge compute capabilities. GSaaS provides managed services for satellite communication without building ground stations. AWS Ground Station allows customers to control satellites and process data in the cloud.*

BODY-344 *AWS Ground Station offers satellite operators access to AWS services for faster, simpler, and cost-effective storage and processing of downloaded data through a network of antenna regions with specific frequencies and link parameters. Operators can easily schedule antenna time on a per-minute basis without long-term commitments.*

BODY-345 *Viasat provides ground station services to NASA and industry, offering real-time communication for LEO satellites with high-speed connectivity and monitoring. Their global network includes 7.3m and 5.4m antennas operating in S, X, and Ka-bands.*

BODY-346 *Specifications for various satellite ground stations in different locations around the world, including details on size, motion capabilities, frequencies, and polarization options.*

BODY-347 *The page discusses satellite ground stations in Japan, Alaska, and Canada, as well as services provided by Viasat and Leaf Space for communication with LEO spacecraft. It highlights the use of cloud-based platforms for scheduling and data delivery, as well as the benefits of real-time earth space relay technology.*

BODY-348 *Leaf Space offers a network with antennas supporting various operations, including upgrades to support Ka-band downlink. Laser Light Communications provides a Global Network platform with high-speed data service and Network On Demand features for cost-effective connectivity.*

BODY-349 *Data Transport as a Service offers secure, high-capacity data delivery with pay-as-you-go pricing. ATLAS Space Operations provides satellite RF communications services through a global network of ground stations and antennas, utilizing the Freedom™ Software Platform for mission success.*

BODY-350 *ATLAS offers a secure, automated global antenna network for satellite communications with low latency and machine-to-machine capabilities. Freedom Space Technologies supports DoD and IC missions as a subsidiary of ATLAS Space Operations.*

BODY-351 *Information about the National Aeronautics and Space Administration (NASA) on page 312.*

BODY-352 *NASA's space relay network, including the Tracking and Data Relay Satellite System, relays data from satellites to ground stations, benefiting small satellites in low-Earth orbit by providing continuous communication coverage.*

BODY-353 *Satellites can use space relay networks to transmit data to the ground, providing faster identification and health information. Special hardware is required, making it necessary to decide before launch. Low latency, low rate options are available for small satellites.*

BODY-354 *The page discusses the Iridium network for low latency messages, use of FDMA and TDMA communication waveforms, installation of transceivers on spacecraft, data plans linked to IMEI, and new services like Certus 100. It also mentions Analytical Space's LEO relay for data aggregation from GEO satellites.*

BODY-355 *The page discusses the importance of antenna gain in satellite communication, highlighting the trade-off between gain and beamwidth. It also explains how antenna patterns work to minimize interference and optimize signal strength.*

BODY-356 *The page discusses the components and hardware required for ground systems in satellite communication, including antennas, feeds, receivers, modems, and digital processing equipment. The cost of a ground station is influenced by the size of the aperture and type of antenna used.*

BODY-357 *The page discusses the Cortex HDR High Data Rate Receiver used by various stations, the USRP X310 and X410 Open-Source Software Defined Radio for SatCom Applications, Kratos OpenSpace SpectralNet technology, and Celestia Satellite Test & Simulation BV's ground-based solutions.*

BODY-358 *Testing equipment for on-board computers, mass memory units, and transponders used in European Space Agency missions. Options include telemetry processing, high-rate interface systems, integrated modem/baseband unit, and level zero processor software. Ongoing efforts to improve product capability with modular, flexible systems.*

BODY-359 *Radomes protect antennas from weather, increase system lifespan, and reduce maintenance costs. Proper selection can improve performance by providing security, controlled environment, and allowing operation in adverse conditions.*

BODY-360 *Various software options are available for managing ground systems in satellite operations, including open-source and purchased solutions from established companies. Examples include AMERGINT softFEP for data formatting and interface conversion, and Kratos quantumFEP for quick turn programs.*

BODY-361 *Kratos qFEP system architecture offers pure software signal processing, AES encryption, and seamless integration. Gpredict is a real-time satellite tracking application with customizable modules and detailed pass predictions.*

BODY-362 *Gpredict is a customizable satellite tracking program while GNU Radio is a software toolkit for developing radio systems in software, with the ability to work with or without hardware. Both are free and open-source tools.*

BODY-363 *GNU Radio is versatile software for handling digital data in RF signals, supporting various data types and acceleration techniques for processing large bandwidths. It is used in ground stations, commercial services, and small satellites for prototyping and testing.*

BODY-364 *HWCNTRL is a satellite ground station control program used in over 30 sites worldwide, supporting multiple antennas and instruments. The Mission and Science Operations Center is where satellite commanding, telemetry archiving, and ground station control take place, with experts managing operations and resolving anomalies as needed.*

BODY-365 *Mission Operations Center submits data for satellite commands and parameter settings, with rehearsals and simulated telemetry before launch. Science*

Operations Center analyzes mission data, with a virtual SOC replacing physical locations for secure missions. Software manages ground segment functions and data flow for communication.

BODY-366 *The page discusses the software used in satellite operations, including flight software for receiving commands, ground station software for tracking and monitoring satellites, and mission planning software for complex tasks.*

BODY-367 *The page discusses the process of optimizing satellite plans, handling mission science data, and conducting end-to-end communication and compatibility testing for SmallSats. Various software and testing methods are used to ensure proper functionality.*

BODY-368 *The page discusses end-to-end testing for ground stations, including network connectivity and hardware solutions provided by various vendors for small satellites. It emphasizes the importance of initial testing before compatibility testing and highlights key components such as antennas, controllers, RF feeds, and software-defined radios.*

BODY-369 *Advanced ground station hardware and software system for satellite communication, including high-power radio capabilities, lightning protection, rotor control, and automatic TLE download. Features include Doppler shift corrections, weather monitoring, lightning detection, and email alerts for non-nominal conditions.*

BODY-370 *Various companies offer innovative ground station kits for communicating with satellites in low-Earth orbit, featuring GUI command recording, manual override, and integrated data handling. ISISPACE offers a low-cost solution with SDR flexibility, GomSpace provides unique end-to-end solutions using their own radios, and Surrey Satellite Technology Ltd. offers complete turnkey ground segment solutions.*

BODY-371 *Surrey and Kratos offer ground station solutions for satellite control and data processing, with Surrey focusing on integration and training, and Kratos specializing in digital processing products for RF signals.*

BODY-372 *Kratos introduced the OpenSpace Platform in 2021, offering a virtual, software-defined architecture for SmallSat operations. The system allows for scalability, upgrades, and enhanced security through digital IF over IP and software defined networks. Cybersecurity is crucial to protect space systems from remote attacks and vulnerabilities.*

BODY-373 *The page discusses the importance of cybersecurity in space systems, including securing external dependencies, on-board security, and supply chain risk management. It also mentions U.S. national guidance and regulations, such as Space Policy Directive 5, and the availability of frameworks and standards to address cybersecurity concerns.*

BODY-374 *Guidance on cybersecurity for small satellites, including encryption, PNT resilience, and system security engineering practices from NASA, NIST, FFRDCs, and CCSDS. MSPA technique discussed for maximizing contact availability of SmallSat clusters at the Annual Small Satellite Conference.*

BODY-375 *The page discusses the increase in antenna size for improved communication with satellites, automation in ground station operations, and the introduction of new large antennas by Viasat to support future space missions. Additionally, it mentions NASA's Lunar Exploration Ground Sites (LEGS) for communication and navigation services.*

BODY-376 *The page discusses the use of ground aggregators in the space industry, focusing on RBC Signals as a provider of global satellite data communication solutions. RBC Signals offers secure and dynamic communication services through a network of company-owned and partner-owned systems worldwide.*

BODY-377 *Table listing service providers for DTE ground system network, including MSPA, KSAT, SSC Infinity, Amazon AWS Ground Station, Viasat, Leaf Space, NASA Near Space Network, NASA/JPL Deep Space Network, and RBC Signals Global Ground Station Network.*

BODY-378 *Scheduling and mission operations software, such as InfoStellar, offer communication services in various frequency bands for enhanced mission operations in the growing commercial small-satellite market. These software products are highly customizable and feature functions related to scheduling, tasking, commanding, telemetry, monitoring, and control.*

BODY-379 *OpenC3 COSMOS is an open-source command and control system for testing and operations, with real-time commanding, telemetry visualization, and offline analysis capabilities. It is fully containerized and microservice based, allowing for integration with any embedded system.*

BODY-380 *Galaxy and Major Tom are mission control software systems used for satellite operations, telemetry monitoring, and ground station scheduling. Galaxy is customizable and can communicate with multiple spacecraft simultaneously, while Major Tom offers cloud-based tools for integrating ground segment applications and services.*

BODY-381 *Orbit Logic offers software for mission planning and space situational awareness, with applications for aerial and satellite imaging. Their suite includes configurable systems, constraints, goals, algorithms, scheduling plans, and visualizations. Major Tom's dashboard allows operators to oversee missions in a comfortable way.*

BODY-382 *The page discusses advanced ground data systems for small satellite technology, including Orbit Logic's CPAW, ACE CrtlPoint from Parsons, and Bright Ascension's Mission Control Software. It also explores the future of ground system technology, focusing on Free Space Optical communications for increased data rates.*

BODY-383 *Optical communication uses lasers to transmit data from spacecraft to ground terminals, offering high data rates with small equipment. Photon detection and nanowire technology are used for signal processing, providing advantages over traditional RF methods. Laser wavelengths allow for smaller capture antennas compared to RF transmissions.*

BODY-384 *NASA is making advancements in optical communication technology, with smaller antennas and higher data transmission rates. Optical Ground Stations are crucial for successful deep space communication, requiring partnerships with large telescopes. Adhering to common standards ensures interoperability between SmallSats and ground stations.*

BODY-385 *JPL is enhancing DSN RF antennas by adding optical segments, creating dual-purpose hybrid antennas. A prototype system has been installed at Goldstone, with plans for a full 8m optical ground aperture on DSS-23 in the early 2030s. This technology will support high-rate optical communications for deep space missions.*

BODY-386 *NASA faces a challenge with FSO technology due to lack of ground segment. Europe's Optical Nucleus Network aims for an automated, cost-efficient multi-mission optical ground station service using modified components and complying with CCSDS standards. KSAT's OGS in Nemea is a pioneering example.*

BODY-387 *KSAT established an OGS in Nemea to offer cost-effective optical ground station services compliant with CCSDS standards, using modified COTS components for cost reduction.*

BODY-388 *Various organizations in different countries are involved in optical communications, using telescopes and ground stations to transmit data between commercial optical terminals. Germany's DLR German Aerospace Center has developed transportable optical ground stations for tracking instruments on satellites. WORK Microwave provides satellite communication technologies and products for ground stations.*

BODY-389 *The page discusses the development of optical ground stations in Australia, New Zealand, and the United States for future space missions, highlighting the use of lasers for high-definition footage on the Moon.*

BODY-390 *Cailabs offers beam shaping products to improve laser communication by counteracting atmospheric turbulence, enhancing free-space optical links for high-throughput data transmission.*

BODY-391 *Cailabs offers turnkey Optical Ground Stations for high-speed laser communication, while BridgeComm's technology enables cost-effective point-to-multipoint transmissions in satellite networks. WarpSpace is developing inter-satellite communication systems based on laser technology for low latency data delivery. The ground segment is crucial for mission success and requires early planning.*

BODY-392 *The page discusses the importance of ground systems in satellite operations, focusing on data efficiency and the transition to optical communications. It highlights challenges and advancements in adapting ground systems for small satellites and high data volumes.*

BODY-393 *Various companies and technologies related to ground stations, optical networks, and space services are discussed in this page.*

BODY-394 *This page provides resources and standards for cybersecurity in space systems, including reports on security protocols, best practices for small satellites, and communication antennas.*

BODY-395 *Various online resources and companies are advancing laser communication technology for space operations, including NASA's Lunar Laser Communication Demonstration and collaborations between Thales and Goonhilly Earth Station.*

BODY-396 *This page discusses identification and tracking systems used by NASA, including ground systems, tracking aids, radio communication devices, coded light signals, and future efforts in this area.*

BODY-397 *Glossary of acronyms related to space defense and surveillance programs, technologies, and organizations.*

BODY-398 *The page discusses challenges in identifying and tracking small satellites, particularly CubeSats, due to the increase in multi-manifest launches. It highlights the difficulty in distinguishing between objects and offers solutions such as coordination with tracking agencies and using tracking aids.*

BODY-399 *The page discusses the United States Space Force Delta 2's role in space surveillance and Space Situational Awareness, including tracking objects in Earth orbit, maintaining the space catalog, and providing data for conjunction assessments. NASA's Conjunction Assessment Risk Analysis program acts as an intermediary between Delta 2 and NASA satellite missions.*

BODY-400 *Commercial entities like ExoAnalytic and LeoLabs provide spacecraft tracking services using telescopes and radars, with plans for more radars to track objects down to 2 cm. The Department of Commerce is working on a space traffic coordination system to merge data from multiple sources for conjunction assessment.*

BODY-401 *Tracking aids are essential for spacecraft to prevent collisions and ensure trackability. Various technologies, such as CubeSat position receivers and coded light signals, can provide augmented tracking data for conjunction assessment. Commercial companies like SpaceNav offer services to process this data.*

BODY-402 *Small satellites can communicate position and ID via radio using devices like the Black Box system or NovAtel OEM7700. These systems provide independent tracking and communication capabilities, with some external versions including solar cells for power generation. The Aerospace Corporation's 'Blinker' concept offers similar functionality.*

BODY-403 *Systems using coded light signals, like the ELROI beacon, offer simple satellite identification but currently rely on host satellite power. Future systems may be smaller and self-powered. Ground equipment is needed to track satellites with these devices.*

BODY-404 *Methods for tracking and identifying small satellites include using LED lights and RF interrogation receivers. LED-based systems require clear skies and ground equipment, while RF systems like CUBIT respond to specific radio signals for identification. Both methods have been successfully demonstrated in space.*

BODY-405 *RF-interrogated devices such as CUBIT and Van Atta arrays increase radar cross-section for tracking, but unique identification is difficult. Laser-interrogated corner cube reflectors provide a return signal when illuminated by a laser, allowing for differentiation between CubeSats based on the number of reflectors.*

BODY-406 *Increasing satellite visibility through high-albedo paint and unique identifiers, such as wire antennas, is a simple method to improve tracking. Efforts are being made to address CubeSat confusion, with plans for advanced identification technologies and a resilient satellite constellation in the future.*

BODY-407 *Tracking and identifying small satellites in space, particularly those deployed in batches, is crucial to avoid confusion and collisions. Multi-manifest launches offer a cost-effective way to deploy numerous satellites. Evaluating tracking options based on cost and capability is essential for mission success.*

BODY-408 *This page lists various publications, conferences, and online resources related to orbit determination, satellite tracking, space debris mitigation, and space domain awareness.*

BODY-409 *Various studies and partnerships related to satellite tracking, traffic management, and mitigating interference in space.*

BODY-410 *This page discusses deorbit systems for spacecraft, including regulations, passive systems like drag sails and deployable booms, and active systems like the TechEdSat Series Exo-Brake. Various companies and organizations involved in deorbiting technology are also mentioned.*

BODY-411 *Glossary of acronyms related to NASA and space technology.*

BODY-412 *Space debris in low-Earth orbit poses a threat to spacecraft due to collisions at high velocities. Regulations require spacecraft to deorbit within 25 years after mission completion. Designers are exploring strategies such as launching in lower orbits or using deorbit systems for faster decay.*

BODY-413 *The page discusses the importance of deorbit methods for small satellites in orbits beyond 400 km, comparing passive and active systems. It also highlights the need for space situational awareness and provides a guide to existing technologies and their development status.*

BODY-414 *The page discusses regulations and guidelines for limiting orbital debris, including the 25-year orbital lifetime guideline, established by NASA, IADC, and FCC. Recent updates include the FCC adopting a new 5-year lifetime rule for satellites in LEO to address current orbital debris issues.*

BODY-415 *Satellite operators urged to remove spacecraft within five years of end-of-life to reduce space debris. FAA proposes new rule for commercial launch vehicles to limit debris growth and collision risk. Solar weather conditions impact deorbit performance in LEO, influencing orbital lifetimes.*

BODY-416 *Passive deorbit methods, such as drag devices, are effective in increasing spacecraft drag area for faster deorbiting in compliance with new requirements. Recent developments in this technology provide solutions to the space debris problem for missions lacking resources for an active system.*

BODY-417 *Table listing various drag sail products and missions, including manufacturer, mission host, launch mass, device mass, initial orbit details, launch and deployment years, drag area, and technology readiness level.*

BODY-418 *Various CubeSats and satellites with different specifications, launch dates, and missions are listed, including CanX-7, NABEO-1, ADEO series, ARTICA series, LightSail-2, ACS3, and Gama ALPHA.*

BODY-419 *Small spacecraft missions have successfully tested passive deorbit technologies using drag sails and booms, demonstrating increased altitude decay rates. The Exo-Brake system, funded by NASA, is a tension-based atmospheric braking device used for accurate deorbit targeting.*

BODY-420 *Various missions and technologies, such as DragSail and Icarus, have been developed to demonstrate drag augmentation for satellite deorbiting, with varying levels of success and innovation.*

BODY-421 *Various deorbit systems, including drag sails and solar sails, are being developed to safely remove spacecraft from orbit. These systems can effectively deorbit satellites of different sizes and weights, improving space debris management.*

BODY-422 *Various companies and organizations have developed and launched drag sail systems for small satellites, CubeSats, and larger spacecraft to aid in deorbiting and extending mission lifetimes in low Earth orbit. Notable examples include Spinnaker 2 and 3, ARTICA system, LightSail-2, and ADEO by HPS.*

BODY-423 *Various space missions, including the deployment of a deorbit mechanism for small satellites, testing of solar sails, and development of a drag sail to reduce space debris, showcase innovative technologies in space exploration and sustainability efforts.*

BODY-424 *Deployable booms are essential components of deorbit systems, such as the ROC-FALL device and D3 CubeSat, offering scalability and versatility for spacecraft sizes. The RODEO device utilizes lightweight film and a roll-out composite boom structure for self-deployment.*

BODY-425 *The page discusses the use of electromagnetic tethers and active spacecraft-based deorbit systems for removing space debris, with examples of technologies developed by companies like Tethers Unlimited and NASA.*

BODY-426 *The page discusses the successful deployment and testing of new technologies for targeted deorbiting of space debris by NASA, RemoveDebris Consortium, and Astroscale. This includes the use of Exo-Brake, CubeSats, harpoons, and magnetic capture mechanisms to address the growing issue of space debris.*

BODY-427 *Astroscale and ClearSpace are developing missions to remove space debris, with Astroscale focusing on satellite deorbiting and ClearSpace planning to capture a non-cooperative upper stage. Both companies have secured funding and partnerships to advance their missions.*

BODY-428 *Momentus and D-Orbit provide space transportation services for satellite deployment and deorbiting. Momentus launched Vigoride transfer vehicles successfully, while D-Orbit offers precision deployment with ION CubeSat carrier platform and a Decommissioning Device for end-of-life maneuvers. Voyage Space partners with OneWeb for sustainable space efforts.*

BODY-429 *Altius DogTags launched into space on OneWeb satellites to address space debris regulations. Passive and active deorbit technologies have matured, with companies developing systems for spacecraft rendezvous and removal. TRL of technologies has increased, expected to grow further with demand for deorbiting services.*

BODY-430 *Various organizations and agencies are working on solutions to mitigate space debris, including drag sails, de-orbit systems, and educational satellites.*

BODY-431 *Various companies and organizations are developing innovative technologies such as solar sails, drag sails, and deorbit modules to address the issue of space debris and improve satellite reentry capabilities.*

BODY-432 *Various articles and press releases about companies and organizations working on developing technology to remove space debris, including Astroscale, Tethers Unlimited, Iridium, and ClearSpace.*

BODY-433 *Various space companies, such as Momentus and D-Orbit, have successfully completed missions deploying payloads and providing space logistics services. Altius Space Machines also celebrated a successful launch of DogTags aboard OneWeb's satellites.*

BODY-434 *The report discusses advancements in small spacecraft technology, including larger CubeSats and smaller SmallSats with increased capabilities. The industry is moving towards more complex missions, with a focus on maximizing spacecraft volume and design. Recent successful missions have contributed to the rapid acceleration of SmallSat technology.*

BODY-435 *NASA is working with commercial providers for flight opportunities and developing policies to mitigate space debris. They are also collaborating with American companies for lunar surface deliveries through the CLPS initiative. The report will be updated annually as technologies evolve.*

BODY-436 *NASA NPR 7123.1D outlines Technology Readiness Levels for hardware and software development, from basic principles to practical application. Compliance is mandatory for NASA employees. This document does not bind the public unless authorized by law or contract.*

BODY-437 *This page outlines the Technology Readiness Levels (TRL) for hardware and software development in space technology, including examples of experiments and success criteria at each level.*

BODY-438 *This page discusses the validation of components and systems in relevant environments to demonstrate overall performance and predict operational environment success. Examples include testing a fiber optic gyroscope and a bi-liquid chemical propulsion engine.*

BODY-439 *Testing and verification of additive manufacturing facility on ISS, with examples of TRL 6 prototypes for remote sensing camera and vacuum pressure integrated suit test.*

BODY-440 *The page discusses the completion and qualification of systems through test and demonstration, including hardware, software, and documentation. It emphasizes the importance of thorough testing, integration, and validation for successful operational use in various environments.*

BODY-441 *This page discusses the Technology Readiness Level (TRL) definitions and examples, emphasizing the importance of testing and operational success criteria for hardware and software in space missions.*

BODY-442 *Instructions to verify the current version of NPR 7123.1D - Appendix E on the NASA Online Directives Information System (NODIS) Library before use.*

NOTABLE PASSAGES

BODY-13 *This report focuses on the spacecraft system in its entirety, provides current best practices for integration, and then presents the state of the art for each specific spacecraft subsystem. Certain chapters have a particular emphasis on CubeSat platforms, as nanosatellite applications have expanded due to their high market growth in recent years.*

BODY-14 *CubeSats are now a common form of small spacecraft that can weigh only a few kilograms and are based on a form factor of a 10 cm square cube, or unit (U).*

BODY-15 *"An accurate TRL assessment requires a high degree of technical knowledge on a subject device, and an in-depth understanding of the mission (including interfaces and environment) on which the device was flown."*

BODY-16 *"At first glance, the TRL descriptions in figure 1.4 appear to be straightforward. It is in the process of trying to assign levels that problems arise. A primary cause of difficulty is in terminology, e.g., everyone knows what a breadboard is, but not everyone has the same definition."*

BODY-17 *"If the new architecture is sufficiently close to the old architecture, then the resulting evaluation could be TRL 6 or 7, but the most important thing to realize is that it is no longer at TRL 9."*

BODY-20 *This chapter addresses the state of the art in the small spacecraft bus offerings and provides the reader with a programmatic overview for small spacecraft mission development.*

BODY-21 *"Hosted payloads, also referred as "satellite-as-a-service," "hitchhiking" or "piggybacking," is increasing in popularity due to its cost savings. The idea is to share the spacecraft bus platform with other payloads and still achieve mission success."*

BODY-24 *"The market has grown considerably over the last 5 years with complete spacecraft bus solutions including I&T and operations options. The addition of I&T and operations gives missions flexibility in implementation, allowing the mission to focus on unique or challenging aspects of the project as needed."*

BODY-36 *The ESPA ring typically separates the primary payload with the upper stage of the launch vehicle, permitting additional mounting allocations for secondary payloads. Multiple rings can be stacked without a primary payload on the top to launch multiple payloads.*

BODY-41 *Design selection can be driven by unique mission constraints, manufacturing lead time, and documented reliability. All of these and many more considerations should be well understood for each trade space option prior to a down-select. Given mission system performance requirements for key performance parameters like mass, volume, power, data link, data budget, and pointing, a functional importance rating and risk-based trade study should be used to screen the many options available. In addition to functional performance, relevant flight heritage or TRL, production lead time, and any available reliability data should be included in the trades. These, as well as cost, could drive the design to be done via COTS or commercial support.*

BODY-42 *"SmallSat subsystem technology will continue to mature and gain flight heritage, to produce improved next generation platforms offered by vendors. Platforms with increased performance will spark the interest of newer vendors as they emerge into the market."*

BODY-47 *"In this chapter, the terms SmallSat and CubeSat are often used in the same context, however, the reader needs to be aware of the distinctions between the two types of spacecraft. Please refer to the introduction of this report for more information on the categories of SmallSats. CubeSats fall under the category of both microsatellites and nanosatellites, and CubeSat missions commonly use COTS parts for space applications."*

BODY-48 *Power generation on SmallSats is a necessity typically governed by a common solar power architecture (solar cells + solar panels + solar arrays). As the SmallSat industry drives the need for lower cost and increased production rates of space solar arrays, the photovoltaics industry is shifting to meet these demands. The standardization of solar array and panel designs.*

BODY-49 *EPS engineers should note beginning-of-life (BOL) vs end-of-life (EOL) performance of the systems as well as their planned testing hours for such systems while on the ground prior to operations. Typically, EPS for SmallSats is over-engineered to handle a dynamic thermal environment, eclipse durations while in LEO, or any other operational scenarios or mission needs while in eclipse at varying sun-angles. Figure 3.1 captures actual space system performances given such wide varying operational conditions for 389 space missions (not only SmallSats) launched since 1989 to 2021 reviewed in the publication (Peretz et. al 2024). The dotted horizontal line indicated where more of the 389 missions surveyed fall under*

BODY-50 *Deployed solar arrays are often the largest structure on a satellite; the ratio between the size of the deployed solar array and the size of the SmallSat may be much higher compared to other conventionally large spacecraft. The size and fundamental frequency of the solar arrays impact spacecraft pointing, propulsion, and delta-V needed for station keeping. Important considerations for SmallSat solar*

BODY-51 *Solar array comparison can be challenging because SmallSat/CubeSat manufacturers who make solar arrays specific to their bus and payload designs often do not report solar array power using the same metrics. Their reported "power" can mean multiple things: power available to the payload, peak power provided by a combination of solar array and battery, or an orbital-specific average power. Reported solar array power (Peak BOL) mainly refers to the peak power of the solar array at the beginning of life, 28°C which is mission-independent. Panel stiffness and moment of inertia usually need to be calculated for a specific spacecraft because they are dependent on multiple factors such as the size and mass of the panel, as well as spacecraft size and weight distribution.*

BODY-54 *"Space missions using solar arrays, regardless of spacecraft mass, are strongly clustered around ~30 W/kg (red dashed line) and are strongly bounded: no missions fly with SP lower than 1 W/kg (lower dashed line), and the maximum empirical SP of this dataset is 200 W/kg (upper dashed line)."*

BODY-55 *"However, even if triple-junction solar cell efficiency improves to the theoretical limit of 68%, the surface area, mass, and storage volume required to support median power requirements for exploration of deep space are beyond the point of feasibility."*

BODY-56 *"A collaboration between the Air Force Research Laboratory (AFRL) and SolAero has developed Metamorphic Multi-Junction (IMM-α) solar cells that are less costly with increased power efficiency for military space applications. The process for developing IMM-α cells involves growing them upside down, where reversing the growth substrate and the semiconductor materials allow the materials to bond to the mechanical handle, resulting in the more effective use of the solar spectrum. A single cell can leverage up to 32% of captured sunlight into available energy."*

BODY-57 *"In 2018, Chinese researchers in organic photovoltaics were able to reach 17% power conversion energy using a tandem cell strategy. This method uses different layers of material that can absorb different wavelengths of sunlight, which enables the cells to use more of the sunlight spectrum, which has limited the performance of organic cells."*

BODY-58 *Due to small spacecraft mass and volume requirements, the batteries and cells in this section will be arranged according to specific energy, or energy per unit mass.*

BODY-60 *"In general, for space applications, high energy density is important because a battery with high gravimetric energy density will be cheaper to launch into orbit (higher battery capacity per unit mass). However, for some high pulse applications, high-power cells would meet mission needs with less weight. However, energy density is not the only factor to investigate during cell selection."*

BODY-61 *"Cylindrical 18650s have become the most commonly used building blocks for many SmallSats today, although prismatic and pouch formats are also available. The lithium-ion industry has seen incremental increases in energy density via the inclusion of silicon in the anode, high voltage cathodes, new electrolyte additives, and improved cell designs."*

BODY-62 *Although 21700 cells are slightly larger than 18650 cells, they have some of the highest energy densities and could offer some mechanical packaging benefits with fewer cells for certain missions.*

BODY-63 *"The power system integrates carbon nanotubes into lithium-ion batteries that significantly increase available energy density. The energy density has exceeded 300 Wh kg-1 during testing, a roughly two-fold increase from the current state of the art."*

BODY-64 *"While the energy density for supercapacitors, also called ultracapacitors, is low (up to 7 Wh kg-1), they offer a very high-power density (up to 100 kW kg-1) which could be useful for space applications that require power transients."*

BODY-65 *"NASA's Solid-state Architecture Batteries for Enhanced Rechargeability and Safety (SABER) project aims to create solid-state batteries that have significantly higher energy than the current state-of-the-art lithium-ion batteries and do not catch fire or lose capacity over time."*

BODY-66 *Key considerations in determining PMAD device selection often include conversion efficiency, input/output voltage range, output power capabilities, and size, weight, and power (SWaP). These metrics are critical to consider for good SmallSat PMAD designs, but it is important to note that PMAD devices are best chosen to suit the exact application of the SmallSat mission. SmallSat missions are often short and more flexible in terms of risk management than larger satellites, and therefore lend themselves to greater flexibility in design choices. One must leverage the benefits and risks to the mission at hand when choosing COTS PMAD systems.*

BODY-69 *"In the realm of SmallSats, wireless power transfer/detection would be useful as redundant options in dusty environments where physical connectors can be contaminated, or in situations where hardware needs to be swapped around and powered (battery swaps). While wireless power transfer/detection is highly inefficient when compared to conventional means, research and development in this technology for use in space applications has a lot of potential in increasing the reliability and robustness of SmallSat power management and distribution."*

BODY-81 *"In the near-term, the surge in public and private investments in small spacecraft propulsion technologies, combined with the immaturity of the overall small spacecraft market, has resulted in an abundance of confusing, unverified, sometimes conflicting, and otherwise incomplete technical literature. Furthermore, the rush by many device developers to secure market share has resulted in some confusion surrounding the true readiness of these devices for mission infusion."*

BODY-84 *"Readers are strongly encouraged to perform more in-depth technical research on candidate devices based on the most up-to-date information available, as well as to assess risk within the context of their specific mission(s). A thoughtful TMA based on the examination of detailed technical data through consultation with device manufactures can reduce program risk and increase the likelihood of program success."*

BODY-85 *"In-Development devices typically align with the NASA TRL range of 4 to 5. While In-Development devices may have specific applications attributed by their developers, no selection for a specific mission has been publicly announced."*

BODY-87 *Chemical systems have enabled in-space maneuvering since the onset of the space age, proving highly capable and reliable. These include hydrazine-based systems, other mono- or bipropellant systems, hybrids, cold gas systems, and solid propellants. Typically, these systems are sought when high thrust or rapid maneuvers are required. As such, chemical systems continue to be the in-space propulsion technology of choice when their total impulse capability is sufficient to meet mission requirements.*

BODY-88 *"Necessary for their implementation. Depending on the device's intended platform (i.e., NanoSat, MicroSat, SmallSat), the propulsion system may be either highly integrated or distributed within the spacecraft."*

BODY-89 *"Hydrazine thrusters typically achieve a specific impulse between 200 – 235 seconds for 1-N class or larger thrusters."*

BODY-90 *Extensive Safety and Handling Requirements: Hydrazine and its derivatives are corrosive, toxic, and potentially carcinogenic. Its vapor requires the use of Self Contained Atmospheric Protective Ensemble (SCAPE) suits. This overhead must be considered when planning ground processing workflow for spacecraft and may impose undesirable constraints on the spacecraft, the launch provider, or other spacecraft participating in the same launch opportunity. Hydrazine propulsion systems typically incorporate redundant serial valves to prevent spills or leaking vapor, which might harm ground personnel or hardware.*

BODY-91 *"Green hydrazine" would provide the low vapor toxicity and high-density specific impulse of ionic liquids while retaining the low combustion and preheat temperatures of conventional hydrazine."*

BODY-94 *Improvements were seen by increasing fuel pump pressure to clear the suspected obstructions. However, the effort was not sufficient to keep the spacecraft in the vicinity of the Moon and the mission was terminated in May 2023.*

BODY-95 *"Aerojet Rocketdyne continues to develop its GR-M1 Advanced Green Monopropellant CubeSat Thruster. It employs the same advanced techniques, ultra-high-temperature catalyst, and refractory metal manufacture as the GPIM GR-1 thruster, but on a nanosat scale. To partially mitigate thermal management challenges exacerbated at the miniature scale, the GR-M1 is designed to operate on a reduced-flame-temperature variant of the ASCENT propellant containing 10% added water."*

BODY-97 NASA ARC developed a polymethyl methacrylate (PMMA) and nitrous oxide hybrid system that had ethylene and nitrous oxide thrusters. The ethylene and nitrous oxide also function as the hybrid ignition source. The hybrid system had a demonstrated efficiency of 91% and calculated ISP of 247 sec, making it competitive with current small satellite propulsion systems.

BODY-98 Nano OTS leverages Parabilis' proven hybrid engine and small satellite technologies for low-cost, high-performance maneuvers using non-toxic green propellants.

BODY-102 "In-space electric propulsion (EP) is any in-space propulsion technology wherein a propellant is accelerated through the conversion of electrical energy into kinetic energy. The electrical energy source powering in-space EP is historically solar, therefore these technologies are often referred to as solar electric propulsion (SEP), although other energy sources such as nuclear reactors or beamed energy are conceivable."

BODY-103 Electrothermal technologies offer some of the most lenient restrictions on propellant selection for in-space propulsion. Whereas chemical systems require propellants with both the right chemical and physical properties to achieve the desired performance, electrothermal systems primarily depend on acceptable physical properties. For example, electrothermal devices can often employ inert gases or even waste products such as water and carbon dioxide. They also allow use of novel propellants such as high storage density refrigerants or in-situ resources.

BODY-105 "SPHERE-1 EYE, a 6U CubeSat developed by Sony Group Corporation, includes a Pale Blue Water Resistojet Thruster. The satellite was launched aboard a SpaceX Falcon 9 on January 3rd, 2023, and has been orbiting Earth with an altitude between 500 km to 600 km. The water propulsion system is expected to prolong the satellite's life by 2.5 years. The propulsion system operated for approximately 2 minutes on March 3rd, 2023, and the company confirmed the successful generation of thrust from the obtained flight telemetry."

BODY-106 Stable operations in either emission mode can provide very precise impulse bits. Charge Balance: Ionic liquid propellants can support electrospray operations with just cation (i.e., positively charged) emission or bi-polar emission (i.e., both anions and cations). For cation-only emission, as with FEEP thrusters, a separate cathode neutralizer is needed to maintain overall charge balance; such neutralizers do not necessarily need to be tightly integrated with the thruster heads, with resultant mass, volume, and power impacts that must be understood for spacecraft integration.

BODY-107 Precision Thrust: Electrospray devices have the potential to provide very precise thrust during continuous operations. For devices that can operate in pulsed mode via pulsed modulation of the high-voltage input, fine impulse bits (i.e., <10 μN-s) may be achievable. Such operations can permit precise control over spacecraft attitude and maneuvering. Verification test data in a relevant environment should be used to properly assess the degree of thrust precision.

BODY-109 "Grid Erosion: Charge-exchange ions formed in between and downstream of the ion optics can impinge upon and erode the grids. Over time, this erosion can lead to a variety of failure modes, including grid structural failure, an inability to prevent electrons from back streaming into the discharge chamber, or the generation of an inter-grid electrical short due to the deposition of electrically conductive grid material."

BODY-112 *The rapid growth in demand for HETs can be attributed to their simple design, historically well-demonstrated reliability, good efficiency, high specific impulse, and high thrust-to-power ratio. Although, the higher voltage gridded-ion thrusters (GIT) can achieve even higher specific impulse, HETs can achieve higher thrust-to-power ratios because the HET's higher density quasi-neutral plasma is not subject to space-charge limitations. The HET's higher thrust-to-power ratio will typically shorten spacecraft transit time. On the other end of the spectrum, arcjets provide significantly higher thrust than HETs, however material limitations prevent arcjets from matching the HET's electrical efficiency and specific impulse. For many missions, HETs provide*

BODY-114 *Throttling Range: HETs typically throttle stably over a wide range of power and discharge voltage. This makes an HET attractive for missions requiring multiple throttle set-points. However, an HET operates most efficiently at specific throttle conditions. Operating at off-nominal conditions may result in decreased specific impulse and/or electrical efficiency.*

BODY-116 *As of March 2023, Astra has nine ACE in orbit.*

BODY-120 *"Minimum Impulse Bit: A compelling capability of pulsed devices is the ability to generate small, precise, and well-timed impulse bits for precise spacecraft maneuvering. By controlling the discharge voltage, very small impulse bits on the order of micronewton-seconds are easily achieved."*

BODY-121 *"The FPPT can provide a large total impulse primary propulsion for micro-satellites through implementation of a novel PTFE fiber propellant storage and delivery mechanism. A major enhancement of the FPPT technology over classical PPTs is the ability to control both the propellant feed rate and pulse energy, thereby providing control of both the specific impulse and thrust."*

BODY-123 *"In-Space Propellant-less Propulsion*

Propellant-less propulsion systems generate thrust via interaction with the surrounding environment (e.g., solar photon pressure, planetary magnetic fields, solar wind and ionospheric plasma pressures, and planetary atmospheres). By contrast, chemical and electric propulsion systems generate thrust by expulsion of reaction mass (i.e., propellant). Four propellant-less propulsion technologies have undergone in-space demonstrations to date, including solar sails, tethers, electric sails (and plasma brakes), and aerodynamic drag devices."

BODY-124 *"NASA's Advanced Composite Solar Sail System (ACS3) technology demonstration uses composite materials for solar sail propulsion. The unfurled sail, approximately 9-m2, will be deployed from a 12U CubeSat that is planned for launch in 2024."*

BODY-125 *"Data from the Space Surveillance Network indicate that the NSTT caused Prox-1 to deorbit more than 24 times faster than otherwise expected. This rate of orbital decay will enable Prox-1 to meet its 25-year deorbit requirement."*

BODY-126 *"Space debris poses a growing threat to active satellites and human activity in space. Allowing decades for defunct spacecraft to decay naturally from low-Earth orbit is no longer sufficient."*

BODY-132 *Note that all data is documented as provided in the references. Unless otherwise published, do not assume the data has been independently verified.*

BODY-163 *"Miniaturization of existing technologies is a continuing trend in small spacecraft GNC. While three-axis stabilized, GPS-equipped, 100 kg class spacecraft have been flown for decades, it has only been in the past few years that such technologies have become available for micro- and nano-class spacecraft."*

BODY-169 *"Magnetic torquers provide control torques perpendicular to the local external magnetic field. Magnetic torquers are often used to remove excess momentum from reaction wheels. As control torques can only be provided in the plane perpendicular to the local magnetic field, magnetic torquers alone cannot provide three-axis stabilization."*

BODY-171 *"A star tracker can provide an accurate estimate of the absolute three-axis attitude by comparing a digital image to an onboard star catalog. Star trackers identify and track multiple stars and provide three-axis attitude several times a second."*

BODY-175 *Sun sensors are used to estimate the direction of the Sun in a spacecraft body frame. Sun direction estimates can be used for attitude estimation, though to obtain a three-axis attitude estimate at least one additional independent source of attitude information is required (e.g., the Earth nadir vector or the direction to a star). Because the Sun is easily identifiable and extremely bright, Sun sensors are often used for fault detection and recovery. However, care must be taken to ensure the Moon or Earth's albedo is not inadvertently perturbing the measurement.*

BODY-180 *"Gyro behavior is a complex topic and gyro performance is typically characterized by a multitude of parameters."*

BODY-184 *Although the usability of GPS is limited to LEO missions, past experiments have demonstrated the ability of using a weak GPS signal at GSO, and potentially soon to cislunar distances. Development and testing in this fast-growing area of research and development may soon make onboard GPS receivers more commonly available.*

BODY-185 *"In deep space navigation, the distances are much greater from the ground station to spacecraft, and the accuracy of the radio signals needs to be measured within a few nanoseconds."*

BODY-186 *"Light Detection and Ranging (LiDAR) is a new type of sensor that is emerging. The technology has matured in terrestrial applications over the last decade and is used in larger spacecraft capable of proximity operations. This sensor type has applications for small spacecraft altimetry and relative navigation, such as Mars helicopters, rendezvous and docking, and formation flying."*

BODY-187 *"In general, technological progress in guidance, navigation, and control is advancing quickly in automotive research areas but is lagging slightly in the aerospace industry. Given the high maturity of existing GNC components, future developments in GNC are mostly focused on incremental or evolutionary improvements, such as decreases in mass and power, and increases in longevity and/or accuracy."*

BODY-188 *As the interest for deep space exploration with small spacecraft grows, semi-to-fully autonomous navigation methods must advance. It is likely that future deep space navigation will rely solely on fully autonomous GNC methods that require zero ground-based intervention to collect/provide navigation data. This is a desirable capability as the spacecraft's dependence on Earth-based tracking resources (such as DSN) is reduced and the demand for navigation accuracy increases at large distances from Earth.*

BODY-194 *Material selection is of primary importance when considering small spacecraft structures. Requirements for both physical properties (density, thermal expansion, and radiation resistance) and mechanical properties (modulus, strength, and toughness) must be satisfied. The manufacture of a typical structure involves both metallic and non-metallic materials, each offering advantages and disadvantages. Metals tend to be more homogeneous and isotropic, meaning properties are similar at every point and in every direction. Non-metals, such as composites, are inhomogeneous and anisotropic by design, meaning properties can be tailored to directional loads.*

BODY-195 *"Deployable structures enable large structural applications with minimal volume requirements. Actuator and switch mechanisms expand the capabilities of small satellites with motion and deployment applications. These mechanisms enable increased small satellite capabilities beyond original structural volume constraints."*

BODY-196 *"The SmallSat community has witnessed an increase in CubeSat standard configuration over the last 10 years from 1U to 3U, to include 6U and 12U. This was due to a higher demand for more science on a smaller platform, and by the need for more volume to design more complex CubeSats that can handle greater responsibility."*

BODY-197 *"A growing development in building custom small satellites is the use of detailed interface requirement guidelines. These focus on payload designs with the understanding of rideshare safety considerations for mission readiness and deployment methods."*

BODY-198 *The CubeSat Design Specification document by the CubeSat Program at Cal Poly was created to provide CubeSat developers baseline requirements that are compatible with as many CubeSat dispensers and launch opportunities as possible to eliminate launch interface failures.*

BODY-200 *"By classical definition, actuators are devices that convert electrical, thermal, hydraulic, and/or pneumatic energy into mechanical motion when said energy is allowed to flow. Active, or commanded, actuators use onboard data links and electrical transistors to determine the transfer of energy; whereas passive, or reactive, actuators allow the spacecraft environment (including external launch systems) to dictate actuator energy transfer."*

BODY-203 *"SmallSat deployable structures are common and are associated with high reliability. Engineers have started developing deployables with different materials to decrease the stowage area, mass, and power."*

BODY-204 *"The need for in-space servicing is receiving more attention from the SmallSat community with the increasing demand of more complex SmallSat with greater capability and longer mission life. These types of challenges are being solved with robotic manipulations that can perform intricate actions in space."*

BODY-205 *"Mechanisms add capabilities and complexities to small satellite design. Additional integration and testing are required. NASA Reliability and Maintainability Standard (36) describes maintainability, and 'test as you fly,' in addition to multiple other mitigation strategies and considerations."*

BODY-206 *"In practice, additive manufacturing has a separate design space and design process, which has seen tighter integration into computer-aided design, computer-aided manufacturing, and modal and structural analysis packages in the past few years. Such tools can enable quicker turnaround times for SmallSat development, and have been instrumental in mass optimization, using AM materials in radiation shielding, and enabling high-throughput, high-quality manufacturing."*

BODY-207 *Surface discharge, or electrostatic discharge (ESD), is a result of in-space charging effects and is caused by interactions between the in-flight plasma environment and spacecraft materials and electronic subsystems. The field buildup and ESD can negatively affect the spacecraft and there are design precautions which must be considered depending on the spacecraft's operational environment. Per ESD guidelines from NASA Spacecraft Charging Handbook 4002A, dielectric materials* above 10^12 Ohm cm should be avoided because charge accumulation occurs regardless. Please refer to the NASA Handbook 4002A, 5.2.1.5 Material Selection for more information. Historically, ESD due to faulty grounding has been a leading cause of spacecraft or subsystem failures.

BODY-208 Unless the application has a very short-term exposure to harsh conditions, and if the conditions are well characterized and controlled, it is not recommended to use PLA for an application beyond TRL 3-4.

BODY-209 ABS has flown as the complete structure for KickSat-2, a FemtoSat deployer for chip-scale satellites. The single-use, short mission duration, and intricate dispenser frame made a conventionally machined deployer mass- and cost-prohibitive.

BODY-215 "Compared to other AM processes, photopolymers and their manufacturing processes allow for superior isotropic material properties, very high resolution, and the ability to manufacture optical quality parts."

BODY-216 "Infill Patterns

Due to the flexibility that AM offers, new methods of lightweighting are now possible. 'Lightweighting' refers to the reduction of mass of structural elements, without compromising structural integrity. The best examples of well-proven heritage methods of lightweighting are 'honeycomb' sandwiched aluminum panels, subtractive machining, and truss structures."

BODY-217 "In some cases, increased shielding is more detrimental than if none was used, owing to the secondary particles generated by highly penetrating energetic particles. Therefore, it is important to analyze both the thickness and type of materials used to shield all critical parts of the spacecraft."

BODY-218 Adding a two-fold increase for the trapped belt radiation uncertainty brings the total radiation near the TID lifetime of many commercial parts, even before estimating a SPE TID contribution. The uncertainty of radiation model results of low-Earth orbit below 840 km has been estimated as at least two-fold; Van Allen Belt models are empirical and rely on data in the orbital environment. The NASA Preferred Reliability Series "Radiation Design Margin Requirements" also recommends a radiation design margin of 2 for reliability. By reducing the total ionizing dose on commercial parts, the mission lifetimes can be increased by reducing the risk of electronic failures on sensitive semiconductor parts.

BODY-219 "Z-shielding enables a low volume shielding solution for CubeSat and SmallSat applications where reduced volume is important. AlTiTa, Z-shielding, at 2.08 g cm-2 reduces the dose from a SPE by half when compared to a standard 0.2 cm aluminum structure."

BODY-220 Ad hoc shielding is more efficient per unit mass than inherent mass shielding because it can be optimized for the spacecraft's intended radiation environment while loosening the structural constraints. The most recent methods include multiple layer shields with layer-unique elemental atomic numbers which are layered advantageously (often in a low-high-low Z scheme), known as "graded-Z" shielding, and advanced low-Z polymer or composite mixtures doped with high-Z, metallic micro-particles. Low-Z elements are particularly capable at shielding protons and ions while generating little secondary radiation, where high Z elements scatter electrons and photons much more efficiently.

BODY-221 The XP Charge Dissipation Coating has reduced resistance compared to typical commercial conformal coatings, which reduces surface charging risk on electronic boards. LUNA XP Coating on an electronic board has transparency for visual parts inspection. For extreme radiation environments, a combination of radiation *shielding and charge dissipation coating reduces the ionizing radiation that contributes to charging and provides a surface pathway for removing charge to ground.*

BODY-231 *Challenges of designing a thermal control system for a SmallSat stem from several intrinsic properties, summarized in table 7-1. Due to the small size and volume limitations inside the deployer or around deployables, there is often no room for multi-layer insulation (MLI) for CubeSats. The thermal solution must be worked out as a coatings problem, exposing the CubeSat to more transient thermal behaviors.*

BODY-233 *"Thermal radiation heat transfer is controlled by using materials that have specific optical surface properties, namely: solar absorptivity and IR emissivity. Solar absorptivity governs how much incident heating from solar radiation a spacecraft absorbs, while IR emissivity determines how much heat a spacecraft emits to space, relative to a perfect blackbody emitter, and what fraction of thermal radiation from IR sources (e.g., the Earth, Moon, any particularly hot spacecraft components) are absorbed by that spacecraft surface."*

BODY-234 *Alternatively, matte white paint has a low solar absorptivity and high IR emissivity for surfaces required to absorb a low percentage of solar heating and emit a high percentage of spacecraft heat (e.g., radiator).*

BODY-235 *"MLI is delicate and performance drops drastically if compressed (causing a thermally conductive "short circuit"), so it should be used with caution or avoided altogether on the exterior of small satellites that fit into a deployer (e.g., P-POD, NLAS)."*

BODY-236 *Thermal straps can be made of metal, traditionally copper or aluminum, or high conductivity carbon materials, such as graphite. They can be formed of multiple foil sheets or wound cables (also referred to as ropes and braids), with end blocks at each end to hold the sheets/cables in place and to mount or otherwise attach to the needed surfaces. Straps with more than two end blocks and multiple material combinations can also be produced and have been used on large spacecraft. Advances in thermal straps are being developed to further increase heat transfer capability and custom thermal straps are now commonly fabricated and tested using graphite material due to improved thermal conductivity.*

BODY-239 *"Thermal louvers are thermally activated shutters that regulate how much heat the louvered surface can dissipate. As the louvers open, the average IR emissivity of the surface changes, changing how much heat the surface dissipates."*

BODY-240 *"A novel deployable radiator is being developed by JPL, California Polytechnic San Luis Obispo, and California State Los Angles. At the core of this technology is an Additively Manufactured Deployable Radiator with embedded Oscillating Heat Pipes (AMDROHP) that enables heat to be efficiently transported across moving interfaces."*

BODY-241 *"Traditional constant conductance heat pipes are cylindrical in shape with a grooved inner wick, like those used on Bi-Spectral Infrared Detection (BIRD), a 92 kg satellite launched in 2001, to join satellite segments. Heat pipes can also be configured as flat plates with tubing sandwiched between two plates and charged with a working fluid inside."*

BODY-245 *"Cryocoolers are used on instruments or subsystems requiring cryogenic cooling, such as high precision IR sensors. Instruments such as imaging spectrometers, interferometers and midwave infrared (MWIR) sensors require cryocoolers to function at extremely low temperatures."*

BODY-246 *TECs are miniature solid-state heat pumps which provide localized cooling via the Peltier effect, which is cooling resulting from passing electric current through a junction formed by two dissimilar metals. TECs have been used to cool star trackers, IR sensors and low noise amplifiers on large spacecraft. Advantages of TECs are that they have no moving parts, are reliable, noiseless, lightweight, and compact. Their use is limited by low efficiency below temperatures of 130K and low performance with large temperature differences. Furthermore, the TECs are fragile to mount and highly sensitive to thermal expansion stresses. External stresses can be mitigated by adding a conductive strap on the cold side.*

BODY-247 *The Active Thermal Architecture (ATA) system is an advanced, active thermal control technology for small satellites in support of advanced missions in deep space, helio-physics, earth science, and communications. The ATA technology is capable of high-power thermal rejection and zonal temperature control of satellite busses, payloads, and high-energy density components. The ATA system includes integrated heaters and a PID-based control algorithm to dynamically tune the satellite's thermal rejection and zonal temperature control as a function of payload and mission parameters.*

BODY-248 *"Many technologies that have to date only been integrated on larger spacecraft are being designed, evaluated, and tested for small spacecraft to meet the growing needs of SmallSat developers as small satellites become more and more advanced."*

BODY-253 *"The avionics system is essentially the foundation for all components and their functions integrated on the spacecraft. As the nature of the mission influences the avionics architecture design, there is a large degree of variability in avionics systems."*

BODY-255 *Modular avionics architectures for small spacecraft can be characterized as either federated or integrated. In a federated avionics architecture, each subsystem of the spacecraft is considered an independent, dedicated autonomous element, with the avionic components performing all functions independently and exchanging data over standardized communications protocols and interfaces. An integrated avionics architecture is a shared, distributed functionality, that can be configured with distributed, heterogeneous and/or mixed criticality elements. In either case, modular avionics architectures can be configured with smart subsystem capabilities, redundant fault tolerant radiation, and anomaly mitigation procedures.*

BODY-257 *System developers are gravitating towards ready-to-use hardware and software development platforms that can provide seamless migration to higher performance architectures.*

BODY-267 *In the context of SSA, a FSW framework can be described as a hierarchal architecture, sometimes referred to as a set of lego-like building block constructs, partitions, and functions. This emerging system-of-systems concept describes the large-scale integration of many independent, self-contained systems that work together to satisfy a global need.*

BODY-268 *"In keeping with trends seen in other disciplines and industries, the Industry 4.0 and "digitally managed everything" is absolutely of critical importance for technological and programmatic efficiencies in SSA systems development."*

BODY-269 *It is essential that FSW be as simple as possible. It is critical to survey options and plan early in any FSW effort. Wherever possible, early development and testing should be performed. Efforts to add additional features should be looked at very critically with a strong effort to stick to the existing plan. With good planning and careful execution, a favorable outcome can be achieved.*

BODY-270 *The 2022 Small Spacecraft Avionics chapter has been updated with a broader, interrelated framework, where CDH, FSW, and smart payloads are not just independent space platform subsystems but are part of an integrated avionics ecosystem which includes all electronic elements of a space platform, now primarily digitally based and or managed. Also, SSA should not be considered as an isolated spaceflight technology component, but rather as a core digital engineering technology emphasis area, capable of taking advantage of and integrating products, processes, and technologies from other disciplines.*

BODY-278 *At JPL, the Integrated Solar Array and Reflectarray Antenna (ISARA) mission demonstrated high bandwidth Ka-band CubeSat communications with over 100 Mbps downlink rate. The back of the 3U CubeSat was fitted with a high gain reflectarray antenna integrated into an existing solar array. The successful demonstration of the reflectarray on ISARA became the basis for the Mars Cube One (MarCO) mission to Mars. The MarCO mission uses two twin CubeSats for a communications relay between the InSight lander and Earth. Using a X-band reflectarray they were able to successfully complete their mission.*

BODY-280 *"Spread-spectrum communication applies a known frequency spreading function to the signal, which helps reduce interference from other transmitters, and provides more secure communications; as such, it is often used for multi-way communication networks."*

BODY-281 *"With the increased proliferation of small satellites in low-Earth orbit comes an increase in vulnerabilities. Many SmallSats are comprised of COTS hardware and/or open-source software. While this strategy allows for a more flexible design approach, adversaries can gain insight into the design. Encryption of data in transit prevents other actors from commanding satellites or intercepting transmissions."*

BODY-282 *Increasing the aperture or diameter of an antenna increases the link margin, which can allow designers to increase the data rate of the system or reduce the necessary transmit power.*

BODY-283 *This report recommends efficient modulation and coding schemes for spacecraft power and bandwidth to increase the data rate and meet bandwidth constraints with the limited power and mass for CubeSat spacecraft. Advanced coding, such as the CCSDS low-density parity-check code (LDPC) family, with various code rates is a powerful technique to provide bandwidth and power with high-order modulation to achieve high data rate requirements for CubeSat missions.*

BODY-284 *Developing networked swarms is less of a hardware engineering problem than a systems and software engineering problem in that one must manage multiple dynamic communication links.*

BODY-295 *"Laser communications have a low probability of intercept, are difficult to jam, and encounter very little interference because of the narrow beamwidth."*

BODY-296 *"Successful optical communications links typically require high pointing accuracy. The optical communication terminal on a spacecraft typically has a two-stage pointing system, with a coarse-pointing stage and a fine-pointing stage. The optical communication system often relies heavily on the spacecraft attitude determination and control system (ADCS) for coarse-pointing, and may use a second pointing mechanism such as a gimbal as additional support for coarse pointing."*

BODY-297 "Lasercom crosslinks can provide a high bandwidth connection between two satellites, as well as perform ranging between the satellites, potentially with high ranging precision. Connecting two satellites across different orbit planes helps with data routing and can reduce how long it takes to route data to the end use."

BODY-298 "Given the early stages of development for optical communication systems, both policy and regulatory approaches are still evolving. In the policy realm, there is an initial draft CCSDS Pink Book in process (CCSDS 141.0-P-1.1) with a goal to facilitate interoperability and cross-support between different communication systems."

BODY-299 The German Aerospace Center (DLR) has been developing LCTs as part of its Optical Space Infrared Downlink System (OSIRIS) program to support lasercom from small satellites. The first, OSIRISv1 is capable of 200 Mbps downlinks and is integrated into the University of Stuttgart's Flying Laptop satellite. This LCT uses a body pointing-only approach. The OSIRISv1 LCT, launched in 2017, has completed commissioning and is being used by DLR to test their optical ground stations.

BODY-300 "While free space optical communication technology development has been making strides towards fielding operational systems, other avenues of research have also been explored. Quantum key distribution is a protocol that shares a secret cryptographic key through entangled photons."

BODY-307 The significant increase in SmallSats launched in the past few years is due to an expansion in SmallSat and CubeSat constellations by operators like SpaceX, OneWeb, and Planet. With more SmallSat missions currently being planned, the demand for SmallSat launches is expected to continually rise.

BODY-308 Launch integration with the launch vehicle is a complex and critical portion of the mission. The launch integration effort for a primary spacecraft typically includes the launch service provider, the spacecraft manufacturer, the spacecraft customer, the launch range operator, and sometimes a launch service integration contractor.

BODY-310 "Dedicated launches for SmallSats have many advantages. A SmallSat on a dedicated launch controls the mission requirements in whole—what they need, when they want to launch, and where they want to go. They generally have a readiness "go/no-go" call on launch day in case something goes wrong with their satellite pre-launch."

BODY-313 "Given the stiffness and fundamental frequency requirements of traditional rideshare missions, many companies are also shifting to 4-point separation systems for MicroSats and SmallSats as a viable alternative to traditional MLB or clamp band systems. These systems function in a similar way to the systems above and are typically rated for microsatellites (≤100 kg), however boast less complexity than a traditional MLB or Clampband."

BODY-314 "A main driver for CubeSat utility is their adhesion to a standard that can be integrated into several different launch configurations. The physical hardware that attaches both a containerized and non-containerized small spacecraft and keeps it insulated from a rocket body include deployers, adapters, dispensers, and launchers."

BODY-315 "The ability for small spacecraft to reach new orbits will enable a much wider range of mission designs for destinations both near and far."

BODY-317 "The D-Orbit ION system is one of the most used OTV/OMV platforms on the market. The system was first used in 2020 to deploy Planet Labs constellation satellites. The system has persisted and has been used nine times as of writing to deploy small satellites, the most of any of the described OMV/OTV platforms."

BODY-318 *The first commercial version of Outpost is scheduled to fly in 2024. Atomos has plans to offer many in-space services to small spacecraft, including deployment and orbit raising, phasing, and inclination changes. The first version of their OTV/OMV, Quark, is planned to launch in 2024. Momentus Space has developed an in-space orbit transfer service for SmallSats, named Vigoride. The maximum payload mass on Vigoride is 750 kg to LEO, and it can be launched from an ESPA or ESPA Grande ring, from ISS airlocks, or a launch vehicle.*

BODY-320 *"Nanoracks CubeSat Deployer (NRCSD) (figure 10.9) is a self-contained CubeSat dispenser system that mechanically and electrically isolates CubeSats from the ISS, cargo resupply vehicles, and ISS crew."*

BODY-322 *"Dedicated launch vehicles take advantage of rapid integration and mission design flexibility, enabling small spacecraft to dictate mission parameters. New integration systems will greatly increase the mission envelope of small spacecraft riding as secondary spacecraft. Advanced systems used to host secondary spacecraft in-orbit or support a rideshare to a dedicated orbit, can increase mission lifetime, expand mission capabilities, and enable orbit maneuvering for smaller spacecraft."*

BODY-329 *There are exciting changes in the government and commercial sector ground stations and services, and the shifting synergy between these. From its inception in 1958, whenever NASA needed to receive data from one of its Earth observing satellites or talk to its astronauts in orbit, it used equipment and services it had needed to develop and build itself. Over time, commercial enterprise acquired the proficiencies necessary to reliably and securely communicate with objects in low-Earth orbit (LEO), services NASA is now pursuing to purchase as any other near-Earth space customer.*

BODY-330 *The ground system is responsible for collecting and distributing the most valuable asset of the mission: the data. Using the proper ground system is key to mission success.*

BODY-333 *Maintaining a 3 dB link margin is adequate for data return from a satellite in low-Earth orbit at a slant range of 1,500 km. Usually commanding to a Near Earth orbit has plenty of margin because of the high power and aperture size of the ground station, and the lower required data rate on the account of the commands' low volume. When considering deep space communication, a 3 dB link margin is desired, but for distant spacecraft, such as New Horizons at 7 billion kilometers from Earth, 1 dB or less margin may be all that is practically possible.*

BODY-334 *Challenges are usually associated with the initial satellite-to-ground station link closure. Typically, two-line elements (TLE) or state vectors are established and shared by the launch provider after deployment. This information can be used to create an initial orbit solution for ground station antenna pointing. Low-Earth orbit missions can use North American Aerospace Defense Command (NORAD) TLE data for satellite location. However, it could take up to a week or more for NORAD to add the new object to their tracking list. This process could be delayed further if multiple spacecraft are ejected in close proximity, and it may not be clear which NORAD element set corresponds to which spacecraft.*

BODY-335 *Another critical time in the life of a spacecraft is commissioning; either commissioning of the spacecraft bus, or commissioning of science instruments, including in-space calibration. During commissioning phases, additional time and support personnel are typically scheduled.*

BODY-336 *"The NSN provides direct-to-earth telemetry, commanding, ground-based tracking, and data and communications services to a wide range of customers. The network consists of NASA, commercial, and partner S-band, X-band, and Ka-band ground stations supporting spacecraft in low-Earth orbit, GEO, Highly Elliptical Orbit (HEO), Lunar orbit, and Lagrange point L1/L2 orbit up to one million miles from Earth."*

BODY-340 *Two missions (Voyager I and Voyager II) have reached interstellar space and still communicate with the DSN.*

BODY-341 *The use of data service interface standards enable interoperability with similar services from other providers.*

BODY-343 *Ground Station as a Service (GSaaS) is a managed service which enables customers to communicate, downlink, & process data from their satellites/spacecrafts on as a pay-as-you go basis without needing them to build their own satellite ground stations. These services are usually scalable and use edge cloud services as an intermediate for customers data.*

BODY-344 *Other AWS services include Amazon VPC, Amazon Rekognition, and Amazon Kinesis Data Streams. These services allow operators to reduce data processing and analysis times for use cases like weather prediction or natural disaster imagery from hours to minutes or seconds. This also enables operators to quickly create business rules and workflows to organize, structure, and route the satellite data before it can be analyzed and incorporated into key applications such as imaging analysis and weather forecasting.*

BODY-345 *Viasat's Real-Time Earth (RTE) ground segment service enables communications for next-generation and legacy LEO satellites using S, X, and Ka-bands. RTE offers downlinks from low megabits per second to multiple gigabits per second empowered by cutting edge software-defined radios. The service includes high-speed connectivity for backhaul, real-time data streaming, and real-time monitoring of passes.*

BODY-347 *Viasat's Real-Time Earth Space Relay is designed to leverage the ViaSat-3 global constellation with a newly developed terminal that will enable on-demand and cost-effective communications for LEO spacecraft anywhere and at any time in their global orbit. Viasat won a portion of the NASA Communications Service Provider contract with its integrated RTE ground solution, which will include on-orbit demonstrations in 2025 that allow users to select between the low latency space relay and the high throughput global ground network.*

BODY-348 *Leaf Space works with its customers to execute both routine operations and complex tasks such as LEOP or MSPA data downlink for formation flying satellite clusters. Leaf Line is characterized by ~3.7m antennas supporting operations in S-band uplink (2025-2110 MHz, EIRP: 50 dBW) and downlink (2200-2290 MHz, G/T: 12.8 dB/K) and X-band downlink (8025-8400 MHz, G/T: 25.4 dB/K), as well as by UHF antennas (uplink and downlink). In 2024, the Leaf Line network will be upgraded to support Ka-band downlink (25.5-27 GHz) at*

BODY-349 *"Through geographical dispersion of its ground stations and cloud services, ATLAS provides a resilient capability that delivers dependable low latency data."*

BODY-352 *"NASA's Tracking and Data Relay Satellite System (TDRSS), shown in figure 11.12, is one of the most well-known space relay networks."*

BODY-353 *Another benefit is the ability to obtain real-time event notifications without the need for prior scheduling. Scientists are interested in using this technology to alert the science community when certain phenomenon are observed.*

BODY-355 *There is a practical compromise where the beamwidth is so small that tracking is difficult and when the antenna gets so large that it is difficult to procure or manage. A typical antenna pattern is shown in figure 11.14. There is a center lobe where most of the transmitted energy is contained. The remaining energy is stored in the sidelobes on either side of the main lobe. The diminished side lobes are intentional so that ground noise from other emitters on Earth are not collected when receiving and so that interference to terrestrial systems is not created when transmitting.*

BODY-356 *"Directional antennas point towards the satellite as it moves over the ground station. Pointing adjustments are necessary in both the vertical (elevation) and horizontal (azimuth) directions. These movements are accomplished using motors and gears."*

BODY-357 *OpenSpace SpectralNet® eliminates the distance constraints of RF transport by digitizing RF signals for transport over IP networks in a way that preserves both frequency and timing characteristics, and then uniquely restores the RF signals at their destination. By eliminating the distance constraints between antennas and signal processing equipment, this technology enables operators to deploy new ground architectures with numerous advantages, such as the ability to mitigate the effects of rain fade for Ku/Ka satellites, reduce costs by centralizing operations, simplify disaster recovery and system maintenance, optimize antenna placement, and develop a migration path toward virtual ground systems. SpectralNet does all of this while protecting the operator's current investment in existing equipment.*

BODY-358 *Efforts are on-going to improve product capability with a focus on modular, flexible, scalable multichannel systems that take advantage of the latest technologies.*

BODY-359 *"A radome can increase the useful life of the antenna and decrease overall maintenance costs for the system. Consideration for a radome should be given early in the design phase of the system, as a radome will allow for lighter duty and less expensive components such as drive motors and foundations due to the elimination of wind loads on the antenna."*

BODY-361 *"Gpredict is different from other satellite tracking programs in that it allows the satellites to be grouped into visualization modules. Each of these modules can be configured independently from others, allowing unlimited flexibility in the look and feel of the modules."*

BODY-362 *"As free software licensed under the terms and conditions of the GNU General Public License, it can be freely used, learned from, modified, and re-distributed."*

BODY-363 *"GNU Radio supports heterogeneous computing, where some of the blocks run on an FPGA or GPU. These acceleration techniques are particularly important for processing large bandwidths or data rates, and also in embedded platforms, where size and power consumption are usually constrained."*

BODY-364 *"A control/status screen is accessible for each instrument in the system, and the user can view and change the settings of any instrument through these screens."*

BODY-365 *"When the contact is complete, the data will be sent back to the MOC by the ground station. Prior to a launch, there will be rehearsals with everyone at their stations, and simulated telemetry with anomalous readings inserted will be used to test the team. This ensures that they are ready with the proper analysis software or integration test data available to quickly diagnose the problem and propose a plan of action."*

BODY-372 *"Effective space system security efforts begin early in the lifecycle and continue through to mission termination. Early architecture and design decisions will have the most significant impact on the overall system's security outcomes and can help avoid rework in later phases."*

BODY-375 *"In 2022 Viasat introduced new 19/24m aperture antennas at their Antenna Systems campus in Duluth, GA supporting several ongoing programs. The size and architecture of these larger apertures support current programs while offering the flexibility and scalability to support future and forward planning missions."*

BODY-376 *"Since this is very similar to the business model of Uber, these aggregator services represent the ongoing "Uber-ization" of ground station services within the space industry. The downstream service markets are observing new players with new products and services. With increasing competition, the differentiating factors are shrinking in number."*

BODY-378 *"Scheduling is emerging as the single most important enabler. As individual providers may have their own scheduling formats, for seamless operations, a common scheduler is critical."*

BODY-379 *OpenC3 COSMOS is a free, open-source, open-architecture, command, control and communications system providing commanding, scripting, and data visualization capabilities for embedded systems and systems of systems. With the release of version 5, COSMOS is now a fully containerized and microservice based architecture, with a web frontend. COSMOS is intended for use during all phases of testing (board, box, and integrated system) and during operations. OpenC3 COSMOS is made up of a set of applications that can be grouped into four categories: real-time commanding and scripting; telemetry visualization; offline analysis; and utilities.*

BODY-380 *Galaxy is a command and telemetry system that is derived directly from the Integrated Test and Operations System (ITOS) telemetry and command system developed by the Hammers Company with NASA GSFC. It has been available commercially since 2000. Galaxy can accept telemetry from, and send commands to, multiple spacecraft and ground stations simultaneously. Users can customize Galaxy for a particular mission via a database in which they provide telemetry and command specifications. Users can design telemetry displays, plots, sequential prints, configuration monitors, and spacecraft commands and table loads in simple text files stored on the computer's file system.*

BODY-381 *Orbit Logic specializes in mission planning, scheduling, and space situational awareness software. The software suite consists of multiple applications that support analysis and operations for aerial and satellite imaging and space-to-ground networking. The mobile, web, desktop, and onboard scheduling applications have a variety of features, including: configurable systems, constraints, and goals; high performance algorithms; deconflicted scheduling plans; visualizations.*

BODY-382 *"Advancements in onboard processing and data storage will demand more capability in getting data to the ground. Mass production of small satellites is quickly becoming a reality and large constellations are now starting to find their way to orbit. This will require ground system technology that can communicate with multiple satellites simultaneously."*

BODY-383 *"Optical communication is attractive for mission designers using small, resource-constrained spacecraft, because it offers a path to relatively high data rates with relatively small, low-power spacecraft equipment. The same volume and power savings can be experienced on the ground terminal side as well."*

BODY-384 *"The demand now is not for just one gigabit per second, not 10 gigabits per second, but tens if not hundreds of gigabits per second. And it's growing exponentially. The only way to achieve that is by starting to use optical communications or laser communications to augment or to complement RF communications" (Barry Matsumori, CEO of BridgeComm, 2022)*

BODY-385 *The operational RF-Optical hybrid will ultimately include 64 mirrors each of 1m diameter, installed as a segmented 8 m optical receive aperture/mirror physically inside one of the new DSN 34 m radio frequency ground terminals (DSS-23, in California). The final phase includes completion of the full 64-segment aperture on DSS-23, as illustrated in figure 11.34, including a full year of field tests. This 8 m equivalent optical ground aperture will be operational in the early 2030s.*

BODY-386 *"Optical downlinks have the potential to become an essential part of data downlinks, especially in the new-space domain."*

BODY-387 *The goal was to bring all building blocks together into an automated, cost-efficient, operational, multi-mission optical ground station service.*

BODY-388 *The First Data Link Between Commercial Optical Terminals has been validated through a temporary Sony optical terminal placed on the ISS, with a channel data-rate of 150 Mbit/s with BER varying from 1e-3 and <1e-6.*

BODY-389 *"It's likely to be how we'll see high-definition footage of the first woman to walk on the Moon," Dr Schediwy said (abc.net.au).*

BODY-390 *"Laser communication is essential for future telecom networks to supplement RF communications and enable: Very high throughput links (> 10 Gb/s and up to Tb/s), Communication without frequency band limitation, Highly secure, stealthy, non-interceptable links. It is essential for operational use cases: Multispectral observation of Earth from space (very bandwidth intensive), Securing sovereign communication, Telecommunication constellations that rely on very broad bandwidth links."*

BODY-391 *"Optical inter-satellite links are 'critical to the success of the Space Development Agency's low Earth orbit constellation' known as Transport Layer. SDA-Funded Laser Terminal Technology could connect to multiple satellites simultaneously. Each satellite in the Pentagon's 'planned mesh network of communications satellites could have as many as many as four laser links so they can talk to other satellites, airplanes, ships and ground stations.'"*

BODY-392 *Efforts are ongoing to keep pace, but only time will tell whether ground systems will advance or impede the small satellite revolution.*

BODY-398 *"Small satellites can improve their chances of being identified and tracked through good coordination with tracking agencies pre-launch, through community sharing of position and covariance data in clearly defined, consistent, standard formats (such as Orbit Ephemeris Message (OEM), and through careful selection of deployment direction and timing. Good spacecraft design choices can also improve the chances of small satellites surviving launch and early orbit and can even make use of in-space commercial radio networks as a 'back-up' method of communicating should primary systems fail."*

BODY-400 *LeoLabs is another commercial entity providing spacecraft tracking and support services. They use a group of distributed Earth-based, phased-array radars to make a commercial-off-the-shelf (COTS) satellite tracking service targeted to the specific requirements of SmallSat operators in low-Earth orbit. They currently have two radar stations in New Zealand, Costa Rica, the Azores, Poker Flat, AK, Midland, TX, Western Australia, and Argentina. There are currently seven functioning radars as of 2023, with plans for more radars strategically located around the world to robustly track objects down to 2 cm in size. The predicted performance also includes a revisit time of over 10 observations per day for specific objects, and a low-E*

BODY-401 *For spacecraft that cannot be routinely tracked by the SSN, it is important to ensure trackability by another means to enable other owner/operators to know where your spacecraft is to prevent debris-producing collisions. This is especially important for SmallSats that have orbital lifetimes that exceed operational lifetimes, and the risk to orbital neighbors remains after tracking activities have stopped.*

BODY-402 *The advantages to such a system are that it provides complete data on a satellite's position and requires no specialized ground equipment (other than the equipment used by the communications tracker, monitor key parameters of a satellite and download data using an omnidirectional antenna in any orientation.*

BODY-405 *"When interrogated at the proper frequency range, the incident RF field received by each antenna is fed to a corresponding antenna via a passive transmission line, where it is re-radiated. This significantly increases the radar cross-section of the object, allowing it to be more easily tracked."*

BODY-406 *"On the horizon, High Earth Robotics plans to create the Argus constellation – twelve optical 6U HERO-1 nanosatellites with space telescope payloads in GEO that can identify objects, take high resolution images of damaged satellites, and help identify solutions to avoid further decomposition. The constellation is intended to be resilient to interference and communications link interruption."*

BODY-407 *Improving the ability to identify and track small satellites in space – especially those deployed in batches from a single launch vehicle – can help both small satellite owners and the entire space enterprise avoid the pitfalls of "CubeSat confusion." Tracking SmallSats from launch through demise is critical to avoid collisions and preserve the space environment from orbital debris.*

BODY-413 *The influx of small spacecraft in LEO has also developed space situational awareness and space traffic management data. For information on this, please see the Identification and Tracking Systems chapter.*

BODY-415 *The Federal Aviation Administration (FAA) announced on September 20th, 2023, a proposal to create a new rule to limit the growth of debris from commercial launch vehicles in order to reduce collision risk and limit space debris in populated LEO environments. The new regulation will give commercial launch operators specific options for orbital debris countermeasures, requiring disposal of their rocket upper stages by performing a controlled reentry within 30 days after mission completion, moving to a less congested or graveyard orbit within 30 days, placing them in an Earth escape trajectory within 30 days, retrieving them with active debris removal within five years after launch, or performing an uncontrolled atmospheric disposal within 25 years.*

BODY-416 *"Drag devices are the most common deorbit device for satellites orbiting in LEO. They are advantageous due to simplicity and small stowed volumes. For certain area-to-mass ratios in altitudes equal to or lower than 800 km, drag devices can be deployed to increase the drag area for faster deorbiting in compliance with the new 5-year requirement."*

BODY-421 *"The sail has an area of 14 m2 that can effectively deorbit a 180 kg spacecraft at an altitude of 850 km in less than 10 years. In October 2022, the dragNET deorbit system was launched as part of the General Atomics GAzelle satellite."*

BODY-422 *In June 2022, China launched a Long March 2D rocket that carried a 25 m2 drag sail attached to the payload adapter on the rocket upper stage. The 300 kg object could deorbit within two years due to this technology.*

BODY-423 *SBUDNIC, a CubeSat designed and built by Brown University students with support from D-Orbit shown in figure 13.8, AMSAT-Italy, La Sapienza-University of Rome, and NASA Rhode Island Space Grant, demonstrated a practical, low-cost method to cut down on space debris. Rather than taking debris out of orbit after it becomes a problem, this $30 drag device can be added onto satellites to radically reduce how long they're in space. SBIDNIC was launched on a SpaceX rocket May 2022 as part of the Transporter 5 ridesharing mission. The plastic drag sail made from Kapton polyimide was deployed at about 520 kilometers, well above the orbit of the International Space*

BODY-424 *"The University of Florida has developed the Drag Deorbit Device (D3) 2U CubeSat which provides attitude stabilization and modulation of the satellite drag area at the same time, making the overall solution an alternative to regular ADCS units. Four 3.7 m long tape spring booms form the D3, which can deorbit a 15 kg satellite from an altitude of 700 km."*

BODY-427 *ClearSpace One will use a group of robotic arms to grab the upper stage, and then both spacecraft will be deorbited together to a lower orbit for final disintegration in the atmosphere. The mission is planned to launch in 2025 to help establish a market for in-orbit servicing and debris removal.*

BODY-434 *The pace of SmallSat technology advancement overall is rapidly accelerating and varies per subsystem. There has been significant subsystem growth in enhanced ground station support, improved technical efficiency, emerging sensor technology, and in rideshare opportunities. Recent flight missions have demonstrated innovative SmallSat technologies; the successful flights of Starling, CAPSTONE, PTD-3 and CLICK A spacecraft have each significantly contributed to SmallSat technology development. Starling successfully demonstrated intersatellite communication; CAPSTONE completed its six-month primary mission of testing the stability of the near-rectilinear halo orbit for Lunar Gateway; PTD-3 achieved a downlink of 200 gigabits per second via optical communication; and CLICK A tested the optical communication hardware that will be implemented on the second CLICK B*

BODY-435 *To comply with new orbital lifetime requirements, satellite operators are employing strategies such as decreasing the spacecraft ballistic coefficient or mass to area ratio. Deorbit technologies such as drag devices that can effectively increase the spacecraft's drag area may become even more important for future spacecraft operations in LEO.*

BODY-438 *This document does not bind the public, except as authorized by law or as NPR 7123.1D -- AppendixE incorporated into a contract. This document is uncontrolled when printed. Check the NASA Online Directives Information System (NODIS) Library to verify that this is the correct version before use: https://nodis3.gsfc.nasa.gov.*

NASA/TP—2024–10001462

State-of-the-Art
Small Spacecraft Technology

Small Spacecraft Systems Virtual Institute

Ames Research Center, Moffett Field, California

February 2024

NASA STI Program ... in Profile

Since its founding, NASA has been dedicated to the advancement of aeronautics and space science. The NASA scientific and technical information (STI) program plays a key part in helping NASA maintain this important role.

The NASA STI program operates under the auspices of the Agency Chief Information Officer. It collects, organizes, provides for archiving, and disseminates NASA's STI. The NASA STI program provides access to the NASA Aeronautics and Space Database and its public interface, the NASA Technical Reports Server, thus providing one of the largest collections of aeronautical and space science STI in the world. Results are published in both non-NASA channels and by NASA in the NASA STI Report Series, which includes the following report types:

- TECHNICAL PUBLICATION. Reports of completed research or a major significant phase of research that present the results of NASA Programs and include extensive data or theoretical analysis. Includes compilations of significant scientific and technical data and information deemed to be of continuing reference value. NASA counterpart of peer-reviewed formal professional papers but has less stringent limitations on manuscript length and extent of graphic presentations.

- TECHNICAL MEMORANDUM. Scientific and technical findings that are preliminary or of specialized interest, e.g., quick release reports, working papers, and bibliographies that contain minimal annotation. Does not contain extensive analysis.

- CONTRACTOR REPORT. Scientific and technical findings by NASA-sponsored contractors and grantees.

- CONFERENCE PUBLICATION. Collected papers from scientific and technical conferences, symposia, seminars, or other meetings sponsored or co-sponsored by NASA.

- SPECIAL PUBLICATION. Scientific, technical, or historical information from NASA programs, projects, and missions, often concerned with subjects having substantial public interest.

- TECHNICAL TRANSLATION. English-language translations of foreign scientific and technical material pertinent to NASA's mission.

Specialized services also include organizing and publishing research results, distributing specialized research announcements and feeds, providing information desk and personal search support, and enabling data exchange services.

For more information about the NASA STI program, see the following:

- Access the NASA STI program home page at http://www.sti.nasa.gov

- E-mail your question to help@sti.nasa.gov

- Fax your question to the NASA STI Information Desk at 443-757-5803

- Phone the NASA STI Information Desk at 443-757-5802

- Write to:
  STI Information Desk
  NASA Center for AeroSpace Information
  7115 Standard Drive
  Hanover, MD 21076-1320

NASA/TP—2024–0001462

State-of-the-Art
Small Spacecraft Technology

Small Spacecraft Systems Virtual Institute

Ames Research Center, Moffett Field, California

National Aeronautics and
Space Administration

Ames Research Center
Moffett Field, CA 94035-1000

February 2024

The use of trademarks or names of manufacturers in this report is for accurate reporting and does not constitute an official endorsement, either expressed or implied, of such products or manufacturers by the National Aeronautics and Space Administration.

Available from:

NASA Center for AeroSpace Information
7115 Standard Drive
Hanover, MD 21076-1320
443-757-5802

This report is also available in electronic form at

http://www.nasa.gov/smallsat-institute/sst-soa

NASA STI Support Services
Mail Stop 148
NASA Langley Research Center
Hampton, VA 23681-2199
757-864-9658

National Technical Information Service
5301 Shawnee Rd.
Alexandria, VA 22312
webmail@ntis.gov
703-605-6000

National Aeronautics and
Space Administration

SMALL SPACECRAFT SYSTEMS VIRTUAL INSTITUTE (S3VI)

Small Spacecraft Technology
State-of-the-Art Report

2023 Edition

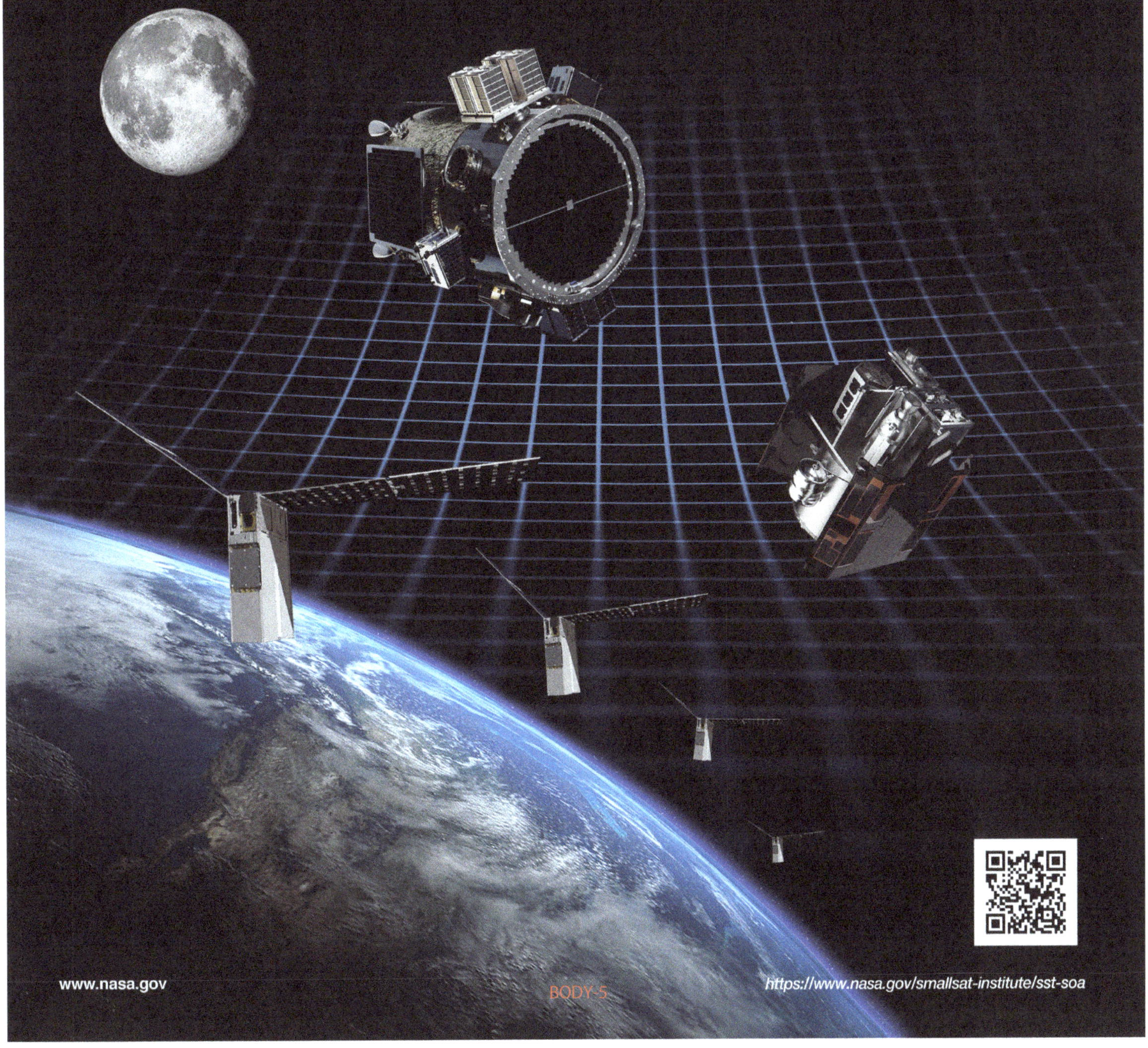

www.nasa.gov

BODY-5

https://www.nasa.gov/smallsat-institute/sst-soa

2023 Cover Page Image Credits

Top Left Vehicle: Elytra on-orbit vehicle
Credit: Firefly Aerospace, Inc.

Top Right Vehicle: ION vehicle Credit: D-Orbit, Inc.

Four Lower Spacecraft: Artist depiction of Starling CubeSat swarm
Credit: NASA

NASA Ames Research Center, Small Spacecraft Systems Virtual Institute

February 2024

Small Spacecraft Systems Virtual Institute Director:
Bruce Yost

Editor:
Sasha Weston

Contributors and Reviewers:

Helpful suggestions and contributions were also received from numerous people across NASA. In particular, the following are acknowledged for their participation as contributors or reviewers on the 2023 SOA report:

From NASA Ames Research Center: Craig Burkhard, Andres Dono, Julianna Fishman, Scott Miller, Marc Murbach and the TechEdSat team, Shang Wu, and Senior Technical Editor Teague Soderman

From NASA Glenn Research Center: Marc Abadie, Drew Ahern, Gabriel Benavides, Jonathan Valenzuela Brok, William Fabanich, James Gilland, Yao Kovach, Thomas Liu, Brian Reed

From NASA Goddard Spaceflight Center and Wallops Flight Facility: Rebekah Austin, Eric Golliher, Eliad Peretz, Luis Santos Soto, and David Steinfeld

From NASA Marshall Space Flight Center: Les Johnson, Jodi Turk, and Hunter Williams

From NASA Kennedy Space Center: Liam Cheney, Shaun Daly, Norman Phelps, Jorge Piquero

From NASA Langley Research Center: Laurence Thomsen

From NASA Headquarters: Lauri Newman

From Jet Propulsion Laboratory: Colleen Marrese-Reading and Frank Picha

From Bryce Space and Technology: Joseph Zimo and Anh Nguyen

The authors would like to also thank all of the companies, universities and organizations who provided information for this report.

Contents

Change Summary

Published Date	Edition	Chapter	Description of Changes
February 2024	2023	Complete Spacecraft Platforms	All technology tables updated.
		Power	The Solar Panel section in the Power chapter was updated.
		In-space Propulsion	All three major sections (Chemical, Electric, and Propellant-less) were updated to reflect the surge of commercial propulsion technologies.
		GNC	Edits pending next edition.
		Structures, Materials, and Mechanisms	The Mechanisms and Primary Structures sections updated.
		Thermal Control	New passive and active technology tables included.
		SmallSat Avionics	On-board Computing Systems table updated.
		Communications	Edits pending next edition.
		Launch, Integration, Deployment, and Orbital Transport	A new OMV section included with information on reusable in-space servicing vehicles.
		Ground Data Systems and Mission Operations	Updated content in Ground Segment Services, Ground Station Components, Ground Data and Supporting Systems sections.
		ID and Tracking	Minor edits throughout chapter.
		Deorbit Systems	New Orbital Debris Regulations section included; Drag Sail table and Passive and Active sections updated.

Preface

When the first edition of NASA's *Small Spacecraft Technology State-of-the-art* report was published in 2013, 247 CubeSats and 105 other non-CubeSat small spacecraft under 50 kilograms (kg) had been launched worldwide, representing less than 2% of launched mass into orbit over multiple years. Since 2013, the fight heritage for small spacecraft has greatly increased as they have become the primary source to space access for commercial, government, private, and academic institutions. Since the last edition of this report released in 2022, there has been an influx of constellations of mini-class small spacecraft with a mass 201 – 600 kg (1).

This report is updated annually to capture the wealth of new information on publicly available small spacecraft systems from NASA and other sources. While updates in all chapters reflect growth in the small spacecraft market, a concerted effort was made to update areas with recent technology developments that may ultimately bridge existing technology gaps. The organizational approach for each chapter has matured over the years to capture not only the development status of current state-of-the-art SmallSat technologies, but to also distill design considerations for the reader to consider when identifying components for their mission. Chapter organization includes an introduction of the technology, current development status of the technology's procurable systems, and summary tables of technologies surveyed. The content in each chapter is uniquely organized to present a mini-stand-alone report on the spacecraft subsystem, and information from previous editions are updated with new and maturating technologies and reference missions if applicable. Lastly, the authors tried to use the terms "SmallSat," "microsatellite," "nanosatellite," and "CubeSat" in a consistent manner, even as these terms are often used interchangeably in the space industry.

Content in this edition is based on data available by September 2023; it only contains information on SmallSat technology and does not include instrumentation or science payloads. Information presented in this report is limited to publicly available material and cannot reflect major advances in development that are not publicly disclosed. We encourage any opportunity to publish mission outcomes and technology development milestones (e.g., via conference papers, press releases, company website) so they can be reflected in this report. Overall, this report is a survey of small spacecraft technologies sourced from open literature; it does not endeavor to be an original source, and only considers literature in the public domain to identify and classify devices. Commonly used sources for data include manufacturer datasheets, press releases, conference papers, journal papers, public filings with government agencies, news articles, presentations, the compendium of databases accessed via NASA's Small Spacecraft Systems Virtual Institute (S3VI) Information Search, and engagement with companies. Data not appropriate for public dissemination, such as proprietary, export controlled, or otherwise restricted data, are not considered. As a result, this report includes many dedicated hours of desk research performed by subject matter experts reviewing resources noted above.

This report should not be considered as a comprehensive overview of all the technologies but a general overview for the current state-of-the-art SmallSat technologies and their development status. It should be noted that technology maturity designations may vary with change to payload, mission requirements, reliability considerations, and/or the environment in which performance was demonstrated. Readers are highly encouraged to reach out to companies for further information regarding the performance and maturity of the described technology.

A central element of this report is to list state-of-the-art technologies by NASA standard Technology Readiness Level (TRL) as defined by the 2020 NASA Engineering Handbook, found in NASA NPR 7123.1C *NASA Systems Engineering Processes and Requirements*. The authors have endeavored to independently verify the TRL value of each technology by reviewing and citing published test results or publicly available data to the best of their ability. Where test results and data disagree with vendors' own advertised TRL, the authors have attempted to engage the vendors to discuss the discrepancy. Readers are strongly encouraged to follow the references cited in the literature describing the full performance range and capabilities of each technology. Readers of this report should reach out to individual companies to further clarify information. It is important to note that this report takes a broad system-level view. To attain a high TRL, the subsystem must be in a flight-ready configuration with all supporting infrastructure—such as mounting points, power conversion, and control algorithms—in an integrated unit.

Future editions of this report may include content dedicated to the rapidly growing fields of assembly, integration, and testing services, and mission modeling and simulation–all of which are now extensively represented at small spacecraft conferences. Many of these subsystems and services are still in their infancy, but as they evolve and reliable conventions and standards emerge, the next iteration of this report may also evolve to include additional chapters.

References

(1) Bryce and Space Technology. "SmallSat by the Numbers, 2023." [Online] Accessed: September 28, 2023. https://brycetech.com/reports/report-documents/Bryce_Smallsats_2023.pdf

Chapter Glossary

(EELV)	Evolved Expendable Launch Vehicle
(ESPA)	EELV Secondary Payload Adapter
(FASTSAT)	Fast, Affordable, Science and Technology Satellite
(LADEE)	Lunar Atmosphere and Dust Environment Explorer
(LCROSS)	Lunar Crater Observation and Sensing Satellite
(NODIS)	NASA Online Directives Information System
(SST)	Small Spacecraft Technology
(STMD)	Space Technology Mission Directorate
(TMA)	Technology Maturity Assessment
(TRL)	Technology Readiness Level
(U)	Unit

1.0 Introduction

1.1 Objective

The objective of this report is to assess and provide an overview of the state of the art in small spacecraft technologies for mission designers, project managers, technologists, and students, connecting current small spacecraft missions to available technologies. This report focuses on the spacecraft system in its entirety, provides current best practices for integration, and then presents the state of the art for each specific spacecraft subsystem. Certain chapters have a particular emphasis on CubeSat platforms, as nanosatellite applications have expanded due to their high market growth in recent years.

This report is funded by NASA's Space Technology Mission Directorate (STMD). It was first commissioned by the Small Spacecraft Technology (SST) program within NASA's STMD in mid-2013 in response to the rapid growth in interest in using small spacecraft for low-Earth orbit, low-cost missions. The report was subsequently updated in 2015, 2018, 2020, 2021, and 2022 to capture smallsat technology growth and maturation. In addition to reporting currently available state-of-the-art technologies that have achieved TRL 5 or above, a prognosis is provided describing technologies as "on the horizon" if they are being considered for future application.

1.2 Scope

The SmallSat mission timeline began at NASA Ames Research Center with the launch of Pioneer 10 and 11 that launched in March 1972 and April 1973, respectively, where both spacecraft weighed < 600 kg. To address the increase in mass and associated cost with the high launch cadence, NASA initiated the Small Explorer (SMEX) program in 1988 to encourage the development of small spacecraft with masses in the range of ~60–350 kg. In 1998, Ames' SmallSat program then focused on lunar exploration and launched Lunar Prospector (< 700 kg), followed by the Lunar Crater Observation and Sensing Satellite (LCROSS), (< 630 kg) in 2009, and the Lunar Atmosphere and Dust Environment Explorer (LADEE), (~380 kg) which was launched in September 2013. In late 2010, NASA launched its first minisatellite called Fast, Affordable, Science and Technology Satellite (FASTSAT), which had a launch mass ~180 kg. This decrease in spacecraft mass, reduced overall cost, and increase in science capabilities ignited interest in miniaturization and maturity of aerospace technologies which have proven to be capable of producing more complex missions for less cost.

The Evolved Expendable Launch Vehicle (EELV) Secondary Payload Adapter (ESPA) payloads provided up to 180 kg mass allocation to six payload slots in 2012 when this report was first being written. As this report is focused on smaller platforms, the "180 kg mass limit" served as a good indicator to further classify the maximum "SmallSat" mass. SmallSats are generally grouped according to their mass, and this report adopts the following five small spacecraft mass categories (1):

- minisatellites are spacecraft with a total mass of 100 – 180 kg;
- microsatellites have a total spacecraft mass of 10-100 kg;
- nanosatellites have a total mass of 1 – 10 kg;
- picosatellites have a mass of 1 – 0.01 kg; and
- femtosatellites have a total spacecraft mass 0.01 – 0.09 kg.

Figure 1.1 offers examples of the various categorized spacecraft. On the lower mass end, there are projects such as KickSat-2, which deployed 100-centimeter (cm) scale "ChipSat" spacecraft, or Sprites, from a 2U femtosatellite deployer in March 2019. These femtosatellite ChipSats are the size of a large postage stamp and have a mass below 10 grams.

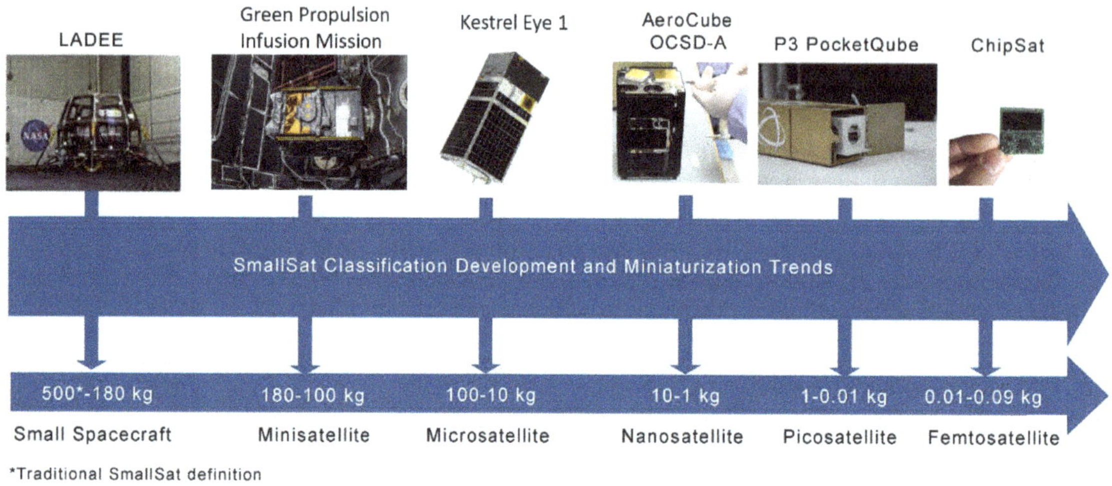

Figure 1.1: Overview of small spacecraft categories. Credit: NASA, SpaceX, Redwire Space, and Alba Orbital.

In 1999, a collaboration between California Polytechnic State University (Cal Poly) in San Luis Obispo and Stanford University in Stanford, California, developed a small educational platform called a "CubeSat" which was designed for space exploration and research for academic purposes. CubeSats are now a common form of small spacecraft that can weigh only a few kilograms and are based on a form factor of a 10 cm square cube, or unit (U) (1). The original CubeSat was composed of a single cube, a 1U, and it is now common to combine multiple cubes to form, for instance, 3U or 6U units as shown in figure 1.2. These larger CubeSat sizes have become more standardized and popular in the past five years as much more science can be achieved at less cost with the additional volume, power, and overall increase in capability.

Figure 1.2: CubeSats are a class of nano- and microsatellites that use a standard size and form factor. Credit: NASA.

It is common to interchange the terms "CubeSat" and "NanoSat" (short for nanosatellite) as the original 1-3U CubeSat platforms fall under the nanosatellite category. Since the physical expansion of CubeSats in 2014 with the 6U form factor, CubeSats now fall into both nanosatellite and microsatellite categories, and this report refers to a nanosatellite as a spacecraft with mass under 10 kg; a microsatellite as a spacecraft with mass greater than 10 kg; and a CubeSat as the accepted form factor. Figure 1.3 illustrates the three smaller SmallSat categories: microsatellites, nanosatellites, and picosatellites.

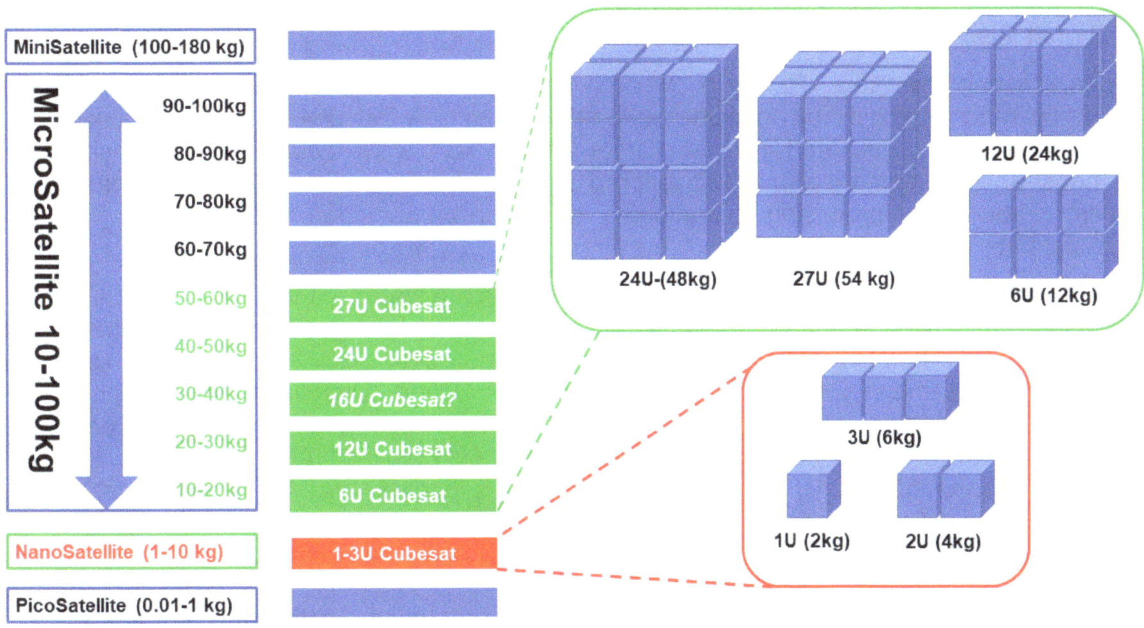

Figure 1.3: Nanosatellite sizes compared to CubeSat containerized sizes. Credit: NASA.

1.3 Assessment

While "state-of-the-art" may be defined as the most recent development stage of technology, this report considers NASA's Technology Readiness Level (TRL) scale (figure 1.4) when assessing SmallSat technology. A technology may be deemed state-of-the-art whenever its TRL is larger than or equal to 5. A TRL of 5 indicates that the component and/or brassboard with realistic support elements was built and operated for validation in a relevant environment so as to demonstrate overall performance in critical areas. Success criteria include documented test performance demonstrating agreement with analytical predictions and documented definition of scaling requirements. Performance predictions are made for subsequent development phases (2).

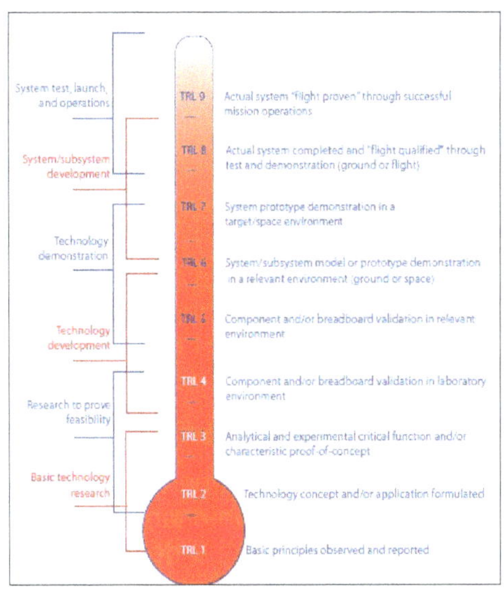

Figure 1.4: NASA's standard TRL scale. Credit: NASA.

An accurate TRL assessment requires a high degree of technical knowledge on a subject device, and an in-depth understanding of the mission (including interfaces and environment) on which the device was flown. TRL values vary depending on design factors for a specific technology. For example, differences in TRL assessment based on the operating environment may result from mechanical loads, mission duration, the thermal environment, or radiation exposure. The authors believe TRLs are most accurately determined when assessed within the context of a program's unique requirements. If a technology has flown on a mission without success, or without providing valid confirmation to the operator, such claimed "flight heritage" is discounted. Some older technologies may still be well suited to certain mission needs and still be regarded as "state-of-the-art." For a technology to be considered obsolete, "retired", or no longer "state-of-the-art", it's performance must have been surpassed by newer technology such that it is no longer used.

While a technology with a TRL value lower than or equal to 4 may not be state of the art, in some cases these technologies may considered "on the horizon." A TRL of 4 is defined as a component and/or breadboard validated in a laboratory environment with documented test performance demonstrating agreement with analytical predictions and a documented definition of the relevant environment. These promising technologies may soon be considered state-of-the-art for small spacecraft.

NASA standard TRL requirements for this report edition are stated in the NPR 7123.1C, Appendix E, which is effective through February 14, 2025. The criteria for selection of appropriate TRL are described in the NASA Systems Engineering Handbook 6105 Rev 2 Appendix G: Technology Assessment/Insertion. Please refer to the NASA Online Directives Information System (NODIS) website https://nodis3.gsfc.nasa.gov/ for NPR documentation. The following paragraphs in sections 1.3.1 and 1.3.2 of this introduction are excerpts from the NASA Engineering Handbook 6105 Rev 2 (pp. 252 – 254). They highlight important aspects of NASA TRL guidelines in hopes of eliminating confusion on terminology and heritage systems.

1.3.1 Terminology

"At first glance, the TRL descriptions in figure 1.4 appear to be straightforward. It is in the process of trying to assign levels that problems arise. A primary cause of difficulty is in terminology, e.g., everyone knows what a breadboard is, but not everyone has the same definition. Also, what is a "relevant environment?" What is relevant to one application may or may not be relevant to another. Many of these terms originated in various branches of engineering and had, at the time, very specific meanings to that particular field. They have since become commonly used throughout the engineering field and often acquire differences in meaning from discipline to discipline, some differences subtle, some not so subtle. "Breadboard," for example, comes from electrical engineering where the original use referred to checking out the functional design of an electrical circuit by populating a "breadboard" with components to verify that the design operated as anticipated. Other terms come from mechanical engineering, referring primarily to units that are subjected to different levels of stress under testing, e.g., qualification, protoflight, and flight units. The first step in developing a uniform TRL assessment (see figure 1.5) is to define the terms used. It is extremely important to develop and use a consistent set of definitions over the course of the program/project."

1.3.2 Heritage Systems

"Note the second box particularly refers to heritage systems (figure 1.5). If the architecture and the environment have changed, then the TRL decreases to TRL 5—at least initially. Additional testing may need to be done for heritage systems for the new use or new environment. If in subsequent analysis the new environment is sufficiently close to the old environment or the new

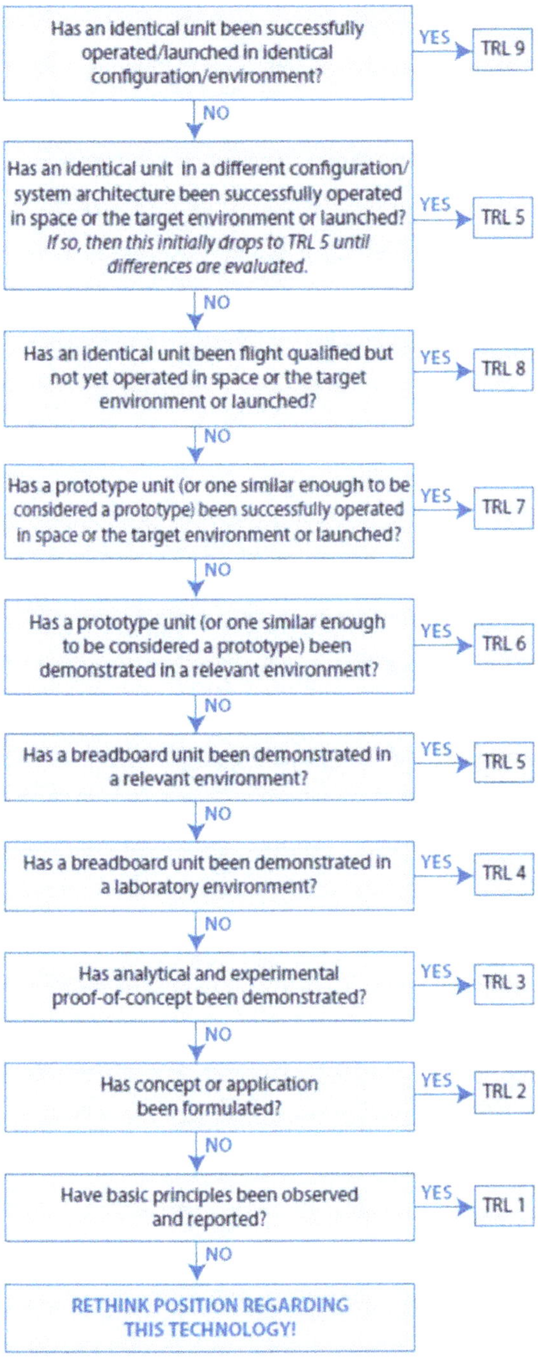

Figure 1.5: Technology Maturity Assessment (TMA) thought process. Credit: NASA.

architecture is sufficiently close to the old architecture, then the resulting evaluation could be TRL 6 or 7, but the most important thing to realize is that it is no longer at TRL 9. Applying this process at the system level and then proceeding to lower levels of subsystems and components identifies those elements that require development and sets the stage for the subsequent phase, determining the new TRL."

References

(1) NASA. What are SmallSats and CubeSats? February 26, 2015. Revised August 6, 2017. https://www.nasa.gov/content/what-are-smallsats-and-cubesats

(2) NASA Systems Engineering Handbook. NASA/SP-2016 6105 Rev. 2. https://www.nasa.gov/feature/release-of-revision-to-the-nasa-systems-engineering-handbook-sp-2016-6105-rev-2

Chapter Contents

Chapter Glossary

(COTS)	Commercial-off-the-Shelf
(EELV)	Evolved Expendable Launch Vehicle
(ESPA)	EELV Secondary Payload Adapter
(GEO)	Geostationary Equatorial Orbit
(I&T)	Integration and Test
(kg)	Kilogram
(LEO)	Low Earth Orbit
(MEO)	Medium Earth Orbit
(MTBF)	Mean Time Between Failures
(NASA)	National Aeronautics and Space Administration
(SHF)	Super High Frequency
(SmallSat)	Small Satellite
(SPA)	Secondary Payload Adapter
(STMD)	Space Technology Mission Directorate
(UHF)	Ultra High Frequency
(UK)	United Kingdom
(Unk)	Unknown
(USA)	United States of America
(VLEO)	Very Low Earth Orbit
(VHF)	Very High Frequency
(W)	Watts
(xGEO)	Beyond Geostationary Equatorial Orbit

2.0 Complete Spacecraft Platforms

2.1 Introduction

The SmallSat market provides a variety of mission-enabling components. Along with a large variety of new and proven components, companies are offering entire spacecraft bus solutions. Spacecraft bus refers to the side of the mission flight segment that provides essential services to the payload. This chapter addresses the state of the art in the small spacecraft bus offerings and provides the reader with a programmatic overview for small spacecraft mission development.

There are 2 distinct types of SmallSat market options in terms of complete spacecraft platforms. One option is not superior to the other and selection may depend on the needs of each individual mission.

- **Hosted payloads** – Also known as "satellite-as-a-service," integrates multiple payloads from different and independent customers into the same platform with some form of resource sharing (cost, autonomy, concept of operations, etc.). Hosted payload configurations and performance vary by provider. Two examples of hosted payloads are:
    - Service provider brokers multiple independent customer payloads into a single spacecraft bus (no primary payload)
    - Service provider intends to launch their own satellite with its own primary goals but have unused resources and allows secondary payloads to be added
- **Dedicated spacecraft bus** – the entirety of the spacecraft bus is at the disposal of a single customer or mission

This chapter organizes the state-of-the-art small spacecraft platforms into these two main categories. The dedicated small spacecraft bus section is further divided by PocketQube, CubeSat, and ESPA-Class offerings. Each subsection contains a summary table with a non-exhaustive list of commercially available small spacecraft platforms.

1. Hosted Payloads (2.2.1)
2. Dedicated Spacecraft Bus (2.2.2)
    a. PocketQubes (2.2.2.1)
    b. CubeSats (2.2.2.2)
    c. ESPA-Class (2.2.2.3)

Following Section 2.2 is a brief explanation on systems engineering considerations that introduces newcomers to the design selection process and highlights specific resources for mission development. On the Horizon is a section that describes upcoming technology considered low maturity and revolutionary in small spacecraft platform with the potential to advance the state-of-the-art.

The list of organizations/companies in this chapter is not all-encompassing and does not constitute an endorsement from NASA. The information is for awareness and guidance only. The performance advertised may differ from actual performance since the information has not been independently verified by the State-of-the-Art document staff and relies on information provided directly from the manufacturers or available public information.

Section 2.6 includes a list of providers with contact information and the source used to complete the tables. It is recommended to contact the organizations/companies directly for further clarification and application to your specific needs.

2.2 State-of-the-Art – Spacecraft Platforms

2.2.1 Hosted Payloads

Hosted payloads, also referred as "satellite-as-a-service," "hitchhiking" or "piggybacking," is increasing in popularity due to its cost savings. The idea is to share the spacecraft bus platform with other payloads and still achieve mission success. The terms of the agreement are negotiated in advance with the provider to ensure necessary on-orbit time, power, pointing and data volume (among other resources) are adequate for the mission.

Figure 2.1: Representation of NASA's FASTSAT minisatellite. Credit: NASA.

Configurations of a hosted payload platform are typically scalable, and several spacecraft platform vendors provide hosted payload services. Larger spacecraft bus hosted options offer deployable capability/mechanisms for smaller nanosatellite missions. NASA's Fast, Affordable, Science and Technology Satellite (FASTSAT) is an example of a minisatellite that hosted smaller science and technology flight missions. It carried several low-TRL experiments and deployed NanoSail-D. See figure 2.1 for an illustration of FASTSAT. Figure 2.2 is from Loft Orbital Hosted Payload Services.

Hosted payload services are becoming more appealing for academic and government scientific missions. This option provides a cost-effective and timely solution to those missions going to the same destination.

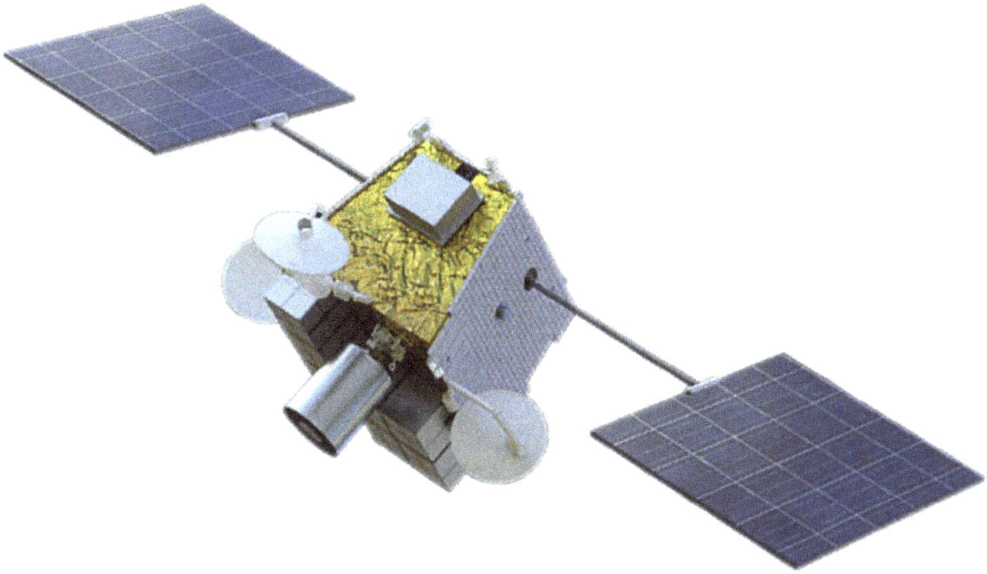

Figure 2.2: A rendering of a generic Longbow-class Loft Orbital satellite. Credit: Loft Orbital.

Table 2-1: Hosted Payload Providers

(The fields indicate maximum capability, organizations may offer multiple options including smaller capabilities within the Hosted Payloads category)

Organization	Max Volume	Max Mass (kg)	Peak Power (W)	3-σ Pointing Control/ Knowledge	Destination	US Office
Artemis Space Technologies UK	0.58 m^3	500	1,500	0.01°/0.01°	LEO, MEO, GEO, Lunar and Deep Space	No
Astranis Space Technologies Corp. USA	0.02 m^3	10	300	<0.1°/ <0.09°	GEO	Yes
Axelspace Japan	0.2 m^3	30	184	<0.05°/ <0.04°	LEO	No
Berlin Space Technologies Germany	1 m^3	200	3,000	<0.017°/< 0.017°	LEO	Yes
Bradford Space USA	0.38 m^3	220	1,500	1.5°/ 0.006°	LEO, GEO, GTO, Cislunar, Lunar, Deep Space	Yes
C3S Electronics Development Hungary	16.5U	18.5	155	0.2°/ 0.2°	LEO, MEO	No
EnduroSat Bulgaria	10U	20	60	0.1°/ 0.05°	LEO	Yes
General Atomics EMS USA	0.46 m^3	200	450	0.03°/ 0.02°	LEO	Yes
German Orbital Systems Germany	4U	8	50	<1°/< 1°	LEO	No
Gran Systems Taiwan	6U	9	25	5°/ 5°	LEO, Lunar	No
Hemeria France	0.1 m^3	35	250	0.03°/0.01°	LEO, GTO, GEO	No
Innova Space Argentina	0.5U	0.5	4	<15°/< 15°	LEO	Yes
In-Space Missions UK	Unk	Unk	Unk	Unk	LEO	Unk
Loft Orbital USA	0.44 m^3	85	>1,000	<0.035°/<0.03°	LEO	Yes
Momentus USA	1 m^3	350	3,000	0.05°/ 0.05°	LEO	Yes
Muon Space USA	60U	30	200	0.03°/ 0.012°	LEO	Yes
NanoAvionics Lithuania	0.7 m^3	150	378	0.15°/ 0.03°	LEO	Yes
NearSpace Launch USA	8U	16	160	0.5°/ 0.2°	LEO	Yes
Northrop Grumman USA	0.37 m^3	50	420	<4°/<1°	LEO	Yes
NovaWurks USA	1 m^3	150	1000	0.002°/0.0004°	LEO, GEO, xGEO	Yes
NPC SPACEMIND Italy	12U	24	100	<0.1°/<0.1°	LEO, MEO	No
OHB LuxSpace Luxembourg	0.3 m3	90	600	<0.022°/ 0.01°	LEO	No

Open Cosmos [UK]	12U	18	160	0.03°/0.02°	LEO	No
Orbital Astronautics [UK]	0.163 m^3	100	5,000	<0.05°/<0.01°	LEO, MEO, GEO, Deep Space	No
Orion Space Solutions [USA]	14U	45	400	<1°/<1°	LEO, GEO, Lunar	Yes
Quantum Space [USA]	0.5 m^3	100	400	0.006°/0.006°	LEO, GEO, Cislunar, Lunar, Deep Space	Yes
Redwire [USA]	0.94 m^3	104	500	0.005°/0.0017°	LEO, MEO, GEO and Deep Space	Yes
SatRev [Poland]	3U	3	25	1°/0.6°	LEO	No
Sierra Space [USA]	0.48 m^3	250	500	0.001°/ <0.001°	LEO, MEO, GEO	Yes
SITAEL [Italy]	0.54 m^3	100	1,000	0.018°/ 0.010°	LEO	No
Space Inventor [Denmark]	24U	50	400	<0.008°/ <0.008°	LEO, GEO, MEO	No
Spacemanic [Czech Republic]	12U	18	500	0.1°/ 0.05°	LEO, MEO, GEO, Lunar	No
Spire Global [USA]	12U	32	300	0.1°/ 0.05°	LEO	Yes
Xplore [USA]	0.125 m^3	55	210	0.17°/ 0.018°	VLEO, LEO, Cislunar	Yes
York Space Systems [USA]	-	300	1,500	0.008°/ 0.004°	LEO, GEO, Lunar	Yes

2.2.2 Dedicated Spacecraft Bus

The market has grown considerably over the last 5 years with complete spacecraft bus solutions including I&T and operations options. The addition of I&T and operations gives missions flexibility in implementation, allowing the mission to focus on unique or challenging aspects of the project as needed. Mission implementation solutions are shown in table 2-2. A complete vendor solution can allow the mission organization to focus primarily on payload development, however this may not be appropriate for all missions. For example, an organization may decide to perform their own mission operations if the vendor offerings do not meet the requirements for the project.

Table 2-2: Mission Implementation Flexibility

Option	Product or Service		
	Spacecraft Bus	**System-Level Integration and Testing**	**Operations**
1	Vendor	Vendor	Vendor
2	Vendor	Vendor	Mission Organization
3	Vendor	Mission Organization	Mission Organization
4	Mission Organization	Mission Organization	Mission Organization

2.2.2.1 PocketQubes

PocketQubes refer to small satellites that conform to a form factor of 5 cm cubes. PocketQubes use a standard deployer and follow a unit nomenclature of P. In this case 1P refers to a single 5 cm cube (see figure 2.3). Consequently, 2P refers to 2 of these single units. A typical PocketQube deployer can deploy up to a 3P satellite but larger deployers may allow additional capability. PocketQube providers have developed spacecraft busses to simplify mission implementation; a list of providers is included in this section; table 2-3 provides available commercial PocketQube products. Figure 2.4 is an example of a PocketQube deployer at Alba Orbital.

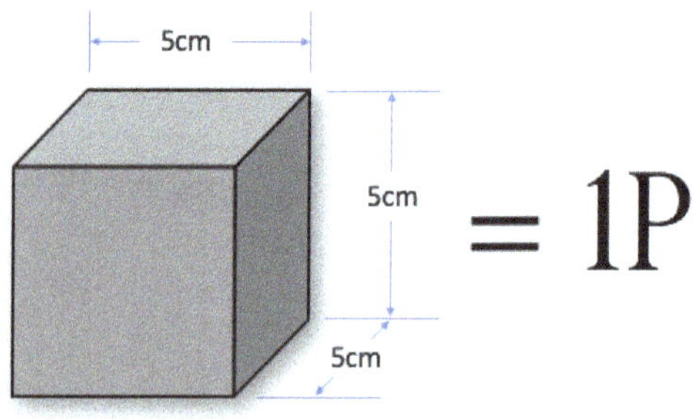

Figure 2.3: PocketQube Dimensions.

Figure 2.4: Alba Orbital Integration of PocketQubes into the Deployers. Credit: Alba Orbital.

Table 2-3: PocketQubes Market Solutions (The fields indicate maximum capability, organizations may offer multiple options including smaller capabilities within the PocketQube category)						
Organization	**Peak Power (W)**	**3-σ Pointing Control/ Knowledge**	**Comm Options**	**Intended Destination**	**Maturity**	**US Office**
Alba Orbital UK	15	5°/2°	UHF, S	LEO	Flown LEO	Yes
Citadel Space Systems UK	20	Unk	UHF, S	Unk	Unk	Unk
DIYSATELLITE Argentina	9	<5°/<5°	VHF, UHF, SHF	LEO, GEO, Lunar	Flown LEO	No
FOSSA Systems Spain	10	<5°/<5°	UHF, S	LEO	Flown LEO	No
Innova Space Argentina	3.9	N/A - Magnetic Passive	UHF	LEO	Flown LEO	Yes
Quub, Inc. USA	26	5°/2°	UHF, S	LEO, Lunar	Flown LEO	Yes

2.2.2.1 CubeSats

CubeSats refer to small satellites that conform to a form factor of 10 cm cubes. The CubeSat standard was created by California Polytechnic State University, San Luis Obispo, and Stanford University's Space Systems Development Lab in 1999 to facilitate access to space for university students. When launch providers started adopting this standard as a secondary payload service it enabled increased, low-cost opportunities for space access. Many organizations are currently using the standard including academia, private industry, and government. For more information on the history of CubeSats, the reader is encouraged to review the Introduction of this report.

CubeSat sizes follow a unit nomenclature in which 1 unit or 1U refers to a single 10 cm cube (see figure 2.5). Consequently, 2U refers to 2 of these single units, 3U is a set of 3 single units, and so forth. CubeSat providers have developed spacecraft busses to accommodate missions from 1U to 27U satellites. This section provides a list of providers separated by satellite size: 0.25U-3U, 6U, 12U and 16U+ in tables 2-4, 2-5, 2-6, and 2-7.

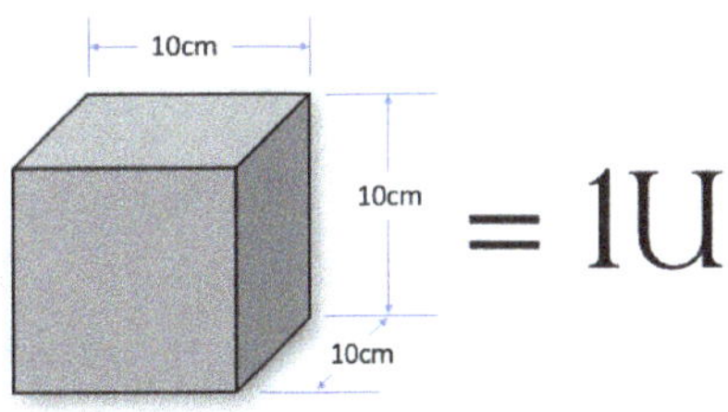

Figure 2.5 - CubeSat Dimensions.

Multiple companies have developed deployers for CubeSats with different dimensions and external volume allocations. Contact your sponsoring organization and/or launch provider for specifics on which deployer is used in your mission. Many CubeSat deployers exist in the market but the primary 2 interfaces follow the classic corner rails or the tabs (clamped and unclamped), as seen in figure 2.6. Most spacecraft bus providers in this chapter can adapt to different interfaces. Please refer to the Launch, Integration, and Deployment chapter for further information on SmallSat deployers. Figure 2.7 includes images of CubeSat missions that have

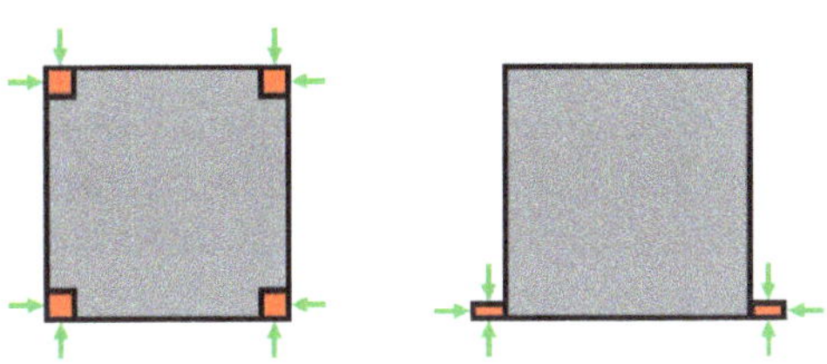

Figure 2.6: Rails vs. Tabs Restraint System Cross-Section.

been successfully flown in space, figure 2.8 provides examples of CubeSat deployers' location on a rocket, and figure 2.9 provides examples for 6U and 16U satellites from Spire Global.

Figure 2.7: Examples of flown CubeSats. (Top left) 1U PhoneSat spacecraft, (top right) 12U CAPSTONE spacecraft, (lower left) 3U CLICK spacecraft, (lower right) 6U PTD-3 spacecraft. Credits: NASA and Terrain Orbital.

Figure 2.8: (left) Location of Artemis CubeSat deployers in between the Orion Crew Vehicle and the Interim Cryogenic Propulsion Stage (ICPS); (right) NASA Nodes mission deployment from ISS. Credit: NASA.

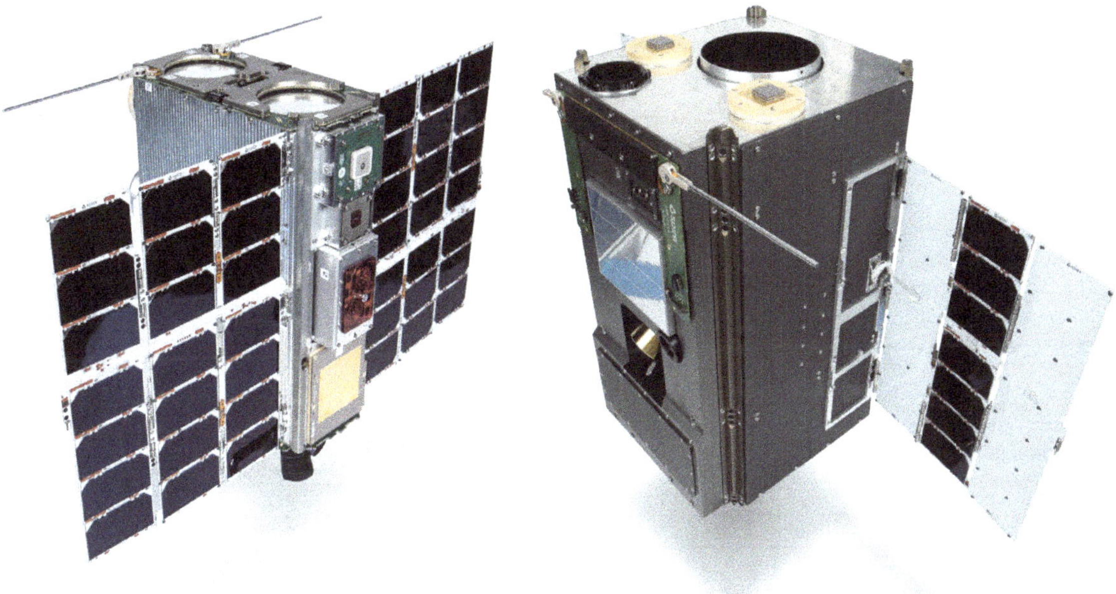

Figure 2.9: Examples of a 6U and 16U CubeSat. Credit: Spire Global.

Table 2-4: 0.25U-3U Market Solutions

(The fields indicate maximum capability, organizations may offer multiple options including smaller capabilities within the 0.25U-3U category)

Organization	Peak Power (W)	3-σ Pointing Control/ Knowledge	Comm Options	Intended Destination	Maturity	US Office
AAC Clyde Space Sweden	90	<0.1°/<0.01°	VHF, UHF, S, X	LEO	Flown LEO	Yes
Alén Space Spain	180	0.2°/0.1°	VHF, UHF, S	LEO	Flown LEO	No
Artemis Space Technologies UK	50	0.01°/0.01°	UHF, S, X, Ka, Ku	Designed for LEO	Flown LEO	No
Blue Canyon Technologies USA	27	0.003°/0.003°	L, S, X	LEO, GEO, Deep Space	Flown LEO Qualified GEO and Deep Space	Yes
C3S Electronics Hungary	35	0.2°/0.2°	UHF, S	LEO, MEO	Flown LEO	No
EnduroSat Bulgaria	30	<1°/<0.6°	UHF, S, X	LEO	Flown LEO	Yes
General Atomics EMS USA	10	0.28°/0.08°	UHF	LEO, MEO, GEO, xGEO	Under Development	Yes
German Orbital Systems Germany	24	<1°/<1°	UHF, S	LEO	Flown LEO	No
GomSpace Denmark	35	2.5°/2°	S	LEO	Flown LEO	Yes
Gran Systems Taiwan	25	5°/ 5°	VHF, UHF	LEO	Flown LEO	No
GUMUSH AeroSpace Turkey	80	<1°/ <0.005°	VHF, UHF, S, X	LEO	Flown LEO	No
Hex20 Australia	25	0.003°/ 0.003°	UHF, S	LEO, MEO, GEO, Lunar	Flown LEO	No
IMT Italy	>5	10°/5°	VHF, UHF	LEO	Under Development	No
Innova Space Argentina	7.5	<15°/<15°	UHF	LEO	Flown LEO	Yes
ISISPACE The Netherlands	50	<15°/<15°	VHF, UHF, S	LEO	Flown LEO	No
NanoAvionics Lithuania	175	13.20°/12.93°	UHF, S, X	LEO	Flown LEO	Yes
NearSpace Launch USA	80	0.5°/0.2°	L, UHF, S, X	VLEO, LEO	Flown LEO	Yes
NPC SPACEMIND Italy	51.6	<0.1°/<0.1°	UHF, S, X, Ka	LEO, MEO, GEO, Lunar	Flown LEO and MEO	No
Open Cosmos UK	160	2.4°/0.67°	UHF, S	LEO	Flown LEO	No

Orbital Astronautics UK	400	0.1°/ 0.01°	S, X, K, Ka, Optical	LEO, MEO	Flown LEO	No
Orion Space Solutions USA	8	1°/1°	L, S, X	LEO	Qualified LEO	Yes
Pumpkin Space Systems USA	200	0.05°/<0.05°	UHF, S, X, Ka	LEO	Flown LEO	Yes
Quub, Inc. USA	44	5°/2°	UHF, S	LEO, Lunar	Flown LEO	Yes
SatRev Poland	36	1°/0.6°	UHF, S	LEO	Flown LEO	No
SkyLabs Slovenia	100	0.3°/0.06°	VHF, UHF, S	LEO, MEO	Flown LEO and MEO	No
Space Flight Laboratory Canada	93	0.009°/0.004°	UHF, S, X, Ka	LEO, GEO, Lunar	Flown LEO Qualified GEO and Lunar	No
Space Inventor Denmark	100	0.01° / 0.01°	VHF, UHF, S, X, L	LEO	Flown LEO	No
Spacemanic Czech Republic	30	0.1°/0.05°	VHF, UHF, S	LEO, GEO, Lunar	Flown LEO Qualified GEO	No
Spire Global USA	35	0.1°/0.05°	UHF, L, S, X, Ka, Ku	LEO	Flown LEO	Yes

Table 2-5: 6U Market Solutions
(The fields indicate maximum capability, organizations may offer multiple options including smaller capabilities within the 6U category)

Organization	Peak Power (W)	3-σ Pointing Control/ Knowledge	Comm Options	Intended Destination	Maturity	US Office
AAC Clyde Space Sweden	150	<0.1°/<0.01°	VHF, UHF, S, X	LEO	Flown LEO	Yes
Alén Space Spain	180	0.2°/0.1°	VHF, UHF, S	LEO	Flown LEO	No
Argotec Italy	100	<0.03°/<0.01°	UHF, S, X, K	LEO, GEO, Lunar, Deep Space	Flown Deep Space Flown Lunar	Yes
Artemis Space Technologies UK	100	0.01°/0.01°	UHF, S, X, Ka, Ku, Optical	LEO, MEO, GEO, Lunar, Deep Space	Flown LEO Qualified MEO, GEO, Lunar, and Deep Space	No
Astro Digital USA	240	<0.1°/<0.05°	UHF, S, X, Ka	LEO	Flown LEO	Yes
Blue Canyon Technologies USA	108	0.003°/0.003°	L, S, X	LEO, GEO, Deep Space	Flown LEO and Lunar Qualified GEO and Deep Space	Yes
C3S Electronics Development Hungary	165	<0.2°/<0.2°	UHF, S	LEO, MEO	Under Development	No
EnduroSat Bulgaria	60	<0.021°/<0.021°	UHF, S, X	LEO	Flown LEO	Yes
General Atomics EMS USA	10	0.28°/0.08°	UHF, S	LEO	Flown LEO	Yes
German Orbital Systems Germany	72	<1°/<1°	UHF, S, X	LEO	Flown LEO	No
GomSpace Denmark	102	0.07°/0.056°	S, X	LEO, Deep Space	Flown LEO Qualified Deep Space	Yes
Hex20 Australia	45	0.003°/0.003°	UHF, S, X	LEO, MEO, GEO, Lunar	Flown LEO	No
IMT Italy	115	0.1°/0.1°	VHF, UHF, S, C, X	LEO	Under Development	No
ISISPACE The Netherlands	100	<0.3°/<0.3°	UHF, S, X	LEO, Lunar	Flown LEO Qualified for Lunar	No
Millennium Space Systems USA	100	<0.03°/<0.014°	UHF, S	LEO	Flown LEO	Yes

NanoAvionics Lithuania	175	0.18°/0.12°	UHF, S, X	LEO	Flown LEO	Yes
NearSpace Launch USA	160	0.5°/0.2°	L, UHF, S, X	LEO	Flown LEO	Yes
NPC SPACEMIND Italy	85.2	<0.1°/<0.1°	UHF, S, X, Ka	LEO, MEO, GEO, Lunar	Flown LEO	No
Open Cosmos UK	160	0.02°/0.01°	UHF, S, X	LEO	Flown LEO	No
Orbital Astronautics UK	1,000	0.1°/0.01°	S, X, K, Ka, Optical	LEO, MEO	Flown LEO	No
Orion Space Solutions USA	15	1°/1°	L, S, X	LEO	Flown LEO	Yes
Pumpkin Space USA	200	0.05°/<0.05°	UHF, S, X, Ka	LEO, Lunar	Flown LEO Qualified Lunar	Yes
Quub, Inc. USA	50	5°/2°	UHF, S, Ku	LEO, Lunar	Under Development	Yes
SatRev Poland	36	1°/0.6°	UHF, S	LEO	Qualified LEO	No
SkyLabs Slovenia	200	0.3°/0.06°	VHF, UHF, S	LEO, MEO	Flown LEO and MEO	No
Space Dynamics Lab USA	80	0.021°/0.021°	UHF, S, X, Ka	LEO, GEO, GTO, Cislunar, Deep Space	Qualified LEO and GEO	Yes
Space Flight Laboratory Canada	240	0.009°/0.004°	UHF, S, X, Ka	LEO, GEO, Lunar	Flown LEO Qualified GEO and Lunar	No
Space Inventor Denmark	200	<0.008°/<0.008°	VHF, UHF, S, X	LEO	Flown LEO	No
Spacemanic Czech Republic	500	0.1°/0.05°	VHF, UHF, S	LEO, GEO, Lunar	Flown LEO Qualified GEO	No
Spire Global USA	200	0.1°/0.05°	UHF, L, S, X. Ka, Ku	LEO	Flown LEO	Yes
Terran Orbital USA	180	0.008°/0.007°	UHF, S, X, C	LEO, GEO, Deep Space	Flown LEO and Lunar Qualified GEO and Deep Space	Yes

Table 2-6: 12U Market Solutions

(The fields indicate maximum capability, organizations may offer multiple options including smaller capabilities within the 12U category)

Organization	Peak Power (W)	3-σ Pointing Control/ Knowledge	Comm Options	Intended Destination	Maturity	US Office
AAC Clyde Space [Sweden]	400	<0.01°/<0.0075°	VHF, UHF, S, X, K, Ka, Ku, Optical	LEO	Qualified LEO	Yes
Argotec [Italy]	180	<0.03°/<0.01°	UHF, S, X, K	LEO, GEO, Lunar, Deep Space, DRO	Under Development	Yes
Artemis Space Technologies [UK]	150	0.01°/0.01°	UHF, S, X, Ka, Ku, Optical	LEO, MEO, GEO, Lunar, Deep Space	Flown LEO Qualified GEO, MEO, Lunar, and Deep Space	No
Blue Canyon Technologies [USA]	108	0.0025°/0.0025°	L, S, X	LEO, GEO, Deep Space	Flown LEO and GEO Qualified Deep Space	Yes
C3S Electronics Development [Hungary]	165	<0.2°/<0.2°	UHF, S	LEO, MEO	Under Development	No
EnduroSat [Bulgaria]	70	0.1°/0.05°	UHF, S, X, K	LEO	Flown LEO	Yes
General Atomics EMS [USA]	115	0.3°/0.024°	UHF, S	LEO	Qualified LEO	Yes
GomSpace [Denmark]	102	0.07°/0.056°	S, X	LEO	Qualified LEO	Yes
Hex20 [Australia]	110	0.003°/0.003°	UHF, X	LEO, MEO, GEO, Lunar	Flown LEO	No
ISISPACE [The Netherlands]	190	<0.03°/<0.03°	UHF, S, X, Ka	LEO	Under Development	No
NanoAvionics [Lithuania]	175	0.18°/0.09°	UHF, S, X	LEO	Flown LEO	Yes
NearSpace Launch [USA]	500	0.5°/0.2°	L, UHF, S, X	LEO, MEO	Under Development	Yes
NPC SPACEMIND [Italy]	96	<0.1°/<0.1°	UHF, S, X, Ka	LEO, MEO, GEO, Lunar	Flown LEO	No
Open Cosmos [UK]	160	0.031°/0.027°	UHF, S, X	LEO	Flown LEO	No
Orbital Astronautics [UK]	1,000	0.05°/0.01°	S, X, K, Ka, Optical	LEO, MEO, GEO	Flown LEO	No
Orion Space Solutions [USA]	40	<1°/<1°	L, S, X, Ka	LEO, GEO	Qualified LEO and GEO	Yes
Pumpkin Space [USA]	400	0.05°/<0.05°	UHF, S, X, Ka	LEO, Lunar	Qualified LEO	Yes
SkyLabs [Slovenia]	500	0.3°/0.06°	VHF, UHF, S	LEO, MEO	Flown LEO and MEO	No

Space Dynamics Lab [USA]	80	0.021°/0.021°	UHF, S, X, Ka	LEO, GEO, GTO, Cislunar, Deep Space	Flown LEO Qualified GTO and GEO	Yes
Space Flight Laboratory [Canada]	322	0.009°/0.004°	UHF, S, X, Ka	LEO, GEO, Lunar	Flown LEO Qualified GEO and Lunar	No
Space Information Laboratories [USA]	180	0.008°/0.008°	S, X, Ka	LEO, GEO, Lunar	Under Development	Yes
Space Inventor [Denmark]	200	<0.008°/<0.008°	VHF, UHF, S, X	LEO	Flown LEO	No
Spacemanic [Czech Republic]	500	0.1°/0.05°	VHF, UHF, S, X	LEO, GEO, Lunar	Flown LEO Qualified GEO	No
Spire Global [USA]	300	0.1°/0.05°	UHF, L, S, X, Ka, Ku	LEO	Qualified LEO	Yes
Terran Orbital [USA]	180	0.008°/0.007°	UHF, S, X, C	LEO, GEO, Lunar, Deep Space	Flown LEO and Lunar Qualified GEO and Deep Space	Yes

Table 2-7: 16U+ Market Solutions
(The fields indicate maximum capability, organizations may offer multiple options including smaller capabilities within the 16U+ category)

Organization	Format	Peak Power (W)	3-σ Pointing Control/ Knowledge	Comm Options	Intended Destination	Maturity	US Office
AAC Clyde Space Sweden	16U	400	<0.01°/<0.0075°	VHF, UHF, S, X, K, Ka, Ku, Optical	LEO	Qualified LEO	Yes
Argotec Italy	16U+	250	<0.07°/<0.03°	UHF, S, X, K	GEO, Lunar, Mars, Deep Space	Under Development	Yes
Artemis Space Technologies UK	16U	200	0.01°/0.01°	UHF, S, X, Ka, Ku, Optical	LEO, MEO, GEO, Lunar, Deep Space	Flown LEO Qualified GEO, MEO, Lunar, and Deep Space	No
Astro Digital USA	16U+	500	<0.05°/<0.01°	UHF, S, X, Ku, Ka, V, W, Optical	LEO	Flown LEO	Yes
Blue Canyon Technologies USA	16U	108	0.0025°/0.0025°	L, S, X	LEO, GEO, Deep Space	Qualified LEO, GEO and Deep Space	Yes
C3S Electronics Hungary	16U+	165	<0.2°/<0.2°	UHF, S	LEO, MEO	Under Development	No
EnduroSat Bulgaria	16U	80	0.1°/0.05°	UHF, S, X, K	LEO	Qualified LEO	Yes
German Orbital Systems Germany	16U	164	<1°/<1°	UHF, S, X	LEO	Qualified LEO	No
GomSpace Denmark	16U	150	0.07°/0.056°	S, X	LEO	Qualified LEO	Yes
Hex20 Australia	27U	150	0.003°/0.003°	UHF, S, X	LEO, MEO, GEO, Lunar	Flown LEO	No
ISISPACE The Netherlands	16U	190	<0.03°/<0.03°	UHF, S, X, Ka	LEO	Under Development	No
NanoAvionics Lithuania	16U	175	0.18°/0.09°	UHF, S, X	LEO	Flown LEO	Yes
Nara Space Korea	16U	232	0.006°/0.006°	S, X	LEO	Qualified LEO	No
NPC SPACEMIND Italy	16U	120	<0.1°/<0.1°	UHF, S, X, Ka	LEO, MEO, GEO, Lunar	Under Development	No
Open Cosmos UK	16U	160	0.031°/0.027°	UHF, S, X	LEO	Flown LEO	No
Orbital Astronautics UK	16U, 27U	1,000	0.05°/0.01°	S, X, K, Ka, Optical	LEO, GEO, Lunar	Qualified LEO	No
Orion Space Solutions USA	16U+	400	<1°/<1°	L, S, X, Ka, Optical	LEO, GEO, Lunar	Qualified LEO, GEO and Lunar	Yes

Pumpkin Space USA	16U, 27U	400	0.05°/<0.05°	UHF, S, X, Ka	LEO, Lunar	Qualified LEO	Yes
SkyLabs Slovenia	20U+	500	<0.005°/<0.003°	VHF, UHF, S	LEO, MEO	Flown LEO and MEO	No
Space Flight Laboratory Canada	16U	322	0.009°/0.004°	UHF, S, X, Ka	LEO, GEO, Lunar	Flown LEO Qualified GEO and Lunar	No
Space Information Laboratories USA	27U	180	0.008°/0.008°	S, X, Ka	LEO, GEO, Lunar	Under Development	Yes
Space Inventor Denmark	16U	200	<0.008°/<0.008°	VHF, UHF, S, X, L, Ka, Ku, QV	LEO, GEO, MEO	Flown LEO and GEO	No
Spacemanic Czech Republic	16U, 27U	1,000	0.1°/0.05°	VHF, UHF, S, X	LEO, GEO, Lunar	Flown LEO Qualified GEO and Lunar	No
Spire Global USA	16U	300	0.1°/0.05°	UHF, L, S, X, Ka, Ku	LEO	Flown LEO	Yes

2.2.2.2 ESPA-Class

The term ESPA-class refers to the Evolved Expendable Launch Vehicle (EELV) Secondary Payload Adapter (SPA) or similar configurations. The ESPA ring typically separates the primary payload with the upper stage of the launch vehicle, permitting additional mounting allocations for secondary payloads. Multiple rings can be stacked without a primary payload on the top to launch multiple payloads.

For this document, the ESPA-class table 2-8 includes options that may not be designed for the ESPA ring, but its mass and volume permit adaptability to rideshare opportunities. The information in this chapter is limited to offerings with mass under 500 kg even though some variants of the ESPA ring can support higher mass. Variants of the ESPA ring include, but are not limited to, ESPA-Heavy and ESPA-Grande. Examples of ESPA Rideshare are provided in figures 2.10 and 2.11, while figure 2.12 shows an example for an ESPA satellite from Muon Space.

Figure 2.10: Example mission configuration using Rideshare Plates. Credit: SpaceX.

Figure 2.11: LandSat-9 ESPA Ring populated with payloads and mass ballasts. Credit: NASA/Randy Beaudoin.

Figure 2.12: ESPA-Class satellite bus from Muon Space during integration at SpaceX facility for Transporter 8 rideshare mission. Credit: Muon Space via ExoLaunch and SpaceX.

Table 2-8: ESPA-Class Market Solutions

(The fields indicate maximum capability; organizations may offer multiple options including smaller capabilities within the ESPA-Class category)

Organization	Peak Power (W)	3-σ Pointing Control/ Knowledge	Comm Options	Intended Destination	Maturity	US Office
Airbus US Space & Defense USA	2,200	0.3°/0.3°	S, Ka, Optical	LEO	Flown LEO	Yes
Argotec Italy	250	<0.005°	UHF, S, X, K	LEO	Under Development	Yes
Artemis Space Technologies UK	1,250	0.01°/0.01°	UHF, S, X, Ka, Ku, Optical	LEO, MEO, GEO, Lunar, Deep Space	Qualified LEO, MEO, GEO, Lunar and Deep Space	No
Astranis Space Technologies Corp. USA	2,500	<0.1°/<0.01°	MIL-Ka, Ka, Ku, Q, V, X	MEO, GEO, Cislunar, Deep Space, Polar, High Inclination	Flown GEO	Yes
Astro Digital USA	2,000	<0.05°/<0.01°	UHF, S, X, Ku, Ka, V, W, Optical	LEO, GEO, Deep Space	Flown LEO	Yes
Ball Aerospace USA	1,000	<0.007°/<0.006°	L, S, X, Ka	LEO, MEO, GEO, Deep Space	Flown LEO	Yes
Berlin Space Technologies Germany	3,000	<0.017°/<0.017°	UHF, S, X	LEO	Flown LEO	Yes
Blue Canyon Technologies USA	1,082	0.0025°/0.0025°	L, S, X	LEO, GEO, Deep Space	Flown LEO and GEO Qualified Deep Space	Yes
Bradford Space USA	1,500	1.5°/0.006°	S, K	LEO, GEO, GTO, Cislunar, Lunar, Deep Space	Under Development	Yes
CesiumAstro USA	4,500	<0.1°/<0.01°	S, L, Ku, Ka, Optical	LEO	Under Development	Yes
EnduroSat Bulgaria	170	0.1°/<0.05°	UHF, S, X, K	LEO	Under Development	Yes
General Atomics EMS USA	450	0.03°/0.02°	S, X	LEO	Qualified LEO	Yes
Hemeria France	250	<0.03°/<0.01°	S, X	LEO, GEO, GTO	Flown LEO Qualified GEO and GTO	No
LeoStella USA	2,000	0.013°/0.009°	UHF, S, X	LEO	Flown LEO	Yes
Lockheed Martin USA	500+	<0.1°/<0.1°	S, X, Ka	LEO, GEO, Lunar, Deep Space	Flown LEO	Yes

					Qualified GEO, Lunar and Deep Space	
Loft Orbital USA	>1,000	<0.035°/<0.03°	S, X, L	LEO	Flown LEO	Yes
Magellan Aerospace Canada	200	0.01°/0.01°	S, X	LEO	Flown LEO	No
Malin Science Space Systems USA	918	<0.015°/<0.015°	UHF, X, Ka	Mars	Under Development	Yes
Millennium Space Systems USA	500	<0.013°/<0.008°	S, X, Ka	LEO, MEO, GEO, Deep Space	Flown LEO and GEO	Yes
Momentus USA	750	0.05°/0.025°	S, Ka, Optical	LEO	Flown LEO	Yes
Muon Space USA	500	0.03°/0.012°	S, X	LEO	Flown LEO	Yes
NanoAvionics Lithuania	660	0.24°/0.09°	UHF, S, X	LEO	Flown LEO	Yes
Northrop Grumman USA	400	<0.01°/<0.008°	S, X, Ka	LEO, GEO, HEO	Flown LEO, GEO, and HEO	Yes
NovaWurks USA	>5,000	0.002°/0.0004°	UHF, S, L, X, Ka, Ku and Optical	LEO, MEO, GEO, GTO, HEO, Lunar and Deep Space	Flown LEO and GTO	Yes
OHB LuxSpace Luxembourg	834	<0.022°/ 0.01°	S, X	LEO	Qualified LEO	No
Open Cosmos UK	2,200	<0.033°/0.03°	S, X	LEO	Under Development	No
Orbital Astronautics UK	5,000	0.05°/0.01°	S, X, K, Ka, Optical	LEO, MEO, GEO, Deep Space	Qualified LEO	No
Quantum Space USA	1,000	0.006°/0.006°	S, X, Ka	LEO, GEO, Cislunar, Lunar, Deep Space	Under Development	Yes
Redwire USA	600	0.005°/0.0017°	X	LEO	Qualified LEO	Yes
Sierra Space USA	500	0.001°/ <0.001°	UHF, S, X	LEO, MEO, GEO	Flown LEO	Yes
SITAEL Italy	1,000	0.018°/0.010°	S, X, Ka	LEO	Under Development	No
Southwest Research Institute USA	1,550	0.015°/0.002°	S, X, Ka	LEO, GEO	Flown LEO Under Development GEO	Yes
Space Dynamics Lab USA	1,600	0.021°/0.021°	UHF, S, X, Ka, Optical	LEO, GEO, GTO, Cislunar, Deep Space	Flown LEO	Yes
Space Flight Laboratory Canada	1,200	0.009°/0.004°	UHF, S, X, Ka	LEO, GEO, Lunar	Flown LEO Qualified GEO and Lunar	No

Space Inventor [Denmark]	400	<0.008°/<0.008°	VHF, UHF, S, X, L, Ka, Ku, QV	LEO, GEO, MEO	Qualified LEO	No
Surrey Satellite Technology Limited [UK]	400	0.01°/0.01°	S, X	LEO	Flown LEO	No
Terran Orbital [USA]	4,000	0.003°/0.002°	UHF, S, X, C	LEO, GEO, Lunar and Deep Space	Under Development	Yes
XPLORE [USA]	950	0.17°/ 0.018°	S, X	VLEO, LEO, Cislunar	Under Development	Yes
York Space Systems [USA]	1,500	0.008°/0.004°	UHF, S, X, Ka, Ku, Optical	LEO, GEO, Lunar	Flown LEO Qualified GEO and Lunar	Yes

2.3 Programmatic and Systems Engineering Considerations

To make an appropriate decision on which design path to take, small satellite mission developers should consider the programmatic and Systems Engineering factors most important to them, such as:

- What are the environments the system will be exposed during development and in flight?
- Are the Concept of Operations well defined and understood?
- How well do the systems meet functional and performance requirements?
- What are the mission's key performance parameters (e.g., mass, volume, power, data link, data budget, pointing) and how much margin do they offer?
- What is the software development environment, and how much flight and ground software can be re-used? Are emulators, simulators, Engineering Development Units (EDUs) and/or flatsats available to aid that development?
- What are the systems'/components' flight heritage, Technology Readiness Level (TRL), and reliability? What is the remaining Research and Development (R&D) level of effort to integrate the system with existing and/or planned systems?
- What is the mission's risk posture? How much development risk and performance risk are acceptable to the mission?
- Is it most important to meet performance requirements, cost, and/or schedule? What is the system's/components' production/lead time, and what are the contractual mechanisms that will be used to procure the systems and ensure timely delivery if delays are encountered?

Design selection can be driven by unique mission constraints, manufacturing lead time, and documented reliability. All of these and many more considerations should be well understood for each trade space option prior to a down-select. Given mission system performance requirements for key performance parameters like mass, volume, power, data link, data budget, and pointing, a functional importance rating and risk-based trade study should be used to screen the many options available. In addition to functional performance, relevant flight heritage or TRL, production lead time, and any available reliability data should be included in the trades. These, as well as cost, could drive the design to be done via COTS or commercial support.

Mission developers may want to take into consideration the following guides to help them in their selection and design process:

- NASA CubeSat 101 Book https://www.nasa.gov/content/cubesat-launch-initiative-resources
- NASA Systems Engineering Handbook https://www.nasa.gov/connect/ebooks/nasa-systems-engineering-handbook
- NASA Small Spacecraft Technology program Guidebook for Technology Development Projects https://www.nasa.gov/sites/default/files/atoms/files/smallsattechdevguidebook_rev-508d1.pdf

2.4 On the Horizon

As spacecraft buses are combinations of the subsystems described in later chapters, it is unlikely there will be any revolutionary changes in this chapter that are not preceded by revolutionary changes in some other chapter. As launch services become less expensive and commonplace with the rise of dedicated SmallSats launches, the market will continue to expand allowing interested universities and researchers to purchase COTS spacecraft platforms as an alternative to developing and integrating SmallSats themselves. Another option is to use numerous turnkey solutions offered by SmallSat vendors who can customize and cater to customer constraints.

SmallSat subsystem technology will continue to mature and gain flight heritage, to produce improved next generation platforms offered by vendors. Platforms with increased performance will spark the interest of newer vendors as they emerge into the market. This was demonstrated in the PocketQube industry: the requirement to satisfy ultra-low mass and volume constraints enabled high-performance capabilities. As the industry grows, there will likely be key technological advancements in SmallSat in-space propulsion, pointing and navigation control, optical communications, radiation tolerance, and radiation hardening. Subsystems described in other chapters in this report include details on radiation testing, but a subsystems' mean time between failures (MTBF) and overall system reliability will become a key design criterion as the sample groups become large enough to be statistically significant.

2.5 Summary

Several vendors have pre-designed fully integrated small spacecraft buses that are space rated and available for purchase. The market ranges from companies that are willing to heavily modify their systems to fit the customer's needs to companies attempting to standardize their system with little to no customization in favor of a better cost proposition. This chapter consolidated a long list of providers with key characteristics to facilitate the research and down-selection process for SmallSat practitioners.

For feedback about this chapter, email: arc-sst-soa@mail.nasa.gov. Please include a business email in case of follow up questions.

2.6 References

The references in this section are provided to facilitate the process in which practitioners can obtain information from the providers. The source indicates how the information provided in this chapter was obtained.

Source definition:

Direct New = organization provided the information through direct communication with the State-of-the-Art team for the current edition of the document.

Direct Old = organization provided the information through direct communication with the State-of-the-Art team on a previous edition of the document, and the team was unable to communicate with the organization to update the current edition of the document.

Website = the team was unable to directly communicate with the organization and limited information was obtained from the organization's website.

Table 2-9: List of Contact Information for Organizations in this Chapter

Organization	Source	Contact Email	Website
AAC Clyde Space	Direct New	enquiries@aac-clydespace.com	www.aac-clyde.space
Airbus US Space & Defense	Direct New	deborah.horn@airbusus.com	www.airbusus.com
Alba Orbital	Direct New	contact@albaorbital.com	www.albaorbital.com
Alen Space	Direct New	sales@alen.space	www.alen.space
Argotec	Direct New	info@argotecgroup.com	www.argotecgroup.com
Artemis Space Technologies	Direct New	info@spaceartemis.com	www.spaceartemis.com
Astranis	Direct New	scott@astranis.com	www.astranis.com
Astro Digital	Direct New	brian@astrodigital.com	www.astrodigital.com

Axelspace	Direct New	Contact Page	www.axelspace.com
Ball Aerospace	Direct New	General Inquiry Form	www.ballaerospace.com
Berlin Space Technologies	Direct New	info@berlin-space-tech.com	www.berlin-space-tech.com
Blue Canyon Technologies	Direct New	info@bluecanyontech.com	www.bluecanyontech.com
Bradford Space	Direct New	info@bradford-space.com	Bradford-Space.com
C3S Electronics Development	Direct Old	info@c3s.hu	www.c3s.hu
CesiumAstro	Direct New	info@cesiumastro.com	www.cesiumastro.com
Citadel Space Systems	Website	contact@citadel.space	Citadel.space
DIYSATELLITE	Direct New	gus@diysatellite.com	www.diysatellite.com
EnduroSat	Direct New	Contact Page	www.endurosat.com
FOSSA Systems	Direct New	contact@fossa.systems	Fossa.systems
General Atomics EMS	Direct New	Chris.white@ga.com	www.ga.com/EMS
German Orbital Systems	Direct New	info@orbitalsystems.de	www.orbitalsystems.de
GomSpace	Direct New	info@gomspace.com	gomspace.com
Gran Systems	Direct New	info@gransystems.com	www.gransystems.com
GUMUSH AeroSpace	Direct New	gumush@gumush.com.tr	www.gumush.com.tr
Hemeria	Direct New	contact@hemeria-group.com	www.hemeria-group.com/en
Hex20	Direct New	lloyd@hex20.com.au	www.hex20.com.au
IMT	Direct New	giovanni.cucinella@imtsrl.it	imtsrl.it
Innova Space	Direct New	info@innova-space.com	innova-space.com/en
In-Space Missions	Website	info@in-space.co.uk	in-space.co.uk
ISISPACE	Direct Old	sales@isispace.nl	www.isispace.nl
LeoStella	Direct New	mike.kaplan@leostella.com	leostella.com
Lockheed Martin	Direct New	timothy.m.linn@lmco.com	-
Loft Orbital	Direct New	gautier@loftorbital.com	www.loftorbital.com
Magellan Aerospace	Direct New	rushi.ghadawala@magellan.aero	www.magellan.aero
Malin Space Science Systems	Direct New	yee@msss.com	www.msss.com
Millennium Space Systems	Direct New	Contact Webpage	www.millennium-space.com
Momentus	Direct New	sales@momentus.space	Momentus.space
Muon Space	Direct New	info@muonspace.com	www.muonspace.com
Nanoavionics	Direct New	info@nanoavionics.com	www.nanoavionics.com
NearSpace Launch	Direct New	nsl@nearspacelaunch.com	www.nearspacelaunch.com
Northrop Grumman	Direct New	John.Dyster@ngc.com	-
NovaWurks	Direct New	info@NovaWurks.com	www.novawurks.com
NPC SPACEMIND	Direct New	info@npcspacemind.com	www.npcspacemind.com
OHB LuxSpace	Direct New	info@luxspace.lu	OHB LuxSpace
Open Cosmos	Direct New	partnerships@open-cosmos.com	open-cosmos.com
Orbital Astronautics	Direct New	hello@orbastro.com	orbastro.com
Orion Space Solutions	Direct New	contact@orionspace.com	orionspace.com
Pumpkin Space Systems	Direct New	sales@pumpkininc.com	www.pumpkinspace.com

Quantum Space	Direct New	sales@quantumspace.us	www.quantumspace.us
Quub, Inc.	Direct New	info@quub.space	quub.space
Redwire	Direct New	sales@redwirespace.com	www.redwirespace.com
SatRev	Direct New	engage@satrev.space	www.satrev.space
Sierra Space	Direct New	spaceapps@sierraspace.com	www.sierraspace.com
SITAEL	Direct New	sales.space@sitael.com	www.sitael.com
SkyLabs	Direct New	sales@skylabs.si	www.skylabs.si
Southwest Research Institute	Direct New	spacecraft-info@swri.org	-
Space Dynamics Lab	Direct New	info@sdl.usu.edu	www.sdl.usu.edu
Space Flight Laboratory	Direct New	info@utias-sfl.net	www.utias-sfl.net
Space Information Laboratories	Direct New	sales@spaceinformationlabs.com	www.spaceinformationlabs.com
Space Inventor	Direct New	sales@space-inventor.com	space-inventor.com
Spacemanic	Direct New	sales@spacemanic.com	www.spacemanic.com
Spire Global	Direct New	joe.carroll@spire.com	www.spire.com
Surrey Satellite Technology Limited	Direct New	info@sstl.co.uk	www.sstl.co.uk
Terran Orbital	Direct Old	info@terranorbital.com	terranorbital.com
Xplore	Direct New	inquire@xplore.com	www.xplore.com
York Space Systems	Direct New	BD@yorkspacesystems.com	www.yorkspacesystems.com

Chapter Contents

Chapter Glossary

(AFRL) Air Force Research Laboratory
(BMS) Battery Management System
(BOL) Beginning-of-Life
(CFRPs) Composite Fiber Reinforced Panels
(CIGS) Cu(In,Ga)Se2
(COTS) Commercial-off-the-Shelf
(EOL) End-of-Life
(EPS) Electrical Power System
(ESA) European Space Agency
(GaN) Galium Nitride
(GRC) NASA Glenn Research Center
(KSC) Kennedy Space Center
(Li-ion) Lithium-ion
($LiCF_x$) Lithium carbon monofluoride
(LiPo) Lithium polymer
($LiSO_2$) Lithium sulfur dioxide
($LiSOCl_2$) Lithium thionyl chloride
(MIL) Military
(QML) Qualified Manufacturers List
(NiCd) Nickel-cadmium
(NiH_2) Nickel-hydrogen
(OPV) Organic Photovoltaic
(OSCAR) Optical Sensors based on carbon materials
(PCB) Printed Circuit Board
(PEASSS) Piezoelectric Assisted Smart Satellite Structure
(PET) polyethylene terephthalate
(PMAD) Power management and distribution
(RHUs) Radioisotopic Heater Units
(RTGs) Radioisotope Thermoelectric Generators
(SABER) Solid-state Architecture Batteries for Enhanced Rechargeability and Safety
(SP) Specific Power
(SWaP) Size, Weight, and Power
(TPV) Thermophotovoltaic
(TR) Thermoradiative
(TRL) Technology Readiness Level
($Wh\ kg^{-1}$) Watt hours per kilogram

3.0 Power

3.1 Introduction

The electrical power system (EPS) encompasses electrical power generation, storage, and distribution. The EPS is a major, fundamental subsystem, and commonly comprises a large portion of volume and mass in any given spacecraft. Power generation technologies include photovoltaic cells, panels and arrays, and radioisotope or other thermonuclear power generators. Power storage is typically applied through batteries; either single-use primary batteries or rechargeable secondary batteries. Power management and distribution (PMAD) systems facilitate power control to spacecraft electrical loads. PMAD takes a variety of forms and is often custom-designed to meet specific mission requirements. EPS engineers often target a high specific power or power-to-mass ratio (Wh kg^{-1}) when selecting power generation and storage technologies to minimize system mass. The EPS volume is most likely to be the constraining factor for nanosatellites.

CubeSats and SmallSats typically operate in a mild radiative environment for short periods in low-Earth orbits, so stringent qualification standards and high Technology Readiness Level (TRL) tend to not be as relevant on those missions unlike in deep space. Therefore, EPS engineers should note some fundamental differences between commercial-off-the-shelf (COTS) parts and space-qualified parts while weighing those differences against spacecraft requirements. Typically, Military or Space (MIL/QML) qualified parts go through a series of specific tests, while COTS go through less stringent ones. For example, Military or Space parts are typically tested and qualified to survive -55°C to 125°C, while the alternative COTS requirement is -40°C to 85°C. The same trend is true for other factors that are a part of the MIL/QML qualification process like radiation, reliability, etc. COTS parts are typically known to have higher performance, while space qualified parts typically have relatively higher reliability. Another key limitation in QML parts is their lack of availability and slow revision timeline. All in all, we find that COTS parts are in many cases more suitable for use in SmallSat designs.

In this chapter, the terms SmallSat and CubeSat are often used in the same context, however, the reader needs to be aware of the distinctions between the two types of spacecraft. Please refer to the introduction of this report for more information on the categories of SmallSats. CubeSats fall under the category of both microsatellites and nanosatellites, and CubeSat missions commonly use COTS parts for space applications. Due to their nearly exclusive use in low-Earth orbit applications, CubeSats are more likely to incorporate COTS parts as they typically feature shorter mission lengths, more favorable environmental conditions, and as a result, need less stringent standards when qualifying parts. Knowing the distinction between a CubeSat and a SmallSat is necessary for determining the potential for incorporating COTS parts in a SmallSat design.

The information described below is not intended to be exhaustive but provides an overview of current state-of-the-art technologies and their development status for a particular small satellite subsystem. It should be noted that TRL designations may vary with changes specific to the payload, mission requirements, reliability considerations, and/or the environment in which performance was demonstrated. Readers are highly encouraged to reach out to companies for further information regarding the performance and TRL of the described technology. There is no intention of mentioning certain companies and omitting others based on their technologies or relationship with NASA.

In this chapter we will review the following categories:

- Power Generation-- including solar cells, panels and arrays (Sections 3.2 & 3.3),
- Energy Storage-- including Li-ion, Lipo, supercapacitors and solid-state batteries (Sections 3.4 & 3.5), and
- Power Management-- including modular architectures and wireless power transfer and telemetry (Sections 3.6 & 3.7).

3.2 State-of-the-Art – Power Generation

Power generation on SmallSats is a necessity typically governed by a common solar power architecture (solar cells + solar panels + solar arrays). As the SmallSat industry drives the need for lower cost and increased production rates of space solar arrays, the photovoltaics industry is shifting to meet these demands. The standardization of solar array and panel designs,

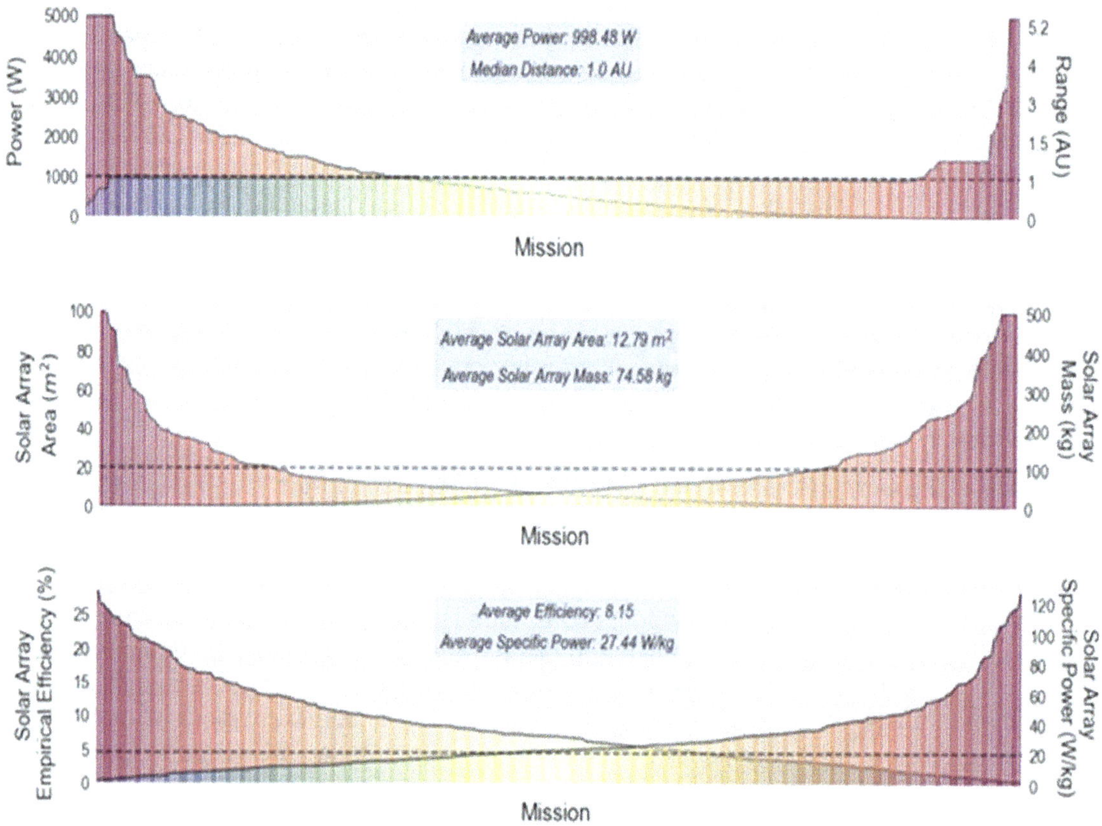

Figure 3.1: (Top) Distribution of mission range, or the furthest point from the sun that the spacecraft reaches, and mission power levels [power capped at 5 KW]. (Middle) Distribution of solar array surface area and solar array mass [mass capped at 500 Kg]. (Bottom) Distribution of solar array empirical efficiency (calculated at Earth) and specific power (for the entire array measured at the destination of the mission), Peretz et al. 2024 (92). X axis represents 389 missions and color scheme represents highest to lowest values (red to blue). Credit: NASA.

deployment mechanisms, and power integration will be critical to meet the desire for large, proliferated constellations.

EPS engineers should note beginning-of-life (BOL) vs end-of-life (EOL) performance of the systems as well as their planned testing hours for such systems while on the ground prior to operations. Typically, EPS for SmallSats is over-engineered to handle a dynamic thermal environment, eclipse durations while in LEO, or any other operational scenarios or mission needs while in eclipse at varying sun-angles. Figure 3.1 captures actual space system performances given such wide varying operational conditions for 389 space missions (not only SmallSats) launched since 1989 to 2021 reviewed in the publication (Peretz et. al 2024). The dotted horizontal line indicated where more of the 389 missions surveyed fall under

When possible, choosing a pre-designed and qualified panel is preferred over designing unique solar panels to reduce the cost and schedule as well as unforeseen design and manufacturing issues. Companies that have the capacity for mass production and automation are rare because space solar arrays, cells, and panels have always been a 'boutique' business; however, standardized designs like the OneWeb and StarLink constellations have been appearing more often to meet the demands of highly proliferated constellations.

The following subsections aim to capture the current state of the art and assist EPS engineers, mission designers, system engineers, etc., in designing, reviewing and ultimately constructing and operating such power flight systems.

3.2.1 Solar Cells

Solar power generation is the predominant method of power generation on small spacecraft. As of 2021, over 90% of all nanosatellite/SmallSat form factor spacecraft were equipped with solar panels and rechargeable batteries (92). Limitations to solar cell use include diminished efficacy in deep-space applications, no generation during eclipse periods, degradation over mission lifetime (due to aging and radiation), high surface area, mass, and cost. To pack more solar cells into the limited volume of SmallSats and NanoSats, mechanical deployment mechanisms can be added, which may increase spacecraft design complexity, reliability, as well as risk. Photovoltaic cells, or solar cells, are made from thin semiconductor wafers that produce an electric current when exposed to light. The light available to a spacecraft solar array, also called solar intensity, varies as the inverse square of the distance from the Sun. The projected surface area of the panels exposed to the Sun also affects power generation and varies as a cosine of the angle between the panel and the Sun.

While single-junction cells are cheap to manufacture, they carry a relatively low efficiency, usually around 20%, and are not included in this report. Modern spacecraft designers favor multi-junction solar cells made from multiple layers of light-absorbing materials that efficiently convert specific wavelength regions of the solar spectrum into energy, thereby using a wider spectrum of solar radiation (1). The theoretical efficiency limit for an infinite-junction cell is 86.6% in concentrated sunlight (2). However, in the aerospace industry, triple-junction cells are commonly used due to their high efficiency-to-cost ratio compared to other cells.

The current state of the art for space solar cells are multi-junction cells ranging from 3 to 5 junctions based on Group III-V semiconductor elements (like GaAs). SmallSats and CubeSats typically use some of the highest performing cells that provide efficiencies over 32%, even though they have a substantially higher cost than terrestrial silicon solar cells (~20% efficient). Ultimately the size, weight, and volume of smaller satellites may be the determining factor in choosing solar cell technology, rather than solar cell efficiency. Being a life-limiting component on most spacecraft, the EOL performance at operating temperature is critical in evaluating their performance. Common factors that degrade the functionality of solar cells include radiation exposure, coverglass/adhesive darkening, contamination, and mechanical or electrical failure.

This section individually covers small spacecraft targeted cells, fully integrated panels, and arrays. Table 3-1 itemizes small spacecraft solar cell efficiency per the available manufacturers. Note the efficiency may vary depending on the solar cells chosen.

Table 3-1: Solar Cells Product Table

Company	Cell Name	BOL Efficiency	Voc (V)	Vmp (V)	Jsc (mA/cm²)	Jmp (mA/cm²)	Pmp (W/m²)	Ref
AZUR Space Germany	Silicon S 32	16.8	0.628	0.528	45.8	43.4	229.2	(3)
	3G30-Adv	29.5	2.7	2.411	17.2	16.71	403	(3)
	4G32-Adv	31.5	3.426	2.999	15.2	14.37	431	(3)
	TJ 3G28C	28	2.667	2.37	16.77	16.14	1367	(3)
Rocket Lab USA	ZTJ	29.5	2.726	2.41	17.4	16.5	397.7	(10)
	ZTJ+	29.4	2.69	2.39	17.1	16.65	397.9	(10)
	ZTJ Omega	30.2	2.73	2.43	17.4	16.8	408.2	(10)
	Z4J	30.0	3.95	3.54	12	11.5	407.1	(10)
	IMMα	32.0	4.78	4.28	10.7	10.12	433.1	(10)
	ZTJM	29.5	2.72	2.38	17.1	16.5	392	(10)
SpectroLab USA	XTJ	29.5	2.633	2.348	17.76	17.02	399.6	(6)
	XTJ-Prime	30.7	2.715	2.39	18.1	17.4	415.9	(6)
	XTE-SF	32.2	2.75	2.435	18.6	17.8	433.4	(5)
	XTE-HF	32.1	2.782	2.49	18	17.4	427.9	(5)
	XTE-LILT	31.6	2.755	2.459	18.1	17.4	427.9	(5)
	UTJ	28.4	2.66	2.35	17.14	16.38	384.93	(7)
	TASC	27	2.52	2.19	32	28	270	(8)
	ITJ	26.8	2.565	2.27	16.9	16	1353	(9)
Emcore Corporation USA	BTJ	28.5	2.7	2.37	17.1	16.3	386	(4)
	ZTJ	29.5	2.726	2.41	17.4	16.5	397	(4)

3.2.2 Solar Panels & Arrays

Solar panels & arrays are constructed from individual solar cells connected in series to form strings and in parallel to form circuits mounted on a substrate backing (e.g., figure 3.2). While very low-power CubeSats and SmallSats may only need body-mounted solar panels, most will require more power from deployed solar arrays. The deployed solar arrays for CubeSats and SmallSats are mostly on rigid substrates made of either a printed circuit board (PCB), composite fiber reinforced panels (CFRPs), or an aluminum honeycomb panel.

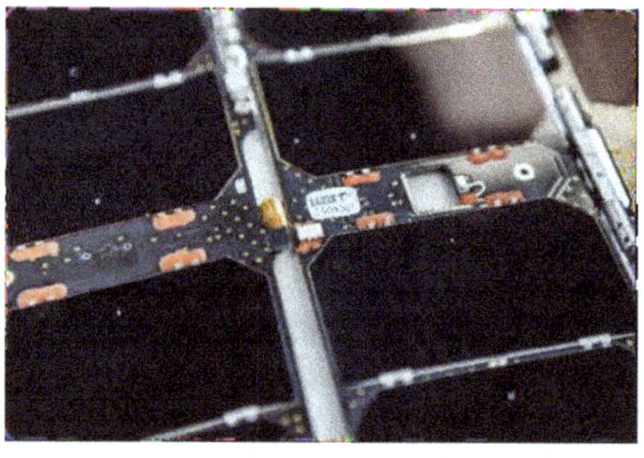

Figure 3.2: AAC Clyde Space solar arrays. Credit: AAC Clyde Space.

Deployed solar arrays are often the largest structure on a satellite; the ratio between the size of the deployed solar array and the size of the SmallSat may be much higher compared to other conventionally large spacecraft. The size and fundamental frequency of the solar arrays impact spacecraft pointing, propulsion, and delta-V needed for station keeping. Important considerations for SmallSat solar arrays are deployment mechanisms, deployed frequency, panel specific power, and power density, as well as stowed volume. Most of these metrics are not listed on the manufacturer's datasheets.

Solar array comparison can be challenging because SmallSat/CubeSat manufacturers who make solar arrays specific to their bus and payload designs often do not report solar array power using the same metrics. Their reported "power" can mean multiple things: power available to the payload, peak power provided by a combination of solar array and battery, or an orbital-specific average power. Reported solar array power (Peak BOL) mainly refers to the peak power of the solar array at the beginning of life, 28°C which is mission-independent. Panel stiffness and moment of inertia usually need to be calculated for a specific spacecraft because they are dependent on multiple factors such as the size and mass of the panel, as well as spacecraft size and weight distribution. Examples of commercial solar array and panel products are shown in table 3-2.

Table 3-2: Solar Array/Panel Products					
Company	**Product**	**Panel Type**	**Specific Power (W/kg)**	**Peak BOL Solar Array Power (W)**	**Ref**
AAC Clyde Space Sweden	Photon	Body Mount + Deployed Rigid	*	9.25 per 3U-12 Face	(11)
Agencia Espacial Civil Ecuatoriana Italy	DSA/1	Deployed Rigid	107	7.2	(18)
Airbus Defense and Space The Netherlands	Sparkwing Solar Panel	Deployed Rigid	165	66	(17)
Blue Canyon Technologies USA	BCT Solar Array	Body Mount + Deployed Rigid	*	28 – 42 for 3U / 48-118 for 6U-12U	(12)
DHV Technologies Spain	CubeSats Solar Panels	Body Mounted (Polyimide)	50	2 for 1U Face	(13)
	CubeSats Solar Panels	Body Mounted (Polyimide)	49	4 for 2U Face	
	CubeSats Solar Panels	Body Mounted (Polyimide)	75	8 for 3U Face	
	CubeSats Solar Panels	Body Mounted (Polyimide)	68	18 for 6U Face	
	CubeSats Solar Panels	Deployed Rigid (Polyimide)	42	12 for 3U Double Deployable and Body Mounted	
	CubeSats Solar Panels	Deployed Rigid (Polyimide)	69	57 for 6/12U Double Deployable and Body Mounted	
	CubeSats Solar Panels	Deployed Rigid (Polyimide)	108	34 for 3U Quadruple Deployable	
	CubeSats Solar Panels	Deployed Rigid (CFRP)	69	68 for 6U Quadruple Deployable	
	CubeSats Solar Panels	Body Mounted (Polyimide)	50	2 for 1U Face	
	CubeSats Solar Panels	Body Mounted (Polyimide)	49	4 for 2U Face	
	Body mounted solar array panel	Sandwich CFRP substrate	84	179	
	Body mounted solar array panel	Sandwich CFRP substrate	90	171	
	Body mounted solar array panel	Low thickness monolithic CFRP substrate	140	96	
	Multiple deployable solar array wing	Sandwich CFRP substrate	57	697	
EnduroSat Bulgaria	1U Solar Panel	Deployed Rigid	50	2.4	(35)
	1.5U Solar Panel	Deployed Rigid	55	2.4	
	3U Solar Panel/Array	Deployed Rigid	66	8.4	

	6U Solar Panel/Array	Deployed Rigid	64	19.2	
	8U Solar Panel/Array	Deployed Rigid	65	24.0	
ExoTerra Resource USA	Fold Out Solar Arrays (FOSA)	Deployed Flexible	140	150	(14)
GomSpace Germany	Nanopower DSP	Deployed Rigid	*	1.2	(19)
	NanoPower P110	Fixed	36.9 – 92 Dependent on variant	1.2 per cell 2.4	(95)
	NanoPower MSP 16 cell	Fixed	42	1.2 per cell 19.2	(96)
	NanoPower TSP Per wing	Deployed Rigid	60	45	
ISISPACE The Netherlands	Smallsat Solar Panels	Body Mount + Deployed Rigid	46	2.3 per U	(20)
Lockheed Martin USA	SmallSat Solar Array	Deployed Rigid with Additively Manufactured Substrate	53.6	170 per panel BOL at 28C	
MMA Design USA	Hawk	Deployed Rigid (PCB)	121	36-112	(15)
	zHawk	Deployed Rigid (PCB)	95	36	(16)
	Next-Gen HaWK	Deployed Rigid (PCB)	100	45 - 310	
NanoAvionics Lithuania	CubeSat GaAs Solar Panel	Deployed Rigid	Unk	Unk	(89)
Pumpkin Space Systems USA	Dual Articulated Deployable Solar Array	Deployed Rigid	31	135	(97)
	Dual-Quad Articulated Deployable Solar Array	Deployed Rigid	30	240	
	Deployable Clamshell Solar Array (DCSA)	Deployed Rigid	44	220	(98)
	Deployable Clamshell Solar Array (DCSA)	Deployed Rigid	38	350	
Redwire Space USA	ROSA	Flexible PV blanket	100	1000	(21)
	Aladdin SmallSat Array	Hybrid Array: Flex Rigid	80	300	
Spacemanic Slovakia	RA_Solar_Panels	Triple Junction GaAs	46	2.4	
Space Dynamics Laboratory USA	SDL Modular Solar Panels	Carbon fiber/Al honeycomb composite	84.5	190 per panel	

* Available with inquiry to manufacturer

** For SmallSat use

While solar arrays efficiency has been the prevailing way to characterize solar array performance, discrepancies between theoretical and empirical data indicate that specific power (SP) of the solar array fundamentally governs space mission feasibility and flexibility. Figure 3.3 visualizes the landscape of mission architectures, parameterized by power (adjusted by a factor of squared distance) and mass; the relation between these axes is the SP.

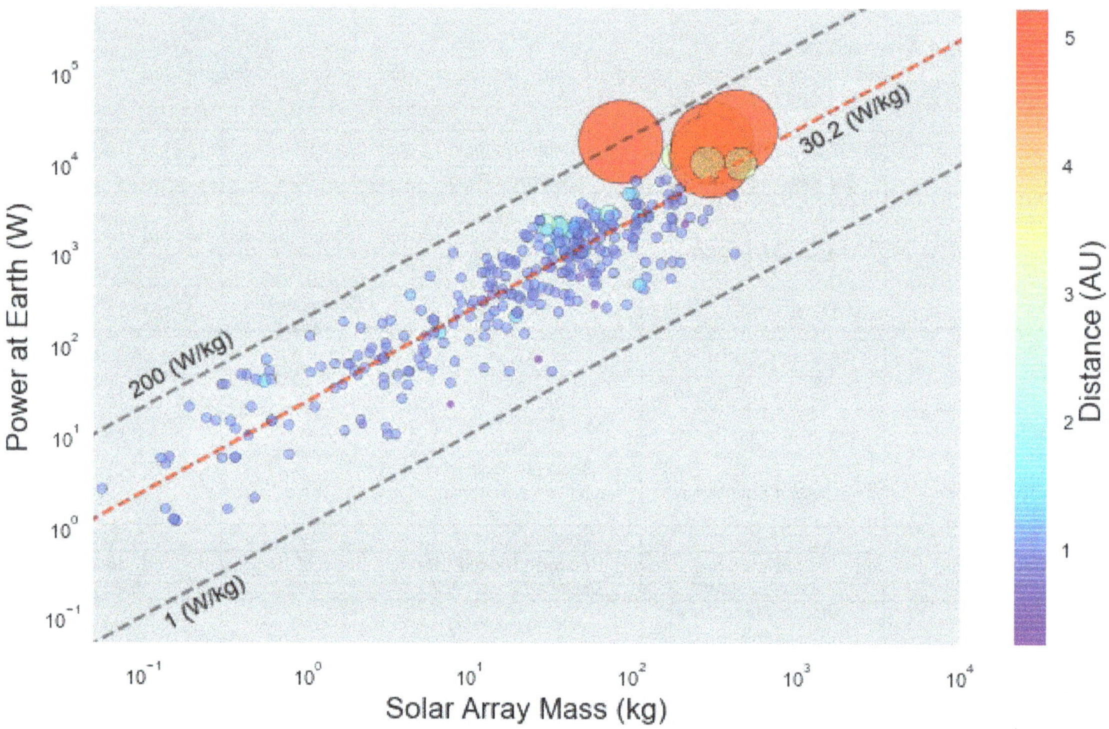

Fig. 3.3: Specific power of a solar array at Earth. A minimum of 1 W/kg is plotted as a dashed black line, showing that no missions fly with any smaller specific power. The larger red circles represent the missions farther from Earth corresponding to the red color on the distance scale. Peretz et al (2024). Credit: NASA.

Space missions using solar arrays, regardless of spacecraft mass, are strongly clustered around ~30 W/kg (red dashed line) and are strongly bounded: no missions fly with SP lower than 1 W/kg (lower dashed line), and the maximum empirical SP of this dataset is 200 W/kg (upper dashed line). There are two clearly unoccupied regions: the empty region of high-mass and low-power is of little interest since this is not generally desirable; however, it is interesting that the high-power low-mass regime is empty, indicating that while this is a highly desirable region, it is technologically inaccessible.

With the average mission power consumption of 1000 W and a medium value of 600 W, figure 3.4 shows what maximal ranges can be achieved with three hypothetical solar array technologies with specific power levels of 10, 100 and 1600 W/Kg (the lines show how the Empirical Specific power change as a function of range until it reaches the lower limit of 1 W/Kg). The color-coded regions are where maximal surface area of 100 m^2 could be contained for power levels of 100,

300, and 1000 W. The areas under the line and within the colored region indicate where current (empirical evidence) and future (theoretical) missions could exist.

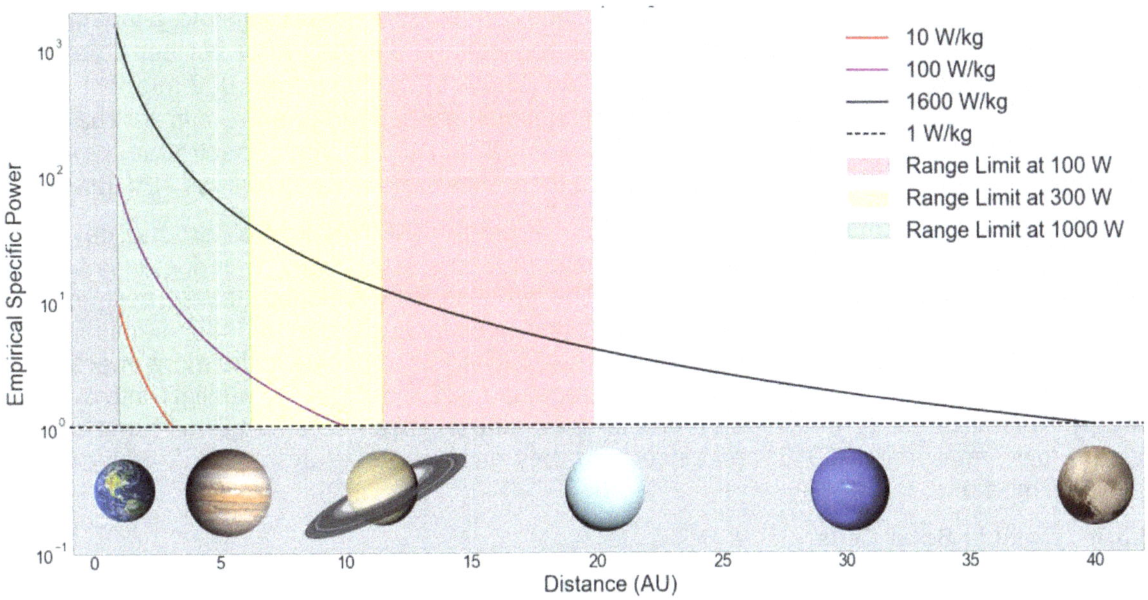

Fig. 3.4: Practical power operation limit for spacecraft. Horizontal Axis is range in AU, Vertical is Specific Power in W/Kg. Lines represent a hypothetical solar array Specific Power. Colored regions show where a solar array with surface area smaller than 100 m^2 could exist (green is more spacecraft power and red is less spacecraft power). Peretz et al (2024). Credit: NASA.

There are two possible approaches to improving solar array specific power: increase generated power or decrease array mass. The former has been the focus of the community for the past 40 years through improving efficiencies. However, even if triple-junction solar cell efficiency improves to the theoretical limit of 68%, the surface area, mass, and storage volume required to support median power requirements for exploration of deep space are beyond the point of feasibility. The mass required from such solar array structures would be measured in thousands of kilograms; clearly a reduction in solar array mass is needed for deep space exploration missions to be feasible.

Future work is needed to reduce solar array mass; deployment mechanism mass could be reduced with light-weight components or alternate configurations (for example, using a tension-based deployment instead of a motor-driven system) to eliminate the requirement for a motor and many spacecraft integration components. Mass reduction for the solar cells themselves can come by reducing cell thickness or increasing flexibility to increase launch vehicle stowage volume. Creating highly foldable arrays may be a desirable solution, with mass decreases possible in both deployment mechanisms and in the design of solar cells.

3.3 On the Horizon – Power Generation

New technologies continue to be developed for space-qualified power generation. Promising technologies applicable to small spacecraft include advanced multi-junction, flexible and organic solar cells, hydrogen fuel cells, and a variety of thermo-nuclear and atomic battery power sources.

3.3.1 Multi-junction Solar Cells

Fraunhofer Institute for Solar Energy Systems has developed different four-junction solar cell architectures that currently reach up to 38% efficiency under laboratory conditions, although some designs have only been analyzed in terrestrial applications and have not yet been optimized (Lackner). Fraunhofer ISE and EV have achieved 33.3% efficiency for a 0.002 mm thin silicon-based multi-junction solar cell, and future investigations are needed to solve current challenges of the complex inner structure of the sub-cells (22). Additionally, SpectroLab has been experimenting with 5- and 6-junction cells with a theoretical efficiency as high as 70% (23).

A collaboration between the Air Force Research Laboratory (AFRL) and SolAero has developed Metamorphic Multi-Junction (IMM-α) solar cells that are less costly with increased power efficiency for military space applications (1). The process for developing IMM-α cells involves growing them upside down, where reversing the growth substrate and the semiconductor materials allow the materials to bond to the mechanical handle, resulting in the more effective use of the solar spectrum (1). A single cell can leverage up to 32% of captured sunlight into available energy. This also results in a lighter, more flexible product. These cells had their first successful orbit in low-Earth orbit in 2018, and since then they have operated in low-Earth orbit on other CubeSat missions.

3.3.2 Flexible Solar Cells

Flexible and thin-film solar cells have an extremely thin layer of photovoltaic material placed on a substrate of glass or plastic. Traditional photovoltaic layers are around 350 microns thick, while thin-film solar cells use layers just one micron thick. This allows the cells to be flexible, lightweight, and cheaper to manufacture because they use less raw material. The performance of commercial flexible CIGS was investigated and reported with the potential for deep space applications at the University of Oklahoma. The authors found promising thin-film solar material using Cu(In, Ga)Se2 (CIGS) solar cells with recorded power conversion efficiencies up to 22.7% (24).

3.3.3 Organic Solar Cells

Another on the horizon photovoltaic technology uses organic or "plastic" solar cells. These use organic electronics or organic polymers and molecules that absorb light and create a corresponding charge. A small quantity of these materials can absorb a large amount of light making them cheap, flexible, and lightweight.

Toyobo Co., Ltd. and the French government research institute CEA have succeeded in making trial organic photovoltaic (OPV) small cells on a glass substrate. Trial OPV modules on a lightweight and thin polyethylene terephthalate (PET) film substrate were demonstrated during their joint research project. Toyobo and CEA succeeded in making the OPV small cells on a glass substrate with the world's top-level conversion efficiency by optimizing the solvents and coating technique. In a verification experiment under neon lighting with 220 lux, equivalent to the brightness of a dark room, the trial product was confirmed to have attained a conversion efficiency of about 25%, or 60% higher than that of amorphous silicon solar cells commonly used for desktop calculators (25).

In October 2016, the Optical Sensors based on carbon materials (OSCAR) stratospheric-balloon flight test demonstrated organic-based solar cells for the first time in a stratospheric environment. While more analysis is needed for terrestrial or space applications, it was concluded that organic solar energy has the potential to disrupt "conventional" photovoltaic technology (26). Since then, a joint collaborative agreement between the German Aerospace Center and the Swedish National

Space Board REXUS/BEXUS has made the balloon payload available for European university student experiments collaborating with the European Space Agency (ESA) (27).

No standardized stability tests are yet available for organic-based solar cell technology, and challenges remain in creating simultaneous environmental influences that would permit an in-depth understanding of organic photovoltaic behavior, but these achievements are enabling progress in organic-based solar cell use. In 2018, Chinese researchers in organic photovoltaics were able to reach 17% power conversion energy using a tandem cell strategy. This method uses different layers of material that can absorb different wavelengths of sunlight, which enables the cells to use more of the sunlight spectrum, which has limited the performance of organic cells (28).

3.3.4 Fuel Cells

Hydrogen fuel cells are appealing due to their small, light, and reliable qualities, and high energy conversion efficiency. They also allow missions to launch with a safe, storable, low-pressure, and non-toxic fuel source. An experimental fuel cell from the University of Illinois that is based on hydrogen peroxide rather than water has demonstrated an energy density of over 1000 Wh kg^{-1} with a theoretical limit of over 2580 Wh kg^{-1} (29). This makes them more appealing for interplanetary missions and during eclipse periods, however unlike chemical cells, they cannot be recharged on orbit. Carrying a large fuel tank is not feasible for small or nanosatellite missions. Regenerative fuel cells are currently being researched for spacecraft applications. Today, fuel cells are primarily being proposed for small spacecraft propulsion systems rather than for power sub-systems (30).

3.4 State-of-the-Art – Energy Storage

Solar energy is not always available during spacecraft operations; the orbit, mission duration, distance from the Sun, or peak loads may necessitate stored, onboard energy. Primary and secondary batteries are used for power storage and are classified according to their different electrochemistry. As primary-type batteries are not rechargeable, they are typically used for short mission durations. Silver-zinc is typically used as they are easier to handle and discharge at a higher rate, however, there are also a variety of lithium-based primary batteries that have a higher energy density, including lithium Sulfur dioxide ($LiSO_2$), lithium carbon monofluoride ($LiCF_x$) and lithium thionyl chloride ($LiSOCl_2$) (36).

Secondary-type batteries include nickel-cadmium (NiCd), nickel-hydrogen (NiH_2), lithium polymer (LiPo) and lithium-ion (Li-ion), which have been used extensively in the past on small spacecraft. Lithium-based secondary batteries are commonly used in portable electronic devices because of their rechargeability, low weight, and high energy, and have become ubiquitous on spacecraft missions. They are generally connected to a primary energy source (e.g., a solar array) and can provide rechargeable power-on-demand. Each battery type is associated with certain applications that depend on performance parameters, including energy density, cycle life, and reliability (36).

Figure 3.5 shows some popular 18650 Lithium-Ion cells and their specific energy densities. While legacy cells had a specific energy of less 200 Wh/kg, latest cells have all exceeded 240 Wh/kg. Traditionally, vendors pack these 18650 cells in various configurations to meet customer needs. Table 3-3 shows a list of battery pack assemblers with their products and TRLs.

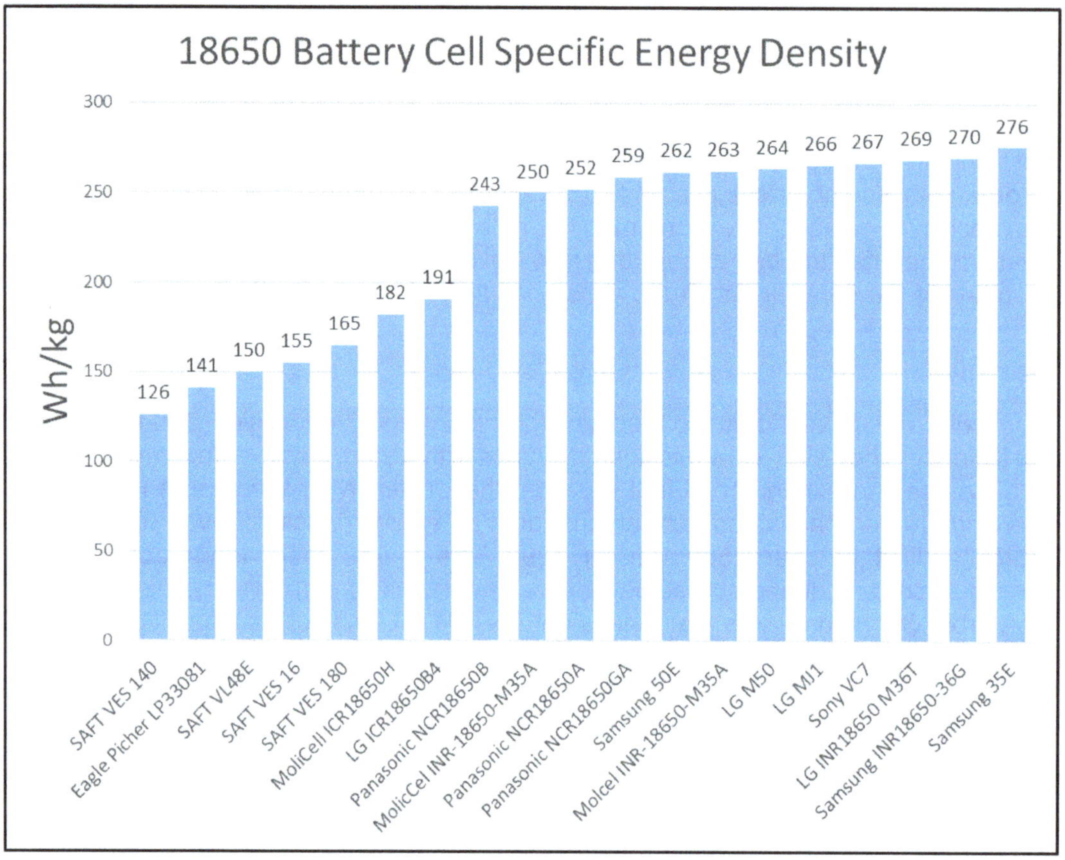

Figure 3.5: Battery cell energy density. Credit: NASA.

This section will discuss the individual chemical cells as well as pre-assembled batteries of multiple connected cells offered from multiple manufacturers. Due to small spacecraft mass and volume requirements, the batteries and cells in this section will be arranged according to specific energy, or energy per unit mass. However, several other factors are worth considering, some of which will be discussed below (37).

Table 3-3: Battery (Pack) Product Table								
Company	**Product**	**Volumetric Energy Density [Wh L^{-1}]**	**Specific Energy [Wh kg^{-1}]**	**Typical Capacity [Ah]**	**Max Discharge Rate [A]**	**Cells Used**	**TRL**	**Ref**
EaglePicher Technologies USA	NPD-002271	271	153.5	14.5	15	EaglePicher Li-ion	7-9	(39)
GomSpace Germany	Nanopower BPX 3000mAh (e.g., 4S-2P)	262.2	172	6.0	4	GomSpace NanoPower Li-ion	7-9	(42)
	Nanopower BP4 3000mAh (e.g., 2S-2P)	239.8	168.8	6.0	4	GomSpace NanoPower Li-ion	7-9	(43)
	Nanopower BP8 (8S-1P)	227.36	177.3	3.0	4	GomSpace NanoPower Li-ion	7-9	
AAC Clyde Space Sweden	Optimus	169.5	119	4.84	2.6	Clyde Space Li-Polymer	7-9	(44)
Ibeos USA	28V Modular Battery	151.1	109.8	9.82	20	*	N/A	(45)
Saft France	VES16 4S1P	109.2	91	4.5	4.5 – Cont. 9 - Pulse	SAFT Li-ion	7-9	(46)
Vectronic Aerospace GmbH Germany	VLB-X	101.96	74.6	12	10 – Cont. 20 – Pulse	SAFT Li-Ion	7-9	(47)
Berlin Space Technologies Germany	BAT-110 Modular Battery (Nominal 3 strings)	69.73	57.75	7.5	3	Li-Fe	7-9	(48)
GUMUSH AeroSpace Istanbul	n-ART BAT	184.5	155.1	6.01	8	Li-Ion	7-9	
Argotech Italy	ELEKTRA	228.5	190.4	3.4	4 – Cont. 10 – Peak	Li-Ion	5	
SkyLabs Slovenia	SKY-NANOeps-BMM	103	89	up to 18Ah	up to 54 A	LiFePO4	9	(100)
Space Dynamics Laboratory USA	SDL 12V Battery Pack	144	75	12.0	96	EaglePicher LP32975	9	

* Available with Inquiry to Manufacturer

The chemistry and cell design impacts the volumetric and specific energy densities. This limit represents the total amount of energy available per unit volume or weight, respectively. Current top-of-the-line Li-ion energy cells exhibit ~270 Wh kg^{-1}. Li-ion batteries exhibit lower energy densities due to the inclusion of a battery management system (BMS), interconnects, and sometimes thermal regulation.

There are generally two groups of cells – high energy or high power. High power cells use a low resistance design, such as increasing coating surface area, or multiple points of contact for the current collector to the cell which can allow for lower overall resistance values and a higher rate of discharge. High energy cells work to optimize gravimetric energy densities to obtain the most energy from the cell. Some common methods to increase gravimetric energy densities are via the addition of silicon to the anode, the use of high voltage cathodes, or using a metallic lithium anode. However, these methods can significantly reduce the cyclability of the battery system in exchange for increased energy density.

In general, for space applications, high energy density is important because a battery with high gravimetric energy density will be cheaper to launch into orbit (higher battery capacity per unit mass). However, for some high pulse applications, high-power cells would meet mission needs with less weight. However, energy density is not the only factor to investigate during cell selection. For non-space commercial applications, faster degradation (lower cyclability) of the battery can be beneficial as the electronic device often lasts as long as the battery, and faster turnover of a device may lead to increased revenue.

While space-designed cells typically underperform in energy density, they over-perform in cyclability with many space-designed cells used for longer (~5-15 year) missions. However, not all degradation modes for the lithium-ion trend in a linear fashion, and trends often take time to settle, thus the test results don't necessarily show the best performing cell until others are further along in testing.

Due to the extremely short mission durations with primary cells, the current state-of-the-art energy storage systems use lithium-ion (Li-ion) or lithium-polymer (LiPo) secondary cells, so this subsection will focus only on these electrochemical compositions, with some exceptions.

3.4.1 Secondary Li-ion and Lipo Batteries

Typically, Li-ion cells deliver an average voltage of 3.6 V, while the highest specific energy obtained is well over 150 Wh kg^{-1} (37). Unlike electronics, battery cells do not typically show significant damage or capacity losses due to radiation. However, in an experiment done by JPL, some capacity loss is seen among these latest lithium-ion battery cells under a high dosage of Cobalt-60. The results are shown below in figure 3.6 (62).

In Lithium-ion batteries, repeated charging cycles of the battery eventually result in aging or degradation that affects the overall energy (Watt-hours) that the battery may provide. Many variables impact aging, such as temperature, charge/discharge rate, depth of discharge, storage conditions, etc. Due to the numerous variables that impact aging, lithium-ion batteries are typically put under life test in mission conditions before launch to ensure the battery will meet the specific mission life requirements.

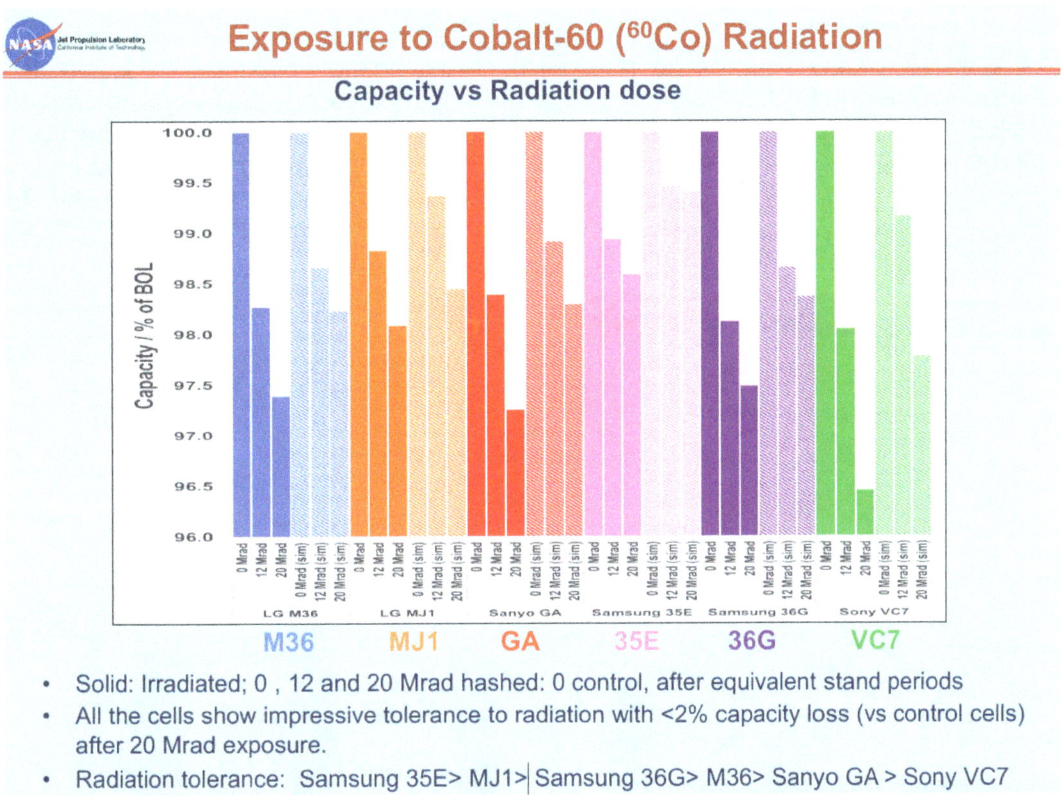

Figure 3.6: Capacity vs. radiation dose. Credit: JPL.

18650 Cells

18650 cylindrical cells (18 x 65 mm) have been an industry standard for lithium-ion battery cells. Many manufacturers have staple high-performance 18650 cells, some of which have flown on multiple spacecraft and are documented in table 3-4 below.

Table 3-4: 18650 Cylindrical Cells

Cell	Specific Energy (Wh kg^{-1})	Flight Heritage
LG ICR18650 B3 (2600 mAh)	191	NASA's PhoneSat, NoDES
Panasonic NCR18650B (3350 mAh)	243	MarCO, ADAPT (Sept 2022*: BioSentinel, Lunar Flashlight, NeaScout)
Molicel ICR18650H (2200 mAh)	182	NASA's EDSN mission
Canon BP-930s (3000 mAh)	112	NASA's TechEdSat missions
LG MJ1 (3500 mAh)	260	NASA's PACE mission

Cylindrical 18650s have become the most commonly used building blocks for many SmallSats today, although prismatic and pouch formats are also available. The lithium-ion industry has seen incremental increases in energy density via the inclusion of silicon in the anode, high voltage cathodes, new electrolyte additives, and improved cell designs.

21700 Cells

21700 (21 x 70 mm) is another type of cylindrical cell that is getting more popular in the automotive industry. Samsung 50E and LG M50 both offer 5000 mAh of energy while the Samsung cells are slightly heavier. The specific energy densities are 262 Wh kg^{-1} and 264 Wh kg^{-1} respectively. Although 21700 cells are slightly larger than 18650 cells, they have some of the highest energy densities and could offer some mechanical packaging benefits with fewer cells for certain missions. Figure 3.7 shows various 21700 battery cell specific densities.

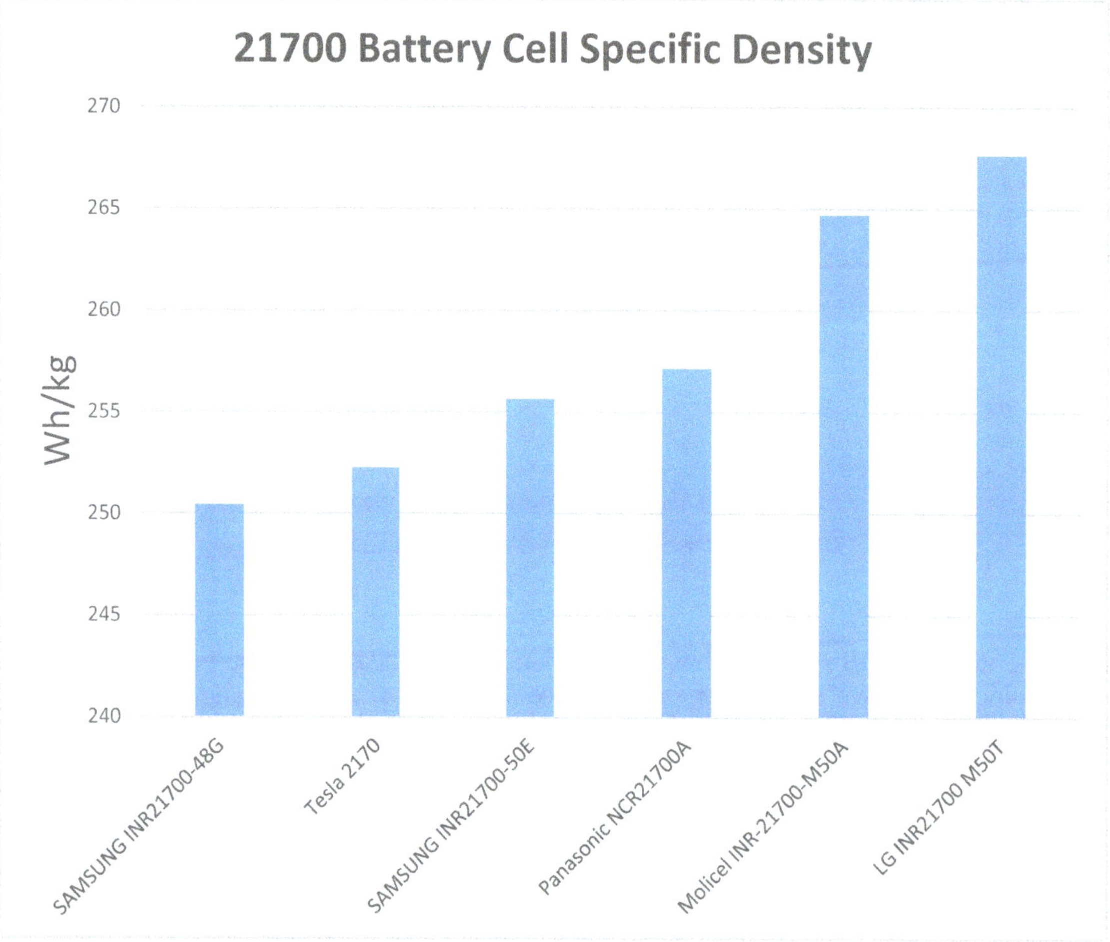

Figure 3.7: 21700 Battery Cell Specific Density. Credit: NASA.

4680 Cells

4680 (46 x 80 mm) cylindrical cells are a battery cell form factor that has been introduced to the energy storage scene by Tesla. The larger format cell potentially exacerbates several of the thermal management drawbacks (particularly internal temperature gradients and heterogeneity in current distribution) associated with other common smaller cells, however, to address these drawbacks, Tesla has a "tabless current collection" method where the current collector foil is used in conjunction with an array of current collectors to reduce ohmic losses and the temperature increases that those losses can cause (63).

When it comes to the manufacturing of Li-ion batteries and battery cells, these companies are at the forefront for their respective sectors listed in table 3-5. Although China has some of the largest consumer electronics and EV battery manufacturers, their products are rarely used in the space industry which requires high performance and high reliability. Therefore, we do not include these products in this report at this time.

Table 3-5: Commercial and Space Li-ion Manufacturers

Commercial Li-ion Manufacturing		Space Li-ion Manufacturing	
Company	**Headquarters**	**Company**	**Headquarters**
Panasonic	Japan	EaglePicher Technologies	USA
LG Chem	South Korea	Enersys	USA
Samsung	South Korea	GS Yuasa	Japan
E-one Moli	Taiwan	Saft	France
Sony	Japan	Tesla	USA

3.5 On the Horizon – Energy Storage

In the area of power storage, there are several ongoing efforts to improve storage capability and relative power and energy densities; a Ragone Chart shown in figure 3.8 illustrates different energy devices (64). For example, the Rochester Institute of Technology and NASA Glenn Research Center (GRC) developed a nano-technology enabled power system on a CubeSat platform. The power system integrates carbon nanotubes into lithium-ion batteries that significantly increase available energy density. The energy density has exceeded 300 Wh kg^{-1} during testing, a roughly two-fold increase from the current state of the art. The results in this program were augmented from a separate high-altitude balloon launch in July 2018 organized through NASA GRC which showed typical charge and discharge behavior on the ascent up to an altitude of 19 km (65). A collaborative project between the University of Miami and NASA Kennedy Space Center (KSC) is aiming to develop a multifunctional structural battery system that uses an electrolytic carbon fiber material that acts as both a load-bearing structure and a battery system. This novel battery system will extend mission life, support larger payloads, and significantly reduce mass. While several panel prototypes have shown successively increased electrochemical performance, further testing of the individual components can improve the accuracy of the computational models (66).

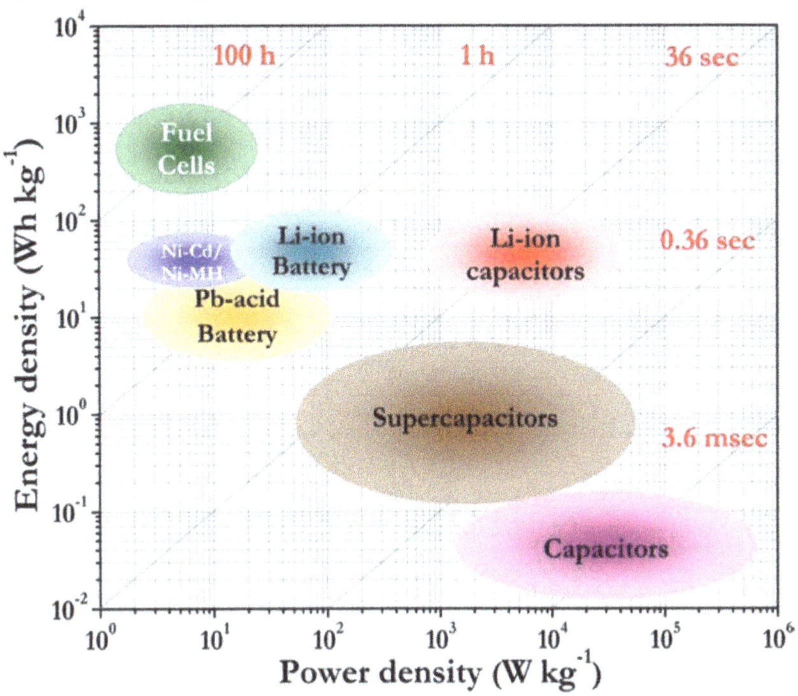

Figure 3.8 Relative power and energy densities of different energy devices. Ragone chart illustration reprinted with permission from Aravindan et al. Copyright (2014) American Chemical Society.

3.5.1 Supercapacitors

While the energy density for supercapacitors, also called ultracapacitors, is low (up to 7 Wh kg^{-1}), they offer a very high-power density (up to 100 kW kg^{-1}) which could be useful for space applications that require power transients. Their fast charge and discharge time, their ability to withstand millions of charge/discharge cycles, and wide range of operational temperatures (-40°C to +70°C), make them a perfect candidate for several space applications (launchers and satellites). This was demonstrated in an ESA Study entitled "High Power Battery Supercapacitor Study" completed in 2010 by Airbus D&S (67). The Nesscap 10F component and a bank of supercapacitors based on the Nesscap 10F component were space-qualified in 2020 after the completion of the ESA Study entitled "Generic Space Qualification of 10F Nesscap Supercapacitors." Although not likely to replace Li-ion batteries completely, supercapacitors could drastically minimize the need for a battery and help reduce weight while improving performance in some applications. Figure 3.8 shows a comparison chart (68), and table 3-6 lists differences in Li-ion batteries and supercapacitors (69).

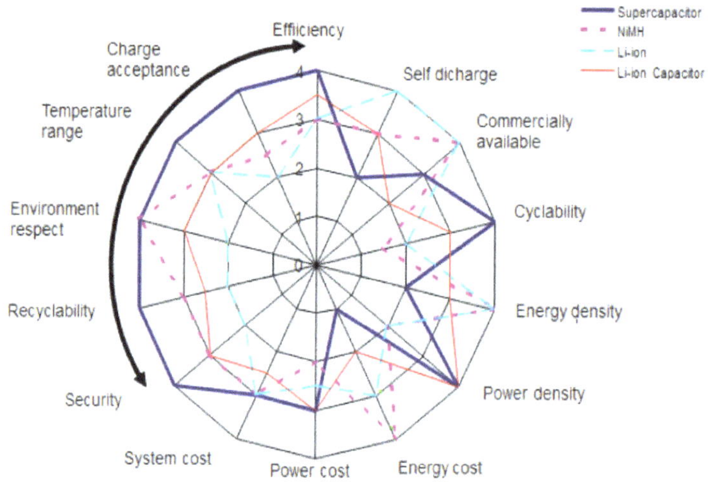

Figure 3.8: Supercapacitor comparison chart. Credit: Airbus Defense and Space and ESA (2016).

Table 3-6: Battery-vs-Supercapacitor Specifications

Feature	Li-Ion Battery	Supercapacitor
Gravimetric energy (Wh kg^{-1})	100 – 265	4 – 10
Volumetric energy (Wh L^{-1})	220 – 400	4 – 14
Power density (W kg^{-1})	1,500	3,000 – 40,000
Voltage of a cell (V)	3.6	2.7 – 3
ESR (mΩ)	500	40 - 300
Efficiency (%)	75 – 90	98
Cyclability (nb charges)	500 – 1,000	500,000 – 20, 000,000
Life (years)	5 – 10	10 – 15
Self-discharge (% per month)	2	40 – 50 (descending)
Charge temperature	0 to 45°C	-40 to 65°C
Discharge temperature	-20C~60°C	-40 to 65°C
Deep discharge pb	yes	no
Overload pb	yes	no
Risk of explosion	yes	no
Charging 1 cell	complex	easy
Charging cells in series	complex	complex
Voltage on discharge	stable	decreasing
cost ($) per kW h	235 – 1,179	11,792

The lithium-ion capacitor is a promising recent development in the world of energy storage, combining the energy storage capabilities of both lithium-ion batteries as well as double-layered capacitors; they provide a middle ground between power density and energy density, but suffer from limited life-cycles. Some lithium-ion capacitors have minimum specific energy of 200 Wh kg^{-1} but are limited by a maximum specific power of <350 W kg^{-1} (88).

3.5.2 Solid-State Batteries

A majority of the batteries being used in contemporary space applications are lithium-ion batteries that use liquid electrolytes. However, these batteries carry an inherent risk of combustion from physical damage as well as thermal runaway due to overcharge. As a result, spacecraft often carry parasitic weight in the form of cooling systems and housing units. There is a long-standing need for battery designs that improve on energy and power density, as well as safety. NASA's Solid-state Architecture Batteries for Enhanced Rechargeability and Safety (SABER) project aims to create solid-state batteries that have significantly higher energy than the current state-of-the-art lithium-ion batteries and do not catch fire or lose capacity over time. Current strides in this project include examination and testing on unique battery chemistries including sulfur-selenium and "holey graphene" (70). See table 3-7 for examples of solid-state batteries.

Table 3-7: Solid-State Batteries

Manufacturer	Product	Wh/kg	Wh/L
Solid Power	Silicon EV Cell	390	930
Solid Power	Lithium Metal	440	930
Solid Power	Conversion Reaction Cell	560	785
QuantumScape	LFP (projected)	230	600
QuantumScape	NMC (projected)	300	1000

3.5.3 Batteries for Low-Temperature Applications

Typical Li-Ion batteries have an operating temperature range of -20°C to 60°C (3). This may not meet the requirements for missions what require lower operating temperature. See table 3-8 for batteries with low-temperature applications.

Table 3-8: Batteries for Low-Temperature Applications

Company/Chemistry	Package	Temperature	Specific Energy
EEMB/Li-Ion (93)	Custom	-40°C ~ 60°C	193.5 (Wh kg^{-1})
Tadiran/$LiSOCl_2$	Custom	-80°C ~ 125°C -40°C ~ 85°C	1420 Wh/l 1420 Wh/l
GREPOW/LiPo	Custom/Pouch	-40°C ~ 60°C -55°C ~ 50°C	N/A
GREPOW/Li-Ion ($LiFePO_4$)	Custom/Pouch	-40°C ~ 50°C	N/A

3.6 State-of-the-Art – Power Management and Distribution

Power management and distribution (PMAD) systems control the flow of power to spacecraft subsystems and instruments and are often custom-designed by mission engineers for specific spacecraft power requirements, however, several manufacturers have begun to provide a variety of PMAD devices for inclusion in small spacecraft missions. PMAD not only delivers power coming from energy sources (typically solar arrays in SmallSat applications) but also conditions energy as well, mitigating harmful transient disturbances and fault conditions from propagating downstream and hurting connected loads.

Several manufacturers supply EPS which typically have a main battery bus voltage of 8.2 V but can distribute a regulated 5.0 V and 3.3 V to various subsystems. The EPS also protects the electronics and batteries from off-nominal current and voltage conditions. As the community settles on standard bus voltages, PMAD standardization may follow. Well-known producers of PMAD systems that focus on the small spacecraft market include Pumpkin, GomSpace, Stras Space, and AAC Clyde Space. However, several new producers have begun to enter the PMAD market with a variety of products, some of which are listed below. Table 3-9 lists PMAD system manufacturers; it should be noted that this list is not exhaustive.

Key considerations in determining PMAD device selection often include conversion efficiency, input/output voltage range, output power capabilities, and size, weight, and power (SWaP). These metrics are critical to consider for good SmallSat PMAD designs, but it is important to note that PMAD devices are best chosen to suit the exact application of the SmallSat mission. SmallSat missions are often short and more flexible in terms of risk management than larger satellites, and therefore lend themselves to greater flexibility in design choices. One must leverage the benefits and risks to the mission at hand when choosing COTS PMAD systems, which may include the following:

- COTS PMAD may require less intensive integration and testing but have drawbacks to be addressed in a custom PMAD build
- Unnecessary features and peripherals (e.g., excess switching, fusing, current capability) can greatly increase SWaP metrics on a SmallSat
- Variability in designs of COTS PMAD devices means that important features and protections are not available in all devices (MPPT, Dead-bus protections, redundancy mechanisms, etc.)

Due to the variability of COTS PMAD options, many choice considerations, from internal power management topologies/materials to telemetry and protection options, are either included or omitted from products depending on the manufacturer. Internal power regulation topologies have traditionally been silicon-based, but relatively recent research into the performance improvements of Gallium Nitride (GaN) topologies has increased the number of GaN-based PMAD options in the consumer market with the following benefits over their silicon counterparts:

- Ability to achieve high switching rates and lower switching losses, allowing for the downsizing of inductors and capacitors, and improving SWaP metrics
- Lack of gate oxide layer in GaN-based field-effect transistors yields improvements in overall efficiency

It must also be noted that GaN-based PMAD options are not to be considered as drop-in replacements for silicon-based PMAD options. Despite the number of performance improvements, GaN architectures come with a variety of drawbacks including high complexity of control circuitry and lack of flight heritage.

In looking at the table below, one must note that there is no single COTS PMAD solution that can fit all needs of a mission at hand. In appealing to a broad range of applications, most COTS PMAD devices make sacrifices that can impact important metrics for SmallSats, including SWaP as well as the efficiency and quality of the power being managed. In choosing to use COTS PMAD devices, designers and system architects should be aware of, and try to minimize, unnecessary features not beneficial to the mission.

Table 3-9: Power Management and Distribution System Products									
Company	**Product**	**Mass (kg)**	**Volume (cm^3)**	**Peak Power Output (W)**	**Input Voltages (VDC)**	**Output Voltages (VDC)**	**Max Efficiency (%)**	**TRL**	**Ref**
AAC Clyde Space Sweden	Starbuck Micro	2.45	3968	120	28	28 / 5	97	9	(72)
	Starbuck Mini	5.90	13133	1200	*	22-34 / 5 / 8/ 12 / 15	*	9	(73)
	Starbuck Nano	0.086	140	*	*	3.3 / 5/ 12	*	9	(74)
Argotec Italy	PCDU VOLTA	0.97	600	100	18-22	1x 3.3V, 1x 5V and 2x 12V	75	9	(94)
	PCDU ZEUS	0,5	500	136 (reg.) 400 (unreg.)	10-24	4x12 V 4x 5 V 8x 28 ± 6 V	80	5	(94)
Berlin Space Technologies Germany	PCU-110	0.960	1191	*	20-25	3.3 / 5/ 12 / 24 / 1.8-28	*	9	(83)
Bradford Space, Inc. USA	SuperNova modular PCDU (1500W config)	2.9	3045	1500	22 - 34	3.3 / 5 / 12 / unreg. batt	95	8	
C3S Electronics Development LLC Budapest	EPS1000	~0.860	~731	90	6…25V 6ch SA	3.3V, 5V, 9.9…12.3V	90%	9	
DHV Technologies Spain	MicroEPS	0.285-1.135 (+0.170 ⁑)	392-1045	592 in eclipse/ 693 in sunlight	10-40 (X/Y) / 9-28 Z	3.3 / 5/ 12 / Batt	93	5	
	NanoEPS	0.155-0.402 (+0.109 ⁑)	283-600	59 in eclipse/ 124 in sunlight	9-28 (X/Y) / 3-18 (Z)	3.3 / 5/ 12 / Batt	93	9	
	PicoEPS	0.110-0.190 (+0.1⁑)	140-197	29 in eclipse/ 74 in sunlight	3-18	3.3 / 5/ 12 / Batt	93	8	
Ecarver GmBH Germany	PCU-SB7	1.500	1800	250	0-24	0-24	85	N/A	(82)
EnduroSat Bulgaria	EPS I	0.208	183	10-20	0-5.5	3.3 / 5 / Batt	86	9	(79)
	EPS I Plus	0.292	259	30	0-5.5	3.3 / 5 / Batt	86	9	(80)

EnduroSat Bulgaria	EPS II	1.280	742	250	10-36	3.3 / 5 / 6-12 / Batt	89	9	(81)
	EPS III	1-3 kg	860-4000	1152	0-45	3.3 / 5 / 12 / 28 / Batt	87	6	-
GomSpace Germany	P31U	0.100	127	30	0-8	3.3 / 5	96	9	(75)
	P60	**	**	100	16/32 V	3.3/5/8/12/18/ 24	Refer to datasheet	-	-
	P80	360-610g ¥	350 – 586	300	0-25	3.3/5/12/18V & Vbat	Refer to datasheet	-	-
GUMUSH AeroSpace Istanbul	n-ART EPS	0.098	160	100	4.5-30	3.3 / 5 / 8-36 / Batt	94	6	
Ibeos USA	150W CubeSat EPS	0.140	124	150	18-42	3.3 / 5 / 12 / Unreg Batt	95	8	(84)
	200W, 28V CubeSat EPS	0.14	124	200	12-34	3.3 / 5 / 12/ Unreg Batt	96	8	-
	Modular EPS (500W – 2,000W)	Startin g at <1	Starting at 1150	500 – 2,000	12-26	5 / 12 / Unreg Batt	98	6	-
ISISPACE The Netherlands	iEPS Type C	0.360	14.13	13	12.8-16	3.3 / 5 / Unreg	95	9	(76)
Nanoavionics Lithuania	CubeSat EPS	*	*	175	2.6-18	3.3 / 5 / 3-18	96	N/A	(85)
Pumpkin Space Systems USA	EPSM 1	0.300	180	300	4-32	3.3-28	99.0	9	(71)
	AMPS	1.3	360	1200	5-32	3.3-28	99	9	-
SkyLabs Slovenia	SKY-NANOeps-PCDU-23c-5d	0.2	216.125	72	10V Unreg	3V3, 5V, 12V, 10V Unreg.	99%	9	(99), (100)
Space Dynamics Laboratory USA	PMU	7	6480	1000	30-50	3.3, 5, 12, 28	97	5	-
Spacemanic Slovakia	AMUN_PSU	0.2	173	50	~8V	3.3-5V	80	9	(101) (102)

* Available with inquiry to manufacturer
** Configuration dependent
† Standard Configuration
‡ Optional radiation shielding case
¥Flexible stacking options Standard options

3.7 On the Horizon – Power Management and Distribution

Power management and distribution have been steadily improving each year due to changes in technology, as well as from different approaches to maximizing the use of these systems, including modular architectures, wireless telemetry, and power transmission options.

3.7.1 Modular Architecture

For small spacecraft, traditional EPS architecture is centralized (each subsystem is connected to a single circuit board). This approach provides simplicity, volume efficiency, and inexpensive component cost. However, a centralized EPS is rarely reused for a new mission, as most of the subsystems need to be altered based on new mission requirements. A modular, scalable EPS for small spacecraft was detailed by Timothy Lim and colleagues, where the distributed power system is separated into three modules: solar, battery, and payload. This allows scalability and reusability from the distributed bus, which provides the required energy to the (interfaced) subsystem (86).

ISISPACE has a modular EPS for CubeSat missions (3U+) that includes a large amount of flexibility in output bus options with adjustable redundancy for certain parts of the device. The modular EPS consists of a power conditioning unit for solar panel input, secondary power storage, a battery holder with an integrated fuse, and a power regulation and distribution unit for subsystem loads. Each unit is designed to be independent, allowing for daisy-chaining and flexibility in redundancy and subsystem upgrades. This device is based on heritage from the Piezoelectric Assisted Smart Satellite Structure (PEASSS) CubeSat flown in 2016, with the device itself successfully flown in 2018 (76).

3.7.2 Wireless Power Transfer and Telemetry

In the commercial world, the technology already exists for wireless sensing and power transmission from the order of microwatts, all the way up to kilowatts. In the realm of SmallSats, wireless power transfer/detection would be useful as redundant options in dusty environments where physical connectors can be contaminated, or in situations where hardware needs to be swapped around and powered (battery swaps). While wireless power transfer/detection is highly inefficient when compared to conventional means, research and development in this technology for use in space applications has a lot of potential in increasing the reliability and robustness of SmallSat power management and distribution.

3.8 Summary

Driven by weight and mostly size limitations, small spacecraft are using advanced power generation and storage technology such as >32% efficient solar cells and lithium-ion batteries. The higher risk tolerance of the small spacecraft community has allowed both the early adoption of technologies like flat lithium-polymer cells, as well as COTS products not specifically designed for spaceflight. This can dramatically reduce cost and increase mission-design flexibility. In this way, power subsystems are benefiting from the current trend of miniaturization in the commercial electronics market as well as from improvements in photovoltaic and battery technology.

Despite these developments, the small spacecraft community has been unable to use other, more complex technologies. This is largely because the small spacecraft market is not yet large enough to encourage the research and development of technologies like miniaturized nuclear energy sources. Small spacecraft power subsystems would also benefit from greater availability of flexible, standardized power management and distribution systems so that every mission need not be designed from scratch. In short, today's power systems engineers are eagerly adopting certain innovative Earth-based technology (like lithium polymer batteries) while, at the same time, patiently waiting for important heritage space technology (like fuel cells and RTGs) to be adapted and miniaturized. Despite the physical limitations and technical challenges these power generation technologies have, most small nanosatellites in the foreseeable future will still likely carry batteries to support the transient load.

For feedback solicitation, please email: arc-sst-soa@mail.nasa.gov. Please include a business email so someone may contact you further.

References

(1) M. Alia-Novobilski. "Advanced Multi-junction Solar Cells Deliver High Efficiency, Reduced Costs for Space." [Online] January 4, 2018. Accessed July 15, 2018. Available at: https://phys.org/news/2018-01-advanced-multi-junction-solar-cells-high.html#jCp

(2) M.A. Green. "Third Generation Photovoltaics: Advanced Solar Energy Conversion." Springer. Vol. 12. 2003.

(3) AzurSpace. "Space Products, Solar Cells." [Online] 2020. Accessed May 15, 2021. Available at: http://www.azurspace.com/index.php/en/products/products-space/space-solar-cells

(4) EMCORE. "BTJ Space Solar Cells." Technical Datasheet. [Online] September 2021. Available at: https://dtsheet.com/doc/1248164/btj-photovoltaic-cell

(5) SpectroLab. "XTE-SF (Standard Fluence) Space Qualified Triple Junction Solar Cell." Datasheet. [Online] May 16, 2020. Accessed: May 16, 2021. Available at: https://www.spectrolab.com/photovoltaics/XTE-SF%20Data%20Sheet%20_12.23.19.pdf

(6) SpectroLab. "30.7% Triple Junction Space Grade Solar Cell." Datasheet. [Online] 2018. Accessed July 15, 2018. Available at: https://www.spectrolab.com/DataSheets/cells/XTJ_Prime_Data_Sheet_7-28-2016.pdf.

(7) SpectroLab. "Ultra-Triple Junction (UTJ) Cells". Datasheet. [Online] 2018. Accessed: July 16, 2021. Available at; http://www.spectrolab.com/DataSheets/cells/2015%20UTJ%20CIC%20Datsheet.pdf

(8) SpectroLab. "Triangular Advanced Solar Cells (TASC)." Datasheet. [Online] Accessed July 16, 2021. Available at: https://www.spectrolab.com/DataSheets/PV/PV_NM_TASC_ITJ.pdf

(9) SpectroLab. "26.8% Improved Triple Junction (ITJ) Solar Cells." Datasheet. [Online] Accessed July 16, 2021. Available at: https://www.spectrolab.com/DataSheets/TNJCell/tnj.pdf

(10) SolAero. "Space Solar Cells/CICs." [Online]. 2020. Accessed: June 30, 2021. Available at: https://solaerotech.com/space-solar-cells-cics/

(11) AAC Clyde Space. "Photon." Datasheet. [Online]. Accessed July 17, 2021. Available at: https://www.aac-clyde.space/assets/000/000/078/PHOTON_original.pdf?1564954830

(12) Blue Canyon Technologies. "Power Systems." Datasheet. [Online] Accessed: July 19, 2021. Available at: https://www.bluecanyontech.com/static/datasheet/BCT_DataSheet_Components_Power Systems.pdf

(13) DHV Technology. “Products - Solar Panels for CubeSats.” [Online] 2020. Available at: https://dhvtechnology.com/products

(14) Exoterra. Fold Out Solar Arrays (FOSA) Datasheet [Online]. Available at: https://exoterracorp.com/products/power/

(15) MMA Design. “Space Power Solutions.” [Online] Available at: https://mmadesignllc.com/products/solar-arrays/

(16) MMA_Design. rHaWK Solar Array. [Online] May 16, 2020. Accessed May 16, 2020. Available at: https://mmadesignllc.com/product/r-hawk-solar-array/.

(17) Airbus Defense and Space Netherlands: Datasheet, “Sparkwing: The plug and play solution in power your small satellite.” [Online] Accessed: July 16, 2021. Available at: https://sparkwing.space/

(18) Agencia Espacial Civil Ecuatoriana. Datasheet, “DSA: 1U Deployable Solar Arrays.” [Online] Accessed: July 18, 2021. Available at: https://satsearch.co/products/exa-1u-deployable-solar-panels-dsa-1a

(19) GomSpace Datasheet, “Nanopower Deployable Solar Panel for 3U or 6U satellite.” [Online] Accessed: July 18, 2021. Available at: https://gomspace.com/UserFiles/Subsystems/datasheet/GS-DS-NanoPower_DSP_1018088_25.pdf

(20) Innovative Solutions in Space: “Small satellite solar panels.” 2021. [Online] Accessed: July 18, 2021. Available at: https://www.isispace.nl/product/isis-cubesat-solar-panels/

(21) Redwire Space. “Redwire’s Innovative Solar Array Technology Powering the Future of Space Exploration.” [Online] May 25, 2021. Accessed: July 18, 2021. Available at: https://redwirespace.com/2021/05/25/redwires-innovative-solar-array-technology-powering-the-future-of-space-exploration/

(22) Seeger, Daniel. "Fraunhofer ISE Announces 33% Efficiency for Multi-junction Solar Cell. [Online] 2018. https://www.pv-magazine.com/2018/04/04/fraunhofer-ise-announces-33-efficiency-for-multi-junction-solar-cell/

(23) King, R. "Raising the Efficiency Ceiling in Multijunction Solar Cells.” Stanford Photonics Research Center Symposium, 2009. 2015.

(24) C. Brown et al. "An Investigation of Flexible Cu(In,Ga)Se2 Solar Cells Under Low Intensity Low Temperature for Potential Applications for Outer Planetary Missions." IEEE Journal of Photovoltaics. 2019.

(25) Toyobo Co., Ltd. "Toyobo to practicalize power-generating material for organic photovoltaics with world’s top-level conversion efficiency under room light." May 23, 2020.

(26) L. Cardinaletti et al. "Organic and Perovskite Solar Cells for Space Applications." Solar Energy Materials and Solar Cells, p. 182. 2018.

(27) S. Nagels. "OSCAR: the First Hasselt University Aerospace Project." 2020.

(28) L. Meng et al. "Organic and Solution-processed Tandem Solar Cells with 17.3% efficiency." 6407, Vol. 361. 2018.

(29) N. Luo et al. “Hydrogen-Peroxide-Based Fuel Cells for Space Power Systems.” Vol. 24, no. 3. 2008.

(30) Ethier, V, et al. "Development of a CubeSat Water-Electrolysis Propulsion System." 2013.

(31) ~~"Radioisotope Power Systems (RPS) Program." National Aeronautics and Space Administration. 2015.~~

(32) ~~Light-Weight Radioisotope Heater Unit | Thermal Systems –. (n.d.). NASA RPS: Radioisotope Power Systems, from https://rps.nasa.gov/power-and-thermal-systems/thermal-systems/light-weight-radioisotope-heater-unit/~~

(33) ~~Joannopoulos, Aristeidis Karalis and J. D.. "'Squeezing' near-field thermal emission for ultra-efficient high-power thermophotovoltaic conversion." 2016.~~

(34) ~~Nighttime Photovoltaic Cells: Electrical Power Generation by Optically Coupling with Deep Space. Tristan Deppe and Jeremy N. Munday. ACS Photonics 2020 7 (1), 1-9. DOI: 10.1021/acsphotonics.9b00679~~

(35) EnduroSat. CubeSat 3U Solar Panel X/Y MTQ. [Online] 2017. Available at: https://www.endurosat.com/products/cubesat-3u-solar-panel-x-y-mtq/

(36) R.A Nelson. "Spacecraft Battery Technology." VIA SATELLITE, pp. 104 - 118, February 1999.

(37) Jung, D. S. and Manzo, M. A.. "NASA Aerospace Flight Battery Program." NASA/TM-2010-216728/Vol I. 2010.

(38) Eagle Pitcher. LP34100 Datasheet [Online] Available at: https://www.eaglepicher.com/sites/default/files/LP34100%205Ah%20021820.pdf

(39) EaglePitcher Technologies. "Lithium-Ion Space Battery." Data Sheet Catalog_Part2_Part5. Page 21

(40) EaglePitcher Technologies. LP33330 Datasheet [Online] Available at: https://www.eaglepicher.com/sites/default/files/LP%2033330%206Ah%20Space%20Cell%200319.pdf

(41) EaglePitcher Technologies. LP32975 Datasheet [Online] Available at: https://www.eaglepicher.com/sites/default/files/LP%2032975%2012%20Ah%20Space%20Cell%20%20040219.pdf

(42) GomSpace. Datasheet, "NanoPower BPX." [Online]. Accessed: July 19, 2021. Available at: https://gomspace.com/shop/subsystems/power/nanopower-bpx.aspx

(43) GomSpace. Datasheet, "NanoPower BP4." [Online]. Accessed: July 19, 2021. Available at: https://gomspace.com/shop/subsystems/power/nanopower-bp4.aspx

(44) AAC Clyde Space. Datasheet, "Optimus Battery." July 28. 2020.

(45) Ibeos. Cubesat Product Overview. [Online] 2020. Accessed: July 16, 2021. Available at: https://ibeos.com/products.html.

(46) Saft. 4s1p VES16 Battery [Online]. 2019. Available at: https://www.saftbatteries.com/products-solutions/products/4s1p-ves16-battery

(47) Vectronic Aerospace. Li-Ion Battery Block VLB-4/ -8 /-16. 2014.[Online] Accessed July 18, 2021. Available at: https://satsearch.co/products/vectronic-aerospace-li-ion-battery-block-vlb-4-8-16

(48) Berlin Space Technologies. "Battery BAT 110," Technical Datasheet, [Online] Available at: https://www.berlin-space-tech.com/portfolio/battery-bat-110/

(49) Samsung. INR18650-35E Datasheet [Online]. Accessed: July 18, 2021. Available at: https://www.orbtronic.com/content/samsung-35e-datasheet-inr18650-35e.pdf

(50) Sony: Murata. US18650VC7 Battery Datasheet [Online]. Accessed: July 18, 2021. Available at: https://www.batterypacks.com/content/sony_us18650vc7.pdf

(51) Panasonic. NCR18650B Datasheet [Online] Available at: https://www.orbtronic.com/content/NCR18650B-Datasheet-Panasonic-Specifications.pdf

(52) Panasonic. NCR18650A Datasheet [Online] Available at: https://www.orbtronic.com/product_images/uploaded_images/panasonic-ncr18650a-datasheet-specs-sanyo-18650-battery-specifications.png

(53) LG Chem. "Product Specification: Rechargeable Lithium Ion Battery Model ICR18650 B3 2600mAh." 2007

(54) Molicel. ICR18650M Rechargebale Battery [Online] Available at: https://www.custompower.com/documents/ICR18650M.pdf

(55) Molicel. IHR18650C Datasheet [Online] Available at: http://www.molicel.com/wp-content/uploads/DM_IHR18650C-V4-80073.pdf

(56) Molicel. ICR18650H Datasheet [Online] Available at: https://www.custompower.com/documents/ICR18650H.pdf

(57) Molicel. ICR18650J Datasheet [Online] Available at: https://www.custompower.com/documents/ICR18650J.pdf

(58) Molicel. ICP103450DA Datasheet [Online]. Available at: https://www.custompower.com/documents/ICP103450DAProductDataSheetF.pdf

(59) Molicel. INR18650A Datasheet [Online] Available at: http://www.molicel.com/wp-content/uploads/DM_INR18650A-V4-80078.pdf

~~(60) Molicel. Panasonic Lithium Ion NCR18650B. 2012.~~

(61) A. Zimmerman and E. Jung. "Verification Sciences & Engineering Case Study of Counterfeit COTS Li-ion Cells." 2019. NASA Aerospace Battery workshop.

(62) JPL. "NASA Battery Workshop." [Online] November 19, 2019. Accessed May 16, 2020. Available at: https://www.nasa.gov/sites/default/files/atoms/files/3-nasa_battery_workshop_nov_2019_high_power_li-ion_cells_final.pdf

(63) T. G. Tranter et al. "Communication—Prediction of Thermal Issues for Larger Format 4680 Cylindrical Cells and Their Mitigation with Enhanced Current Collection." 2020 J. Electrochem. Soc. 167 160544

(64) Aravindan, V., Gnanaraj, J., Lee, Y., & Madhavi, S. (2014). Insertion-type electrodes for nonaqueous Li-ion capacitors. Chemical reviews, 114 23, 11619-35.

(65) R. Raffaelle. "Demonstration of a Nano-Enabled Space." [Online] 2018. Available at: https://ntrs.nasa.gov/archive/nasa/casi.ntrs.nasa.gov/20170011060.pdf

(66) R.L. Karkkainen. "Development of Lightweight CubeSat with Multi-Functional Structural Battery Systems." [Online] 2018. Available at: https://ntrs.nasa.gov/archive/nasa/casi.ntrs.nasa.gov/20170011078.pdf

(67) B. Faure, L. Cosqueric, and V. Gineste. "Evaluation and Qualification of Commercial Off-The-Shelf Supercapacitors for Space." 2nd Space Passive Component Days (SPCD) International Symposium. Noordwijk, The Netherlands: s.n., 2016.

(68) SupercapTech. SupercapTech. [Online] May 21, 2016. Accessed: June 1, 2020. Available at: https://www.supercaptech.com/battery-vs-supercapacitor

(69) POWERBOX. "The Silent Power of Supercapacitors." White paper 017. 2019.

(70) Gipson, L. "NASA Seeks to Create a Better Battery with SABERS." [Online] April 7, 2021. Available at: https://www.nasa.gov/feature/nasa-seeks-to-create-a-better-battery-with-sabers/

(71) Pumpkin, Inc. CubeSat Kit Electrical Power System 1 (EPS1). [Online] 2018. Available at:" https://www.pumpkinspace.com/store/p216/High-power_Multi-channel_Electrical_Power_System_Module_%28EPSM1%29.html

(72) AAC Clyde Space. Starbuck Micro Original Datasheet. [Online] 2020. Available at: https://www.aac-clyde.space/wp-content/uploads/2021/11/AAC_DataSheet_Starbuck-Micro.pdf

(73) AAC Clyde Space. Starbuck Mini Original Datasheet. [Online] 2020. Available at: https://www.aac-clyde.space/wp-content/uploads/2021/11/AAC_DataSheet_Starbuck-Mini-updated.pdf

(74) AAC Clyde Space. Starbuck Nano Original Datasheet. [Online] 2020. Available at: https://www.aac-clyde.space/wp-content/uploads/2021/11/AAC_DataSheet_Starbuck-Nano.pdf

(75) GomSpace. NanoPower P31u. [Online] 2018. Accessed: July 16, 2021. Available at: https://gomspace.com/Shop/subsystems/power-supplies/nanopower-p31u.aspx

(76) Innovative Solutions in Space. IEPS Electrical Power System. [Online] 2020. Available at: https://www.ISISPACE.nl/product/ieps-electrical-power-system/

(77) DHV Technology. "Electrical Power Systems." [Online] Available at: https://dhvtechnology.com/products/electrical-power-systems/

(78) Extreme Engineering Solutions: "XPm2020." [Online] 2021. Accessed: July 20, 2021. Available at: https://www.xes-inc.com/products/end-of-life-power-supplies/xpm2020/

(79) EnduroSat. EPS I. [Online] Available at: https://www.endurosat.com/cubesat-store/cubesat-power-modules/eps-power-module/

(80) EnduroSat. EPS I Plus. [Online] Available at: https://www.endurosat.com/cubesat-store/cubesat-power-modules/eps-power-module-i-plus/

(81) EnduroSat. EPS II + Battery Pack. [Online] Available at: https://www.endurosat.com/cubesat-store/cubesat-power-modules/eps-power-module-ii/

(82) Ecarver GmbH: "Battery BP3-SN." [Online] Accessed: July 18, 2021. Available at: https://satellite-power-systems.de/products.html

(83) Berlin Space Technologies. "Power Control Unit PCU-110." [Online]. Accessed: July 19, 2021. https://www.berlin-space-tech.com/portfolio/powercontrolunit-pcu-110/

(84) Ibeos. Datasheet, "150-Watt SmallSat Electric Power Subsystem (EPS)." [Online] 2019. Available at": https://www.ibeos.com/150w-eps-datasheet

(85) NanoAvionics. “CubeSat Electrical Power System EPS”. [Online] 2017. Accessed: July 15, 2018. Available at: https://n-avionics.com/cubesat-components/solar-panels-and-power-systems/cubesat-electrical-power-system-eps/

(86) T.M. Lim et al. "A Modular Electrical Power System." IEEE Transactions on Aerospace and Electronic Systems, Volume: 54, Issue: 4, August 2018.

(87) Canon. BP-930 Material Safety Data Sheet. [Online] 2011. Available at: https://files.canon-europe.com/files/webcontent/safety-data-sheets/EU_BP-Series-Batteries/bp-930_pdf.pdf

(88) M. Soltani and S. H. Beheshti. “A comprehensive review of lithium -ion capacitor: development, modelling, thermal management and applications,” Journal of Energy Storage, Volume 34, 2021.

(89) Nanoavionics. “CubeSat GaAs Solar Panel.” [Online] 2020. Accessed: May 16, 2020. Available at: https://nanoavionics.com/cubesat-components/cubesat-gaas-solar-panel/

(90) NASA. 2020 Mission Perseverance Rover: Power Source.” [Online] 2020. Accessed: September 27, 2021. Available at: https://mars.nasa.gov/mars2020/spacecraft/rover/electrical-power/

(91) A.J. Colozza and R.L Cataldo. “Low Power Radioisotope Conversion Technology and Performance Summary”. NASA/TM—2018-219940.

(92) E. Peretz et al. “The Importance of Solar Power Technology Adaptations to Space Exploration”, IEEE 2024 (In preparation)

(93) EEMB Battery. Military Grade Li-ion Battery [Online] Accessed 2022. Available at: https://www.eemb.com/low-temperature-version.html

(94) Argotec VOLTA. [Online] Accessed September 24, 2023. Available at: https://www.argotecgroup.com/wp-content/uploads/2022/03/Argotec_VOLTA_scheda_prodotto.pdf

(95) GomSpace. “NanoPower P110 Datasheet”. October 22, 2018. Available at: https://gomspace.com/UserFiles/Subsystems/datasheet/gs-ds-nanopower-p110-210.pdf

(96) GomSpace. “NanoPower MSP Datasheet”. February 10, 2021. Available at: https://gomspace.com/UserFiles/Subsystems/datasheet/gs-ds-nanopower-msp_(DS_-_1016340_-_1_-_21)_-_1.pdf

(97) Pumpkin Space Systems. [Online] “135W Dual Articulated Deployable Solar Array”. 2022. Available at: https://www.pumpkinspace.com/store/p215/135W_Dual_Articulated_Deployable_Solar_Array.html

(98) Pumpkin Space Systems. [Online] “220W Deployable Clamshell”. 2022. Available at: https://www.pumpkinspace.com/store/p210/220W_Deployable_Clamshell_Solar_Array_%28DCSA%29.html

(99) Skylabs. [Online] “NANOeps-BMM-172wh”. Available at” https://www.skylabs.si/products/nanoeps-bmm-172wh/

(100) Skylabs. [Online] “TRISAT-R with NANOsky satellite avionics continues to operate flawlessly”. March 16, 2023. Available at: https://www.skylabs.si/trisat-r-with-nanosky-satellite-avionics-continues-to-operate-flawlessly/

(101) Spacemanic. [Online] "AMUN Power Supply Unit" Available at: https://www.spacemanic.com/amun-power-supply-unit/

(102) Spacemanic. [Online] "BDSAT-2 launched aboard Space X's mission: Transporter 6 on Falcon 9 rocket". January 3, 2023. Accessed October 1, 2023. Available at: https://www.spacemanic.com/news/bdsat-2-launched-aboard-space-xs-mission-transporter-6-on-falcon-9-rocket/

Chapter Contents

Chapter Glossary

(ABS)	Acrylonitrile Butadiene Styrene
(AC)	Alternating Current
(ACE)	Apollo Constellation Engine
(ACO)	Announcement for Collaborative Opportunity
(ACS)	Attitude Control System
(ADN)	Ammonium Dinitramide
(AFRL)	Air Force Research Laboratory
(AIS)	Applied Ion Systems
(AOCS)	Attitude and Orbit Control System
(AR)	Aerojet Rocketdyne
(ARC)	Ames Research Center
(BOL)	Beginning of Life
(CHIPS)	CubeSat High Impulse Propulsion System
(CMNT)	Colloid MicroNewton Thrusters
(CNAPS)	Canadian Nanosatellite Advanced Propulsion System
(CNES)	French National Center for Space Studies
(CPOD)	CubeSat Proximity Operations Demonstration
(CUA)	CU Aerospace LLC
(DFMR)	Design for Minimum Risk
(DRM)	Design Reference Mission
(DSSP)	Digital Solid State Propulsion LLC
(EOL)	End of Life
(EMC)	Electromagnetic Compatibility
(EMI)	Electromagnetic Interference
(EP)	Electric Propulsion
(EPL)	Electric Propulsion Laboratory, Inc.
(EPSS)	Enabling Propulsion System for Small Satellites
(ESA)	European Space Agency
(ESPs)	Electrically Controlled Solid Propellant
(FASTSAT)	Fast, Affordable, Science and Technology Satellite
(FEEP)	Field Emission Electric Propulsion
(FPPT)	Fiber-Fed Pulsed Plasma Thruster

(GEO)	Geostationary Equatorial Orbit
(GIT)	Gridded-ion Thrusters
(GOCE)	Gravity Field and Steady-State Ocean Circulation Explorer
(GOX)	Gaseous Oxygen
(GPIM)	Green Propellant Infusion Mission
(GPS)	Global Positioning System
(GRC)	Glenn Research Center
(GSFC)	Goddard Space Flight Center
(HAN)	Hydroxylammonium Nitrate
(HET)	Hall-effect Thruster
(HTP)	High Test Peroxide
(HTPB)	Hydroxyl-terminated Polybutadiene
(IPS)	Integrated Propulsion System
(ISS)	International Space Station
(JANNAF)	Joint Army Navy NASA Air Force
(JHU ERG)	Johns Hopkins University Energetics Research Group
(JPL)	Jet Propulsion Laboratory
(LFPS)	Lunar Flashlight Propulsion System
(LISA)	Laser Interferometer Space Antenna
(MAPS)	Modular Architecture Propulsion System
(MarCO)	Mars Cube One
(MCD)	Micro-cavity Discharge
(MEMS)	Microelectromechanical System
(MEO)	Medium Earth Orbit
(MMH)	Monomethyl Hydrazine
(MPUC)	Monopropellant Propulsion Unit for CubeSats
(MSFC)	Marshall Space Flight Center
(MVP)	Monofilament Vaporization Propulsion
(N_2O)	Nitrous Oxide
(NEA)	Near-Earth Asteroid
(NODIS)	NASA Online Directives Information System
(NSTT)	Nanosat Terminator Tape
(OTS)	Orbital Transfer System
(OTV)	Orbital Transfer Vehicle

(PacSci EMC)	Pacific Scientific Energetic Materials Company
(PBM)	Plasma Brake Module
(PMD)	Propellant Management Device
(PMDs)	Propellant Management Devices
(PMI)	Progress toward Mission Infusion
(PMMA)	Polymethyl Methacrylate
(PPT)	Pulsed Plasma Thrusters
(PPU)	Power Processing Unit
(PTD)	Pathfinder Technology Demonstration
(PTFE)	Polytetrafluoroethylene
(PUC)	Propulsion Unit for CubeSats
(ROMBUS)	Rapid Orbital Mobility Bus
(SAA)	Space Act Agreement
(SBIR)	Small Business Innovative Research
(SCAPE)	Self Contained Atmospheric Protective Ensemble
(SEP)	Solar Electric Propulsion
(SMAP)	Soil Moisture Active Passive
(SMART-1)	Small Missions for Advanced Research in Technology
(SME)	Subject Matter Experts
(SSTL)	Surrey Satellite Technology Ltd.
(SSTP)	Small Spacecraft Technologies Program
(TBD)	To Be Determined
(TCMs)	Trajectory Correction Maneuvers
(TMA)	Technology Maturity Assessment
(TRL)	Technology Readiness Level
(UTIAS)	University of Toronto Institute for Aerospace Research
(VAT)	Vacuum arc thrusters
(VENuS)	Vegetation and Environment monitoring on a New Microsatellite
(WFF)	Wallops Flight Facility

4.0 In-Space Propulsion

4.1 Introduction

In-space propulsion devices for small spacecraft are rapidly increasing in number and variety. Although a mix of small spacecraft propulsion devices have established flight heritage, the market for new propulsion products continues to prove dynamic and evolving. In some instances, systems and components with past flight heritage are being reconsidered to meet the needs of smaller spacecraft. This approach minimizes new product development risk and time to market by creating devices similar to those with existing spaceflight heritage, although accounting for small spacecraft volume, mass, power, safety and cost considerations. Such incremental advancement benefits from existing spaceflight data, physics-based models, and customer acceptance of the heritage technologies, which eases mission infusion. In other instances, novel technologies are being conceived specifically for small spacecraft using innovative approaches to propulsion system design, manufacturing, and integration. While the development of novel technologies typically carries a higher risk and slower time to market, these new technologies strive to offer small spacecraft a level of propulsive capability not easily matched through the miniaturization of heritage technologies. Such novel devices are often highly integrated and optimized to minimize the use of a small spacecraft's limited resources, lower the product cost, and simplify integration. Regardless of the development approach, the extensive investments by commercial industry, academia, and government to develop new propulsion products for small spacecraft suggests long-term growth in the availability of propulsion devices with increasingly diverse capabilities.

In the near-term, the surge in public and private investments in small spacecraft propulsion technologies, combined with the immaturity of the overall small spacecraft market, has resulted in an abundance of confusing, unverified, sometimes conflicting, and otherwise incomplete technical literature. Furthermore, the rush by many device developers to secure market share has resulted in some confusion surrounding the true readiness of these devices for mission infusion. As third parties independently verify device performance, and end-users demonstrate these new devices in their target environments, the true maturity, capability, and flight readiness of these devices will become evident. In the meantime, this report will attempt to reduce confusion by compiling a list of publicly described small spacecraft propulsion devices, identifying publicly available technical literature for further consideration, recognizing missions of potential significance, and organizing the data to improve comprehension for both neophytes and subject matter experts.

This chapter avoids a direct technology maturity assessment (TMA) based on the NASA Technology Readiness Level (TRL) scale, recognizing insufficient in-depth technical insight into current propulsion devices to perform such an assessment accurately and uniformly. An accurate TRL assessment requires a high degree of technical knowledge on a subject device as well as an understanding of the intended spacecraft bus and target environment. While the authors strongly encourage a TMA that is well-supported with technical data prior to infusing technologies into programs, the authors believe TRLs are most accurately assessed within the context of a program's unique requirements. Rather than attempting to assess TRL in the absence of sufficient data, this chapter introduces a novel classification system that simply recognizes Progress toward Mission Infusion (PMI) as an indicator of the efficacy of the manufacturers' approach to system maturation and mission infusion. PMI should not be confused with TRL as PMI does not directly assess technology maturity. However, PMI may prove insightful in early trade studies. The PMI classification system used herein is described in detail in Section 4.4.2.

4.1.1 Document Organization

This chapter organizes the state-of-the-art in small spacecraft propulsion into the following categories:

1. In-Space Chemical Propulsion (4.6.1)
2. In-Space Electric Propulsion (4.6.2)
3. In-Space Propellant-less Propulsion (4.6.3)

Each of these categories is further subdivided by the prevailing technology types. The subsections organize data on each prevailing technology type as follows:

a. Technology Description
b. Key Integration and Operational Considerations
c. Current & Planned Missions
d. Summary Table of Devices
e. Notable Advancements

The organizational approach introduces newcomers to each technology, presents technology-specific integration and operation concerns for the reader's awareness, highlights recent or planned missions that may raise the TRL of specific devices, and tabulates procurable devices of each technology. Some sections also include an incomplete list of notable advancements. While the key integration and operational considerations are not comprehensive, they provide initial insights that may influence propulsion system selection. In the cases where a device has a long history of spaceflight, this chapter reviews only select missions.

4.2 Public Data Sources and Disclaimers

This chapter is a survey of small spacecraft propulsion technologies as discussed in open literature and does not endeavor to be an original source. As such, this chapter only considers literature found in the public domain to identify and classify devices. Commonly used sources for public data include manufacturer datasheets, press releases, conference papers, journal papers, public filings with government agencies, and news articles.

This chapter summarizes device performance, capabilities, and flight history, as presented in publicly available literature. Data not appropriate for public dissemination, such as proprietary, export controlled, or otherwise restricted data, are not considered. As such, actual device maturity and flight history may be more or less extensive than what is documented herein. Device manufacturers should be consulted for the most up-to-date and relevant data before performing a TMA.

This chapter's primary data source is literature produced by device manufacturers. Unless otherwise published, do not assume independent verification of device performance and capabilities. Performance and capabilities described may be speculative or otherwise based on limited data.

The information presented is not intended to be exhaustive but to provide a general overview of current state-of-the-art technologies and their development status. It should be noted that technology maturity designations may vary with change to payload, mission requirements, reliability considerations, and/or the environment in which performance was demonstrated. Readers are highly encouraged to reach out to companies for further information regarding the performance and maturity of the described technology. There is no intention of mentioning certain companies and omitting others based on their technologies or relationship with NASA.

Suggestions or corrections for this document should be submitted to the NASA Small Spacecraft Virtual Institute Agency-SmallSat-Institute@mail.nasa.gov for consideration prior to the publication of future issues. When submitting comments, please cite appropriate publicly accessible references. Private correspondence is not considered an adequate reference.

4.3 Definitions

- *Device* refers to a component, subsystem, or system, depending on the context.
- *Technology* refers to a broad category of devices or intangible materials, such as processes.

4.4 Technology Maturity

4.4.1 Application of the TRL Scale to Small Spacecraft Propulsion Systems

NASA has a well-established guideline for performing TMAs, described in detail in the NASA Systems Engineering Handbook (1). A TMA determines a device's technological maturity, which is usually communicated according to the NASA TRL scale. The TRL scale is defined in NASA Procedural Requirements (NPR) 7123 (2). The NASA Systems Engineering Handbook and NPR 7123 can be accessed through the NASA Online Directives Information System (NODIS) library. Assessment of TRLs for components, systems, or software allows for coherent communication between technologists, program managers, and other stakeholders regarding the maturity of a technology. Furthermore, TRL is a valuable tool to communicate the potential risk associated with the infusion of technologies into programs. For TRLs to be applied across all technology categories, the NASA TRL definitions are written broadly and rely on subject matter experts (SME) in each discipline to interpret appropriately.

Recently, U.S. Government propulsion SMEs suggested an interpretation of the TRL scale specifically for micro-propulsion. The Micro-Propulsion Panel of the JANNAF Spacecraft Propulsion Subcommittee in 2019 published the *JANNAF Guidelines for the Application of Technology Readiness Levels (TRLs) to Micro-Propulsion Systems* (3). The guideline was recently updated in 2022 to reflect the latest community input (4). This guideline suggests an interpretation of TRL for micro-propulsion and reflects both NASA and DOD definitions for TRL. The JANNAF panel consists of participants from the Air Force Research Laboratory (AFRL), Glenn Research Center (GRC), Jet Propulsion Laboratory (JPL), and Goddard Space Flight Center (GSFC). The panel further receives feedback from the non-Government propulsion community. While this JANNAF guideline focuses on micro-propulsion (e.g., CubeSats), the guideline still has relevance to rigorously assessing TRLs for the more general category of small spacecraft in-space propulsion. By establishing a common interpretation of TRL for small spacecraft propulsion, a more coherent and consistent communication of technology maturity can occur between small spacecraft propulsion providers and stakeholders. The JANNAF guideline is open to unlimited distribution and may be requested from the Johns Hopkins University Energetics Research Group (JHU ERG). Ensure the use of the latest JANNAF guideline, as the guideline may continue to evolve with further community input.

A fundamental limitation of the JANNAF guideline for TRL assessment, and TMA in general, is an assumption of in-depth technical knowledge of the subject device. In the absence of detailed technical knowledge, especially in a broad technology survey as presented herein, a TMA may be conducted inaccurately or inconsistently. Furthermore, assessment of TRL assumes an understanding of the end-user application. The same device may be concluded to be at different TRLs for infusion into different missions. For example, a device may be assessed at a high TRL for application to low-cost small spacecraft in low-Earth orbits, while assessed at a lower TRL for application to geosynchronous communication satellites or NASA interplanetary missions due to different mission requirements. Differences in TRL assessment based on the operating environment may result from considerations such as thermal environment, mechanical loads, mission duration, or radiation exposure. Propulsion-specific variances between missions might include propellant type, total propellant throughput, throttle set-points, burn durations, and the total number of on/off cycles. As such, an accurate TRL assessment not only requires an in-depth technical understanding of a device's development history, including specifics on past flight-qualification activities, but also an understanding of mission-specific environments and interfaces. The challenge of assessing an accurate TRL in a broad technology survey poses a significant burden for data collection, organization, and presentation. Such activities are better suited for programs seeking to infuse new technologies into their missions.

Given the rapid evolution of small spacecraft propulsion technologies and the variety of mission environments, as well as generally limited device technical details in open literature, the propulsion chapter implements a novel system to classify technical maturity according to Progress toward Mission Infusion (PMI). This novel classification system is not intended to replace TRL but is a complementary tool to provide initial insight into device maturity when it is not yet feasible to accurately and consistently apply the TRL scale. This novel classification system is discussed in detail below.

Readers are strongly encouraged to perform more in-depth technical research on candidate devices based on the most up-to-date information available, as well as to assess risk within the context of their specific mission(s). A thoughtful TMA based on the examination of detailed technical data through consultation with device manufactures can reduce program risk and increase the likelihood of program success. This survey is not intended to replace the readers' own due diligence. Rather, this survey and PMI seek to provide early insights that may assist in propulsion system down-select to a number of devices where an in-depth TMA becomes feasible.

4.4.2 Progress Toward Mission Infusion (PMI)

Rather than directly assessing a device's technical maturity via TRL, propulsion devices described herein are classified according to evidence of progress toward mission infusion. This is a novel classification system first introduced in this survey. Assessing the PMI of devices in a broad survey, where minimal technical insight is available, may assist with down-selecting propulsion devices early in mission concept development. Once a handful of devices are identified as possible solutions for a specific mission concept, a detailed TMA and rigorous TRL assessment should be conducted. The PMI classification system sort devices into one of four broad technology development categories: Concept, In-Development, Engineering-to-Flight, and Flight-Demonstrated. The following sections describe the PMI classification system in-detail. Furthermore, figure 4.1 summarizes the PMI classifications.

Concept, 'C'

The *Concept* classification reflects devices in an early stage of development, characterized by feasibility studies and the demonstration of fundamental physics. Concept devices typically align with the NASA TRL range of 1 to 3. At a minimum, these devices are established as scientifically feasible, perhaps through a review of relevant literature and/or analytical analysis. These devices may even include experimental verification that supports the validity of the underlying physics. These devices may even include notional designs. While Concept devices are generally not reviewed herein, particularly promising Concept devices will be classified in tables with a 'C'.

In-Development, 'D'

The *In-Development* classification reflects the bulk of devices being actively matured and covered in this survey, where only a modest number of devices may progress to regular spaceflight. In-Development devices typically align with the NASA TRL range of 4 to 5. While In-Development devices may have specific applications attributed by their developers, no selection for a specific mission has been publicly announced. In the absence of a specific mission, device development activities typically lack rigorous system requirements and a process for independent requirement validation. Furthermore, qualification activities conducted in the absence of a specific mission typically require a delta-qualification to address mission-specific requirements. At a minimum, In-Development devices are low-fidelity devices that have been operated in an appropriate environment to demonstrate basic functionality and support prediction of the device's ultimate capabilities. They may even be medium- or high-fidelity devices operated in a simulated final environment, but lacking a specific mission pull to define requirements and a qualification program. They may even be medium- or high-fidelity devices operated in a spaceflight demonstration but lacking sufficient fidelity or demonstrated capability to reflect the anticipated final product. These devices are typically described as a technology push, rather than a mission pull. In-Development devices will be classified in tables with a 'D'.

Engineering-to-Flight, 'E'

The *Engineering-to-Flight* classification reflects devices with a publicly announced spaceflight opportunity. This classification does not necessarily imply greater technical maturity than the In-Development classification, but it does assume the propulsion device developer is receiving mission-specific requirements to guide final development and qualification activities. Furthermore, the Engineering-to-Flight classification assumes a mission team performed due diligence in the selection of a propulsion device, and the mission team is performing regular activities to validate that the propulsion system requirements are met. Thus, while the PMI classification system does not directly assess technical maturity, there is an underlying assumption of independent validation of mission-specific requirements, where a mission team does directly consider technical maturity in the process of device selection and mission infusion. Engineering-to-Flight devices typically align with the NASA TRL range of 5 to 6. At a minimum, these are medium-fidelity devices that have been operated in a simulated final environment and demonstrate key capabilities relative to the requirements of a specific mission. These devices may even be actively undergoing or have completed a flight qualification program. These devices may even include a spaceflight, but in which key capabilities failed to be demonstrated or further engineering is required. These devices may even include a previously successful spaceflight, but the devices are now being applied in new environments or platforms that necessitate design modifications and/or delta-qualification. These devices must have a specific mission pull documented in open literature. A design reference mission (DRM) may be considered in place of a specific mission pull, given detailed documentation in open literature, which includes a description of the DRM, well-defined propulsion system requirements, maturation consistent with the DRM requirements, and evidence of future mission need consistent with the DRM. Engineering-to-Flight devices will be classified in tables with an 'E'.

Flight-Demonstrated, 'F'

The *Flight-Demonstrated* classification reflects devices where a successful technology demonstration or genuine mission has been conducted and described in open literature. Flight-Demonstrated devices typically align with the NASA TRL range of 7 to 9. These devices are high-fidelity components or systems (in fit, form, and function) that have been operated in the target in-space environment (e.g., low-Earth orbit, GEO, deep space) on an appropriate platform, where all key capabilities were successfully demonstrated. These devices may even be final products, which have completed genuine missions (not simply flight demonstrations). These devices may even be in repeat production and routine use for several missions. The devices must be described in open literature as successfully demonstrating key capabilities in the target environment to be considered Flight-Demonstrated. If a device has flown, but the outcome is not publicly known, the classification will remain Engineering-to-Flight. Flight-Demonstrated devices will be classified in tables with an 'F'.

Concept, 'C'

- At minimum, an idea has been established as scientifically feasible.
- May even include experimental verification of the underlying physics.
- May even include notional device designs.
- Approximately aligns to NASA TRL 1-3

In-Development, 'D'

- At minimum, a low-fidelity device that has been operated in an appropriate environment to demonstrate the basic functionality and predict the ultimate capabilities.
- May even be a medium- or high-fidelity device operated in a simulated final environment, but the device lacks a specific mission pull to define requirements and a qualification program.
- May even be a medium- or high-fidelity device operated in a flight demonstration, but the device lacks sufficient fidelity or demonstrated capability to reflect the anticipated final product.
- Approximately aligns to NASA TRL 4-5

Engineering-to-Flight, 'E'

- At minimum, a medium-fidelity device that has been operated in a simulated final environment and demonstrates key capabilities relative to the requirements of a specific mission.
- May even include a qualification program in-progress or completed.
- May even include a spaceflight, but the device fails to demonstrate key capabilities.
- May even include a successful spaceflight, but the device is now being applied in a new environment or platform, necessitating a delta-qualification.
- A specific mission opportunity must be identified in open literature.
- Approximately aligns to NASA TRL 5-6

Flight-Demonstrated, 'F'

- At minimum, a high-fidelity component or system (fit, form, and function) that has been operated in the intended in-space environment (e.g., LEO, GEO, deep space) on an appropriate platform, where key capabilities have been successfully demonstrated.
- May even be a final product that has completed a mission (not strictly a technology demonstration).
- May even be a product in repeat production and routine use for a number of missions.
- A successful spaceflight must be identified and the outcome described in open literature.
- Approximately aligns to NASA TRL 7-9

Figure 4.1: Progress toward mission infusion (PMI) device classifications. Credit: NASA.

4.5 Overview of In-Space Propulsion Technology Types

In-space small spacecraft propulsion technologies are generally categorized as (i) chemical, (ii) electric, or (iii) propellant-less. This chapter surveys propulsion devices within each technology category. Additionally, liquid-propellant acquisition and management devices are reviewed as an important component of in-space propulsion systems. Although other key subsystems have not yet been reviewed, such as small spacecraft propulsion power processing units, they may be included in future updates of this publication. Table 4-1 lists the in-space propulsion technologies reviewed. Figure 4.2 graphically illustrates the range of thrust and specific impulse for these small spacecraft propulsion devices. The thrust and specific impulse ranges provided in table 4-1 and figure 4.1 only summarize the performance of small spacecraft devices covered in this survey and may not reflect the broader capability of the technologies beyond small spacecraft or the limits of what is physically possible with further technology advancement. Furthermore, propulsion systems are often highly throttleable, and the devices surveyed herein may in many cases be capable of offering performance beyond the ranges presented in table 4-1.

Chemical systems have enabled in-space maneuvering since the onset of the space age, proving highly capable and reliable. These include hydrazine-based systems, other mono- or bipropellant systems, hybrids, cold gas systems, and solid propellants. Typically, these systems are sought when high thrust or rapid maneuvers are required. As such, chemical systems continue to be the in-space propulsion technology of choice when their total impulse capability is sufficient to meet mission requirements.

On the other hand, the application of electric propulsion devices has been historically far more limited. While electric propulsion can provide an order of magnitude greater total impulse than chemical systems, research and development costs have typically eclipsed that of comparable chemical systems. Furthermore, electric propulsion generally provides thrust-to-power levels below 75 mN/kW. Thus, a small spacecraft capable of delivering 500 W to an electric propulsion system may generate no more than 38 mN of thrust. Therefore, while the total impulse capability of electric propulsion is generally considerable, these systems may need to operate for hundreds or thousands of hours, compared to the seconds or minutes that chemical systems necessitate for a similar impulse. That said, the high total impulse and low thrust requirements of specific applications, such as station keeping, have maintained steady investment in electric propulsion over the decades. Only in recent years has the mission pull for electric propulsion reached a tipping point where electric propulsion may overtake chemical for specific in-space applications. Electric propulsion system types considered herein include electrothermal, electrospray, gridded ion, Hall-effect, pulsed plasma and vacuum arc, and ambipolar.

Propellant-less propulsion technologies such as solar sails, tethers, electric sails (and plasma brakes), and aerodynamic drag devices have long been investigated, but they have yet to move beyond small-scale demonstrations. However, growing needs such as orbital debris removal may offer compelling future applications.

Some notable categories are not covered in this survey, such as nuclear in-space propulsion technologies. While substantial investment continues in such areas for deep space science and human exploration, such technologies are generally at lower TRL and typically aim to propel spacecraft substantially larger than the 180 kg wet mass limit covered by this report.

Whenever possible, this survey considers complete propulsion systems, which are composed of thrusters, feed systems, pressurization systems, propellant management and storage, and power processing units, but not the electrical power supply. However, for some categories, components (e.g., thruster heads) are mentioned without consideration of the remaining subsystems necessary for their implementation. Depending on the device's intended platform (i.e., NanoSat, MicroSat, SmallSat), the propulsion system may be either highly integrated or distributed within the spacecraft. As such, it is logical to describe highly integrated propulsion units at the system level, whereas components of distributed propulsion systems may be logically treated at the sub-system level, where components from a multitude of manufacturers may be mixed-and-matched to create a unique mission-appropriate propulsion solution.

Table 4-1: Summary of Propulsion Technologies Surveyed

Technology	Thrust Range	Specific Impulse Range [sec]
4.6.1 CHEMICAL PROPULSION TECHNOLOGIES		
Hydrazine Monopropellant	0.25 – 28 N	180 – 285
Alternative Mono- and Bipropellants	50 mN – 22 N	150 – 310
Hybrids	8 – 222 N	215 – 300
Cold Gas	10 μN – 3.6 N	40 – 110
Solid Motors	37 – 461 N	187 – 269
Propellant Management Devices	-	-
4.6.2 ELECTRIC PROPULSION TECHNOLOGIES		
Electrothermal	0.1 mN – 1 N	20 – 350
Electrosprays	20 μN – 20 mN	225 – 3,000
Gridded Ion	0.1 – 20 mN	500 – 3,000
Hall-Effect	0.25 – 55 mN	200 – 1,920
Pulsed Plasma and Vacuum Arc Thrusters	4 – 500 μN	87 – 3,200
Ambipolar	0.5 – 17 mN	400 – 1,100
4.6.3 PROPELLANTLESS PROPULSION TECHNOLOGIES		
Solar Sails	TBD	-
Tethers	TBD	-
Electric Sails	TBD	-
Aerodynamic Drag	TBD	-

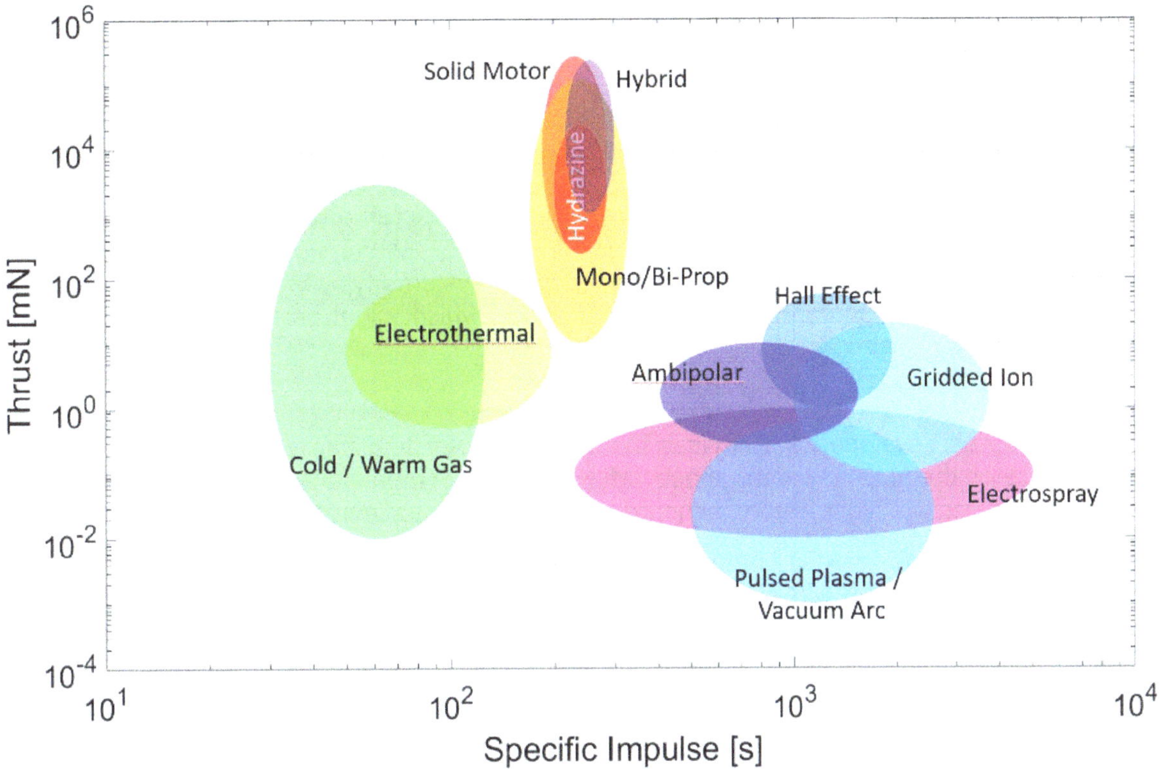

Figure 4.2: Typical small spacecraft in-space propulsion trade space (thrust vs. specific impulse). Credit: NASA.

4.6 State-of-the-Art in Small Spacecraft Propulsion

4.6.1 In-Space Chemical Propulsion

Chemical propulsion systems are designed to satisfy high-thrust impulsive maneuvers. They offer lower specific impulse compared to their electric propulsion counterparts but have significantly higher thrust to power ratios.

Hydrazine Monopropellant

a. Technology Description

Hydrazine monopropellant systems use catalyst structures (such as S-405 granular catalyst) to decompose hydrazine to produce hot gases. Hydrazine thrusters and systems have been in extensive use since the 1960's. The low mass and volume of a significant number of heritage hydrazine propulsion systems makes them suitable for small spacecraft buses, and some hydrazine thrusters used on large spacecraft may be appropriate as the main propulsion system for small spacecraft. Hydrazine thrusters typically achieve a specific impulse between 200 – 235 seconds for 1-N class or larger thrusters.

b. Key Integration and Operational Considerations

- **Extensive Flight Heritage:** Since hydrazine has been used extensively in spaceflight applications, the technology's traits are well understood (5).
- **Extensive Component Ecosystem:** A robust ecosystem of components and experience exists because hydrazine systems are widely used. As such, hydrazine propulsion

systems are frequently customized for specific applications using the available components.

- **Qualified for Multiple Cold Restarts:** These systems have the advantage of typically being qualified for multiple cold starts.
- **Extensive Safety and Handling Requirements:** Hydrazine and its derivatives are corrosive, toxic, and potentially carcinogenic. Its vapor requires the use of Self Contained Atmospheric Protective Ensemble (SCAPE) suits. This overhead must be considered when planning ground processing workflow for spacecraft and may impose undesirable constraints on the spacecraft, the launch provider, or other spacecraft participating in the same launch opportunity. Hydrazine propulsion systems typically incorporate redundant serial valves to prevent spills or leaking vapor, which might harm ground personnel or hardware.

c. Current & Planned Missions

ArianeGroup has developed a 1-N class hydrazine thruster that has extensive flight heritage, including use on the ALSAT-2 small spacecraft (6) (7). The company also offers a 20-N hydrazine thruster with extensive flight history, although not with small spacecraft.

Aerojet-Rocketdyne (L3 Harris company) has several small hydrazine thrusters with extensive flight histories. The 0.09-N MR-401, 1-N MR-103 series, 4-N MR-111G, and the 22-N MR-106L hydrazine thrusters are flight proven (8) (9) (10). In addition to four MR-107S 222-N thrusters, the OSIRIS-Rex spacecraft propulsion system has six MR-106L, sixteen MR-111G, and two MR-401 thrusters (11).

Moog has the 1-N MONARC-1, 5-N MONARC-5, and 22-N MONARC-22 series hydrazine thrusters (12). These thrusters have decades-long flight histories, although primarily as attitude control thrusters in larger satellites. NASA JPL's Soil Moisture Active/Passive (SMAP) spacecraft (a small satellite) used eight MONARC-5 thrusters (13).

Northrop Grumman has a line of hydrazine thrusters with long flight histories as attitude control thrusters for larger satellites. The thrust lines include the 1-N MRE-0.1, 5-N MRE-1.0, and 18-N MRE 4.0 (14).

Rafael offers 1-N, 5-N, and 25-N flight proven hydrazine thrusters (15).

IHI Aerospace offers the 1-N MT-9, 4-N MT-8A, and 20-N MT-2 flight proven hydrazine thrusters (17).

Stellar Exploration supplied their Monopropellant CubeSat System for an EchoStar spacecraft and the CAPSTONE spacecraft. The EchoStar nanosatellite built by Tyvak has been successfully commissioned and placed in the altitude prescribed in EchoStar's license for its S-band frequency (18). The CAPSTONE spacecraft was similarly built by Tyvak and launched on a Rocket Lab Electron from New Zealand on 28 June 2022. The CAPSTONE spacecraft is a 12U, 25 kg CubeSat that will help reduce risk for future lunar spacecraft by validating navigation technologies and verifying the dynamics of a halo-shaped orbit as planned for Gateway, the Moon-orbiting outpost that is part of NASA's Artemis program. Although CAPSTONE experienced communication and propulsion challenges along the way, the spacecraft achieved the expected lunar orbit (19) (20) (21).

d. Summary Table of Devices

See table 4-2 for current state-of-the-art hydrazine monopropellant devices applicable to small spacecraft.

e. Notable Advances

Aerojet-Rocketdyne, in work sponsored by NASA GSFC, is developing a hybridization of ionic liquid and conventional hydrazine constituents to form the green hydrazine propellant blend (GHPB). "Green hydrazine" would provide the low vapor toxicity and high-density specific impulse of ionic liquids while retaining the low combustion and preheat temperatures of conventional hydrazine. This approach would provide a direct drop-in replacement for hydrazine propulsion systems. In testing completed to date, green hydrazine blends have demonstrated long-term thermal stability/storability, low shock/impact sensitivity, and good operational stability. Furthermore, they have demonstrated a 100-fold reduction in vapor pressure/toxicity and a similar low-temperature start capability as compared to pure hydrazine (22).

Alternative Monopropellants and Bipropellants

a. Technology Description

For the past several decades, Earth storable propellants (i.e., storable at room temperature) have dominated in-space chemical propulsion systems, hydrazine for monopropellants and nitrogen tetroxide with either monomethylhydrazine or hydrazine for bipropellants. These propellants have the advantages of being ambient-temperature liquids (thus not requiring extensive thermal management) and easily ignited (catalyst for hydrazine, hypergolic for bipropellants). However, due to the extensive handling and toxicity concerns of the conventional chemical propellants, more propulsion systems using alternate propellants have been developed and adopted. Often described as "green" propellants, these alternative propellants are not necessarily benign, but are not vapor hazards like the conventional Earth storables and thus do not require specialized suits and breathing apparatus for handling.

Alternative monopropellants based on ionic liquids have received extensive technology development in recent years and have been matured to flight status. Ionic liquid monopropellants are ionic salts dissolved in water and blended with a fuel. Catalysts react the aqueous ionic salt and fuel blend for combustion. Thus, these formulations are not true monopropellants, since they have fuel and oxidizer components, but functionally they are no different than monopropellants.

The two matured ionic liquid monopropellant blends are LMP-103S, which is based on ammonium dinitramide (ADN) or ASCENT (Advanced Spacecraft Energetic Non-Toxic), formally referred to as AF-M315E, which is based on Hydroxylammonium Nitrate (HAN). These monopropellants do not present a vapor hazard and can be handled with conventional personal protection equipment (gloves, face shield). Depending on the formulations, they also can offer higher specific impulse and higher density-specific impulse than monopropellant hydrazine. The challenges with ionic liquid monopropellants are that their catalysts require more preheating than hydrazine and their higher combustion temperatures require higher-temperature, with more expensive catalyst and chamber materials. The ionic liquid monopropellants are not drop-in replacements for hydrazine, although their propulsion system components are similar.

Hydrogen peroxide is a high-density liquid that can be catalytically decomposed exothermically like hydrazine. Hydrogen peroxide does not present a vapor hazard, although high concentrations (> 90%) can rapidly react with impurities in storage containers. Hydrogen peroxide was used extensively in the early days of space propulsion before losing favor to the higher performing hydrazine. It is now being examined again as a nontoxic alternative to hydrazine. Development efforts have focused on high-test peroxide (HTP) which have concentrations from 85 to 98%.

Green bipropellant combinations have also been developed, particularly for small satellite applications. Hydrogen peroxide and nitrous oxide have been explored as the oxidizer options with an alcohol as the fuel. A water electrolysis system breaks water down to gaseous hydrogen and gaseous oxygen, which serve as propellants for the engine. An electrolysis system, using water as the source of propellant, could enable the use of in-situ resources.

b. Key Integration and Operational Considerations

- **Improved Hazard Safety Classifications:** Air Force Range Safety AFSPCMAN91-710 (23) requirements state that if a propellant is less prone to external leakage, which is often seen with the ionic liquid 'green' propellant systems due to higher viscosity of the propellant, then the hazardous classification is reduced. External hydrazine leakage is considered "catastrophic," whereas using ionic liquid green propellants reduces the hazard severity classification to "critical" and possibly "marginal" per MIL-STD-882E (Standard Practice for System Safety) (24). A classification of "critical" or less only requires two-seals to inhibit external leakage, meaning no additional latch valves or other isolation devices are required in the feed system (24). While these propellants are not safe for consumption, they have been shown to be less toxic compared to hydrazine. This is primarily due to green propellants having lower vapor pressures, being less flammable, and producing more benign constituent product gases (such as water vapor, hydrogen, and carbon dioxide) when combusted.
- **Simplified Safety and Handling Requirements:** Fueling spacecraft with green propellants, generally permitted as a parallel operation, may require a smaller exclusionary zone, allowing for accelerated launch readiness operations (25). These green propellants are also generally less likely to exothermically decompose at room temperature due to higher ignition thresholds. Therefore, they require fewer inhibit requirements, fewer valve seats for power, and less stringent temperature storage requirements. The reduced hazard associated with some of these propellants may enable projects to take a Design for Minimum Risk (DFMR) approach to address some propulsion system safety concerns, but only with the support of associated range and payload safety entities.
- **Immature Component Ecosystem:** While there are thrusters that are relatively mature (PMI E/F), incorporating them into integrated propulsion systems is challenging, and the maturity of stand-alone propulsion systems has lagged the pace of component development. Historically, research and development efforts, like Small Business Innovative Research (SBIR) efforts, have focused on component development, and not the entire system. Efforts are now being made to focus on the development of system solutions. Most of these non-toxic propellants are still in some phase of development. Additionally, data on the propellants is widely restricted. Therefore, a comprehensive, public, peer-reviewed databased of compatible materials does not currently exist, and would-be system developers using these propellants may have difficultly accessing such data to guide their efforts.
- **Other Considerations for Green Propellants:** Other 'green propellants' such as Hydrogen Peroxide, High Test Peroxide (HTP), and HTP/Alcohol bipropellants also have their own unique handling considerations. For instance, HTP is a strong oxidizer and can exothermically decompose rapidly if improperly stored or handled. Hydrogen Peroxide, however, has been used as a rocket propellant for many decades, and there is a lot of information on safe handling, materials selection, and best practices. Electrolyzed water is another propellant option, wherein water is decomposed into hydrogen and oxygen and combusted as a traditional bi-propellant thruster. However, generating and managing the power required to electrolyze the water in a compact spacecraft presents its own unique challenges. Yet it does provide a safe-to-launch system with very benign constituents.

c. Current & Planned Missions

ECAPS offers a range of High Performance Green Propulsion (HPGP) thrusters, including the LMP-103S (figure 4.3), at 100-mN, 1-N, 5-N, and 22-N thrust levels. The 1-N HPGP thrusters were first demonstrated on orbit in the Prototype Research Instruments and Space Mission technology Advancement (PRISMA) mission completed in June 2011. PRISMA consisted of two small satellites, each carrying hydrazine and HPGP systems to show off a side-by-side comparison of their performance (26).

Figure 4.3: ECAPS HPGP thruster. Credit: Bradford ECAPS.

ECAPS developed an LMP-103S based propulsion system for a constellation of Earth observing satellites, called SkySat. SkySat has a propulsion system that uses four 1-N HPGP thrusters. Thirteen SkySat satellites with the ECAPS propulsion system have been launched and are fully operational (27).

ECAPS LMP-103S propulsion system is also used on the Autoscale demonstration of rendezvous technologies called ELSA-d, which launched in March 2021. ELSA-d has eight 1-N ECAPS HPGP thrusters to provide re-orbiting and de-orbiting capability. A system issue impacted three of eight ECAPS thrusters and an unresolved root cause resulted in the loss of a fourth thruster. Nevertheless, many mission goals were successfully accomplished, improving the providers readiness for offering a commercial deorbit service (28).

VACCO provided a hybrid Micro Propulsion System (MiPS) for the ArgoMoon 6U CubeSat built for the Italian Space Agency. The ArgoMoon MiPS, figure 4.4, has an ECAPS 100mN LMP-103S thruster with four VACCO 25 mN R134a cold gas thrusters. ArgoMoon was successfully deployed in the Artemis I mission in November 2022 (29) (30) (31).

Figure 4.4: ArgoMoon hybrid Micro Propulsion System. Credit: VACCO.

Aerojet-Rocketdyne built the ASCENT-based propulsion system for the NASA Green Propellant Infusion Mission (GPIM). GPIM was an on-orbit demonstration of the ASCENT (then called AF-M315E) propulsion system, using five 1-N thrusters (figure 4.5) for small attitude control maneuvers (32). GPIM was integrated on a small spacecraft bus (Ball Aerospace's BCP-100) and launched in June 2019 as a secondary payload on a Falcon Heavy. The five thrusters were successfully fired in space, across a range of operating modes, testing their ability to control the satellite's attitude, and demonstrating their effectiveness at changing its orbital inclination (33).

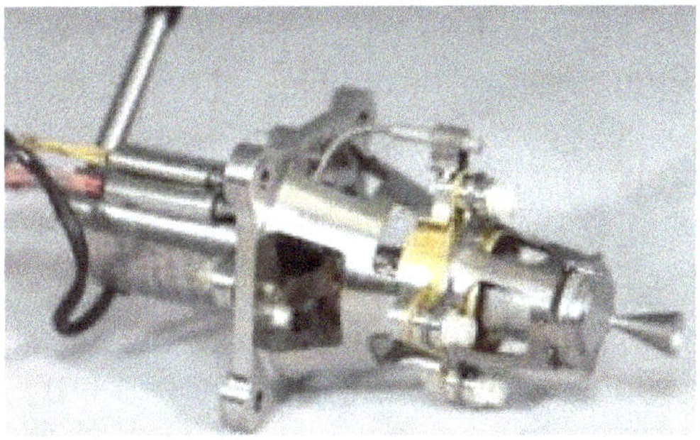

Figure 4.5: GR1 thruster. Credit: Aerojet.

An ASCENT propulsion system, the Lunar Flashlight Propulsion System (LFPS) was developed for the JPL Lunar Flashlight mission (34). Lunar Flashlight (figure 4.6) was launched as a secondary payload in a December 2022 Falcon 9 launch, to map the lunar south pole for volatiles. LFPS was a pump-fed monopropellant system, using four 0.1-N ASCENT thrusters (figure 4.7) built by Rubicon Space Systems (a division of Plasma Processes) and a micro-pump built by Flight Works Inc. The propellant management system was fabricated using additive manufacturing. During the first few days of flight, it was found that 3 of the 4 thrusters were underperforming. Based on ground testing it was thought the underperformance might have been caused by obstructions in the fuel lines that limited propellant flow to the thrusters. Improvements were seen by increasing fuel pump pressure to clear the suspected obstructions. However, the effort was not sufficient to keep the spacecraft in the vicinity of the Moon and the mission was terminated in May 2023 (35).

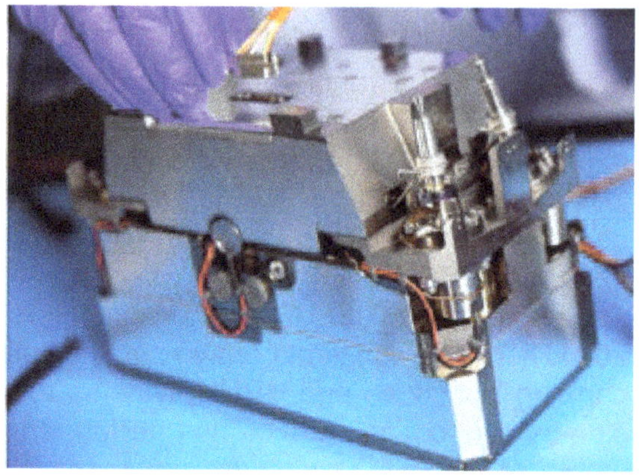

Figure 4.6: Lunar Flashlight Propulsion System. Credit: NASA.

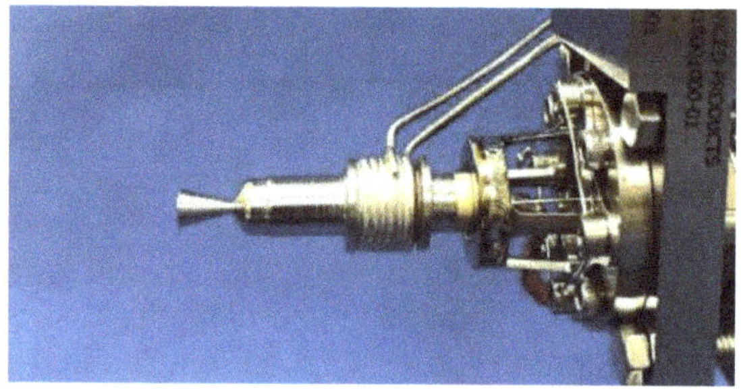

Figure 4.7: Rubicon 0.1N ASCENT thruster. Credit: NASA MSFC.

Benchmark Space Systems offers thrusters that can be used with High Test Peroxide (HTP) as a monopropellant or as an oxidizer in a bipropellant combination with a fuel. Benchmark provided a bipropellant Halcyon Avant propulsion system that uses four 22-N thrusters using HTP and isopropyl alcohol for the Sherpa-LTC2 orbital transfer vehicle (OTV) (36). The Sherpa-LTC2 was launched on a Falcon 9 in September 2022. The OTV was used to elevate the orbit of the Varuna Technology Demonstration Mission satellite (37). In June 2021, three monopropellant HTP Halcyon systems launched aboard the SpaceX Transporter 2 rideshare mission. One of the systems debuted Orbit Fab's RAFTI refueling kit as part of their Tenzing mission. The other two missions supported an undisclosed mission partner (38).

Figure 4.8: Halcyon Avant (Sherpa-LTC2 Configuration). Credit Benchmark Space Systems.

CisLunar Explorer, part of a NASA Centennial Challenges program will use a water electrolysis propulsion system developed by Cornell University's Space Systems Design Studio on a pair of 3U CubeSat. In this system, the water is electrolyzed in the propellant tank. The gaseous hydrogen/gaseous oxygen mixture is flowed through a flame arrestor into the combustion chamber where it is ignited by a glow plug (39). The

CubeSat was originally intended to be launched on Artemis I but difficulties during integration bumped it from the mission (40) (41).

NASA's Small Spacecraft Technology (SST) program at Ames Research Center (ARC) launched the first Pathfinder Technology Demonstration (PTD) mission in January 2021 (42)(43)(44). PTD-1 (figure 4.9) tested the HYDROS-C water electrolysis propulsion system, developed by Tethers Unlimited Inc. With a volume less than 2.4U, the HYDROS-C uses water as propellant. In-orbit, water was electrolyzed into oxygen and hydrogen, then combusted like a traditional bi-propellant thruster. Limited performance data has been evaluated and made public (45). The system requires 10 – 15 minutes of recharge time between pulses. A variant of the HYDROS-C system is the HYDROS-M system, which is intended to be sized for MicroSats. Tethers Unlimited became a subsidiary of ARKA Group LP in 2020.

Figure 4.9: PTD-1 HYDROS-C. Credit: NASA.

NanoAvionics developed an ADN-based monopropellant propulsion system under the Enabling Propulsion System for Small Satellites (EPSS) program. The EPSS monopropellant system was demonstrated on LituanicaSAT-2, a 3U CubeSat, to correct orientation and attitude, avoid collisions, and extend orbital lifetime. LituanicaSAT-2 was launched in June 2017 and successfully separated from the primary payload (Cartosat-2) as part of the European QB50 initiative. According to product literature, multiple missions have since launched, with the latest being in April 2019 (46) (47).

d. Summary Table of Devices

See table 4-3 for the current state-of-the-art in other mono- and bipropellant devices applicable to small spacecraft.

e. Notable Advances

Aerojet Rocketdyne continues to develop its GR-M1 Advanced Green Monopropellant CubeSat Thruster. It employs the same advanced techniques, ultra-high-temperature catalyst, and refractory metal manufacture as the GPIM GR-1 thruster, but on a nanosat scale (48). To partially mitigate thermal management challenges exacerbated at the miniature scale, the GR-M1 is designed to operate on a reduced-flame-temperature variant of the ASCENT propellant containing 10% added water. The heat transfer to surrounding spacecraft structure both during heat up and operation are comparable to conventional hydrazine thrusters.

Rubicon Space Systems (a division of Plasma Processes) is maturing 1N and 5N ASCENT thrusters (49), intended for SmallSat applications. Both offerings are built using the same materials and processes as those used on the 0.1-N thrusters delivered for the Lunar Flashlight Mission. Additionally, Rubicon Space Systems intends to engineer a short-life, lower cost version of the 5N thruster. The prototype thruster accumulated > 1-kg throughput and over 500 seconds before the end of the NASA Phase I SBIR. The Phase II effort will continue to develop the 5N thruster.

CU Aerospace LLC (CUA) has developed the Monopropellant Propulsion Unit for CubeSats (MPUC) system, figure 4.10 (50). The monopropellant is an H_2O_2-ethanol blend denoted as CMP-X. Tests on a thrust stand with a bladder-fed propellant tank have demonstrated continuous constant thrust for >50 minutes at a thrust level of 250 mN at I_{SP} of 179 s with an average input power of ~6 W during catalyst warmup. 1.5U and 2U system designs have an estimated 1400 N-s and 2120 N-s total impulse, respectively. A ~900°C flame temperature allows the thrust chamber to use non-refractory construction materials. CMP-X has low toxicity and was subjected to UN Series 1, 2, 3, and 6 testing; CMP-X demonstrated no detonation propagation when confined under a charge of high explosive, it exhibited thermal stability with no explosion or detonation during bonfire testing, and was not sensitive to drop impact or friction. CMP-X passed the criteria for either a 1.4S or a "Not Class 1" determination and may be excluded from the explosive class. Long-term storage testing shows no degradation over > 1200 days with testing ongoing. This flight-like, additively-manufactured, thruster passed environmental (vibration and thermal vacuum) testing and post-environmental thrust data were within experimental error of the pre-environmental data (51).

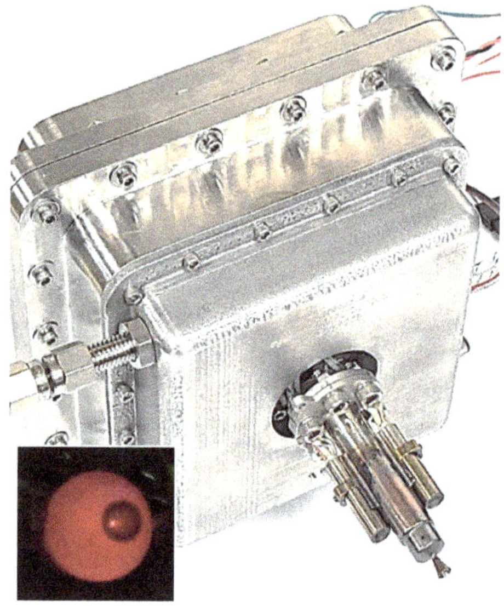

Figure 4.10: MPUC System. Credit: CU Aerospace.

Dawn Aerospace has developed the CubeDrive bipropellant (nitrous oxide and propylene) system that consists of a single B1 thruster, propellant tank, valves, and electronics. The thruster can also be operated in cold gas mode (for smaller impulse bits) by not engaging the spark igniter. The 0.8U (figure 4.11) to 4U configurations provide 425 to 3,500 N-s of total impulse, respectively (52).

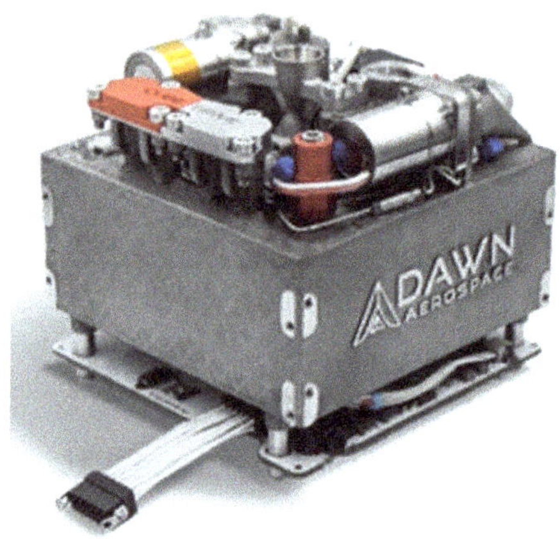

Figure 4.11: CubeDrive. Credit: Dawn Aerospace.

VACCO has integrated four ECAPS 1-N LMP-103S thrusters into their Integrated Propulsion System (IPS) (figure 4.12), a bolt-on propulsion module for delta-V and attitude control applications (53) (54).

Figure 4.12: VACCO Industries IPS. Credit: VACCO Industries.

Hybrids

a. Technology Description

Hybrid propulsion is a mix of both solid and liquid/gas forms of propulsion. In a hybrid rocket, the fuel is typically a solid grain, and the oxidizer (often gaseous oxygen) is stored separately. The rocket is then ignited by injecting the oxidizer into the solid motor and igniting it with a spark or torch system. Since combustion can only occur while the oxidizer is flowing, these systems can readily be started or shut down by controlling the oxidizer flow.

b. Key Integration and Operational Considerations

- **Improved Safety and Handling:** Hybrid systems are inherently safer to handle than solid motor systems because there is no oxidizer pre-mixed into the solid motor, which reduces the risk of pre-mature ignition.
- **Integrates Attributes of Solids and Liquids:** Hybrids achieve many positive attributes of both solid motors (storability & handling) and liquid engines (restart & throttling).
- **Combustion Efficiency:** Combustion efficiency tends to be lower than either solid motors or liquid engines.
- **Other Drawbacks:** Regression rate control and fuel residuals tend to be more problematic in hybrid designs.

c. Current & Planned Missions

An arc-ignition 'green' CubeSat hybrid thruster system prototype was developed at Utah State University and demonstrated in flight under the Undergraduate Student Instrument Project (USIP). The hybrid rocket design used a 3D printed acrylonitrile butadiene styrene (ABS) plastic as the fuel and high-pressure gaseous oxygen (GOX) as the oxidizer. For safety considerations the oxidizer was diluted to 60% nitrogen and 40% oxygen for the demonstration. On March 25, 2018, the system was successfully tested aboard a sounding rocket launched from NASA Wallops Flight Facility (WFF) into space and the motor was successfully re-fired 5 times. During the tests, 8 N of thrust and a specific impulse of 215 s were achieved as predicted (55) (56). The Space Dynamics Lab has miniaturized this technology to be better suited for CubeSat applications (0.25 - 0.5 N). A qualification unit is currently in development for the miniaturized system.

d. Summary Table of Devices

See table 4-4 for current state-of-the-art hybrid devices applicable to small spacecraft.

e. Notable Advances

Utah State University has an ongoing test series with Nytrox, a blend of nitrous oxide and oxygen, and ABS. This testing is focused on a 25-50 N system for a 12U sized vehicle. Investigating different nozzle materials for low erosion in long duration burns is a key concern (57) (58).

JPL is developing a hybrid propulsion system for a 12U CubeSat and a 100 kg SmallSat. Testing included regression rate characterization of clear and black Poly (Methyl MethAcrylate) fuels with GOX to be included in propulsion system sizing. Later vacuum testing included an improvement of the ignition system to a laser operated system that eliminates the need for a separate ignition fuel gas (59) (60) (61) (62).

NASA ARC developed a polymethyl methacrylate (PMMA) and nitrous oxide hybrid system that had ethylene and nitrous oxide thrusters. The ethylene and nitrous oxide also function as the hybrid ignition source. The hybrid system had a demonstrated efficiency of 91% and calculated I_{SP} of 247 sec, making it competitive with current small satellite propulsion systems (63) (64).

Aerospace Corporation and Penn State University developed an "Advanced Hybrid Rocket Motor Propulsion Unit for CubeSats (PUC)". The design used additive manufacturing techniques for the carbon filled polyamide structure including the nitrous oxide tank and a paraffin grain within an acrylic shell, with acrylic diaphragms 3-D printed in-situ in the grain for enhanced performance. This design fits in a 1U space, for a 3 to 6U spacecraft (65).

Parabilis Space Technologies has developed two small satellite propulsion systems. Rapid Orbital Mobility Bus (ROMBUS) is a hybrid rocket-based system with nitrous oxide as the oxidizer and the attitude control system/reaction control system thruster propellant. It provides high-impulse thrust for satellite translational maneuvers which can be used for initial orbit insertion, rapid orbit rephasing, threat/collision avoidance, and targeted re-entry at the satellite's mission end of life (66). Nano Orbital Transfer System (OTS) is a Hydroxyl-terminated polybutadiene (HTPB) and nitrous oxide (N_2O) hybrid system, with N_2O based ACS thrusters. Nano OTS leverages Parabilis' proven hybrid engine and small satellite technologies for low-cost, high-performance maneuvers using non-toxic green propellants. The OTS has a modular design, enabling rapid and low-cost configuration of stages to accommodate 3U size NanoSats up to >50 kg MicroSat-size vehicles (67).

Cold Gas

a. Technology Description

Cold gas propulsion systems are simple, mature, and safe, although they provide relatively limited total impulse. There is no combustion in cold gas systems, with thrust produced by the expulsion of a gaseous propellant through a diverging nozzle. Propellants can be stored as compressed gases, saturated liquids, or solids. Gaseous nitrogen is a commonly used propellant for cold gas systems, although many other gases have been used in the long history of cold gas propulsion. For gases the tradeoff is between performance and storage; lower molecular weight gases offer higher specific impulse but require more voluminous storage. Saturated liquids are stored at low pressure and vaporized when flowed into a low-pressure chamber. The dense storage of saturated liquids and their property of self-pressurization has made them popular as a propellant for CubeSat missions. The liquid propellants may require a pressurant or are self-pressurized, if stored in a two-phase liquid-gas state. Solid propellants can be used in a cold gas system by subliming the solid propellant with adequate heat.

A derivative of cold gas systems is electrothermal or 'warm gas' systems, in which the propellant is somewhat heated without chemical reaction and accelerated through a nozzle. The additional heating results in a modest improvement in thrust and specific impulse compared to a pure cold gas system, but typically burdens the spacecraft with increased power consumption. Electrothermal systems are described in more detail in the Electric Propulsion section.

b. Key Integration and Operational Considerations

- **Low Cost and Complexity:** Cold gas thrusters are often attractive and suitable for small buses due to their relatively low cost and complexity.
- **Safe:** Most cold gas thrusters use inert, non-toxic propellants, which are an advantage for secondary payloads that must adopt "do no harm" approaches to primary payloads.
- **Small Impulse Bit:** Cold gas systems are often well suited to provide attitude control since they can provide very small minimum impulse bits for precise maneuvering.
- **Small Total Impulse:** The low specific impulse of these systems limits them from providing large orbital correction maneuvers.
- **Integrated Systems Optimized for CubeSats:** Designs optimized around a CubeSat's limited resources have improved the capability of these systems for nanosatellite buses.

c. Missions

The Micro-Electromechanical-based PICOSAT Satellite Inspector, or MEPSI, built by the Aerospace Corporation flew aboard STS-113 in 2002 and STS-116 in 2006. The spacecraft included both target and imaging/inspector vehicles connected via a tether. The two vehicles were each 4 x 4 x 5 in^3 in volume and had five cold-gas thrusters, producing approximately 20 mN. The

MEPSI propulsion system was produced using stereo-lithography. It was suited as a propulsion research unit for PicoSats (68).

Marotta developed a cold gas micro-thruster, CGMT-000-9, for fine attitude adjustment maneuvers that flew on the NASA ST-5 mission in 2006. The thruster operated in blowdown mode with gaseous nitrogen, starting at 2.4 N and ending at 0.05 N (69) (70).

In June 2014, Space Flight Laboratory at University of Toronto Institute for Aerospace Research (UTIAS) launched two 15 kg small spacecraft (CanX-4 and CanX-5) to demonstrate formation flying. The Canadian Nanosatellite Advanced Propulsion System (CNAPS), shown in figure 4.13, consisted of four thrusters fueled with liquid sulfur hexafluoride. This propulsion module is a novel version of the previous NanoPS that flew on the CanX-2 mission in 2008. The non-toxic sulfur hexafluoride propellant was selected because it has a high vapor pressure and density, which are important properties for making a compact self-pressurizing system. The propulsion system was successfully used for drift recovery and autonomously maintaining a formation from 1-km range down to 50-m separation (71) (72) (73) (74).

Figure 4.13: CanX-4 and CanX-5 formation flying nanosatellites with CNAPS propulsion systems. Credit: UTIAS SFL.

Microspace Rapid Pte Ltd of Singapore developed a cold gas propulsion system for the POPSAT-HIP1 CubeSat demonstration mission that launched June 2014. It consists of eight micro-nozzles that provide control for three rotational axes with a single thrust axis for translational applications. The total delta-v was estimated from laboratory data to between 2.25 and 3.05 ms^{-1}. Each thruster has 1 mN of nominal thrust using argon propellant. An electromagnetic microvalve with a very short opening time of 1 m-s operates each thruster (75).

GomSpace (acquired NanoSpace in 2016) has developed two related propulsion systems called the NanoProp CGP3 and NanoProp 6U. Both use proportional thrust control of four nozzles to control spacecraft attitude and provide delta-v. The CGP3 was flown on the TW-1 3U CubeSat launched in 2015. The 6U configuration was flown on GOMX-4B in 2018 as a formation flight demonstration (76) (77) (78) (79).

An ACS cold gas propulsion system using R-236fa was produced and tested by Lightsey Space Research for the NASA ARC BioSentinel mission, a 6U CubeSat that launched on Artemis I in November 2022. The propulsion system enables detumbling and pointing for communication back to Earth. The propulsion system uses a 3D-printed propellant tank to reduce part count and make efficient use of the available volume. The system contains six RCS thrusters and one delta-v thruster. The delta-v thruster was included to allow for collision avoidance but was ultimately not needed. One of the RCS thruster valves failed closed during RCS checkout. Rather than further attempting to actuate the valve, and risk the valve failing open, a workaround was identified to perform momentum unloading with the remaining five RCS thrusters. As of May 2023, the propulsion system has accumulated 408 firings and continues to operate as expected. The initial phase of the mission was completed in April 2023. NASA has extended the mission by up to an additional 18 months, or as late as November 2024 (80) (81) (82) (83).

The ThrustMe I2T5 subliming iodine cold gas module, figure 4.14, was the first iodine propulsion system to be spaceflight tested, flown on the Xiaoxiang 1-08 satellite in 2019 (84) (85). Since then, the system was also launched on RTAF's NAPA-2 in June 2021 and on Spire's L3C in January 2023. Additional I2T5 are anticipated to launch on the Robusta-3A satellite, developed by CSUM, which will carry various scientific payloads related to meteorology and technology demonstration (86) (87).

Figure 4.14: I2T5 Iodine Cold Gas Module. Credit: ThrustMe.

The CubeSat Proximity Operations Demonstration (CPOD) is a mission led by Terra Orbital (88) to demonstrate autonomous on-orbit rendezvous and proximity operations using two identical 3U CubeSats. Each spacecraft incorporates a cold gas propulsion system built by VACCO Industries that provides up to 186 N-s of total impulse. This module uses the self-pressurizing refrigerant R236fa propellant, which is exhausted through a total of eight thrusters distributed in pairs at the four corners of the module. CPOD launched in May 2022 as part of SpaceX's Transporter-5 mission. The CPOD mission demonstrated rendezvous of the CubeSats with intersatellite distances closing from 997 km to a minimum of 0.36 km. On-orbit limitations on experimentation were attributed to multiple factors as the launch date slipped due to many years of delay, including obsolete hardware, partial failure in the solar panels, and a propulsion system anomaly suspected to be a plenum leak (89) (90).

The Mars Cube One (MarCO) technology demonstration mission consisted of two identical CubeSats that followed the InSight spacecraft to Mars in loose formation in 2018. The MarCO spacecraft performed five trajectory correction maneuvers (TCMs) during the mission to Mars. The TCMs were conducted using an R236fa cold gas propulsion system developed by VACCO Industries, which contains four thrusters for attitude control and another four for the TCMs. MarCO-B developed a propulsion system leak which required constant maintenance. A tank to plenum leak was identified prior to launch, which the mission accepted. However, a second leak through a MarCO-B thruster valve developed during flight. The combination of both leaks resulted in a continuous moment on the MarCO-B spacecraft. The MarCO spacecraft succeeded in their mission to relay telemetry from the InSight lander during its descent to Mars (91) (92) (93) (94).

Near-Earth Asteroid Scout (NEA Scout) was a joint MSFC and JPL mission that had a VACCO cold gas MiPS (R236FA propellant), with six 25-mN thrusters, to assist the main propulsion system, a solar sail. However, after deployment from Artemis I in November 2022, the project team was not able to communicate with the spacecraft (95) (96) (97).

d. Summary Table of Devices

See table 4-5 for the current state-of-the-art cold gas devices applicable to small spacecraft.

Solid Motors

a. Technology Description

Solid rocket motors have an oxidizer and fuel mechanical mixture stored in solid form (propellant grain). For small satellites, solid rocket motors may be used for impulsive maneuvers such as orbit insertion or quick de-orbiting. They achieve moderate specific impulses and high thrust magnitudes. There are some electrically controlled solid thrusters that operate in the milli-newton (mN) range that are restartable and have steering capabilities. Solid rocket arrays can be compact and suitable for small buses. Composed of several miniature solid rockets, individual units can be fired, alone or together, as needed.

b. Key Integration and Operational Considerations

- **Thrust Vector Control:** Thrust vector control systems can be coupled with existing solid rocket motors to provide controllable high delta-v maneuvering.
- **Usually Single-Burn:** In general, solid motors are considered a single-burn event system. To achieve multiple burns, the system must be either electrically restartable (aka electric solid propellants), or several small units must be matrixed into an array configuration. Because electrically controlled solid propellant (ESPs) are electrically ignited, they are considered safer than traditional solid energetic propellants.

c. Current & Planned Missions

Northrop Grumman offers the STAR 3 motor which has a maximum thrust of 2050 N. The STAR 3 motor was used as the Transverse Impulse Rocket System for the Mars Exploration Rover (MER) program, and was used in 2004 to reduce the lateral velocity of the MER Spirit Lander. The STAR 4G motor was developed and tested by NASA GSFC as an orbit adjust motor for deploying nanosatellite constellations but was not flown, although it is still offered by Northrop Grumman along wth other, higher total impulse motors in the STAR line (98).

The Pacific Scientific Energetic Materials Company (PacSci EMC) Modular Architecture Propulsion System (MAPS) array (figure 4.15) has a 10-plus year in-orbit lifespan. The MAPS system provides three axes capability to control attitude control, deorbit, drag makeup, and plane and attitude changes with a delta-v greater than 50 m s^{-1}. The capability of MAPS "plug-and-play" bolt-on design and clean-burning propellant array is scalable and can be custom fit for a range of interfaces. MAPS was flown in 2017 aboard PacSciSat, which successfully completed all mission objectives (99) (100).

Figure 4.15: PacSci EMC MAPS sealed solid propellant rocket motor array. Credit: PacSci.

d. Summary Table of Devices

See table 4-6 for the current state-of-the-art solid motor devices applicable to small spacecraft.

Propellant Management Devices

a. Technology Description

While not directly a thrust producing device, propellant management devices (PMDs) are frequently used in liquid propulsion systems to reliably deliver propellant to thruster units. PMDs are commonly a critical part of in-space liquid propulsion systems that do not use bellows or membrane type tanks. As small spacecraft look toward more complex propulsion system requirements, PMDs will undoubtedly play an integral role. Historically, small spacecraft have used bellows or membrane tanks to ensure propellant delivery and expulsion. However, there is the potential to incorporate PMD structures into additively manufactured tanks and propulsion systems, permitting more conformal structures to be created and optimized for small spacecraft missions. As such, PMDs are briefly covered here for awareness. A more detailed treatment and explanation can be found in literature. A comprehensive, up-to-date list of the types of PMDs, as well as missions employing PMDs, is available in Hartwig (101).

b. Key Integration and Operational Considerations

The purpose of PMDs is to separate liquid and vapor phases within the propellant storage tank upstream of the thruster, and to transfer vapor-free propellant in any gravitational or thermal environment. PMDs have flight heritage with all classical storage systems, have been flown once with LMP-103S, have no flight heritage with cryogenic propellants, and have been implemented in electric propulsion systems. Multiple PMDs are often required to meet the demands of a particular mission, whether using storable or cryogenic propellants.

c. Current & Planned Missions

The Lunar Flashlight Propulsion System used a PMD sponge and ribbon vane. The sponge was additively manufactured, while the ribbon vane was cut from sheet metal and bent to conform to the required dimensions. Surface tension properties, a necessary parameter for PMD sizing, were determined for the ASCENT propellant by Kent State University, funded and managed by NASA. The design and modelling was a joint effort between MSFC and GRC.

d. Summary Table of Devices

No summary table is included for propellant management devices in this report edition.

e. Notable Advances

Northrop Grumman has made advances in the development of SmallSat and CubeSat scale diaphragm propellant tanks using materials that are compatible with hydrazine and some green monopropellant fuels (102), and demonstrated the utility of additive manufacturing in producing tank shells.

4.6.2 In-Space Electric Propulsion

In-space electric propulsion (EP) is any in-space propulsion technology wherein a propellant is accelerated through the conversion of electrical energy into kinetic energy. The electrical energy source powering in-space EP is historically solar, therefore these technologies are often referred to as solar electric propulsion (SEP), although other energy sources such as nuclear reactors or beamed energy are conceivable. The energy conversion occurs by one of three mechanisms: electrothermal, electrostatic, or electromagnetic acceleration (133) (134). Each of these technologies are covered herein.

This survey of the state-of-the-art in EP does not attempt to review all known devices but focuses on those devices that can be commercially procured or devices that appear on a path toward commercial availability. The intent is to aid mission design groups and other in-space propulsion end-users by improving their awareness of the full breadth of potentially procurable EP devices that may meet their mission requirements.

Instead of detailing the complete operating range for each propulsion device, the authors decided to provide only the metrics associated with the nominal operating condition to improve comprehension of the data and make initial device comparisons more straightforward. When a manufacturer does not specifically state a nominal operating condition in literature, the manufacturer may have been contacted to determine a preferred nominal operating condition, otherwise a nominal operating condition was assumed based on similarity to other devices. For those metrics not specifically found in published literature, approximations have been made when calculable from the available data. Readers are strongly encouraged to follow the references cited to the literature describing each device's full performance range and capabilities.

Electrothermal

a. Technology Description

Electrothermal technologies use electrical energy to increase the enthalpy of a propellant, whereas chemical technologies rely on exothermal chemical reactions. Once heated, the propellant is accelerated and expelled through a conventional converging-diverging nozzle to convert the acquired energy into kinetic energy, like chemical propulsion systems. The specific impulse achieved with electrothermal devices is typically of similar magnitude as chemical devices given that both electrothermal and chemical devices are fundamentally limited by the working temperature limits of materials. However, electrothermal technologies can achieve somewhat higher specific impulses than chemical systems since they are not subject to the limits of chemical energy storage.

Electrothermal devices are typically subclassified within one of the following three categories.

1. *Resistojet devices* employ an electrical heater to raise the temperature of a surface that in turn increases the temperature of a gaseous propellant.
2. *Arcjet devices* sustain an electrical arc through an ionized gaseous propellant, resulting in ohmic heating.
3. *Electrodeless* thrusters heat a gaseous propellant through an inductively or capacitively coupled discharge or by radiation.

b. Key Integration and Operational Considerations

- **Propellant Selection:** Electrothermal technologies offer some of the most lenient restrictions on propellant selection for in-space propulsion. Whereas chemical systems require propellants with both the right chemical and physical properties to achieve the desired performance, electrothermal systems primarily depend on acceptable physical properties. For example, electrothermal devices can often employ inert gases or even waste products such as water and carbon dioxide. They also allow use of novel propellants such as high storage density refrigerants or in-situ resources. That said, not all propellants can be electrothermally heated without negative consequences. Thermal decomposition of complex molecules may result in the formation of polymers and other inconvenient byproducts. These byproducts may result in clogging of the propulsion system and/or spacecraft contamination.
- **Propellant Storage:** Electrothermal devices may require that propellants be maintained at a high plenum pressure to operate efficiently. This may require a high-pressure propellant storage and delivery system.
- **High Temperature Materials:** The working temperature limit of propellant wetted surfaces in the thruster head is a key limitation on the performance of electrothermal devices. As such, very high temperature materials, such as tungsten and molybdenum alloys, are often employed to maximize performance. The total mass and shape of these high temperature materials are a safety consideration during spacecraft disposal. While most spacecraft materials burnup on re-entry, the re-entry behavior of devices using these high temperature materials will be scrutinized when assessing the danger of debris to life and property.
- **Power Processing:** While some simple resistojet devices may operate directly from spacecraft bus power, other electrothermal devices may require a relatively complex power processing unit (PPU). For example, a radio-frequency electrodeless thruster requires circuitry to convert the direct current bus power to a high-frequency alternating current. In some cases, the cost and integration challenges of the PPU can greatly exceed those of the thruster.

- **Thermal Soak-back:** Given the high operating temperatures of electrothermal devices, any reliance on the spacecraft for thermal management of the thruster head should be assessed. While the ideal propulsion system would apply no thermal load on the spacecraft, some thermal soak-back to the spacecraft is inevitable, whether through the mounting structure, propellant lines, cable harness, or radiation.

c. Missions

The Bradford (formerly Deep Space Industries) Comet water-based electrothermal propulsion system (figure 4.16) has been implemented by multiple customers operating in low-Earth orbit, including HawkEye 360, Capella Space, and BlackSky Global (135). All missions use the same Comet thruster head, while the BlackSky Global satellites use a larger tank to provide a greater total impulse capability. The HawkEye 360 Pathfinder mission is a constellation of three small microsatellites built by Space Flight Laboratory (SFL) based on its 15-kg NEMO platform; each spacecraft measures 20 x 20 x 44 cm^3 with a mass of 13.4 kg (136) (137). The Comet provides each HawkEye 360 pathfinder a total delta-v capability of 96 ms^{-1}. The approximate dimensions of the BlackSky Global spacecraft are 55 x 67 x 86 cm^3 with a mass of 56 kg (139).

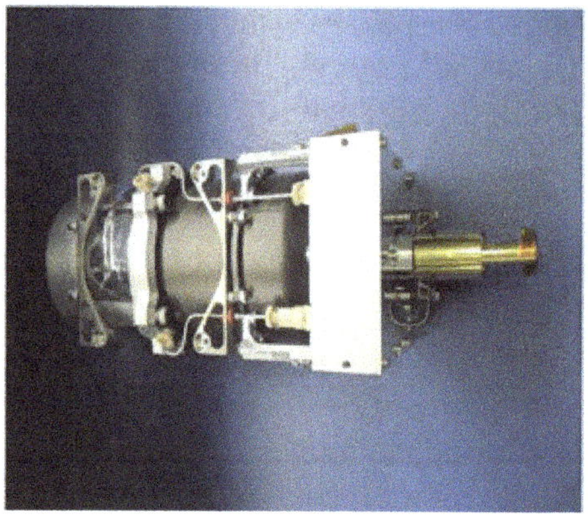

Figure 4.16: Comet-1000. Credit: Bradford Space.

The Propulsion Unit for CubeSats (PUC) system (140), figure 4.17, was designed and fabricated by CU Aerospace LLC (Champaign, IL) and VACCO Industries under contract with the U.S. Air Force to supply two government missions (141). The system was acquired for drag makeup capability to extend asset lifetime in low-Earth orbit. The system uses SO_2 as a self-pressurizing liquid propellant. The propulsion system electrothermally heats the propellant using a micro-cavity discharge (MCD) and expels the propellant through a single nozzle (142). It can alternatively use R134a or R236fa propellants, but only in a cold-gas mode with reduced performance. Eight (8) flight units were delivered to the Air Force in 2014, although it remains unknown if any of the units have flown.

Figure 4.17: PUC module. Credit: CU Aerospace.

In 2019, CU Aerospace was selected for a NASA STMD Tipping Point award to design, fabricate, integrate, and perform mission operations for the Dual Propulsion Experiment (DUPLEX) 6U CubeSat. DUPLEX has two of CU Aerospace's micro-propulsion systems onboard, one Monofilament Vaporization Propulsion (MVP) system (143) (144) (145), figure 4.18, and one Fiber-Fed Pulsed Plasma Thruster (FPPT) system (149) (150) (151) (152) (153), figure 4.43. The MVP is an electrothermal device that vaporizes and heats an inert solid polymer propellant fiber to 725 K. The coiled solid filament approach for propellant storage and delivery addresses common

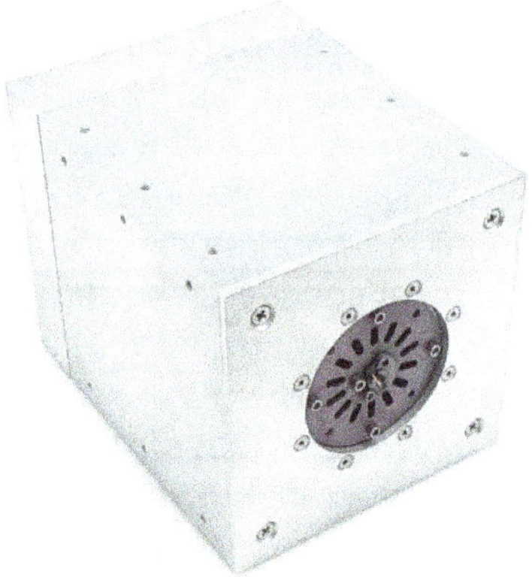

Figure 4.18: MVP module. Credit: CU Aerospace.

propellant safety concerns, which often limit the application of propulsion on low-cost CubeSats. In-orbit operations will demonstrate multiple mission capabilities including inclination change, orbit raising and lowering, drag makeup, and deorbit burns. Launch is manifested in early-2024 (155).

Figure 4.19: Aurora Resistojet Module for Attitude Control. Credit: Aurora Propulsion Technologies.

AuroraSat-1 is a technology demonstration 1.5U CubeSat that is demonstrating multiple propulsion devices by Aurora Propulsion Technologies. AuroraSat-1 carries Aurora's smallest version of Aurora Resistojet Module for Attitude control (ARM-A) (156), figure 4.19, and a demonstration unit of their Plasma Brake Module (PBM) (157). The ARM-A system integrated into AuroraSat-1 has six resistojet thrusters for full 3-axis attitude control and 70 grams of water propellant, providing a total impulse of 70 N-s. AuroraSat-1 is built by SatRevolution with Aurora providing the payloads. The satellite was launched by Rocket Lab in May 2022. (158) (159). See section 4.6.3 for discussion of the PBM module.

The OPTIMAL-1 technology demonstration 3U nanosatellite by ArkEdge Space Inc. launched in late 2022 and was deployed from the International Space Station in early 2023. Among other technology demonstration devices, OPTIMAL-1 contains a Pale Blue water resistojet thruster (160).

SPHERE-1 EYE, a 6U CubeSat developed by Sony Group Corporation, includes a Pale Blue Water Resistojet Thruster. The satellite was launched aboard a SpaceX Falcon 9 on January 3rd, 2023, and has been orbiting Earth with an altitude between 500 km to 600 km. The water propulsion system is expected to prolong the satellite's life by 2.5 years. The propulsion system operated for approximately 2 minutes on March 3rd, 2023, and the company confirmed the successful generation of thrust from the obtained flight telemetry (161).

d. Summary Table of Devices

See table 4-7 for current state-of-the-art electrothermal devices applicable to small spacecraft.

Electrosprays

a. Technology Description

Electrospray propulsion systems generate thrust by electrostatically extracting and accelerating ions or droplets from a low-vapor-pressure, electrically conductive, liquid propellant (figure 4.20). This technology can be generally classified into the following types according to the propellant used:

Ionic-Liquid Electrosprays: These technologies use ionic liquids (i.e., salts in a liquid phase at room conditions) as the propellant. The propellant is stored as a liquid, and onboard heaters may be present to maintain propellant properties within the desired operational temperature range. Commonly used propellants include 1-ethyl-3-methylimidazolium tetrafluoroborate (EMI-BF4) and bis(trifluoromethylsulfonyl)imide (EMI-Im). Thrusters that principally emit droplets are also referred to as colloid thrusters.

Field Emission Electric Propulsion (FEEP): These technologies use low-melting-point metals as the propellant. The propellant is typically stored as a solid, and onboard heaters are used to liquefy the propellant prior to thruster operations. Common propellants include indium and gallium.

Feed systems for electrospray technologies can be actively fed via pressurant gas or passively fed via capillary forces. The ion (high-I_{SP}) or droplet (moderate-I_{SP}) emission can be controlled by modulation of the high-voltage (i.e., >1 kV) input in a closed-loop feedback system with current measurements. Stable operations in either emission mode can provide very precise impulse bits.

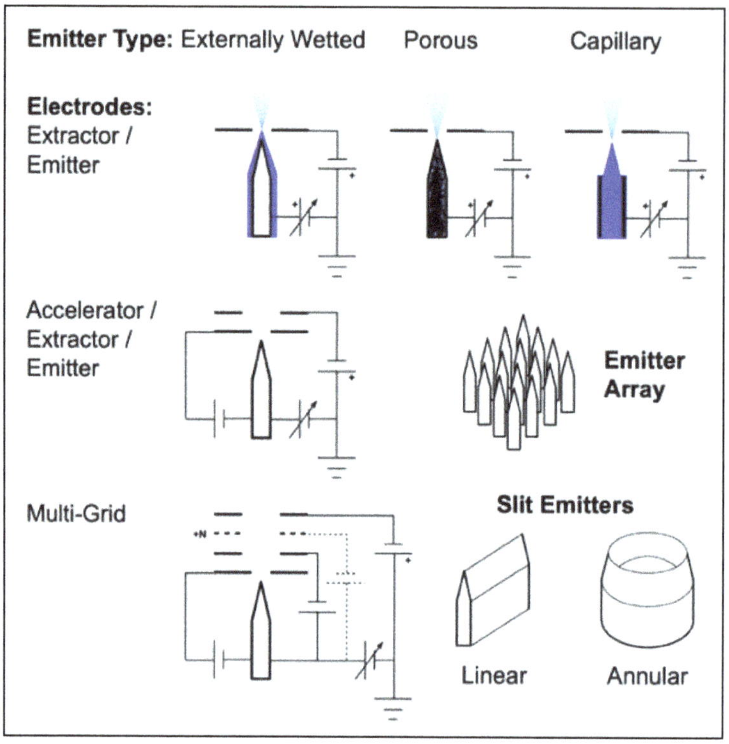

Figure 4.20: Schematic of typical electrospray emitter and electrode configurations. Credit: NASA.

b. Key Integration and Operational Considerations

- **Charge Balance**: Ionic liquid propellants can support electrospray operations with just cation (i.e., positively charged) emission or bi-polar emission (i.e., both anions and cations). For cation-only emission, as with FEEP thrusters, a separate cathode neutralizer is needed to maintain overall charge balance; such neutralizers do not necessarily need to be tightly integrated with the thruster heads, with resultant mass, volume, and power impacts that must be understood for spacecraft integration. For electrospray technologies using bi-polar emission, reliable high-voltage switching in the power electronics becomes a critical consideration.
- **Plume Contamination**: Because propellants for electrospray propulsion systems are electrically conductive and condensable as liquids or solids, impingement of the thruster plume on spacecraft surfaces may lead to electrical shorting and surface contamination of solar panels and sensitive spacecraft components. Consideration of only the primary beam plume may not be sufficient, as droplet emission and molecular fragmentation within the plume can generate off-axis species. Plume shields are frequently employed on spacecraft to protect sensitive surfaces in the absence of high-fidelity electrospray plume models or characterization data.
- **Propellant Handling and Thruster Contamination**: Ionic liquids and metallic propellants can be sensitive to humidity and oxidation, so care is needed if extended storage is required prior to flight. Electrospray technologies can also be sensitive to contamination of the thruster head during propellant loading, ground testing (e.g., backsputter or outgassed materials from the test facility), and handling (i.e., foreign object debris). Precautions should be taken to minimize contamination risks from manufacturing, through test, and to launch. Post-launch, ionic liquids can outgas (e.g., water vapor) when exposed to the space environment, and such behavior should be accounted for in the mission operations. Due to environmental factors such as spacecraft outgassing and atomic oxygen levels, thruster operations following spacecraft deployment may need a conditioning or burn-in period before achieving full propulsive performance.

- **Performance Stability and Lifetime**: As an electrospray propulsion system operates over time, the propulsive performance can degrade as the plume impinges upon and deposits condensable propellant on thruster head surfaces; in time, sufficiently deposited propellant buildup can electrically short out the thruster electrodes and terminate thruster operation. Especially for missions with large total impulse requirements, lifetime testing or validated life models of the electrospray propulsion system in a relevant environment is important for understanding end-of-life behavior.
- **Specific Impulse**: Even for electrosprays that principally emit ions, operational thruster modes and instabilities can result in droplet emission that degrade the specific impulse and thrust efficiency. Caution is advised when considering claimed specific impulse or other propulsive properties (e.g., thrust vector and beam divergence) derived from plume characteristics; verification test data in a relevant environment is important for properly assessing these claims.
- **Precision Thrust**: Electrospray devices have the potential to provide very precise thrust during continuous operations. For devices that can operate in pulsed mode via pulsed modulation of the high-voltage input, fine impulse bits (i.e., <10 μN-s) may be achievable. Such operations can permit precise control over spacecraft attitude and maneuvering. Verification test data in a relevant environment should be used to properly assess the degree of thrust precision.

c. Missions

The ESA Laser Interferometer Space Antenna (LISA) Pathfinder spacecraft was launched in December 2015 on Vega flight VV06. Onboard were two integrated propulsion modules associated with the NASA Space Technology 7 Disturbance Reduction System (ST7 DRS). Each propulsion module contained four independent Busek Colloid MicroNewton Thrusters (CMNT), propellant-less cathode neutralizers, power processing units, digital control electronics, and low-pressure propellant tanks. The propulsion system was successfully commissioned in-orbit in January 2016 after having been fully fueled and stored for almost eight years. The electrospray modules (figure 4.21) were operated at the Earth-Sun Lagrange Point 1 for 90 days to counteract solar disturbance forces on the spacecraft; seven of the eight thrusters demonstrated performance consistent with ground test results, and the full propulsion system met mission-level performance requirements (163).

Figure 4.21: Flight CMNT modules for LISA Pathfinder. Credit: Busek.

The Enpulsion Nano FEEP (figure 4.22), formerly the IFM Nano, was first integrated onboard a 3U Planet Labs Flock 3P CubeSat and launched via PSLV-C40 in January 2018. The indium-propellant propulsion system (with integrated thruster head, propellant storage, and power processing unit) was demonstrated in a 491-km by 510-km orbit. Two thruster firing sequences were reported, with the first a 15-minute firing in non-eclipse and the second a 30-minute firing in eclipse. Global Positioning System (GPS) telemetry data onboard the spacecraft indicated good

Figure 4.22: Nano. Credit: Enpulsion.

agreement with the ~220 µN commanded thrust (164). Since this initial demonstration, the Enpulsion Nano has flown onboard other spacecraft, but limited on-orbit data is publicly available. These missions include the ICEYE-X2 (launched onboard Falcon-9 flight F9-64 in December 2018) to provide low-Earth orbit interferometric synthetic aperture radar observations (165) (166) and the DOD-funded Harbinger technology demonstrator (launched onboard Electron flight STP-27RD in May 2019) (167) (168). The Enpulsion Nano was also integrated onboard the Zentrum für Telematik (Würzburg) NetSat formation-flying demonstrator mission, which launched as a Soyuz-2 rideshare in September 2020 (169) (170). A summary of available on-orbit statistics, anomalies, and lessons learned for the Enpulsion Nano family (Nano, Nano R^3, and Nano AR^3) is available (171).

Figure 4.23: Micro R^3. Credit: Enpulsion.

An Enpulsion Micro R^3 (figure 4.23) was housed onboard the GMS-T mission, which launched in January 2021 onboard a Rocket Lab Electron. The telecommunications satellite uses an OHB Sweden Innosat platform with propulsive capability for collision avoidance. Inaugural on-orbit commissioning of the propulsion system was confirmed in March 2021 (173).

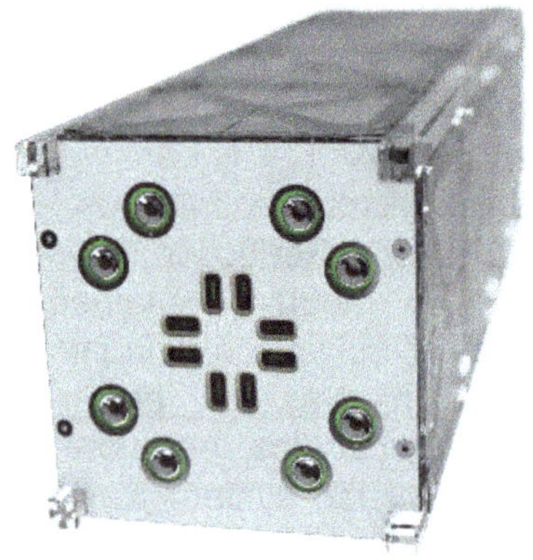

Figure 4.24: Eight NanoFEEP thrusters integrated on 3U-Cubesat bus. Credit: Morpheus Space.

The University Würzburg Experimental Satellite 4 (UWE-4) was launched as a secondary payload onboard the Soyuz Kanopus-V 5 and 6 mission in December 2018. This 1U spacecraft housed two Morpheus Space NanoFEEP systems, with each system consisting of two gallium-propellant thrusters, a power processing unit board for the UNISEC Europe bus, and a propellant-less cathode neutralizer. An experiment using one thruster as an attitude control actuator was reported, with the increased spacecraft rotation rate corresponding to a derived thrust magnitude of ~5 µN; anomalous torque was attributed to unexpected impingement of the thruster plume upon the spacecraft antenna (174) (175). Orbit lowering capability was demonstrated in 2020; of the four individual thrusters, three experienced anomalous behavior during the UWE-4 mission (176). A 3U-Cubsat implementation of the same NanoFEEP technology is shown in figure 4.24.

d. Summary Table of Devices

See table 4-8 for current state-of-the-art electrospray devices applicable to small spacecraft.

Gridded-Ion

a. Technology Description

Gridded-ion propulsion systems ionize gaseous propellant via a plasma discharge, and the resultant ions are subsequently accelerated via electrostatic grids (i.e., ion optics). This technology can be generally classified according to the type of plasma discharge employed:

- **Direct-Current (DC) Discharge**: The propellant is ionized via electron bombardment from an internal discharge cathode (figure 4.25).

- **Radio-Frequency (RF) Discharge**: No internal discharge cathode is present. Instead, the propellant is ionized via RF or microwave excitation from an RF generator (figure 4.26).

Gridded-ion thrusters typically operate at high voltages and include an external neutralizer cathode to maintain plume charge neutrality. High specific impulses can be achieved, but the thrust density is fundamentally limited by space-charge effects. While the earliest thruster technologies used metallic propellants (i.e., mercury and cesium), modern gridded-ion thrusters use noble gases (e.g., xenon, krypton) or iodine.

b. Key Integration and Operational Considerations

- **Performance Prediction:** Due to the enclosed region of ion generation and acceleration, gridded ion thrusters tend to be less sensitive to test-facility backpressure effects than other devices such as Hall thrusters. This allows for more reliable prediction of in-flight performance based on ground measurements. Furthermore, the separation between ion generation and acceleration mechanisms within the device tend to make calculations of thrust and ion velocity (or I_{SP}) more straightforward.
- **Grid Erosion**: Charge-exchange ions formed in between and downstream of the ion optics can impinge upon and erode the grids. Over time, this erosion can lead to a variety of failure modes, including grid structural failure, an inability to prevent electrons from back streaming into the discharge chamber, or the generation of an inter-grid electrical short due to the deposition of electrically conductive grid material. Proper grid alignment must be maintained during thruster assembly, transport, launch, and operations to minimize grid erosion. Random vibration tests at the protoflight level should be conducted to verify the survivability of the ion optics against launch loads. Validated thermal modeling may be needed to assess the impact of grid thermal expansion during thruster operations.
- **Foreign Object Debris**: The grids are separated by a small gap, typically less than 1 mm, to maximize the electric field and thrust capability of the device. As a result, gridded-ion thrusters tend to be sensitive to foreign object debris, which can bridge the inter-grid gap and cause electrical shorting. Precautions should be taken to minimize such contamination risks from manufacturing, through test, and to launch.
- **Cathode Lifetime**: Cathodes for plasma discharge or plume neutralization may be sensitive to propellant purity and pre-launch environmental exposure. Feed system cleanliness, bake-out, and use of a high-purity propellant are key factors in maximizing cathode lifetime. The technology provider may recommend maximum cumulative atmospheric exposure and humidity to reduce risk.
- **Roll Torque**: Misalignments in the ion optics can lead to disturbances in the thrust vector, resulting in a torque around the roll axis that cannot be addressed by the mounting gimbal. For missions requiring extended thruster operations, a secondary propulsion system or reaction wheels may be needed to counter the torque buildup (177).
- **Electromagnetic Interactions**: For RF-discharge thrusters, electromagnetic interference and compatibility (EMI/EMC) testing may be critical to assess the impact of thruster operations on spacecraft communications and payload functionality.
- **Power Electronics:** Operation of gridded-ion thrusters requires multiple high-voltage power supplies for discharge operation (ion generation), ion acceleration, and neutralization, leading to potentially complex and expensive power electronics.
- **Iodine Propellant:** To address the volume constraints of small spacecraft, iodine is an attractive propellant. Compared to xenon, iodine's storage density is three times greater. Furthermore, iodine stores as a solid with a low vapor pressure, which addresses spacecraft integration concerns associated with high-pressure propellant storage. However, iodine is a strong oxidizer, and long-duration impact on the thruster and

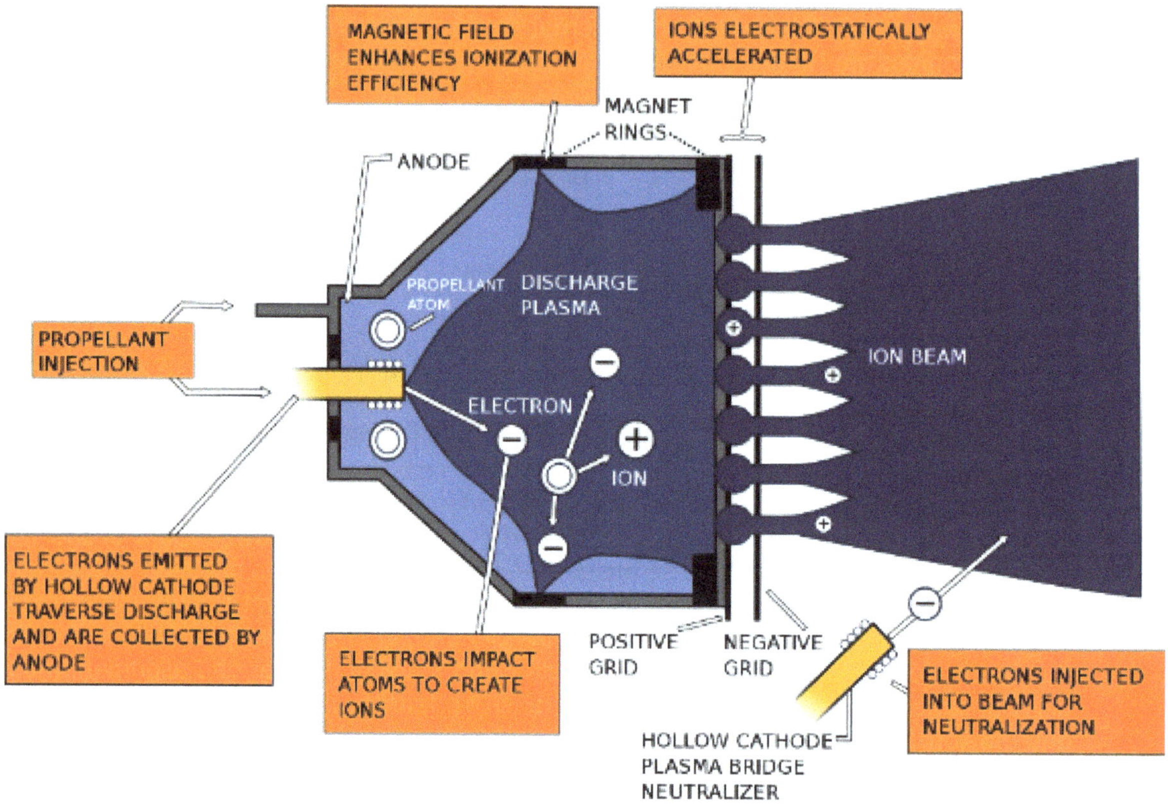

Figure 4.25: Schematic of typical DC-discharge gridded-ion thruster. Credit: NASA.

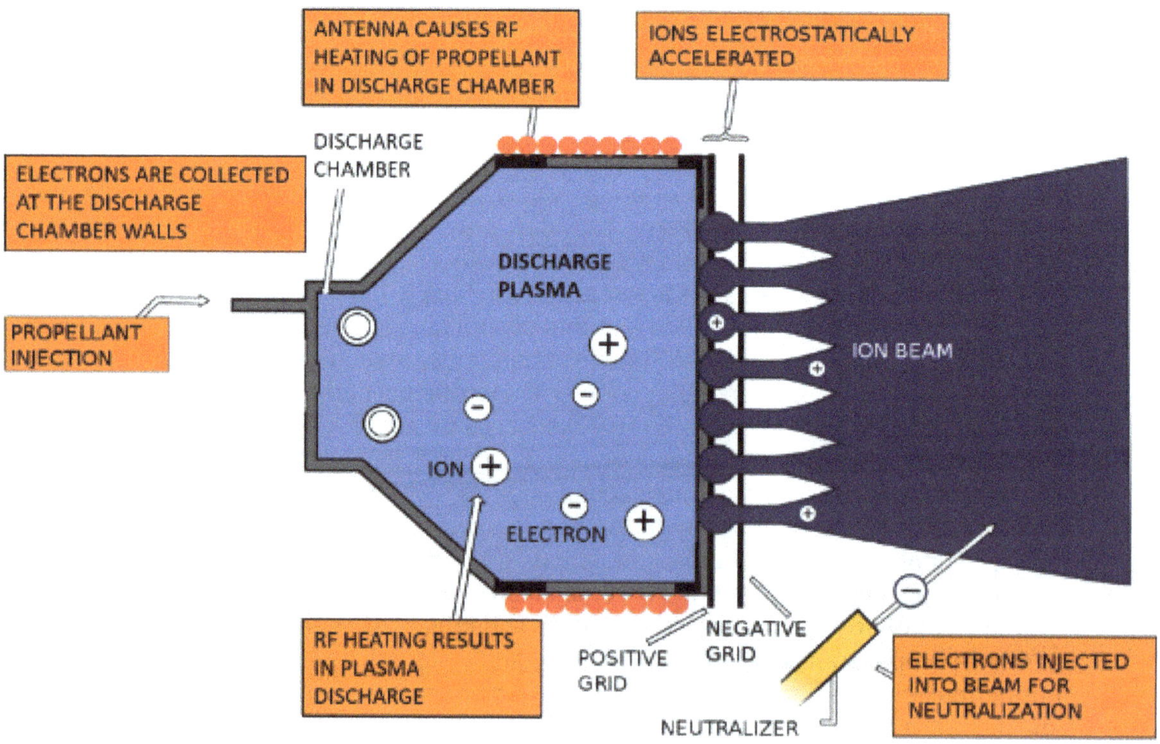

Figure 4.26: Schematic of typical RF-discharge gridded-ion thruster. Credit: NASA.

potential spacecraft interactions and the long-term reliability of feed system and thruster components exposed to iodine.

c. Missions

The ESA Gravity Field and Steady-State Ocean Circulation Explorer (GOCE) was launched in March 2009 onboard a Rokot / Briz-KM to provide detailed mapping of Earth's gravitational field and ocean dynamics from an altitude of ~220-260 km. Two QinetiQ T5 DC-discharge gridded-ion thrusters (figure 4.27), with one serving as a redundant backup, successfully provided drag-free control of the 1000-kg satellite until xenon propellant exhaustion in October 2013 (178) (179) (180) (181).

Figure 4.27: T5 gridded-ion thruster for GOCE mission. Credit: QinetiQ.

ThrustMe's NPT30-I2 1U or 1.5U (figure 4.28), an integrated, RF-discharge gridded-ion propulsion system, has at least two units in space as of July 2023. The Beihangkongshi-1 satellite was launched in November 2020 onboard a Long March 6 rocket. As part of the first on-orbit demonstration of iodine-propellant electric propulsion, two 90-minute burns provided an orbit altitude change of 700 m (182) as described in a Nature publication (183). In April 2023, the 1.5U flew onboard a Space Flight Laboratory of the University of Toronto Institute for Aerospace Studies (UTIAS) 35-kg DEFIANT bus for the Norwegian Space Agency's NorSat-TD mission (184), which included a demonstration of satellite collision avoidance maneuvers (185). With more than 100 systems ordered by clients (186) as of July 2023, upcoming launches include the NPT30-I2 onboard a GomSpace 12U CubeSat for the ESA GOMX-5 technology demonstration mission, NTU Singapore's INSPIRESAT-4, and Turion Space's DROID.002 (187).

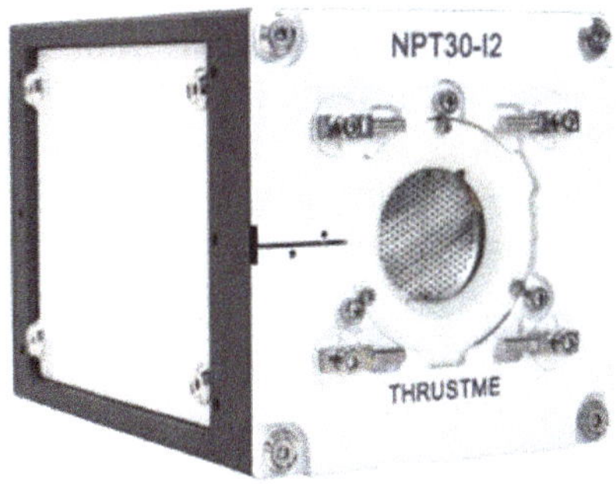

Figure 4.28: NPT30-I2-1U. Credit: ThrustMe.

The Lunar IceCube and LunarH-Map missions both included a Busek BIT-3 propulsion system (figure 4.29) with solid iodine propellant. The BIT-3 system was the primary propulsion for each spacecraft's lunar transfer, orbit capture, orbit lowering, and spacecraft disposal. Each integrated BIT-3 system includes a low-pressure propellant tank with heated propellant-feed system components, a power processing unit to control the RF thruster and RF cathode, and a two-axis gimbal assembly.

Figure 4.29: BIT-3 thruster. Credit: Busek.

The Lunar IceCube mission, led by Morehead State University, aimed to characterize the distribution of water and other volatiles on the Moon from a highly inclined lunar orbit with a perilune less than 100 km. The 6U spacecraft was launched as a secondary payload onboard Artemis I (188) (189). Communication issues with the CubeSat were reported shortly after launch (190).

The Lunar Polar Hydrogen Mapper (LunaH-Map) mission, led by Arizona State University, aimed to map hydrogen distributions at the lunar south pole from a lunar orbit with a perilune less than 20 km. The 6U spacecraft was launched as a secondary payload onboard Artemis I (191). The mission ended in late-2023 as the BIT-3 was inoperable and unable to perform the required mission maneuvers. The mission team concluded that the propulsion system's tank valve was stuck closed (192) (193).

d. Summary Table of Devices

See table 4-9 for current state-of-the-art gridded-ion devices applicable to small spacecraft.

Hall-Effect

a. Technology Description

The Hall-effect thruster (HET) is arguably the most successful in-space EP technology by quantity of units flown. The Soviet Union first flew a pair of EDB Fakel SPT-60 HETs on the Meteor-1-10 spacecraft in 1971. Between 1971 and 2018, over 300 additional HETs flew internationally, although EDB Fakel produced the vast majority. The first flight of a non-Russian HET was on board the European Space Agency (ESA) Small Missions for Advanced Research in Technology (SMART-1) spacecraft in 2003. SMART-1 employed the French PPS-1350 HET, produced by Safran (194). The first flight of a U.S. manufactured HET, the Busek BHT-200, was onboard the TacSat-2 spacecraft (195), a U.S. Air Force Research Laboratory (AFRL) experimental satellite in 2006. In 2010, Aerojet, another U.S. entity, began commercially delivering their 4.5 kW XR5 HET (196), formerly BPT-4000. Launches of HETs greatly accelerated in 2019 with the launch of 120 SpaceX Starlink and 6 OneWeb spacecraft (197), each using an HET. As of November 2023, SpaceX has launched over 5,500 Starlink satellites, and OneWeb has launched over 600 satellites. Suffice to say that HETs have become a mainstream in-space propulsion technology.

The rapid growth in demand for HETs can be attributed to their simple design, historically well-demonstrated reliability, good efficiency, high specific impulse, and high thrust-to-power ratio. Although, the higher voltage gridded-ion thrusters (GIT) can achieve even higher specific impulse, HETs can achieve higher thrust-to-power ratios because the HET's higher density quasi-neutral plasma is not subject to space-charge limitations. The HET's higher thrust-to-power ratio will typically shorten spacecraft transit time. On the other end of the spectrum, arcjets provide significantly higher thrust than HETs, however material limitations prevent arcjets from matching the HET's electrical efficiency and specific impulse. For many missions, HETs provide a good balance of specific impulse, thrust, cost, and reliability.

HETs are a form of ion propulsion, ionizing and electrostatically accelerating the propellant. Historically, all HETs flown in space have relied on xenon propellant, given its high molecular weight, low ionization energy, and ease of handling. The recent exception is the SpaceX Starlink spacecraft using krypton propellant. While HETs typically operate less efficiently with krypton propellant, and krypton has more challenging storage requirements, krypton gas is considerably lower cost than xenon gas. Lower cost is a compelling attribute when the potential number of spacecraft are projected in the thousands, as with constellations. While xenon is generally a superior propellant, krypton fed Hall thrusters will likely become more common in the future, especially for large constellations. SpaceX's 2nd Generation Starlink spacecraft made another first using argon as a Hall thruster propellant. While argon has been commonly tested with lab HETs due to its low cost and good availability, argon is not generally considered a good choice for spaceflight due to its challenging storage requirements. However, given SpaceX's need for such large volumes of propellant for the thousands of satellites planned, supply chain challenges may provide good justification for their use of this generally non-ideal propellant. Many other propellants have been considered and ground tested for Hall-effect thrusters, but to date only Hall-effect thrusters using xenon, krypton, or argon have flown.

As schematically shown in figure 4.30, HETs apply a strong axial electric field and radial magnetic field near the discharge chamber exit plane. The **E x B** force greatly slows the mean axial velocity of electrons and results in an azimuthal electron current many times greater than the beam current. This azimuthal current provides the means by which the incoming neutral propellant is collisionally ionized. These ions are electrostatically accelerated and only weakly affected by the magnetic field. The electron source is a low work function material typically housed in a refractory metal structure (i.e., hollow cathode), historically located external to the HET body. Many recent thruster designs have begun centrally mounting the cathode in the HET body as shown in figure 4.30. The cathode feeds electrons to the HET plasma and neutralizes the plasma plume ejected from the thruster. The high voltage annular anode sits at the rear of the discharge chamber and typically functions as the propellant distribution manifold.

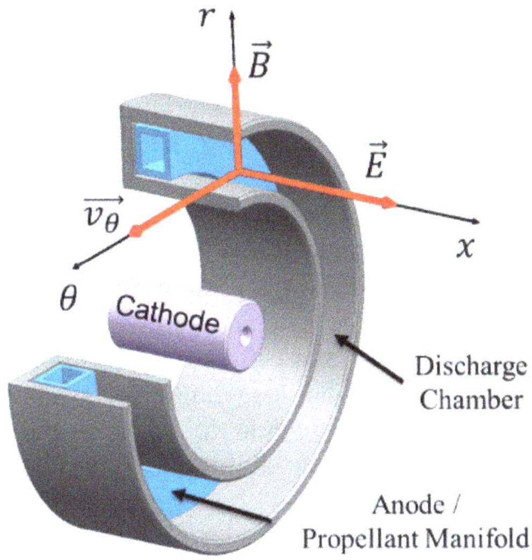

Figure 4.30: Hall-effect Thruster schematic. Credit: NASA.

b. Key Integration and Operational Considerations

- **Ground Facility Effects:** Ground facility effects may result in inconsistencies between ground and flight performance. The significance of the inconsistencies depends on factors such as test facility scale, test facility pumping speed, intrusiveness of diagnostics, and thruster electrical configuration.
- **Contamination:** Plume ions of an HET can affect spacecraft surfaces by erosion or contamination, even at large plume angles. Ground facility measurement of ion density at large angles may under predict flight conditions.
- **Thermal Soak-Back:** HET core temperature may exceed 400°C with the cathode exceeding 1000°C. Most HET waste heat radiates directly from the HET surfaces. However, some thermal soak-back to the spacecraft will occur through the mounting structure, propellant feed lines, electrical harness, and radiation.
- **Survival Heaters:** Given the thermal isolation between the HET and spacecraft, the HET may require a survival heater depending on the qualification temperature and flight environments.
- **Performance:** HET performance may vary over the life of the device due to erosion and contamination of the plasma wetted HET surfaces. Magnetically shielded thrusters demonstrate less time dependency to their performance than classical HETs.
- **Thruster Lifetime:** Classical HETs are primarily life-limited by erosion of the discharge chamber wall. Magnetically shielded HETs are primarily life-limited by erosion of the front pole covers.
- **Cathode Lifetime:** Cathode lifetime may be sensitive to propellant purity and pre-launch environmental exposure. Feed system cleanliness, bake-out, and use of a high purity propellant are key factors in maximizing cathode lifetime. The HET manufacturer may recommend a maximum cumulative atmospheric exposure and humidity. Some cathode

emitter formulations are less sensitive to propellant impurities and atmospheric exposure, but these formulations may require other trades such as a higher operating temperature.

- **Roll Torque**: The **E x B** force results in a slight swirl torque. For missions requiring extended thruster operations, a secondary propulsion system or reaction wheels may be needed to counter the torque buildup. The roll torque may largely be countered by periodically reversing the direction of the magnetic field. Field reversal requires switching the polarity of current to the magnet coils. Field reversal is only possible with HETs using electromagnets.
- **Thrust Vector**: Non-uniformity of the azimuthal plasma, magnetic field, or propellant flow may result in slight variations of the thrust vector relative to the HET physical centerline. Temperature variation of the HET, such as during startup, may result in a slight walking of the thrust vector.
- **Heaterless Cathodes:** Heaterless cathode technologies continue to mature. The benefit of a heaterless cathode is elimination of the cathode heater, typically an expensive component due to rigorous manufacturing and acceptance processes. However, the physics of heaterless cathode life-limiting processes require further understanding. Nevertheless, heaterless cathode demonstrations have empirically shown significant promise. Heaterless cathode requirements on the EP system differ from an HET with a cathode heater. Impacts on the power processing unit and feed system should be well understood when trading a heaterless versus heated cathode.
- **Throttling Range:** HETs typically throttle stably over a wide range of power and discharge voltage. This makes an HET attractive for missions requiring multiple throttle set-points. However, an HET operates most efficiently at specific throttle conditions. Operating at off-nominal conditions may result in decreased specific impulse and/or electrical efficiency.

c. Missions

Canopus-V (or Kanopus-V) is a Russian Space Agency spacecraft for Earth observation with a design life of 5 years. The 450 kg spacecraft launched in 2012 employed a pair of EDB Fakel SPT-50 thrusters. Similarly, the Canopus-V-IK (Kanopus-V-IK) launched in 2017 with a pair of SPT-50. The SPT-50 thrusters have a long history of spaceflight dating back to the late 1970s. Although the Canopus bus exceeds 450 kg, the power class and physical scale of the SPT-50 are appropriate for smaller spacecraft. The SPT-50 is nominally a 220 W thruster operated on xenon propellant (198) (199) (200).

The KazSat-1 and KazSat-2 spacecraft produced by Khrunichev Space Center in cooperation with Thales Alenia Space launched in 2006 and 2011, respectively. The KazSat spacecraft are geosynchronous communication satellites. These spacecrafts employ the EDB Fakel SPT-70BR thruster. The SPT-70BR is Fakel's latest version of the SPT-70 product line. EDB Fakel optimized the SPT-70 for operation between 600 and 700 W, but no more than 900 W. Experiments demonstrate a lifetime of 3,100 hours, equating to about 450 kNs. The SPT-70 thrusters have a long history of spaceflight dating back to the early 1980s. Control of KazSat-1 was lost in 2008 (201) (202).

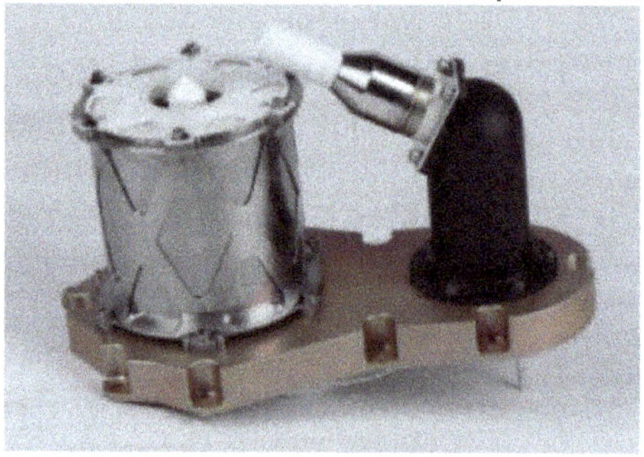

Figure 4.31: BHT-200 thruster. Credit: Busek.

The Busek BHT-200 (figure 4.31) has the distinction of being the first U.S.-made HET to operate in space. The BHT-200 has flight heritage from demonstrations on the TacSat-2 mission launched in 2006, FalconSat-5 mission launched in 2010, and FalconSat-6 mission launched in

2018. A Busek PPU powered the 200 W HET for each of the FalconSat missions (203) (204) (205). Ground testing of the BHT-200 includes multiple propellants, although all spaceflights of the BHT-200 have used xenon. Busek developed an iodine compatible derivative of the BHT-200 for the NASA iSat mission. It was determined during the course of the iSat project that additional development related to iodine compatible cathodes was required before conducting an in-space demonstration of the technology (206) (207).

The Israel Space Agency and the French National Center for Space Studies (CNES) jointly developed the Vegetation and Environment monitoring on a New Microsatellite (VENuS) spacecraft launched in 2017. The 268 kg VENuS spacecraft includes a pair of Rafael IHET-300 thrusters (figure 4.32) and 16 kg of xenon propellant. Inflight operations have demonstrated operation between 250 and 600 W. Rafael developed the IHET-300, nominally operating at 300 W, specifically for small spacecraft (208) (209) (210) (211) (212).

Figure 4.32: IHET-300 thruster. Credit: Rafael.

The European and Italian space agencies selected the SITAEL HT100 (figure 4.33) for an in-orbit validation program to evaluate the device's capabilities for orbital maintenance and accelerated reentry of a small spacecraft. The uHETSat mission will be the first in-orbit demonstration of the HT100. The mission will use the thruster to raise and lower its orbit over a series of verification tests conducted during a planned two-year lifespan. The mission seeks to conduct at least one thousand ignition cycles. SITAEL recently completed ground qualification of the complete propulsion system. The HT100 is nominally a 175 W device operating on xenon propellant. The uHETSat will use the SITAEL S-75 microsatellite platform that is 75 kg with dimensions of 60 x 40 x 36 cm^3. The spacecraft launched December 1, 2023 (213) (214) (215) (216). The HT100 is also baselined for the Italian Space Agency Platino-1 and Platino-2 spacecraft (217).

Figure 4.33: HT100 thruster. Credit: SITAEL.

The Astra Spacecraft Engine (ASE), figure 4.34, successfully achieved orbital ignition onboard the Spaceflight Sherpa-LTE1 orbital transfer vehicle, which launched on SpaceX's Transporter-2 mission on June 30, 2021 (218). This single-string system is sized to achieve a controlled de-orbit of Sherpa-LTE1 (219). On-orbit performance was demonstrated by operating the system for 5-minute durations. The first 54 maneuvers have been reported (220). After outgassing, performance metrics were nominal within one standard deviation of ground test data. On-orbit thrust averaged 22.4 mN, and specific impulse for each 5-minute thrust maneuver averaged 1108 seconds. Total propulsion system power processing efficiency averaged 94%, including feed system power,

Figure 4.34: Astra Spacecraft Engine (ASE). Credit: Astra.

circuit efficiency, and housekeeping circuits (221). As of November 2023, the ASE aboard the Sherpa-LTE1 is continuing mission operations and has operated for more than 800 five-minute maneuvers (i.e., accumulated total duration of ~70 hours) (218). The ASE, (formerly the Apollo Constellation Engine) is a propulsion system that was acquired in Astra's purchase of Apollo Fusion in 2021. The ASE is designed for operation with xenon and krypton propellants and sized to fit ESPA-class missions. The propulsion system includes several key technologies, including permanent magnets, a heatless instant start cathode, and a radiation hardened PPU. Astra has also reportedly sold ASE units to OneWeb (222) and LeoStella (223), Apex (224), Astroscale (225), and Maxar (226). As of March 2023, Astra has nine ACE in orbit (224).

Exotrail's first in-orbit propulsion demonstration mission was in November 2020. NanoAvionics and Exotrail partnered to integrate Exotrail's spaceware - nano (figure 4.35) into a NanoAvionics M6P nanosatellite 6U bus, which was built, integrated, and qualified in 10 months. Following this, Exotrail signed a contract with AAC Clyde Space to provide propulsion for the Eutelsat ELO 3 (2022) and ELO4 (2023) 6U CubeSats (227) (228) (229) (230) (231), and a contract with Aerospace Lab to provide its spaceware - nano for Aerospace Lab's Risk Reduction Flight (RRF) mission. The Aerospace Lab spacecraft, known as "Arthur," was launched in 2021. The propulsion system will be used to demonstrate the spacecraft maneuver capabilities (232) (233). Blue Canyon Technologies (BCT) selected the Exotrail's spaceware system for the company's Venus-class microsatellite platform, which will be used for the NASA's INCUS mission, which is expected to launch in 2026 (234) (235).

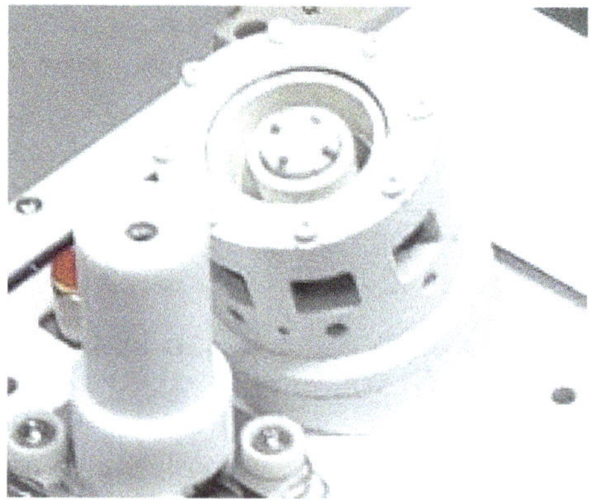

Figure 4.35: spaceware-nano thruster. Credit: Exotrail.

A spaceware - micro cluster² (figure 4.36) will be integrated onboard York Space Systems S-Class platform for a satellite mission scheduled for launch in late 2023, aiming to orbit the Moon and deliver Earth-to-Moon telecommunication services in support of Intuitive Machines' lunar south pole mission. With Exotrail's spaceware - micro cluster², York Space Systems will be able to execute maneuvers such as a lunar transfer orbit (236). Exotrail also has agreements to provide the spaceware – micro for missions by OHB Luxspace (237), Satrec Initiative (238), Astro Digital (239), and Muon Space (240).

Figure 4.36: spaceware-micro thruster. Credit: Exotrail.

Exotrail will also launch its spacevan In-Orbit Demonstration (IOD) mission in October 2023. The spacevan will use Exotrail's spaceware - micro cluster² to demonstrate its capabilities (e.g., plane change or altitude change) (241).

Airbus will integrate Exotrail's spaceware - mini thruster, Exotrail's 300W to 600W class electric propulsion system, onto Airbus' Earth Observation satellite platform portfolio. Exotrail anticipates completion of the thruster's qualification activities in 2024 (242).

Blue Canyon, a Raytheon subsidiary, is producing satellites for the DARPA Blackjack program. Blue Canyon selected Exoterra's Halo thruster (figure 4.37), for its Phase 2 and Phase 3 satellites (243). Exoterra has demonstrated Halo on three Blackjack satellites that launched in June 2023 on the SpaceX Transporter-8 rideshare. These are the first flights of ExoTerra's Halo electric propulsion system (244). Additionally, ExoTerra has received a NASA Tipping Point award to perform an in-orbit demonstration of their 12U Courier SEP spacecraft bus, tentatively planned for launch in 2024. The bus includes ExoTerra's Halo thruster, xenon flow control system (XFC), power processing unit (PPU), and deployable solar arrays. The Courier spacecraft provides up to 1 km/s of delta-v, while hosting a 2U, 4 kg payload. The Tipping Point mission objective is to demonstrate the SEP system by spiraling to 800 km from a drop-off orbit of 400 km and then deorbiting. Primary mission objectives include demonstration of the solar array deployment and power generation, PPU efficiency, and 2 kg of thruster propellant throughput. For the Tipping Point mission, the 0.85 kg, 1/3U thruster will nominally operate at 135 W discharge power and produce ~8 mN of thrust (245) (246) (247) (248) (249).

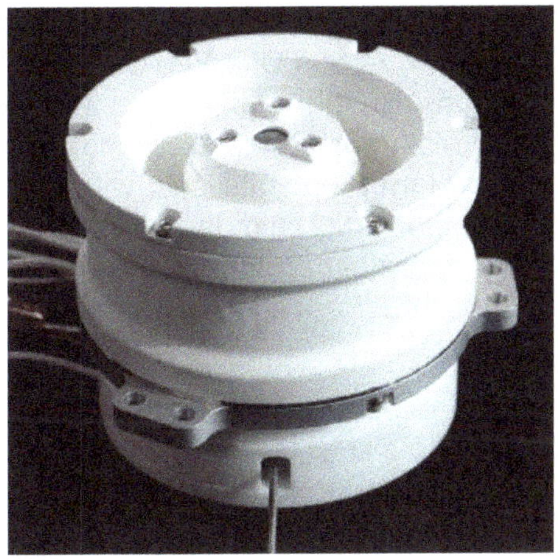

Figure 4.37: Halo thruster. Credit: ExoTerra Resource.

Exoterra Resource established a Space Act Agreement (SAA) with NASA's Jet Propulsion Lab (JPL) in July 2022 to deliver a propulsion system consisting of the Halo12 thruster, power processing unit, and xenon flow control unit to Boeing in support of their RocketR and P1229 programs (250). Halo12 is a commercialized version of JPL's Magnetically Shielded Miniature (MaSMi) Hall thruster technology recently licensed by Exoterra. JPL has performed extensive development of the technology between 2011 and 2021, culminating in a 7,205-hour wear test processing over 100 kg of xenon propellant and producing 1.55 MN-s total impulse. Whereas JPL produced the MaSMi thruster demonstrated in their past life qualification testing, under this SAA Exoterra will fabricate and deliver to JPL the Halo12 for qualification testing. JPL will further update their prior analyses to reflect Boeing requirements and design changes between MaSMi and Halo12. The Halo12 qualification testing will culminate in a life test demonstrating 1.78 MN-s of specific impulse. Nominally 70% of the impulse will be at 1,400 W and 30% at lower power points. The life qualification is anticipated to conclude by mid-2025. It is not yet clear when Boeing might launch their first spacecraft using Halo12 (250) (251).

Orbion's Aurora Hall-effect thruster (figure 4.38) system was selected for a U.S. Space Force 400-kg prototype weather satellite, under contract with General Atomics Electromagnetic Systems (GA-EMS). The Aurora propulsion system consists of a thruster, cathode, power processing unit, propellant flow controller, and cable harness. The system will be used for orbit raising, orbit maintenance, and de-orbit over the 3-5 year mission (252) (253) (254).

Figure 4.38: Aurora HET, PPU and XFC. Credit: Orbion Space.

Busek has supplied its BHT-350, figure 4.39, Hall-effect thruster to Airbus OneWeb Satellites (AOS) for a range of missions. Busek engineered and qualified the thrusters for orbit raising, orbit maintenance, and end-of-life de-orbit. Eighty OneWeb satellites with BHT-350s launched in December 2022 and January 2023. More than 100 units were operating on OneWeb satellites as of August 2023 with all thrusters reported as operational (255) (256) (257) (258).

Figure 4.39: BHT-350 Flight Units. Credit: Busek Co.

Busek shipped its first flight BHT-600 Hall-effect thruster system to a U.S. Government customer in early 2021. The BHT-600 previously demonstrated a 7,000-hour ground test performed at NASA GRC as part of a NASA Announcement for Collaborative Opportunity (ACO) Space Act Agreement (SAA), figure 4.40. The thruster successfully demonstrated 70 kilograms of xenon propellant throughput before the test was terminated (259) (260).

Figure 4.40: BHT-600 Installed in NASA GRC Vacuum Test Facility. Credit: Busek Co.

Northrop Grumman's (NG) Tactical Space Systems Division has developed the NGHT-1X (figure 4.41) thruster for its next generation satellite servicing vehicle known as the Mission Extension Pod (MEP). MEP carries power and propulsion for self-propelled orbit raising as well as station keeping and momentum management when docked to a client vehicle. MEP is designed for a 6-year mission life but can carry a propellant load that permits even longer lifetimes. The NGHT-1X is designed to operate nominally in the 600 to 900 W range, but has demonstrated stable operation from 200 to 1,100 W. Within the nominal power range, the thruster is designed to generate a minimum total impulse of 2.1 MN-s, not including margin, to enable the MEP mission. NG partnered with the NASA Glenn Research Center (GRC) to develop and commercialize the NGHT-1X, licensing NASA's technology for a high propellant throughput, sub-kilowatt hall-effect thruster. NG's SpaceLogistics has sold all three MEPs that are on the manifest for the first launch on a Falcon 9 rocket planned for 2025 (261) (262) (263) (264) (265).

Figure 4.41: NGHT-1X Engineering Model Hall-Effect Thruster. Credit: Northrop Grumman.

The EOS SAT-1 developed by Dragonfly Aerospace for the EOS Data Analytics space mission successfully tested its SETS SPS-25 propulsion system. The EOS SAT-1 launched on a Falcon 9 rocket on January 3, 2023 (266) (267).

The Aliena Pte Ltd MUSIC-SI Hall thruster (figure 4.42) was integrated on the NuX-1 3U nanosatellite in July 2021. In-orbit deployment was achieved in January 2022 during SpaceX Transport 3 Mission. Aliena's Hall thruster became the first to operate onboard such a small satellite, featuring the lowest power consumption for a commercial Hall thruster at 20W (268) (269) (270) (271) (272).

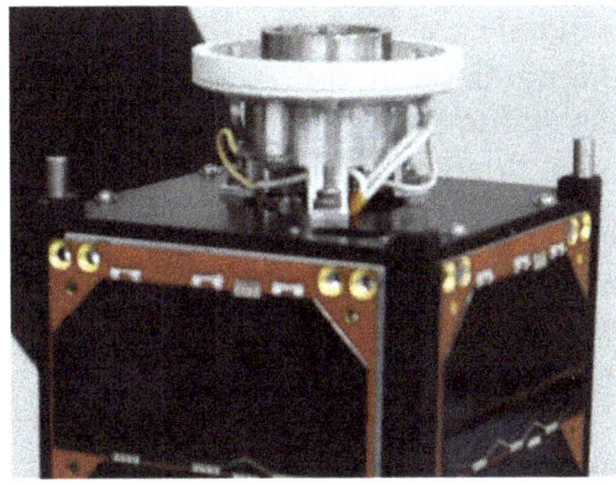

Figure 4.42: NuX-1 satellite with MUSIC-SI Hall Thruster. Credit: Aliena Pte Ltd

The MUSIC-HM Hall thruster from Aliena and four ARM-A resistojets from Aurora Propulsion Technologies form the Multi-modal Electric Propulsion Engine (MEPE). MEPE was deployed on an Orbital Astronautics 12U satellite (ORB-12 Strider) in July 2023 by a PSLV rocket (PSLV C-56 DS-SAR Mission). The mission will demonstrate the capability of these propulsion systems to support a small satellite constellation. MEPE's dual electric propulsion technologies allows both the high thrust, low specific impulse of resistojets and low thrust, high specific impulse of Hall thrusters. Both systems share a common propellant, electronics, and fluidics (268) (272) (273).

d. Summary Table of Devices

See table 4-10 for current state-of-the-art HET devices applicable to small spacecraft.

Pulsed Plasma and Vacuum Arc Thrusters

a. Technology Description

Pulsed Plasma Thrusters (PPT) produce thrust by triggering an electric arc between a pair of electrodes that typically ablates a solid-state propellant like polytetrafluoroethylene (PTFE) or ionizes a gaseous propellant. The plasma may be accelerated by either electrothermal or electromagnetic forces. Whether the mechanism of acceleration is electrothermal, electromagnetic, or often some combination thereof, is determined by the device topology (274).

Electrothermal PPTs characteristically include a chamber formed by a pair of electrodes and solid propellant, wherein propellant ablation and heating occurs. During and immediately following each electric discharge, pressure accumulates and accelerates the propellant through a single opening. Electromagnetic PPTs characteristically do not highly confine the propellant as plasma forms. The current pulse, which may exceed tens of thousands of amps, highly ionizes the ablated material or gas. The current pulse further establishes a magnetic field, where the **j x B** force accelerates the plasma. PPT devices that are predominantly electrothermal typically offer higher thrust, while devices that are predominantly electromagnetic offer higher specific impulse.

The simplest PPTs have no moving parts, which may provide a high degree of reliability. However, as the solid propellant is consumed, the profile of the propellant surfaces constantly changes. Thus, PPTs with static solid propellant demonstrate a change in performance over their life and inherently have a relatively limited lifetime. More complex solid propellant PPTs include a propellant feed mechanism. Typically, the propellant surface profile changes during an initial burn-in period, but then settles into a steady-state behavior where the propellant advancement is balanced by the propellant ablation.

PPT devices are suitable for attitude control and precision pointing applications. PPTs offer small and repeatable impulse bits, which allow for very high precision maneuvering. The complete propulsion system consists of a thruster, an ignitor, and a power processing unit (PPU). Energy to form the pulsed discharge is stored in a high voltage capacitor bank, which often accounts for a significant portion of the system mass. Once the capacitors are charged, resulting in a large differential voltage between the electrodes, the ignitor provides seed material that allows the discharge between the electrodes to form. Various materials and gases (including water vapor) have been tested with PPTs, however PTFE remains most common.

Vacuum arc thrusters (VAT) are another type of pulsed plasma propulsion (275). This technology consists of two metallic electrodes separated by a dielectric insulator. Unlike PPTs, one VAT electrode is sacrificial, providing the propellant source. The mechanism for propellant acceleration is predominantly electromagnetic, resulting in a characteristically high specific impulse and low thrust. One variant of the VAT is predominantly electrostatic, by the inclusion of a downstream electrostatic grid.

b. Key Integration and Operational Considerations

- **Safety**: PPT capacitor banks often store tens of joules of energy at potentially a couple thousand volts. Follow good electrical safety practices when operating and storing PPTs in a laboratory environment.
- **Input Power Range**: PPTs and VATs are pulsed devices, which operate by discharging energy stored in capacitors with each pulse. Thus, the propulsion system's average power draw from the spacecraft bus can be quite low or high depending on the capacitor energy storage and pulse frequency. This flexibility allows PPTs to be applied to spacecraft with limited power budgets of just a few watts, or ample power budgets of hundreds of watts.
- **Minimum Impulse Bit:** A compelling capability of pulsed devices is the ability to generate small, precise, and well-timed impulse bits for precise spacecraft maneuvering. By controlling the discharge voltage, very small impulse bits on the order of micronewton-seconds are easily achieved.
- **Compact and Simple Designs:** PPTs and VATs are typically very simple and compact devices. While the total impulse capability is small compared to other forms of EP, these devices offer a particularly attractive solution for CubeSats, where low cost may be a more significant consideration than total impulse. The systems are also attractive for learning environments where propulsion expertise such as high-pressure feed systems and propellant management may be lacking.
- **Late-Time Ablation**: Although pulsed devices allow for operation over a wide range of pulse frequency, thruster efficiency typically improves with higher pulse rate. Late time ablation is a key inefficiency of solid propellant pulsed devices, where material continues to ablate from the propellant surface well after the discharge pulse. The amount of material accelerated may be maximized through higher frequency pulsing.
- **Thrust-to-Power:** Pulsed devices suffer from several inefficiencies including late time ablation, frozen flow, and wall heating. Propulsion system efficiency is typically below 20% and may be as low as a few percent. Thus, although pulsed devices may have high specific impulse, the thrust-to-power is low. Small spacecraft with limited power for propulsion may find that large propellant loads provide little benefit as there is inherently a limitation to the number of pulses achievable over the life of the power-limited spacecraft.
- **Thermal Soak-back:** The low thruster efficiencies may result in large thermal loads on the spacecraft due to thermal soak-back, especially at high rates of pulsing. The spacecraft's ability to radiate this energy to limit heating may set an upper bound on pulse frequency.
- **Ignitor:** Pulsed devices usually require some form of ignitor to provide seed material to lower the impedance between the electrodes and initiate the discharge pulse. As such, the lifetime of the ignitor may drive the lifetime of the thruster. Ignitors may fail due to

erosion or fouling that prevents sparking. Some devices may include multiple redundant ignitors to increase system lifetime.

- **Shorting:** The electrodes of pulsed devices are separated by isolating elements. Shadow shielding or other physical features are typically necessary to avoid shorting between electrodes as conductive material ejected by the thruster accumulates. While PTFE is an insulator, the PTFE is reduced to carbon and fluorine when ablated, where carbon accumulation provides a potentially conductive path. VATs employ metal propellants that can similarly result in unintended shorting.
- **Spacecraft Contamination:** As with any conductive propellant, spacecraft contamination is a concern. Plume interaction with the spacecraft must be understood to assess the impact of the plume on the operation of critical surfaces such as solar panels, antennas, and radiators.

c. Missions

In 2019, CU Aerospace was selected for a NASA STMD Tipping Point award to design, fabricate, integrate, and perform mission operations for the DUPLEX 6U CubeSat having two of CU Aerospace's micro-propulsion systems onboard, one Monofilament Vaporization Propulsion (MVP) system (143) (144) (145), shown in figure 4.18, and one Fiber-Fed Pulsed Plasma Thruster (FPPT) system (149) (150) (151) (152) (153), shown in figure 4.43. The FPPT can provide a large total impulse primary propulsion for micro-satellites through implementation of a novel PTFE fiber propellant storage and delivery mechanism. A major enhancement of the FPPT technology over classical PPTs is the ability to control both the propellant feed rate and pulse energy, thereby providing control of both the specific impulse and thrust. The FPPT can also provide precision control capability for small spacecraft requiring capabilities such as precision pointing or formation flying. Thrust-vectoring capability of ±10° in the yaw and pitch axes (also with the potential for roll control authority) has been incorporated into the system allowing for limited ACS capability and wheel desaturation for deep space missions. In-orbit operations will demonstrate multiple mission capabilities including inclination change, orbit raising and lowering, drag makeup, and deorbit burns. Launch is manifested in early-2024 (155).

Figure 4.43: FPPT 1.7U module. Credit: CU Aerospace.

The Benchmark Space Systems' Xantus thruster (figure 4.44) is a millinewton-class non-toxic metal plasma thruster that is compatible with multiple solid metal propellants. The baseline molybdenum propellant has a measured specific impulse of 1774 seconds. The Xantus thruster first launched in January 2023 onboard the USSF Rapid Revisit Optical Cloud Imager (RROCI) 12U CubeSat demonstration mission. However, the RROCI spacecraft failed to deploy after launch. Xantus is also manifested in 2024 onboard Orion Space Solutions' EWS (Electro Optical Weather System) mission (276) (277).

Comat's Plasma Jet Pack (PJP) vacuum arc thruster can be configured with up to four nozzles operating on a solid/inert propellant to provide vectorized thrust (278).

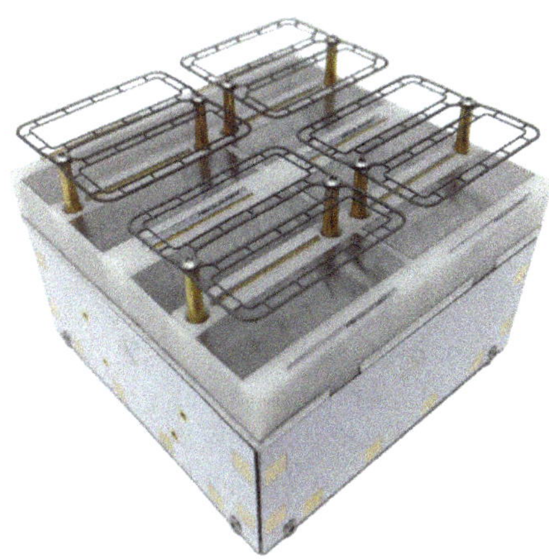

Figure 4.44: Xantus Thruster. Credit: Benchmark Space Systems.

The PJP has launched as part of the Vega VV23 mission on the ∑yndeo-2 6U CubeSat. At the time of this publication, no details have yet been reported on the commissioning of the propulsion system (279).

The Neumann Drive is a center-triggered pulsed cathodic arc propulsion technology developed by Neumann Space using molybdenum propellant. The Neumann Drive is offered in multiple configurations offering a range of propulsive performance depending on available spacecraft power (280). The ND-15 thruster, designed for nanosatellites, is currently in orbit undergoing testing on Australia's Skykraft satellite (281) (282) (283), and another ND-15 will be flown in late 2023 on the University of Melbourne's SpIRIT nanosatellite (284) (287). The ND-50, a more compact system, is planned for missions in 2024 on the 6U EDISON satellite (288) and 150 kg CarbSAR satellite (289).

d. Summary Table of Devices

See table 4-11 for current state-of-the-art pulsed plasma and vacuum arc devices applicable to small spacecraft.

Ambipolar

a. Technology Description

Ambipolar thrusters ionize gaseous propellant within a discharge cavity via various means, including DC breakdown or RF excitation. The escape of high-mobility electrons from the discharge cavity creates a charge imbalance in the plasma discharge, and the subsequent ambipolar diffusion accelerates ions out of the cavity to generate thrust.

Because the thruster plume is charge neutral, no neutralizer assembly is necessary. A variety of propellants are theoretically usable due to the absence of exposed electrodes (and their associated material compatibility concerns).

b. Key Integration and Operational Considerations

- **Propellant Agnostic**: While ambipolar thrusters may be operable on a variety of propellants thanks to the devices' lack of exposed electrodes, different propellants will have different ionization costs (i.e., impact on thruster efficiency), plume behavior, and propellant storage requirements that should be considered during propellant selection.
- **Electromagnetic Interactions**: For RF-discharge thrusters, electromagnetic interference and compatibility (EMI/EMC) testing may be critical to assess the impact of thruster operations on spacecraft communications and payload functionality.
- **Thermal Soakback**: Low thruster efficiencies may result in large thermal loads on the spacecraft due to thermal soakback. Validated thermal modeling should be considered to assess impacts to the host spacecraft.

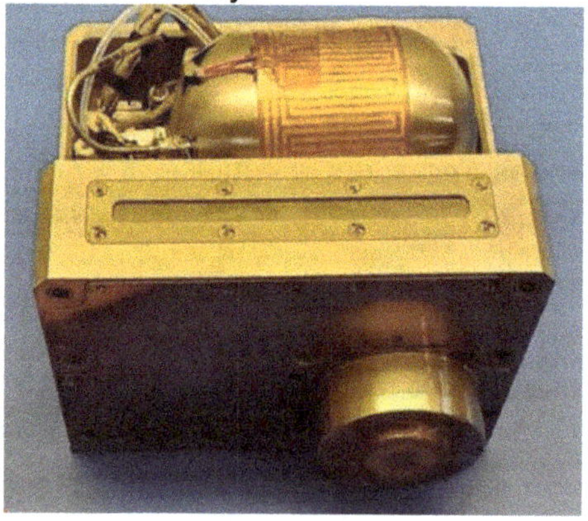

Figure 4.45: Maxwell Block 1. Credit Phase Four.

c. Missions

The SpaceX Falcon 9 Transporter-1 launch in January 2021 included two small satellites with the Phase Four Maxwell Block 1 onboard. This integrated propulsion system (figure 4.45) includes the RF thruster and power electronics along with a xenon propellant tank and feed system (290). Phase Four has further reported in 2023 that its Maxwell Block 2, with improved modularity and performance compared to Block 1, has been demonstrated on orbit (291).

The UniSat-7 mission, led by GAUSS, is a 36-kg microsatellite that launched via Soyuz-2-1a Fregat in March 2021. This technology demonstration mission included a T4i iodine-propellant REGULUS (50-I2-Small) module (figure 4.46), an integrated propulsion system that includes thruster, power processing unit, and heated propellant-feed components. The propulsion demonstration is expected to include orbit raising and lowering around the 600-km orbit (293) (294) (295).

Figure 4.46: REGULUS propulsion module. Credit: T4i.

A 6U CubeSat from Team Miles was awarded a rideshare slot onboard Artemis I, as one of the winning teams in NASA's Cube Quest Challenge. The objective of the mission was to demonstrate deep space communications from beyond a 2.5-million mile range. Twelve ConstantQ water-propellant thrusters (figure 4.47), an earlier version of Team Miles' current M1.4 system, were integrated onboard the CubeSat to provide primary propulsion as well as 3-axis control (296) (297). While the spacecraft was deployed in November 2022, communications contact was not established (403).

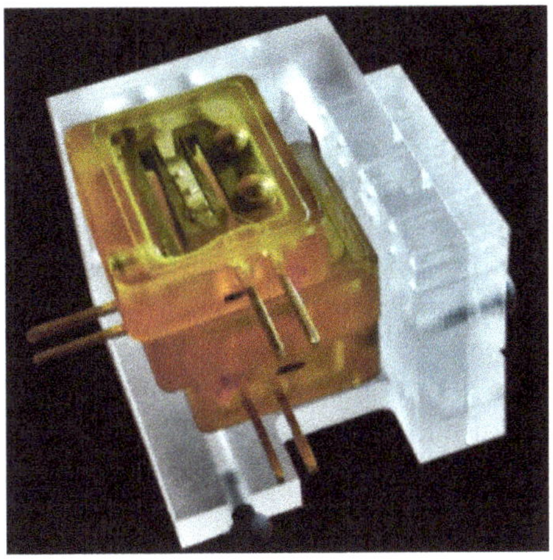

Figure 4.47: ConstantQ thruster head. Credit: Miles Space.

d. Summary Table of Devices

See table 4-12 for current state-of-the-art ambipolar devices applicable to small spacecraft.

4.6.3 In-Space Propellant-less Propulsion

Propellant-less propulsion systems generate thrust via interaction with the surrounding environment (e.g., solar photon pressure, planetary magnetic fields, solar wind and ionospheric plasma pressures, and planetary atmospheres). By contrast, chemical and electric propulsion systems generate thrust by expulsion of reaction mass (i.e., propellant). Four propellant-less propulsion technologies have undergone in-space demonstrations to date, including solar sails, tethers, electric sails (and plasma brakes), and aerodynamic drag devices.

Solar Sails

a. Technology Description

Solar sails use solar radiation pressure to generate thrust by reflecting photons via lightweight, highly reflective membranes. While no commercial products are presently available, a handful of missions have sought to demonstrate the technology using small spacecraft.

b. Missions

NASA's NanoSail-D2, figure 4.48, launched as a 3U CubeSat secondary payload onboard the Fast, Affordable, Science and Technology Satellite (FASTSAT) bus in November 2010. The 10-m^2 sail made of CP-1 deployed from a 650 km circular orbit and de-orbited the spacecraft after 240 days in orbit (298).

Figure 4.48: Deployment test of the NanoSail solar sail. Credit: NASA.

The Planetary Society's LightSail 2 mission launched as a 3U CubeSat secondary payload on the Department of Defense's Space Test Program (STP-2) in June 2019. The 32-m^2 mylar solar sail was deployed at 720-km altitude and demonstrated apogee raising of ~10 km. Its mission was still ongoing as of September 2022 (299).

NASA's Near-Earth Asteroid (NEA) Scout (figure 4.49) mission launched as a secondary payload onboard Artemis I in November 2022. Unfortunately, contact was never made with the spacecraft, and the sail was not able to be tested. The 6U CubeSat was to deploy an 85-m^2 solar sail and conduct a flyby of an asteroid within two years of launch (300) (301).

Figure 4.49: Deployment test of the Near-Earth Asteroid Scout solar sail. Credit: NASA.

NASA's Advanced Composite Solar Sail System (ACS3) technology demonstration uses composite materials for solar sail propulsion. The unfurled sail, approximately 9-m^2, will be deployed from a 12U CubeSat that is planned for launch in 2024 (302).

Tethers

a. Technology Description

NASA has developed tether technology for space applications since the 1960's. A tether is nothing more than a long wire, conducting or nonconducting, deployed from a spacecraft in orbit. Two types of tethers and systems can be used for space transportation, with both having had their fundamental physical principles demonstrated in space:

- **Electrodynamic Tethers (EDT)**: A propulsive force of $F = IL \times B$ (Lorentz force) is generated on a spacecraft-tether system when a current I from an onboard power supply is fed into a tether of length L against the electromotive force (emf) induced in it by the geomagnetic field B. This concept will work near any planet with a magnetosphere (Earth, Jupiter, etc.). The resulting force may be used to accelerate the deorbit of a spacecraft, boost a spacecraft's orbital altitude (by reversing the direction of current flow using an onboard power supply), or change its orbital inclination.
- **Momentum Exchange Tethers**: Using completely different physical principles, long nonconducting tethers can exchange momentum between two masses in orbit to place one body into a higher orbit or a transfer orbit for lunar and planetary missions. Recently

completed system studies of this concept indicate that it would be a relatively low-cost in-space asset with long-term, multi-mission capability.

b. Missions

Important milestones include retrieval of a tether in space on the Tethered Satellite System (TSS) mission (TSS-1, 1992), the accidental momentum-exchange boost of the TSS-1R satellite when the tether broke during flight (TSS-1R, 1996), successful deployment of a 20-km-long tether in space (SEDS-1, 1993), closed loop control of tether deployment (SEDS-2, 1994), and operation of an electrodynamic tether with tether current driven in both directions (i.e., power and thrust modes) (PMG, 1993).

More recently, Georgia Institute of Technology's Prox-1 mission was launched as a secondary payload on the Department of Defense's Space Test Program (STP-2) in June 2019. The 70 kg spacecraft served as the host and deployer for the LightSail 2 mission. The Prox-1 spacecraft housed a Tethers Unlimited Nanosat Terminator Tape (NSTT), shown in figure 4.50, which deployed a 70-m tether in September 2019 to lower the orbit from 717 km. Data from the Space Surveillance Network indicate that the NSTT caused Prox-1 to deorbit more than 24 times faster than otherwise expected. This rate of orbital decay will enable Prox-1 to meet its 25-year deorbit requirement (303) (304) (305).

Figure 4.50: Nanosat Terminator Tape (NSTT). Credit: Tethers Unlimited.

The Naval Postgraduate School's NPSat-1 was launched as a secondary payload on STP-2 and deployed its NSTT in late 2020 (305). TriSept's DragRacer technology demonstration mission, launched as a rideshare onboard an Electron rocket in November 2020, sought to conduct a direct comparison of the deorbiting rates of two Millennium Space Systems satellites, one of which used a 250-m NSTT (305) (306). A comparison of flight data for operation of the NSTT from each of these three missions has been publicly released (307).

Electric Sails

a. Technology Description

An electric sail (also known as an electric solar wind sail or an E-sail) uses the dynamic pressure of the solar wind as a source of thrust. By using positively charged wires deployed from a spacecraft to create electric fields around each wire, a large 'virtual' sail is created that deflects solar wind protons and extracts their momentum. Unlike EDTs, they do not derive thrust using the Lorentz force. Electric sails must be used outside the Earth's magnetosphere.

A closely related cousin to the electric sail is the plasma brake, an electrostatic deorbit propulsion system that, like the electric sail, uses Coulomb collisions between the charged wire and the ions in the surrounding environment (e.g., the Earth's ionosphere) to generate an electrostatic force orthogonal to the tether direction, which lowers the spacecraft's orbital altitude.

These technologies are relatively immature, but researchers in Europe and the USA have been making incremental progress on their development.

b. Missions

The AuroraSat-1 satellite was launched on an Electron rocket on May 5, 2022 (158) (159). The spacecraft was built by SatRevolution with Aurora Propulsion Technologies providing the payloads. The mission serves as a technology demonstration for a Plasma Brake module (157) (figure 4.51) and an Aurora Resistojet Module for Attitude control (ARM-A) (156) (figure 4.19), both produced by Aurora. The Plasma Brake module on AuroraSat-1 is a dual-redundant system for demonstration purposes. A 50-m tether will be deployed to demonstrate its deorbiting capability.

Figure 4.51: Plasma Brake Module (PBM) demo unit. Credit: Aurora Propulsion Technologies.

Aerodynamic Drag

a. Technology Description

Satellites have historically deorbited from low-Earth orbits with the aid of thrusters or passive atmospheric drag. Given the increasing rate of new spacecraft launched and the associated potential for new orbital debris following completion of missions, orbital debris management has gained increasing attention. Space debris poses a growing threat to active satellites and human activity in space. Allowing decades for defunct spacecraft to decay naturally from low-Earth orbit is no longer sufficient, as the Federal Communications Commission in late 2022 adopted a draft rule that requires <2000-km altitude spacecraft to deorbit within 5 years of mission completion (309). Aerodynamic drag devices may provide one method to rapidly remove spacecraft from low-Earth orbits upon mission completion.

Below about 1,000 km altitude, the atmosphere exerts a measurable drag force opposite the relative motion of any spacecraft, which results in a slow orbital decay. The intensity of the drag force exerted on the spacecraft depends on numerous factors such as local atmospheric density, the spacecraft forward facing area, the spacecraft velocity, and a drag coefficient. The drag coefficient accounts for the drag force's dependency on an object's unique geometric profile. While the spacecraft velocity and local atmospheric density are largely mission dependent, a spacecraft's forward-facing area and drag coefficient can be altered by introducing aerodynamic drag devices such as exo-brakes and ballutes. These deployable or inflatable parachutes and balloons can greatly increase the drag force exerted on spacecraft by an order of magnitude or more and significantly increase the rate of orbital decay.

Furthermore, aerodynamic drag devices may be useful to reduce spacecraft propellant mass required for orbit capture and disposal at other planetary bodies, given sufficient atmospheric density exists.

For further details on these devices, see chapter on Deorbit Systems.

Table 4-2: Hydrazine Chemical Propulsion

Manufacturer	Product	Propellant	Thrust per Thruster (Quantity)	Specific Impulse	Total Impulse	Mass	Envelope	Power	ACS	PMI Status	Missions	References
---	---	---	[N]	[s]	[kN-s]	[kg]	[cm³ or U]	[W]	Y/N	C,D,E,F	---	---
Integrated Propulsion Systems												
Aerojet Rocketdyne USA	MPS-120	Hydrazine	0.25 – 1.0 (4)	-	2 (2U) 0.8 (1U)	2.5† (2U) 1.6† (1U)	2U 1U	-	Y	D	-	(104)
Aerojet Rocketdyne USA	MPS-125	Hydrazine	0.25 – 1.0 (4)	-	14 (8U) 9.9 (6U) 5.2 (4U)	12.1† (8U) 9.3† (6U) 6.2† (4U)	8U 6U 4U	-	Y	D	-	(104)
Stellar Exploration USA	Monopropellant CubeSat System	Hydrazine	-	200s	-	-	-	-	Y	F	Echostar Global 3, NASA Capstone	(18) (19) (20) (21) (105)
Stellar Exploration USA	Bipropellant CubeSat system	Hydrazine/ NTO	-	285	-	-	-	-	Y	D	-	(105)
Thruster												
Aerojet Rocketdyne USA	MR-103	Hydrazine	1	202 - 224	183	0.33 - 0.37	-	< 16	-	F	numerous	(8)
Aerojet Rocketdyne USA	MR-106	Hydrazine	22	228 - 235	561	0.59	-	< 36	-	F	numerous	(8)
Aerojet Rocketdyne USA	MR-111	Hydrazine	4	219 - 229	262	0.37	-	< 16	-	F	numerous	(8)
Aerojet Rocketdyne USA	MR-401	Hydrazine	0.08	182	200	0.60	-	-	-	F	Numerous	(8)
ArianeGroup France	1 N	Hydrazine	1	200 - 223	135	0.29	-	-	-	F	ALSAT-2, numerous	(6) (7)
ArianeGroup France	20 N	Hydrazine	20	222 - 230	517	0.65	-	-	-	F	Numerous	(7)
IHI Aerospace Japan	MT-9	Hydrazine	1	208 - 215	(100 kg)*	-	-	-	-	F	Numerous	(17)
IHI Aerospace Japan	MT-8A	Hydrazine	5	212 - 225	(190 kg)*	-	-	-	-	F	Numerous	(17)
IHI Aerospace Japan	MT-2	Hydrazine	20	210 - 226	(115 kg)*	-	-	-	-	F	Numerous	(17)
Moog USA	MONARC-1	Hydrazine	1	227	111	0.38	113x50 mm	18 (valve)	-	F	numerous	(12)
Moog USA	MONARC-5	Hydrazine	4.5	226	613	0.49	203x380 mm	18 (valve)	-	F	NASA SMAP, numerous	(12) (13)
Moog USA	MONARC-22-6	Hydrazine	22	228	533	0.72	203x380 mm	30 (valve)	-	F	numerous	(12)
Moog USA	MONARC-22-12	Hydrazine	22	228	1,173	0.69	229x530 mm	30 (valve)	-	F	numerous	(12)
Northrop Grumman USA	MRE-0.1	Hydrazine	1	216	(34 kg)*	0.5	114x175 mm	15	-	F	numerous	(14)
Northrop Grumman USA	MRE-1.0	Hydrazine	5	218	(544 kg)*	0.5	114x188 mm	15	-	F	numerous	(14)
Northrop Grumman USA	MRE-4.0	Hydrazine	18	217	(249 kg)*	0.5	61x206 mm	30	-	F	numerous	(14)
Rafael Israel	1N	Hydrazine	1	205 - 214	100	0.31	-	9 (valve)		F	numerous	(15) (16)
Rafael Israel	5N	Hydrazine	6	210 - 220	74	0.31	-	9 (valve)		F	numerous	(15) (16)
Rafael Israel	25N	Hydrazine	28	205 - 220	100	0.53	-	15 (valve)		F	numerous	(15) (16)

Note that all data is documented as provided in the references. Unless otherwise published, do not assume the data has been independently verified.
† denotes a wet mass, ‡ denotes a dry mass, "-" = denotes data not available or applicable, * denotes mass of propellant throughput, rather than total impulse

Table 4-3: Alternative Monopropellant and Bipropellant Propulsion

Manufacturer	Product	Propellant	Thrust per Thruster (Quantity)	Specific Impulse	Total Impulse	Mass	Envelope	Power	ACS	PMI Status	Missions	References
---	---	---	[N]	[s]	[kN-s]	[kg]	[cm^3 or U]	[W]	Y/N	C,D,E,F	---	---
Integrated Propulsion Systems												
Aerojet Rocketdyne USA	MPS-130	ASCENT	0.25 – 1.0 (4)	-	2.7 (2U) 1.1 (1U)	2.8† (2U) 1.7† (1U)	2U 1U	-	Y	D	-	(103) (104)
Aerojet Rocketdyne USA	MPS-135	ASCENT	0.25 – 1.0 (4)	-	19.4 (8U) 13.7 (6U) 7.3 (4U)	14.7† (8U) 11.2† (6U) 7.2† (4U)	8U 6U 4U	-	Y	D	-	(104)
Aerospace Corp. USA	HyPer	H_2O_2	-	-	-	-	0.25U	-	-	D	-	(106)
Benchmark Space Systems USA	Halcyon	HTP	0.25 1 5 10	150 - 170	(1.56 kg)* (6.6 kg)* (32.5 kg)* (65 kg)*	-	-	< 4	Y	E	Transporter 2 (Undisclosed missions)	(38) (107) (108) (109)
Benchmark Space Systems USA	Halcyon Avant	HTP + Fuel	2 10 22	290 - 310	(6.8 kg)* (34 kg)* (74.8 kg)*	-	-	< 15	Y	F	Sherpa-TLC2	(36) (37) (109)
Benchmark Space Systems USA	Peregrine	HTP & Alcohol	0.1 - 22 (1-8)	270 - 300	5 - 150	-	-	-	-	D	-	(109)
ECAPS Sweden	SkySat 1N HPGP Propulsion System	LMP-103S	1.0 (4)	> 200	21	22†	55 x 55 x 15	10	Y	F	Skysat, PRISMA, Astroscale ELSA-d	(26) (27) (28) (29) (110) (111) (112) (113)
Busek USA	BGT-X5 System	ASCENT	0.5	220 - 225	0.57	1.5†	1U	20	N	D	-	(114) (115) (116)
Cornell University USA	Cislunar Explorer	Water (Electrolysis)	-	-	-	-	6U total (2-units)	-	-	E	CubeQuest Challenge (removed from Artemis 1)	(39) (40) (41)
CU Aerospace USA	MPUC	(CMP-X) Peroxide/ Ethanol blend	0.23	178	1.4 (1.5U) 2.1 (2U)	2.5† (1.5U) 3.3† (2U)	1.5U 2U	3	N	D	-	(50) (51)
Dawn Aerospace New Zealand	CubeDrive	Nitrous Oxide & Propylene	0.5	-	0.425 1.450	1.17† 2.70†	0.8U 2U	-	N	D	-	(52)
Dawn Aerospace New Zealand	SatDrive	Nitrous Oxide & Propylene	0.5 – 16.7	250 - 280	5 – 500	-	-	-	N	D	-	(52)
Moog USA	Monopropellant Propulsion Module	Green or 'Traditional'	0.5	224	0.5	1.0†	1U	2 x 22.5 W/Thruster	N	D	-	(117)
Rubicon Space Systems USA	Sprite	ASCENT	0.05 - 0.15	215	> 1.2	< 2†	1.5U	2 - 16	N	D	-	(118)
Rubicon Space Systems USA	Phantom	ASCENT	0.3 - 0.5	-	> 9	< 10†	8U	15 - 50	Y	D	-	(119)
NASA MSFC USA	LFPS	ASCENT	0.1 (4)	> 200s	> 3.5	< 5.5†	~2.4U	15 - 47	Y	E	Lunar Flashlight	(34) (35)
NanoAvionics Lithuania	EPSS C1K	IADN-blend	1.0 BOL 0.22 EOL	213	> 0.4	1.2†	1.3U	9.6 (preheat) 1.7 (firing)	N	F	Lituanica-2	(46) (47)
NanoAvionics Lithuania	EPSS C2	IADN-blend	1.0 BOL 0.25 EOL	220	> 1.7	2.6†	2U	-	N	D	-	(46)

Note that all data is documented as provided in the references. Unless otherwise published, do not assume the data has been independently verified.
† denotes a wet mass, ‡ denotes a dry mass, "-" = denotes data not available or applicable, * denotes mass of propellant throughput, rather than total impulse, BOL = Beginning of Life, EOL = End of Life

BODY-128

Manufacturer	Product	Propellant	Thrust per Thruster (Quantity)	Specific Impulse	Total Impulse	Mass	Envelope	Power	ACS	PMI Status	Missions	References
Table 4-3 (cont.): Other Monopropellant and Bipropellant Propulsion												
---	---	---	[N]	[s]	[kN-s]	[kg]	[cm³ or U]	[W]	Y/N	C,D,E,F	---	---
Integrated Propulsion Systems (cont.)												
Tethers Unlimited USA (Subsidiary of ARKA Group LP)	HYDROS-C	Water (Electrolysis)	> 1.2	> 310	> 2.1	2.7†	19 x 13 x 9.2	5 - 25	N	F	Pathfinder Technology Demonstration	(44) (45) (120) (121)
Tethers Unlimited USA (Subsidiary of ARKA Group LP)	HYDROS-M	Water (Electrolysis)	> 1.2 (1)	> 310	> 18	13.7†	38.1 dia. x 19.1	7 - 40	N	D	-	(122)
VACCO USA	ArgoMoon Hybrid MiPS	LMP-103S/ R134a	0.1 (1)	190	0.78	2.07†	1.3U	4.3 20 (max)	Y	F	ArgoMoon (Artemis I)	(29) (30) (31)
VACCO USA	Green Propulsion System (MiPS)	LMP-103S	0.1 (4)	190	3.3	5†	3U	15 (max)	Y	D	-	(123)
VACCO USA	Integrated Propulsion System	LMP-103S	1 (4)	200	12	14.7†	11U	50 (max)	Y	D	-	(53) (54)
Thruster Heads												
Aerojet Rocketdyne USA	GR-M1	ASCENT	0.25 - 0.5	206	(3.4 kg)*	--	--	14 (max)	-	D	-	(48)
Aerojet Rocketdyne USA	GR-1	ASCENT	0.3 - 1.4	231	(12 kg)*	-	-	18 (max)	-	F	GPIM	(8) (32) (33) (48)
Aerojet Rocketdyne USA	GR-22	ASCENT	8 - 25	250	74	-	-	28	-	F	GPIM	(8) (32) (33)
Benchmark Space Systems USA	Felicette	HTP	0.4 - 1.2	161	> 10	-	-	-	-	E	Transporter 2 (Undisclosed missions)	(38) (109)
Benchmark Space Systems USA	Lynx	HTP	2 - 2.5	295	> 20	-	-	-	-	D	-	(109)
Benchmark Space Systems USA	Ocelot	HTP	18 - 22	302	> 240	-	-	-	-	F	Sherpa-TLC2	(36) (37) (109)
ECAPS Sweden	0.1 N HPGP	LMP-103S	0.03 - 0.1	196 - 209	(1 kg)*	0.04 excl. FCV	-	6.3 - 8	-	F	ArgoMoon	(26) (29) (112)
ECAPS Sweden	1 N HPGP	LMP-103S	0.25 - 1	204 - 235	(24 kg)*	0.38	-	8 - 10	-	F	PRISMA, SkySat, ELSA-d	(26) (27) (28) (29) (112) (113)
ECAPS Sweden	1 N GP	LMP-103S/LT	0.25 - 1	194 - 227	(5 - 8 kg)*	0.38	-	8 - 10	-	D		(113)
ECAPS Sweden	5 N HPGP	LMP-103S	1.5 - 5.5	239 - 253	-	0.48	-	15 - 25	-	D	-	(112)
ECAPS Sweden	22 N HPGP	LMP-103S	5.5 - 22	243 - 255	(150 kg)*	1.1	-	25 - 50	-	D	-	(112)
Dawn Aerospace New Zealand	B20 Thruster	Nitrous Oxide & Propylene	6.1 - 16.7	-	-	0.6	17.6 x 8.0 x 7.9	-	-	F	numerous	(52)
Dawn Aerospace New Zealand	B1 Thruster	Nitrous Oxide & Propylene	0.46 - 1.28	-	-	0.26	10.8 x 7.9 x 4.0	-	-	F	numerous	(52)
Rubicon Space Systems USA	0.1N	ASCENT	0.03 - 0.28	215 – 235	(3.1 kg)*	0.06	-	7 - 9	-	E	Lunar Flashlight	(34) (35) (49)
Rubicon Space Systems USA	1N	ASCENT	0.2 - 1.1	236 - 250	(> 7 kg)*	0.18	-	< 15	-	D	-	(49)
Rubicon Space Systems USA	5N LT	ASCENT	1 - 6	221 - 261	(3 kg)*	0.25	-	< 25	-	D	-	(49)
Rubicon Space Systems USA	5N HT	ASCENT	1 - 6	221 - 261	(110 kg)*	0.30	-	< 25	-	D	-	(49)

Note that all data is documented as provided in the references. Unless otherwise published, do not assume the data has been independently verified.
† denotes a wet mass, ‡ denotes a dry mass, "-" = denotes data not available or applicable, * denotes mass of propellant throughput, rather than total impulse

Table 4-4: Hybrid Chemical Propulsion

Manufacturer	Product	Propellant	Thrust (Quantity)	Specific Impulse	Total Impulse	Mass	Envelope	Power	ACS	PMI Status	Missions	References
---	---	---	[N]	[s]	[kN-s]	[kg]	[cm^3 or U]	[W]	Y/N	C,D,E,F	---	---
Aerospace Corp. USA	Propulsion Unit for CubeSats	Paraffin/Nitrous Oxide	-	-	-	-	1U	-	-	D		(65)
JPL USA	Hybrid Rocket	PMMA/GOX	-	> 300	-	-	-	-	-	D	-	(59) (60) (61) (62)
NASA Ames USA	Hybrid Rocket	PMMA/ Nitrous Oxide	25	247	-	-	-	-	-	D		(63) (64)
Parabilis Space Technologies USA	ROMBUS	Various/N2O	222	260	44.6	49	55 dia. x 46	-	Y	D		(66)
Parabilis Space Technologies USA	NanoSat Obrital Transfer System	HTPB/N2O	9.4	245	-	-	-	-	Y	C		(67)
Utah State Univ. USA	Green Hybrid Rocket	ABS/Nytrox	25 - 50	220 - 300	-	-	-	< 30 for 1-2 seconds	-	D		(57) (58)
Utah State Univ. USA	Green Hybrid Rocket	ABS/GOX	8	215	-	-	-	-	-	D	-	(55) (56)

Note that all data is documented as provided in the references. Unless otherwise published, do not assume the data has been independently verified.
† denotes a wet mass, ‡ denotes a dry mass, "-" = denotes data not available or applicable

Table 4-5: Cold Gas Propulsion

Manufacturer	Product	Propellant	Thrust (Quantity)	Specific Impulse	Total Impulse	Mass	Envelope	Power	ACS	PMI Status	Missions	References
---	---	---	[mN]	[s]	[N-s]	[kg]	[cm^3 or U]	[W]	Y/N	C,D,E,F	---	---
Integrated Propulsion Systems												
Aerospace Corp. USA	MEPSI	R236fa	20	-	-	0.188†	-	-	Y	E	STS-113 and STS-116	(68)
Benchmark Space Systems USA	Starling	Nitrogen	10 - 1000	70	-	0.75†	0.5U	< 4	Y	D	-	(109)
GomSpace Sweden	NanoProp CGP3	Butane	0.01 – 1 (4)	60 - 110	40	0.35†	5 x 10 x 10	< 2	Y	E	TW-1	(76) (78) (79) (126)
GomSpace Sweden	NanoProp 6U	Butane	0.1 – 10 (4)	60 - 110	80	0.900†	2 x 10 x 20	< 2	Y	F	GOMX-4B	(76) (77) (127)
Lightsey Space Research USA	BioSentinel Propulsion System	R236fa	20	47	79.8	1.28†	4 x 10 x 20	< 1 idle, < 4 operating	Y	E	BioSentinel	(80) (81) (82) (83)
Microspace Rapid Pte Ltd Singapore	POPSAT-HIP1	Argon	0.083 – 1.1 (8)	43	-	-	-	-	-	E	POPSAT-HIP1	(75)
ThrustMe France	I2T5	Iodine	0.35	-	75	0.9†	0.5U	5	N	F	Xiaoxiang 1-08, NAPA-2, Spire L3C Demo, Robusta-3A (2024**)	(84) (85) (86) (87)
UTIAS/SFL Canada	CNAPS	Sulfur Hexafluoride	12.5 - 50	40	100	-	-	-	N	F	CanX-4/CanX-5	(71) (72) (73) (74)
VACCO USA	NEA Scout	R236fa	25 (6)	-	500	2.54†	2U	< 55 warmup < 9 operating	Y	E	NEA Scout	(95) (96) (97)
VACCO USA	MiPS Standard	R236fa	25 (4)	-	82 - 515	0.85 – 2.46†	0.4 – 1.5U	< 12	Y	D	-	(128)
VACCO USA	MarCO-A and -B MiPS	R236fa	25 (8)	-	755	3.49†	8 x 15 x 20	-	Y	E	MarCO-A & -B	(91) (92) (93) (94)
VACCO USA	C-POD	R236fa	10 (8)	40	186	1.25†	1U	5	Y	E	CPOD	(88) (89) (90)
Thruster Heads												
Marotta USA	CGMT	Nitrogen	105 - 2360	-	-	< 0.06	-	< 7	-	F	NMP ST5	(69) (70)
Moog USA	058E143-146	Nitrogen	10 - 40	60	-	0.04	1.4 x 5.7	< 10	-	F	CHAMP, GRACE	(129)
Moog USA	058E142A	Nitrogen	120	57	-	0.016	1.4 x 2.0	< 35	-	F	Spitzer Space Telescope	(129)
Moog USA	058E151	Nitrogen	120	65	-	0.07	1.9 x 4.1	< 10.5	-	F	Spitzer Space Telescope	(129)
Moog USA	058-118	Nitrogen	3600	57	-	0.023	0.7 x 2.5	< 30	-	F	SAFER, Pluto Fast Flyby	(129)
Moog USA	58E163A	Nitrogen, Xenon, Argon	1300	70 N2, 21 Xe, 54 Ar	-	0.115	2.4 x 5.3	< 10.5	-	F	GEO applications	(129)

Note that all data is documented as provided in the references. Unless otherwise published, do not assume the data has been independently verified.
† denotes a wet mass, ‡ denotes a dry mass, "-" = denotes data not available or applicable, ** anticipated launch date

Table 4-6: Solid Motor Chemical Propulsion

Manufacturer	Product	Propellant	Thrust (Quantity)	Specific Impulse	Total Impulse	Mass	Envelope	Power	ACS	PMI Status	Missions	References
---	---	---	[N]	[s]	[N-s]	[kg]	[cm³ or U]	[W]	Y/N	C,D,E,F	---	---
Integrated Propulsion Systems												
PacSci EMC USA	MAPS	-	-	-	-	-	-	-	-	E	PacSciSat	(99) (100)
Thruster Heads												
Industrial Solid Propulsion USA	ISP 30 sec. Motor	80% Solids HTPB/AP	37	187	996	0.95†	5.7	-	-	D	Optical target at Kirtland AFB	(131)
Northrop Grumman USA	STAR 3	TP-H-3498	461	266	1,250	1.16†	8 dia. x 29	-	-	E	Mars Exploration Rover Spirit lander	(132)
Northrop Grumman USA	STAR 4G	TP-H-3399	258	269	2,650	1.5†	11.3 dia. x 13.8	-	-	D	-	(132)

Note that all data is documented as provided in the references. Unless otherwise published, do not assume the data has been independently verified.
† denotes a wet mass, ‡ denotes a dry mass, "-" = denotes data not available or applicable

Table 4-7: Electrothermal Electric Propulsion

Manufacturer	Product	Propellant	Thrust*	Specific Impulse*	Total Impulse*	Mass	Envelope	Power	ACS	PMI Status	Missions	References
---	---	---	[mN]	[s]	[N-s]	[kg]	[cm³ or U]	[W]	Y/N	C,D,E,F	---	---
Integrated Propulsion Systems												
AIS USA	AIS-SWAG1-PQ	Adamantane	0.12	20	3	0.224†	4.2 x 4.2 x 5.4	5	N	D	-	(308)
AIS USA	AIS-SWAG1-DUO	Adamantane	0.24	20	6	0.449†	8.6 x 4.2 x 5.4	10	N	D	-	(308)
AIS USA	AIS-SWAG1-QUAD	Adamantane	0.48	20	12	0.898†	8.6 x 8.6 x 5.4	20	N	D	-	(308)
Aurora Finland	ARM-A	Water	0.5	100	70	0.280†	0.3U	10£	Y	E	AuroraSat-1 (2022), ORB-12 STRIDER	(156) (158) (159) (273)
Aurora Finland	ARM-C	Water	1	-	-	0.050†	45	12 (max)	N	D	-	(162)
Benchmark Space Systems USA	Starling Ardent	Nitrogen	10 - 1000	70 – 140	-	1.15	1U	10 – 100	Y	D	-	(109)
Bradford Space Netherlands	Comet-1000	H_2O	17	175	1,150	1.55†	2,300	50 (max)	N	F	HawkEye 360, Capella Space, Transporter 7	(135) (136) (137) (138)
Bradford Space Netherlands	Comet-8000	H_2O	17	175	8,348	6.68†	23,760	55 (max)	N	F	BlackSky	(135) (138) (139)
CU Aerospace USA	CHIPS-180	R134a	15	67	176	1.03†	540	20	Y	D	-	(310) (311) (312) (313)
CU Aerospace USA	CHIPS-500	R134a	25	69	505	1.84†	1300	25	Y	D	-	(310) (311) (312) (313)
CU Aerospace USA	CHIPS-1000	R134a	31	70	1,030	3.13†	2500	30	Y	D	-	(310) (311) (312) (313)
CU Aerospace and VACCO USA	PUC	SO_2	4.5	70	184	0.72†	0.35U	15	N	E	8 flight units delivered to AFRL	(140) (141) (142)
CU Aerospace USA	MVP	Delrin Fiber	4.5	66	280	1.06†	0.93U	45	N	E	DUPLEX (launch 2024**)	(143) (144) (145) (146)
EPL USA	APS 100	Ammonia	28	280	2,200	2.63†	2.5U	100	N	D	-	(314)
EPL USA	APS 500	Ammonia	150	350	25,200	13.4†	27,300	450	N	E	Atomos Space (2024**)	(314) (315)
Pale Blue Japan	PBR-9	Water	1.0	45	35	0.6†	0.5U	9	N	D	-	(316)
Pale Blue Japan	PBR-20	Water	1.0	70	200	1.5†	1U	20	N	E	ArkEdge OPTIMAL-1	(160) (316)
Pale Blue Japan	Water Resistojet Propulsion System	Water	2.7	60	170	1.4†	1.25U	< 30	N	E	SPHERE-1 EYE	(161)
SteamJet Space Systems UK	Steam TunaCan Thruster	Water	6	172	219	0.54†	402	< 19.9	N	D	-	(147)
SteamJet Space Systems UK	Steam Thruster One	Water	6	172	-	-	-	19.9	N	D	-	(148)
Thruster Heads												
SITAEL Italy	XR-50	Ar, Xe, N_2	100	55-85	≤ 72,000	0.22‡	45.8	≤ 50	-	D	-	(317) (318)
SITAEL Italy	XR-100	Ar, Xe, N_2	125	63-105	≤ 90,000	0.22‡	45.8	≤ 80	-	D	-	(317) (318)
SITAEL Italy	XR-150	Ar, Xe, N_2	250 (Ar) 100 (Xe)	58-110	≤ 180,000	0.22‡	45.8	≤ 95	-	D	-	(317) (318)

Note that all data is documented as provided in the references. Unless otherwise published, do not assume the data has been independently verified.
*nominal values (see references for full performance ranges), ** anticipated launch date, † denotes a wet mass, ‡ denotes a dry mass, £ per active thruster, "-" = denotes data not available or applicable

Table 4-8: Electrospray Electric Propulsion

Manufacturer	Product	Propellant	Thrust*	Specific Impulse*	Total Impulse*	Mass	Envelope	Power	Neutralizer	PMI Status	Missions	References
---	---	---	[μN]	[s]	[N-s]	[kg]	[cm^3 or U]	[W]	---	C,D,E,F	---	---
Integrated Propulsion Systems												
Busek USA	CMNT (4x heads)	EMI-Im (ionic)	4 x 20	225	980	14.8†	29U	16.5	Carbon Nanotube	F	LISA Pathfinder	(163)
Busek USA	BET-MAX (Config. A)	EMI-Im (ionic)	4 x 55	850	4 x 92§	0.8†	1250	12	Carbon Nanotube	E	US Government	(319) (320) (321) (322) (323) (324) (325)
Busek USA	BET-MAX (Config. B)	EMI-Im (ionic)	4 x 40	2300	4 x 250	0.8†	1250	14	Carbon Nanotube	D	-	(325)
Enpulsion Austria	Nano	Indium (FEEP)	330	1,500 (alpha); 3,000 (gamma)	> 3,000 (alpha); > 5,000 (gamma)	0.90†	10 x 10 x 8.3	40	Thermionic	F	Flock 3p', ICEYE X2, Harbinger, NetSat, and others	(164) (165) (166) (167) (168) (169) (170) (171) (326) (327) (328)
Enpulsion Austria	Nano R^3	Indium (FEEP)	350	1,500 (alpha); 2,700 (gamma)	> 3,000 (alpha); > 5,000 (gamma)	1.4†	9.8 x 9.9 x 9.5	45	Thermionic	E	(Evolution of Nano design)	(171) (172) (329)
Enpulsion Austria	Micro R^3	Indium (FEEP)	1,000	2,100 (alpha); 2,750 (gamma)	> 25,000 (alpha); > 35,000 (gamma)	3.9†	14 x 12 x 13.3	105	Thermionic	F	GMS-T	(172) (173) (330) (331)
Enpulsion Austria	Neo	Indium (FEEP)	20,000	1,500	> 550,000	30†	34 x 34 x 15	800	Thermionic	D	-	(332)
Morpheus Space Germany	NanoFEEP (2x heads)	Gallium (FEEP)	< 40	-	-	0.16‡	9 x 2.5 x 4.3	< 3	Propellant-less	E	UWE-4	(174) (175) (333) (334)
Morpheus Space Germany	MultiFEEP (2x heads)	Gallium (FEEP)	< 140	-	-	0.28‡	9 x 4.5 x 4.5	< 19	Propellant-less	D	-	(333)

Note that all data is documented as provided in the references. Unless otherwise published, do not assume the data has been independently verified.
*nominal values (see references for full performance ranges), ** anticipated launch date, † denotes a wet mass, ‡ denotes a dry mass, § demonstrated, "-" = denotes data not available or applicable

Table 4-9: Gridded-Ion Electric Propulsion

Manufacturer	Product	Type	Propellant	Thrust*	Specific Impulse*	Total Impulse*	Mass	Envelope	Power	Cathode Type	PMI Status	Missions	References
---	---		---	[mN]	[s]	[kN-s]	[kg]	[cm^3 or U]	[W]	---	C,D,E,F	---	---
Integrated Propulsion Systems													
Avant Space Russia	GT-50	RF Ion	Xenon	< 7	-	-	< 8†	< 4U	< 240	Hollow	D	-	(335) (336)
Busek USA	BIT-3	RF Ion	Iodine	1.0	1,960	31.7	2.9†	18 x 8.8 x 10.2	70	RF	E	Lunar IceCube; LunaH-Map	(188) (189) (190) (191) (192) (193) (337) (338) (339) (340) (341) (342)
Pale Blue Japan	PBI-40 Hybrid	RF Ion (Resistojet)	Water (Water)	0.15 (0.9)	> 500 (40)	-	< 2.5†	9 x 12 x 12	28 23	RF	E	RAISE-3 (DDL); RAISE-4 (2024**)	(343) (344) (345) (346)
ThrustMe France	NPT30-I2	RF Ion	Iodine	< 1.3	2,400	5.5 (1U) 14 (1.5U)	1.2† (1U) 1.85† (1.5U)	1U 1.5U	< 81	Thermionic	F	Beihangkongshi-1; NORSAT-TD; INSPIRESAT-4 (2023**); GOMX-5 (2024**); DROID.002 (2024**)	(182) (183) (184) (185) (186) (187) (347) (348) (349)
Thruster Heads													
Ariane Group Germany	RIT µX	RF Ion	Xenon	< 0.5	-	-	0.44‡	7.8 x 7.8 x 7.6	< 50	RF	D	-	(350) (351) (352) (353) (354) (355)
Ariane Group Germany	RIT 10 EVO	RF Ion	Xenon	< 15	-	-	1.8‡	18.6 x 18.6 x 13.4	< 435	Hollow	E	(Identical to flight-heritage RIT-10 with contemporary grid design)	(350) (352) (356)
QinetiQ UK	T5	DC Ion	Xenon	< 20	< 3,000	-	2‡	19 x 19 x 24.2	< 600	Hollow	F	GOCE	(178) (179) (180) (181)

Note that all data is documented as provided in the references. Unless otherwise published, do not assume the data has been independently verified.

*nominal values (see references for full performance ranges), ** anticipated launch date, † denotes a wet mass, ‡ denotes a dry mass, "-" = denotes data not available or applicable, RF = Radio Frequency, DC = Direct Current, DDL = Destroyed During Launch

Table 4-10: Hall-Effect Electric Propulsion Thrusters

Manufacturer	Product	Propellant	Thrust*	Specific Impulse*	Total Impulse*	Mass	Envelope	Thruster Power*	Cathode Type	PMI Status	Missions	References
---	---	---	[mN]	[s]	[kN-s]	[kg]	[cm^3 or U]	[W]	Notes	C,D,E,F	---	---
Aliena Pte Ltd Singapore	MUSIC-SI	Xenon	< 0.25	< 200	1	2†	1.5U	20	None	E	NuX-1	(268) (269) (270) (271) (272)
Aliena Pte Ltd Singapore	MUSIC-HM	Xenon	3	1,000	15	5†	4U	100	EM-HL	E	ORB-12 STRIDER	(268) (272) (273)
Astra USA	ASE	Xenon	25	1,400	300	1.0	-	400‡	CM-HL	F	Sherpa-LTE, Aries (2024**), ELSA-M (2024**)	(218) (220) (219) (220) (221) (222) (223) (224)
Astra USA	ASE	Krypton	18	1,300	300	1.0	-	400‡	CM-HL	D	-	(218)
Busek USA	BHT-100	Xenon	6.3	1,086	150	1.2	275 wo cath.	105	EM-SH	D	-	(203) (357)
Busek USA	BHT-200	Xenon	13	1,390	84§	1.2	675 wo cath.	250‡	EM-SH	F	TacSat-2, FalconSat-5, -6	(203) (204) (205) (206)
Busek USA	BHT-200-I	Iodine	14	1390	-	1.2	675 wo cath.	250	EM-SH	E	NASA iSat (Cancelled)	(204) (206) (207)
Busek USA	BHT-350	Xenon	17	1,244	212§	1.9	-	350	EM-SH	F	OneWeb Satellites	(255) (256) (257) (258)
Busek USA	BHT-600	Xenon	39	1,500	1000§	2.8	1,470 wo cath.	680‡	EM-SH	E	US Government	(203) (259) (260) (358) (359)
Busek USA	BHT-600-I	Iodine	39	-	-	2.8	1,470 wo cath.	600	EM-SH	D	-	(206) (358) (359) (360)
EDB Fakel Russia	SPT-50	Xenon	14	860	126§	1.2	1,092	220	EM-SH	F	Canopus-V	(198) (199) (200) (201) (361)
EDB Fakel Russia	SPT-50M	Xenon	14.8	930	266	1.3	---	220	EM-SH	D	-	(361)
EDB Fakel Russia	SPT-70BR	Xenon	39	1,470	435§	2.0	1,453	660	EM-SH	F	KazSat-1, KazSat-2	(201) (202)
EDB Fakel Russia	SPT-70M	Xenon	41.3	1,580	-	-	-	660	EM-SH	D	-	(202)
EDB Fakel Russia	SPT-70M	Krypton	31.3	1,460	-	-	-	660	EM-SH	D	-	(202)
ExoTerra USA	Halo	Xenon	19.6	1,294	>375	0.83	375	400‡	CM-HL	F	Tipping Point (2024**), Blackjack	(243) (244) (245) (246) (247) (248) (249)
ExoTerra USA	Halo	Krypton	12	900	>175	0.83	375	300‡	CM-HL	D	-	(362)
ExoTerra USA	Halo12	Xenon	50	1,900	> 2,000	3.4	1,700	1,000‡	CM-HL	E[1]	Boeing RocketR and P1229 Programs	(251) (250) (363)
ExoTerra USA	Halo12	Krypton	55	1,575	> 2,000	3.4	1,700	1,000‡	CM-HL	D[1]	-	(364)
Exotrail France	spaceware nano	Xenon	2.5	800	6	-	2.5U	60	EM-SH	F	M6P Demo, Arthur, ELO3 and ELO4, Otter Pup, INCUS (2026)	(227) (228) (229) (230) (231) (232) (233) (234) (235)
Exotrail France	spaceware micro	Xenon	7	1,000	60	-	960	150	EM-SH	E	York Space (2023**), SpaceVan (2023**), Astro Digital (2024), Satrec Initiative (2025), Muon Space (2026)	(227) (230) (236) (237) (238) (239) (240) (241)
Exotrail France	spaceware mini	Xenon	23	1,300	300	-	-	400	EM-SH	E	Airbus' Earth Observation satellite platform portfolio	(227) (242)

Note that all data is documented as provided in the references. Unless otherwise published, do not assume the data has been independently verified.

*nominal values (see references for full performance ranges), ** anticipated launch date, ‡ PPU input power, § demonstrated, CM = Center Mounted, EM = Externally Mounted, SH = Swaged Heater, HL = Heater-less, JPL = Jet Propulsion Laboratory, SETS = Space Electric Thruster Systems, EDB = Experimental Design Bureau, [1]ExoTerra is commercializing the JPL developed MaSMi thruster, "-" = denotes data not available or applicable

Table 4-10 (cont.): Hall-Effect Electric Propulsion Thrusters

Manufacturer	Product	Propellant	Thrust*	Specific Impulse*	Total Impulse*	Mass	Envelope	Thruster Power*	Cathode Type	PMI Status	Missions	References
---	---	---	[mN]	[s]	[kN-s]	[kg]	[cm³]	[W]	Notes	C,D,E,F	---	---
JPL [USA]	MaSMi	Xenon	55	1,920	> 1,550§	3.4	1,700	1,000	CM-HL	D	-	(365) (366) (367) (368) (369) (370) (371) (372) (373) (374) (375) (376) (377)
Northrop Grumman [USA]	NGHT-1X	Xenon	55	1,700	> 2,000	3.1	-	900	CM-SH	E	MEP (2025**)	(261) (262) (263) (264) (265)
Orbion [USA]	Aurora	Xenon	15	1,320	200	1.5	1,500	250	EM-SH	E	GA-EMS (**)	(252) (253) (254)
Rafael [Israel]	R-200	Xenon	13	1,160	200	-	-	250	EM-HL	D	-	(208) (378) (379)
Rafael [Israel]	IHET-300	Xenon	> 14.3	> 1,210	> 135	1.5	1,836	300	EM-SH	F	VENuS	(208) (209) (210) (211) (212)
Rafael [Israel]	R-800	Xenon	-	-	> 600	-	-	800	EM-HL	D	-	(208) (380)
Safran [France]	PPS-X00	Xenon	43	1,530	1,000	< 3.2	-	650	EM-SH	D	-	(381) (382) (383)
SITAEL [Italy]	HT100	Xenon	9	1,300	73	-	407 wo cath.	175	EM-SH	E	uHETSat (2023**)	(213) (214) (215) (216)
SITAEL [Italy]	HT400	Xenon	27.5	1230	1,000	2.77	1,330	615	EM-SH	D		(384) (385) (386)
SETS [Ukraine]	ST25	Xenon	7.6	1,000	82	0.95 (with 2 cathodes)	1,003	140	EM-SH	E	EOS SAT-1	(266) (267) (387) (388)
SETS [Ukraine]	ST40	Xenon	25	1,450	400	1.2 (with 2 cathodes)	1,170	450	EM-HL	D	-	(389)

Note that all data is documented as provided in the references. Unless otherwise published, do not assume the data has been independently verified.
*nominal values (see references for full performance ranges), ** anticipated launch date, ‡ PPU input power, § demonstrated, CM = Center Mounted, EM = Externally Mounted, SH = Swaged Heater, HL = Heater-less, JPL = Jet Propulsion Laboratory, SETS = Space Electric Thruster Systems, EDB = Experimental Design Bureau, [1]ExoTerra is commercializing the JPL developed MaSMi thruster, "-" = denotes data not available or applicable

Table 4-11: Pulsed Plasma and Vacuum Arc Electric Propulsion

Manufacturer	Product	Propellant	Thrust*	Impulse Bit	Specific Impulse*	Total Impulse*	Mass	Envelope	Power*	ACS	PMI Status	Missions	References
---	---	---	[μN]	[μNs]	[s]	[N-s]	[kg]	[cm^3 or U]	[W]	Y/N	C,D,E,F	---	---
Integrated Propulsion Systems													
AIS USA	AIS-VAT1-PQ	Bismuth	26	-	87	0.13	0.056†	37	5	N	D	---	(308)
AIS USA	AIS-VAT1-DUO	Bismuth	52	-	87	0.26	0.084†	74	10	N	D	---	(308)
AIS USA	AIS-VAT1-QUAD	Bismuth	104	-	87	0.52	0.177†	148	20	N	D	---	(308)
Benchmark Space Systems USA	Xantus Metal Plasma Thruster	Molybdenum	400	1	1,774	5,000	1.2†	0.53U	40	N	E	USSF RROCI (not deployed), USSF EWS (2024**)	(390) (276) (277)
Comat France	Plasma Jet Pack	Metal	150	20	2,000	120£	1.2†£	0.66U£	30	N	E	∑yndeo-2	(278) (279) (391)
CU Aerospace USA	FPPT-1.7	PTFE Fiber	500	165	3,200	24,000	3.0†	1.7U	96	Y	E	DUPLEX (2024**)	(149) (150) (151) (152) (153) (154)
Hypernova Space Technologies South Africa	NanoThruster A, Size XS	Solid-state fuel	80	-	≥500	-	0.8†	0.5	10	N	D	---	(392)
Mars Space Ltd UK Clyde Space Sweden	PPTCUP	PTFE	40	40	600	44	0.28†	0.33U	2	N	D	---	(393) (394)
Neumann Space Australia	ND-15	Molybdenum	3.75	45	2,000	880	1.9†	1.5U	15	N	E	Skyride, Apogee	(280) (281) (282) (283) (284) (285) (286) (287)
Neumann Space Australia	ND-50	Molybdenum	100	150	2,000	1,800	1.3†	1U	50	N	E	Edison (2024**), CarbSAR (2024**)	(280) (288) (289)

Note that all data is documented as provided in the references. Unless otherwise published, do not assume the data has been independently verified.
*nominal values (see references for full performance ranges), ** anticipated launch date, † denotes a wet mass, ‡ denotes a dry mass, "-" = denotes data not available or applicable, ESV = Ejector Spring Volume, £ assumes power unit and 2 nozzles

Table 4-12: Ambipolar Electric Propulsion

Manufacturer	Product	Propellant	Thrust*	Specific Impulse*	Total Impulse*	Mass	Envelope	Power	ACS	PMI Status	Missions	References
---	---	---	[mN]	[s]	[kN-s]	[kg]	[cm³]	[W]	Y/N	C,D,E,F	---	---
Integrated Propulsion Systems												
Phase Four [USA]	Maxwell (Block 1)[RF]	Xenon	7	400	5	5.9‡	19 x 13.5 x 19	450	N	F	Capella	(290) (292) (395) (396) (397) (398)
Phase Four [USA]	Maxwell (Block 2)[RF]	Xenon	13	700	-	5.0 (without tank)	22 x 12 x 24 (without tank)	450	N	E	(Mission Not Specified)	(398) (399) (291)
Phase Four [USA]	Maxwell (Kr)[RF]	Krypton	13.6	1,110	-	-	-	550	N	D	-	(291) (400)
T4i [Italy]	REGULUS-50-I2[RF]	Iodine	0.55	550	3 (Small); 7 (Medium); 11 (Large)	2.5† (Small); 4† (Medium); 5.1† (Large)	9.4 x 9.5 x {15, 18, 20} {Small, Medium, Large}	50	N	E	UniSat-7	(293) (294) (401) (295) (402)
T4i [Italy]	REGULUS-150-I2[RF]	Iodine	2	1000	11	6.1†	12 x 12 x 27	150	N	D	-	(402)
Miles Space [USA]	M1.4	Water	17.2	-	-	1.0†	9.5 x 9.5 x 9.5	6	N	E	Team Miles	(296) (297) (403) (403)

Note that all data is documented as provided in the references. Unless otherwise published, do not assume the data has been independently verified.
*nominal values (see references for full performance ranges), ** anticipated launch date, † denotes a wet mass, ‡ denotes a dry mass, "-" = denotes data not available or applicable, RF = Radio Frequency

Table 4-13: Propellant-less Propulsion

Manufacturer	Product	Propellant	Thrust*	Specific Impulse*	Total Impulse*	Mass	Envelope	Power	ACS	PMI Status	Missions	References
---	---	---	[mN]	[s]	[kN-s]	[kg]	[cm³]	[W]	Y/N	C,D,E,F	---	---
Aurora Propulsion Technologies [Finland]	Plasma Brake	-	< 100 mN/m	-	-	< 2	1U	< 4	N	E	AuroraSat-1	(157) (158) (159)
Tethers Unlimited [USA]	NSTT	-	-	-	-	0.81	18 x 18 x 1.8	-	N	F	Prox-1, NPSat-1, DragRacer	(303) (304) (305) (306) (307) (405)

Note that all data is documented as provided in the references. Unless otherwise published, do not assume the data has been independently verified.
*nominal values (see references for full performance ranges), ** anticipated launch date, † denotes a wet mass, ‡ denotes a dry mass, "-" = denotes data not available or applicable
See Chapter on Passive Deorbit Systems for review of aerodynamic drag devices.

BODY-139

References

(1) "NASA Systems Engineering Handbook." NASA/SP-2016-6105, Rev2.
(2) "NASA Systems Engineering Processes and Requirements." NPR 7123.1C, Effective 14 February 2020, Expiration 14 February 2025.
(3) JANNAF Spacecraft Propulsion Subcommittee, Micro-Propulsion Panel, "JANNAF Guidelines for the Application of Technology Readiness Levels (TRLs) to Micro-Propulsion Systems." JANNAF 2019, Tampa, FL, Distribution Unlimited
(4) JANNAF Spacecraft Propulsion Subcommittee, Micro-Propulsion Panel, "JANNAF Guidelines for the Application of Technology Readiness Levels (TRLs) to Micro-Propulsion Systems. 2022 Edition" JANNAF 2022, Huntsville, AL, Distribution Unlimited
(5) Marshall, W. M. and M. Deans. "Recommended Figures of Merit for Green Monopropellants." NASA TM-2013-216560. National Aeronautics & Space Administration. Cleveland, OH. August, 2013.
(6) "AlSat-2". URL:https://directory.eoportal.org/web/eoportal/satellite-missions/a/alsat-2
(7) Arianegroup. "Chemical Monopropellant Thruster Family." Product Brochure. URL:http://www.space-propulsion.com/brochures/hydrazine-thrusters/hydrazine-thrusters.pdf
(8) Aerojet Rocketdyne an L3Harris Technology Company "Monopropellant Rocket Engines Spec Sheet" Product Brochure. URL:https://www.l3harris.com/resources/monopropellant-rocket-engines-data-sheet
(9) Stratton, J. 2004. "The use of the aerojet MR-103H thruster on the New Horizons mission to Pluto." 55th International Astronautical Congress.
(10) Press Release, "Aerojet Rocketdyne Propulsion Guides Cassini to its Grand Finale at Saturn," September 15, 2017.
(11) Press Release, "Aerojet Rocketdyne Propulsion Powers OSIRIS-REx's Approach of Asteroid Bennu", August 24, 2018.
(12) Moog Space and Defense Group. "Monopropellant Thrusters Datasheet", URL: https://www.moog.com/products/propulsion-controls/spacecraft-propulsion/thrusters/monopropellant-thrusters.html
(13) SMAP – Spaceflight 101, URL: https://spaceflight101.com/smap/
(14) Northrop Grumman Website, "Propulsion Products and Services," URL: https://www.northropgrumman.com/space/propulsion-products-and-services/
(15) Rafael datasheet, URL:https://www.rafael.co.il/wp-content/uploads/2022/11/SPACE-FOLDER-2022-2023.pdf
(16) Rafael, "Rafael Space Propulsion Catalogue," 2021, URL: https://www.rafael.co.il/wp-content/uploads/2021/07/RAFAEL-SPACE-PROPULSION-2021-CATALOGUE-2.pdf
(17) IHI datasheet, URL:https://www.ihi.co.jp/ia/en/products/space/satprop/index.html
(18) Werner, D., "Stellar Exploration propels EchoStar nanosatellite to valuable slot," SpaceNews, August 10, 2021, URL: https://spacenews.com/stellar-exploration-propels-echostar-nanosatellite/
(19) "CAPSTONE Launches to Test New Orbit for NASA's Artemis Moon Missions," NASA Press Release, June 28, 2022.
(20) "What is CAPSTONE," NASA website, Accessed 22 November 2023, URL: https://www.nasa.gov/smallspacecraft/capstone/
(21) Foust, J., "CAPSTONE working well more than a year after launch," SpaceNews, 24 July 2023, URL: https://spacenews.com/capstone-working-well-more-than-a-year-after-launch/
(22) Robert K. Masse and Benjamin A. Glassy, Low-Vapor-Toxicity Hydrazine Propellant Blends," Joint Meeting of the 49th Combustion (CS), 37th Airbreathing Propulsion (APS), 37th Exhaust Plume and Signatures (EPSS), and 31st Propulsion Systems Hazards

(PSHS) Subcommittees; and the 66th JANNAF Propulsion Meeting (JPM) and meeting of the Programmatic and Industrial Base (PIB), June 3 7, 2019.
(23) United States Air Force – Air Force Space Command. AFSPCMAN 91-710. "Range Safety User Requirements." 1 July 2004. Certified Current 17 September, 2013.
(24) Spores, R., Masses, R., Kimbrel, S., McLean, C. AIAA-2013-3849. "GPIM AF-M315E Propulsion System." 49th AIAA/ASME/SAE/ASEE Joint Propulsion Conference & Exhibit. San Jose.
(25) Mulkey, H., J. Miller, and C. Bacha. "Green Propellant Loading Demonstration at U.S. Range." AIAA-2016-4576. AIAA Propulsion and Energy Forum. Salt Lake City, UT. July 25-27, 2016.
(26) Dinardi, A. and Persson, M., "High Performance Green Propulsion (HPGP): A Flight-Proven Capability and Cost Game-Changer for Small and Secondary Satellites," 26th Annual AIAA/USU Conference on Small Satellites, August 2023.
(27) Friedhoff, P., K. Anflo, M. Persson, and P. Thormahlen. "Growing Constellation of Ammonium Dinitramide (ADN) Based High Performance Green Propulsion (HPGP) Systems." AIAA-2018-4754. AIAA Propulsion and Energy Forum, Cincinnati, OH, July 9-11, 2018.
(28) "Astroscale Debris-Removal Demo Makes Close Approach Despite Thruster Issues," SpaceNews, May 4, 2022.
(29) "Bradford ECAPS Thrusts Out the Firm's Successes," Satnews, 12 June 2020, URL: https://news.satnews.com/2020/06/12/bradford-ecaps-thrusts-out-the-firms-successes/
(30) VACCO Indurstries, "ArgoMoon Propulsion System," Product Brochure, URL: https://cubesat-propulsion.com/wp-content/uploads/2017/08/X17025000-data-sheet-080217.pdf
(31) eoPortal, "ArgoMoon CubeSat," Accessed 22 November 2023, URL: https://www.eoportal.org/satellite-missions/argomoon#argomoon
(32) McLean, C H, W D Deininger, B M Marotta et al. AIAA-2015-3751. "Green Propellant Infusion Mission Program Overview, Status, and Flight Operations." 51st AIAA/ASME/SAE/ASEE Joint Propulsion Conference, July 2015
(33) "NASA Satellite Set to Conclude Successful Green Propellant Demo Mission," Spaceflight Now, August 17, 2020.
(34) Smith, C., Cheek, N., Burnside, C., et. al., "The Journey of the Lunar Flashlight Propulsion System from Launch through End of Mission," 37th Annual Small Satellite Conference, August 2023.
(35) "NASA Ends Lunar Flashlight Mission Because of Thruster Problems, Space News, May 15, 2023.
(36) "Benchmark Green Propellant Thrusters Installed on Sherpa Tug," Space News, December 21, 2021.
(37) "Sherpa Tug Starts Raising Orit More Than a Month After Launch," Space News, October, 2022.
(38) "Benchmark Space Systems triples propulsion production capacity and doubles personnel to meet demand," Satellite Evolution, 5 April 2022.
(39) Zucherman, A., et. al., "Cislunar Explorers: Lessons Learned from the Development of an Interplanetary CubeSat," 2020 Small Satellite Conference, Logan, UT, SSC20-WP2-23.
(40) Muhlberger, C., "Cislunar Explorers: A Student Cubesat Demonstrating low-Cost Technology for Lunar Exploration," Lunar Exploration Analysis Group (2021).
(41) NASA Space Science Data Coordinated Archive, "Cislunar Explorers," Accessed 22 Nov 2023, URL: https://nssdc.gsfc.nasa.gov/nmc/spacecraft/display.action?id=CISLUNEXP
(42) "Pathfinder Technology Demonstrator, Demonstrating Novel CubeSat Technologies in Low-Earth Orbit", February 2,2021, URL: https://www.nasa.gov/smallspacecraft/pathfinder-technology-demonstrator/

(43) Messier, D. "NASA Selects Tethers Unlimited's HYDROS-C Thruster for First PTD CubeSat Mission." http://www.parabolicarc.com/2018/06/30/nasa-selects-tuis-hydrosc-thruster-ptd-CubeSat-mission/

(44) Satnews, "Rocket Fuel You Can Drink?" 26 January 2021.

(45) Porter, A., Freedman, M., Grist, R., Wesson, C., and Hanson, M., "Flight Qualification of a Water Electrolysis Propulsion System," 35th Small Satellite Conference, Logan, UT, Paper No: SSC21-XI-06, 2021.

(46) NanoAvionics. "CubeSat Propulsion System EPSS." https://nanoavionics.com/CubeSat-components/CubeSat-propulsion-system-epss/

(47) NanoAvionics, "A Successful In-Orbit Test of the First Ever Chemical Propulsion System Running On-Board a CubeSat Was Performed," Press Release, 21 July 2017.

(48) Robert K. Masse, Ronald A. Spores, May Allen, "AF-M315E Advanced Green Propulsion – GPIM and Beyond," AIAA Propulsion & Energy Forum, Virtual Forum, August 24-26, 2020.

(49) Rubicon Space Systems., "Thrusters," Product Brochure, URL: https://www.rubicon.space/products-cat/1/thrusters

(50) CU Aerospace. "Monopropellant Propulsion Unit for CubeSats (MPUC) System", Company Website, Datasheet, July 2023, https://cuaerospace.com/products-services/space-propulsion-systems/monopropellant-propulsion-unit-mpuc

(51) King, D., Woodruff, C., Camp, J., and Carroll, D. "Development and Testing of a Low Flame Temperature, Peroxide-Alcohol-Based Monopropellant Thruster." 35th Small Satellite Conf., Paper # SSC21-S1-41, Logan, UT, Aug. 6-9, 2021.

(52) Dawn Aerospace Website, "Green Cubesat Propulsion," URL: https://www.dawnaerospace.com/green-propulsion-cubedrives

(53) VACCO Industries, "Integrated Propulsion System," Product Brochure, URL: https://www.CubeSat-propulsion.com/wp-content/uploads/2019/08/Integrated-Propulsion-System-datasheet-080519.pdf

(54) Cardin, J., Schappell, T., and Day, C., "Testing of a Green Monopropellant Integrated Propulsion System," 34th Annual Small Satellite Conference, SSC20-IX-02, Logan, UT, Aug 2020.

(55) Whitmore , S A, S L Merkley, Z S Spurrier, and S D Walker. 2015. "Development of a Power Efficient, Restartable, 'Green' Propellant Thruster for Small Spacecraft and Satellites." 29th Annual AIAA/USU Conference on Small Satellites.

(56) Bulcher, A. M. and S. A. Whitmore. "A Green Hybrid Thruster Using Moderately Enriched Compressed Air as the Oxidizer." AIAA-2018-4841. AIAA Propulsion and Energy Forum. Cincinnati, OH. July 9-11, 2018. doi: 10.2514/6.2018-4841

(57) Whitmore, S., Babb, R., Gardner, T., Lloyd, K., Stephens, J.," Pyrolytic Graphite and Boron Nitride as Low-Erosion Nozzle Materials for Long-Duration Hybrid Rocket Testing". AIAA-2020-3740, AIAA Propulsion and Energy Forum, 2020.

(58) Whitmore, S., Babb, R., Stephens, J.,Horlacher, J., "Further Development of Low-Erosion Nozzle Materials for Long-Duration Hybrid Rocket Burns". AIAA Propulsion and Energy Forum, 2021.

(59) Jens, E., Karp, A., Williams, K., Nakazono, B., Rabinovitch, J., Dyrda, D., and Mechentel, F. "Low Pressure Ignition Testing of a Hybrid SmallSat Motor." AIAA-2019-4009, AIAA Propulsion and Energy Forum, 2019. https://doi.org/10.2514/6.2019-4009.

(60) Rabinovitch, J., Jens, E, Karp, A, Nakazono, N, Conte, A, Vaughan, D., "Characterization of PolyMethylMethAcrylate as a Fuel for Hybrid Rocket Motors". AIAA-2018-4530, AIAA Propulsion and Energy Forum, 2018.

(61) Jens, E., Karp, A., Rabinovitch, J., Nakazono, B., Conte, A., Vaughan, D." Design of Interplanetary Hybrid CubeSat and SmallSat Propulsion Systems." AIAA-2018-4668, AIAA Propulsion and Energy Forum, 2018.

(62) Dyrda, D., Mechentel, F., Cantwell, B., Karp, A., Rabinovitch, J., Jens, E., "Diode Laser Ignition Testing for PMMA/GOX Hybrid Motors." AIAA-2019-4095, AIAA Propulsion and Energy Forum, 2019.

(63) Simurda, L., Zilliac, G. and Zaseck, C., "High Performance Hybrid Propulsion System for Small Satellites," AIAA-2013-3635, 49th AIAA/ASME/SAE/ASEE Joint Propulsion Conference, 2013.

(64) Simurda, L, Zilliac, G., "Continued Testing of a High Performance Hybrid Propulsion System for Small Satellites", AIAA-2015-4201, AIAA Propulsion and Energy Forum, 2015.

(65) Desain, J., Brady, B., Curtiss, T., Hsu-Schouten, A., Fuller, J., Boyer, J.E., McKnight, B., Cortopassi, A., "Advanced Hybrid Rocket Motor Propulsion Unit for CubeSats(PUC)". Aerospace Report No. TOR-2015-03522.

(66) ROMBUS Rapid Orbital Mobility Bus." Parabilis Space Technologies, datasheet

(67) Nano OTS Nanosatellite Orbital Transfer System", Parabilis Space Technologies, datasheet

(68) Hinkley, D. "A Novel Cold Gas Propulsion System for Nanosatellites and Picosatellites." AIAA, 22nd Annual Conference on Small Satellites, SSC08-VII-7.

(69) Schappell, D T, E Scarduffa, P Smith, and N Solway. AIAA-2005-4055. "Advances in Marotta Electric and Satellite Propulsion Fluid Control Activities." 41st AIAA/ASME/SAE/ASEE Joint Propulsion Conference.

(70) Marotta, "Cold Gas Microthruster," Datasheet, Accessed 23 Nov 2023, URL: https://marotta.com/products/flow-controls/satellite-propulsion-controls/cold-gas-microthruster/

(71) Pauliukonis, R S. 2017. "Fuel system comprising sulfur hexafluoride and lithium containing fuel." https://patents.google.com/patent/US3325318.

(72) Bonin, G, N Roth, S Armitage, J Newman, B. Risi, and R. Zee., "CanX-4 and CanX-5 Precision Formation Flight: Mission Accomplished!" 29th Annual AIAA/USU Conference on Small Satellites, 2015.

(73) Newman, J and Zee, R. E. "Drift Recovery and Station Keeping Results for the Historic CanX-4/CanX-5 Formation Flying Mission." 29th Annual AIAA/USU Conference on Small Satellites, Logan, UT, 2015.

(74) Risi, B.W. "Propulsion System Development for the CanX-4 and CanX-5 Dual Nanosatellite Formation Flying Mission." Master of Applied Science Thesis, University of Toronto, 2014.

(75) Manzoni, G, and Y L Brama. 2015. "CubeSat Micropropulsion Characterization in Low Earth Orbit." 29th Annual AIAA/USU Conference on Small Satellites.

(76) GOMSpace. "Attitude and Orbit Control Systems." https://gomspace.com/shop/subsystems/attitude-orbit-control-systems/default.aspx

(77) "GomX-4," https://directory.eoportal.org/web/eoportal/satellite-missions/g/gomx-4.

(78) Kvell, U, M Puusepp, F Kaminski, and et al. 2014. "Nanosatellite orbit control using MEMS cold gas thrusters." Proceedings of the Estonian Academy of Sciences, 63, 2S, 279–285.

(79) Palmer, K., Sundqvist, J., Salaverri, A., Gronland, T., Li, Z., and Wu, S.. "In-Orbit Demonstration of a MEMS-based Micropropulsion System for CubeSats." Small Satellite Conference, August 2016.

(80) National Aeronautics & Space Administration. "BioSentinel." Fact Sheet. https://www.nasa.gov/sites/default/files/atoms/files/biosentinel_fact_sheet-16apr2019_508.pdf

(81) Lightsey, E. G., T. Stevenson, and M. Sorgenfrei. "Development and Testing of a 3-D-Printed Cold Gas Thruster for an Interplanetary CubeSat." Proceedings of the IEEE, Vol. 106, No. 3, March, 2018. doi: 10.1109/JPROC.2018.2799898

(82) Napoli, M., et al, "BioSentinel: Mission summary and lessons learned from the first deep space biology CubeSat mission," 37th Small Satellite Conference, Logan, UT, Aug 2023, SSC23-I-02.
(83) Rutter, D., "NASA Extends BioSentinel's Mission to Measure Deep Space Radiation," NASA Website, 7 Aug 2023, URL: https://www.nasa.gov/image-article/nasa-extends-biosentinels-mission-to-measure-deep-space-radiation/
(84) Editorial, "Iodine Impulse for SmallSats Demo'd On-Orbit by ThrustMe and Spacety," SmallSat News, 25 November 2019.
(85) Werner, D., "Thrustme, Spacety report initial success of cold gas thruster," SpaceNews, 23 November, 2019.
(86) Martinez, J., Rafalskyi, D., Aanesland, A., Laurand, X., Martinez, S., and Quinsac, G., "An off-axis iodine propulsion system for the Robusta-3A mission, 34th Small Satellite Conference, SSC20-IX-03, 2020.
(87) ThrustMe, "I2T5 Cold Iodine Thruster," Company Datasheet, Version 02/21, 2021.
(88) Bowen, J, M Villa, and A Williams. 2015. "CubeSat based Rendezvous, Proximity Operations, and Docking in the CPOD Mission." 29th Annual AIAA/USU Conference on Small Satellites.
(89) VACCO Industries. "NASA C-POD Micro CubeSat Propulsion System." Product Brochure. URL: https://cubesat-propulsion.com/reaction-control-propulsion-module/
(90) Spiegel, I., Zhou, B., Goodloe, R., Fox, B., and DiMateo, J., "CubeSat Proximity Operations Demonstration (CPOD) Mission Results," 37th Annual Small Satellite Conference, Logan, UT, Aug 2023, SSC23-XI-01.
(91) Klesh, A, and J Krajewski. 2015. "MarCO: CubeSats to Mars in 2016." 29th Annual AIAA/USU Conference on Small Satellites.
(92) Jet Propulsion Laboratory, National Aeronautics and Space Administration. "Mars Cube One (MarCO)." URL: https://www.jpl.nasa.gov/news/press_kits/insight/launch/appendix/mars-cube-one/
(93) VACCO Industries. "JPL MarCO – Micro CubeSat Propulsion System." Product Brochure. URL: https://cubesat-propulsion.com/jpl-marco-micro-propulsion-system/
(94) Klesh, A., Baker, J., and Krajewski, J., "MarCO: Flight Review and Lessons Learned," 33rd Annual Small Satellite Conference, Logan, UT, 2019, SSC19-III-04.
(95) NEA Scout Propulsion System, Datasheet, VACCO Industries, July 2017
(96) Lockett, T., Castillo-Rogez, J., Johnson, L., Lightholder, J., and Few, A., "Near-Earth Asteroid Scout Flight Mission," IEEE A&E Systems Magazine, March 2020, DOI. No. 10.1109/MAES.2019.2958729
(97) "NEA Scout," NASA website, Accessed 24 Nov 2023, URL: https://science.nasa.gov/mission/nea-scout/
(98) Northrup Grumman Propulsion Product Catalog, April 2016.
(99) BusinessWire.com. 2017. "PacSci EMC Demonstrates First Ever Successful Orbital Maneuvers and Orbit Raising of a CubeSat Using a Commercial Solid Rocket Motor Array." https://www.businesswire.com/news/home/20170925006504/en/PacSci-EMC-Demonstrates-Successful-Orbital-Maneuvers-Orbit
(100) Terran Orbital Mission Summary, "PACSCISAT," Accessed 24 Nov 2023, URL: https://terranorbital.com/missions/pacscisat/
(101) Hartwig, J.W. "Propellant Management Devices for Low Gravity Fluid Management: Past, Present, and Future Applications." AIAA Journal of Spacecraft and Rockets Vol. 54, No. 4, 808 – 824. 2017.
(102) W. Tam, G. Kawahara, K. Wlodarczyk, "Review of ATK Diaphragm Tanks-An Update", Space Propulsion 2018, SP2018_00024.

(103) Carpenter, C, D Schmuland, J Overly, and R Masse. 2014. "Test Results for the MPS-120 and MPS-130 CubeSat Propulsion Systems." 28th Annual AIAA/USU Conference on Small Satellites. Logan.
(104) Aerojet Rocketdyne. “Modular Propulsion Systems.” Product Brochure. https://www.rocket.com/sites/default/files/documents/CubeSat%20Mod%20Prop-2sided.pdf
(105) Stellar Exploration, “Propulsion for Small Spacecraft.” Product Brochure.
(106) Aerospace Corporation. “Propelling the Field of SmallSats Forward.” https://aerospace.org/article/propelling-field-small-sats-forward
(107) Rainbow, J., “Benchmark Space Systems unveils in-space mobility-as-a-service business,” SpaceNews, 23, June, 2021.
(108) Gagne, K. R., M. R. McDevitt, and D. L. Hitt. "A Dual Mode Propulsion System for Small Satellite Applications." Aerospace. Vol. 5, No. 52. 2018. doi: 10.3390/aerospace5020052
(109) Benchmark Space Systems. “Products.” Product Datasheets, URL: https://www.benchmarkspacesystems.com/products
(110) Anflo, K. and B. Crowe. “In-Space Demonstration of an ADN-based Propulsion System.” AIAA-2011-5832. 47th AIAA/ASME/SAE/ASEE Joint Propulsion Conference & Exhibit. San Diego, CA. July 31 – August 3, 2011.
(111) “SkySat: Quick Facts,” eoPortal, July 26, 2016.
(112) Bradford ECAPS. “High Performance Green Propulsion.” Product Brochure. https://www.ecaps.space/assets/pdf/Bradford_ECAPS_Folder_2017.pdf
(113) Bradford ECAPS. “1N GP Thruster.” https://www.ecaps.space/products-1ngp.php
(114) Busek Co., Inc. “BGT-X5 Green Monopropellant Thruster.” Product Brochure.
(115) Busek Co., Inc. “Green Monopropellant Thrusters.” http://busek.com/technologies__greenmonoprop.htm
(116) Tsay, M., C. Feng, and J. Zwahlen. “System-Level Demonstration of Busek’s 1U CubeSat Green Propulsion Module ‘AMAC’.” AIAA-2017-4946. 53rd AIAA/SAE/ASEE Joint Propulsion Conference. Atlanta, GA. 10-12 July, 2017.
(117) Moog, Inc. “Monopropellant Propulsion Module.” Product Brochure. https://www.moog.com/content/dam/moog/literature/sdg/space/propulsion/moog-monopropellant-propulsion-module-datasheet.pdf
(118) Rubicon Space Systems., “Sprite Propulsion System,” Product Brochure, URL: https://www.rubicon.space/product/2/sprite-propulsion-system
(119) Rubicon Space Systems., “Phantom Propulsion System,” Product Brochure, URL: https://www.rubicon.space/product/20/phantom-propulsion-system
(120) Tethers Unlimited. “Hydros-C” Product Brochure. URL: https://www.tethers.com/wp-content/uploads/2021/01/HYDROS-C-1.pdf
(121) NASA, 2015, “Pathfinder Technology Demonstrator.” National Aeronautics & Space Administration.
(122) Tethers Unlimited. “Hydros-M” Product Brochure.
(123) VACCO Industries. “Green Propulsion System.” Product Brochure. https://www.CubeSat-propulsion.com/wp-content/uploads/2019/08/X19041000-Green-Propulsion-System-data-sheet-073019.pdf
(124) Benchmark Space Systems., “Products”., Accessed Sept 2023. URL: https://www.benchmarkspacesystems.com/products#slide-1-6
(125) Jens, E., Karp, A., Rabinovitch, J., Nakazono, B., Conte, A., Vaughan, D.” Design of Interplanetary Hybrid CubeSat and SmallSat Propulsion Systems.” AIAA-2018-4668, AIAA Propulsion and Energy Forum, 2018.
(126) GOMspace. “NanoProp CGP-3.” Product Brochure. https://gomspace.com/UserFiles/Subsystems/flyer/gomspace_nanoprop_cgp3.pdf

(127) GOMspace. "NanoProp 6U." Product Brochure. https://gomspace.com/UserFiles/Subsystems/flyer/gomspace_nanoprop_cgp6_flyer.pdf
(128) VACCO Industries. "Standard Propulsion System." Product Brochure. URL: https://cubesat-propulsion.com/standard-propulsion-system/
(129) Moog, "Cold Gas Thrursters," Moog Datasheet, 2021
(130) Thrasher, J. and M. McPherson. "Design and Qualification of DSSP's CubeSat Delta-V Motor (CDM-1)." Digital Solid State Propulsion, Inc. Poster presented at Small Satellites Conference. Logan, UT. 2015. https://digitalcommons.usu.edu/cgi/viewcontent.cgi?article=3261&context=smallsat
(131) Industrial Solid Propellant, Inc. "Product Portfolio." Product Brochure. http://www.specificimpulse.com/
(132) Northrop Grumman, "Propulsion Products Catalog." Product Brochure. 2016, URL: https://cdn.prd.ngc.agencyq.site/-/media/wp-content/uploads/NG-Propulsion-Products-Catalog.pdf
(133) Jahn, R. G. and Choueiri, E. Y. "Electric Propulsion." Academic Press, Encyclopedia of Physical Science and Technology, Third Edition, Volume 5, 2002.
(134) Goebel, D. M. and Katz, I. "Fundamentals of Electric Propulsion: Ion and Hall Thrusters." California Institute of Technology, Jet Propulsion Laboratory, JPL Space Science and Technology Series, 2008.
(135) Bradford Space, "Comet: Water-based Propulsion for Small Satellites." Company Website, Datasheet, Copyright 2019.
(136) Sarda, K., Roth, N., Zee, R. E., Cajacob, D., and Orr, N. G. "Making the Invisible Visible: Precision RF-Emitter Geolocation from Space by the HawkEye 360 Pathfinder Mission." 32nd AIAA/USU Conference on Small Satellites, Logan, UT, 2018.
(137) eoPortal, "HawkEye 360 Pathfinder Cluster Mission to identify RFI locations." June, 2020. URL: https://directory.eoportal.org/web/eoportal/satellite-missions/h/hawkeye
(138) Bradford Space, "Missions," Company website, Accessed Sept. 2023, URL: https://www.bradford-space.com/missions
(139) BlackSky Global, "Global-4 Orbital Debris Assessment Report (ODAR)." FCC Licensing, 18 December, 2018.
(140) CU Aerospace, "Propulsion Unit for CubeSats." Company Website, URL: https://cuaerospace.com/products-services/space-propulsion-systems/propulsion-unit-for-cubesats-puc, Datasheet, 2020.
(141) Carroll, D. L., Cardin, J. M., Burton, R. L., Benavides, G. F., Hejmanowski, N., Woodruff, C., Bassett, K., King, D., Laystrom-Woodard, J., Richardson, L., Day, C., Hageman, K., and Bhandari, R. "Propulsion Unit for CubeSats (PUC)." 62nd JANNAF Propulsion Meeting, Nashville, TN, 1-5 June, 2015, Paper Tracking # 4059, 2015, Dist. A.
(142) Burton, R.L., Benavides, G.F., and Carroll, D.L. "Space Thruster using Robust Microcavity Discharge." U.S. Patent No. 9,242,747 B1, 2016.
(143) Woodruff, C., Carroll, D., King, D., Burton, R., and Hejmanowski, N. "Monofilament Vaporization Propulsion (MVP) – CubeSat propulsion system with inert polymer propellant." Small Satellite Conf., Paper # SSX18-III-09, Logan, UT, Aug. 6-9, 2018.
(144) CU Aerospace, "Monofilament Vaporization Propulsion System Solid Inert Polymer Propellant." Company Website, URL: https://cuaerospace.com/products-services/space-propulsion-systems/monofilament-vaporization-propulsion-mvp, Datasheet, 2020.
(145) Woodruff, C., Parta, M., King, D., Woodruff, A., Burton, R., and Carroll, D. "Monofilament Vaporization Propulsion (MVP) Flight-like System Performance." 37th International Electric Propulsion Conference (IEPC) 2022, Boston, Massachusetts, Paper # IEPC 2022-575, 2022.

(146) CU Aerospace, "Space Systems Hardware: Hardware for the Future," Company website, Accessed Sept. 2023, URL: https://cuaerospace.com/technologies/space-systems-hardware
(147) Steam TunaCan Thruster," SteamJet company website, Accessed Sept. 2023, URL: https://steamjet.space/
(148) Steam Thruster One," SteamJet company website, Accessed Sept. 2023, URL: https://steamjet.space/
(149) Woodruff, C., King, D., Burton, R., Bowman, J., and Carroll, D. "Development of a Fiber-Fed Pulsed Plasma Thruster for Small Satellites." Small Satellite Conference, Logan, UT, Paper # SSC19-WKVIII-06, 2019.
(150) Woodruff, C., King, D., Burton, R., and Carroll, D. "Fiber-fed Pulsed Plasma Thruster for Small Satellites." 36th International Electric Propulsion Conference (IEPC) 2019, Vienna, Austria, Paper # IEPC 2019-A899, 2019.
(151) Woodruff, C., King, D., Burton, R., and Carroll, D. "Fiber-Fed Advanced Pulsed Plasma Thruster (FPPT)." U.S. Patent # 10,570,892, Feb. 25, 2020, and 11,242,844, Feb. 8, 2022.
(152) Woodruff, C., Parta, M., King, D., Burton, R., and Carroll, D. "Fiber-Fed Pulsed Plasma Thruster (FPPT) with Multi-axis Thrust Vectoring." 37th International Electric Propulsion Conference (IEPC) 2022, Boston, Massachusetts, Paper # IEPC 2022-558, 2022.
(153) CU Aerospace, "Fiber-Fed Pulsed Plasma Thruster System Solid Inert Polymer Propellant." Company Website, URL: URL: https://cuaerospace.com/products-services/space-propulsion-systems/fiber-fed-pulsed-plasma-thruster-fppt, Datasheet, 2022.
(154) CU Aerospace, "Space Systems Hardware: Hardware for the Future", Accessed Nov. 2023. URL: https://cuaerospace.com/technologies/space-systems-hardware
(155) "NASA Funds Long-Standing Partners in CubeSat Development." Press Release, University of Illinois at Urbana/Champaign, Aerospace Engineering, 9 Oct., 2019.
(156) Aurora Propulsion Technologies, "ARM-A Datasheet." Company Website, URL: https://aurorapt.fi/thrusters/, Datasheet, Copyright May, 2021.
(157) Aurora Propulsion Technologies, "Deorbiting Old Satellites & Used Rockets with Aurora Plasma Brake." Company Website, URL: https://aurorapt.fi/deorbiting/#plasmabrake, Brochure, Copyright March, 2022
(158) "Aurora Successfully Launches First Sustainable Propulsion Tech Demonstration," Aurura Propulsion Technologies, Press Release, May 3, 2022.
(159) "The Flying Object (AuroraSat-1, Aurora CubeSat)," Nanosats Database, URL: https://www.nanosats.eu/sat/aurorasat-1, Updated July 23, 2022.
(160) Spacewatch.Global, "Space BD Successfully Deploys OPTIMAL-1 into Orbit," Feb. 3, 2023, URL: https://spacewatch.global/2023/02/space-bd-successfully-deploys-optimal-1-into-orbit/
(161) Nakagawa, Y., Iwakawa, A., Kameyama, and S., Kikuchi, T., "On-orbit Demonstration of the Water Resistojet Propulsion System on Commercial 6U-Sat Sphere-1 EYE," 37th Annual Small Satellite Conference, Logan, UT, August 2023, SSC23-VI-02.
(162) "Collision Avoidance for CubeSats and SmallSats with ARM-C," Aurora Brochure, Company Website, Dated March 21, 2022.
(163) Ziemer, J., Marrese-Reading, Colleen, Cutler, C., Dunn, C., Romero-Wolf, A., Javidnia, S., Le, T., Li, I., Barela, P., Demmons, N., Hruby, V., Slutsky, J., Thorpe, J., Maghami, P., Hsu, O., and O'Donnell, J. "In-Flight Verification and Validation of Colloid Microthruster Performance." AIAA Propulsion and Energy Forum, Cincinnati, OH, AIAA-2018-4643.
(164) Krejci, D., Reissner, A., Seifert, B., Jelem, D., Horbe, T., Plesescu, F., Friedhoff, P., and Lai, S. "Demonstration of the IFM Nano FEEP Thruster in Low Earth Orbit." 4S Symposium, 2018.

(165) Krejci, D., Reissner, A., Schonherr, T., Seifert, B., Saleem, Z., and Alejos, R. "Recent Flight Data from IFM Nano Thrusters in a Low Earth Orbit." 36th IEPC, Vienna, Austria, IEPC-2019-A724.
(166) Amos, J. "ICEYE's Small Radar Satellites Achieve Big Capability." BBC News, 6 May, 2020, URL: https://www.bbc.com/news/science-environment-52560809
(167) Werner, D. "Key Hurdle Cleared for York Space Systems and U.S. Army Small Satellite Launch." SpaceNews, 22 February, 2018, URL: https://spacenews.com/key-hurdle-cleared-for-york-space-systems-and-u-s-army-small-satellite-launch/
(168) Clark, S. "Rocket Lab Deploys Experimental U.S. Military SmallSats on First Night Launch." Spaceflight Now, 5 May, 2019. URL: https://spaceflightnow.com/2019/05/05/rocket-labs-deploys-experimental-u-s-military-SmallSats-on-first-night-launch/
(169) Wörner, N. "Sensor Networks of Satellites." ElektronikNet, 26 May, 2020. URL: https://www.elektroniknet.de/international/sensor-networks-of-satellites-176831.html
(170) "NetSat – Pioneering Research in Formation Control", website accessed June 2021, https://www.telematik-zentrum.de/en/projects/netsat/
(171) D. Krejci and A. Reissner, "The First 100 FEEP Propulsion Systems in Space: A Statistical View and Lessons Learnt of 4 Years of ENPULSION", IEPC-2022-199, 37th International Electric Propulsion Conference, Cambridge, MA, 19-23 June 2022.
(172) Grimaud, L., Schonherr, T., Vasiljevich, I., Reissner, A., Krejci, D., and Seifert, B., "Qualification status update of the MICRO R3 and NANO R3 FEEP thrusters," IEPC-2022-200, 37th International Electric Propulsion Conference, Cambridge, MA, 19-23 June 2022.
(173) "First Use of the Enpulsion Micro R3 Thruster in the GMS-T Mission", Space Daily, 18 March 2021.
(174) Bangert, P., Kramer, A., and Schilling, K. "UWE-4: Integration State of the First Electrically Propelled 1U CubeSat." 31st Annual AIAA/USU Conference on Small Satellites, SSC17-WK-47, 2017.
(175) Kramer, A., Bangert, P., and Schilling, K. "Hybrid Attitude Control On-Board UWE-4 Using Magnetorquers and the Electric Propulsion System NanoFEEP." 33rd Annual AIAA/USU Conference on Small Satellites, SSC19-WKI-02, 2019.
(176) D. Werner, "Miniature Thruster Propels Single CubeSat to Lower Orbit", SpaceNews, 22 July 2020, https://spacenews.com/uwe-4-lowers-altitude/
(177) N. Arthur. "Ion Thruster Produced Roll Torque." AIAA-2019-4166.
(178) N. Wallace et al., "The GOCE Ion Propulsion Assembly – Lessons Learnt from the First 22 Months of Flight Operations", IEPC-2011-327.
(179) "ESA's GOCE Mission Comes to an End", ESA press release #33-2013, 21 Oct 2013, https://www.esa.int/Newsroom/Press_Releases/ESA_s_GOCE_mission_comes_to_an_end
(180) C.H. Edwards et al., "The T5 Ion Propulsion Assembly for Drag Compensation on GOCE", 2nd International GOCE User Workshop, ESA-ESRIN Frascati Italy, 8-10 March 2004.
(181) P.N. Randall et al., "T5 Performance, Industrialization and Future Applications", IEPC-2019-688.
(182) A. Jones, "French Startup Demonstrates Iodine Propulsion in Potential Boost for Space Debris Mitigation Efforts", SpaceNews, 22 January 2021, https://spacenews.com/french-startup-demonstrates-iodine-propulsion-in-potential-boost-for-space-debris-mitigation-efforts/
(183) Rafalskyi, D., Martínez, J.M., Habl, L. et al. In-orbit demonstration of an iodine electric propulsion system. Nature 599, 411–415 (2021). https://doi.org/10.1038/s41586-021-04015-y

(184) "ThrustMe NPT30-I2 iodine electric propulsion system launched on board the NorSat-TD satellite", TURUSTME web post,17 April 2023, URL: https://www.thrustme.fr/post/89-thrustme-npt30-i2-iodine-electric-propulsion-system-launched-on-board-the-norsat-td-satellite
(185) "Trial Integration Completed By Norwegian Space Agency Of ThrustMe's Propulsion System For NorSat-TD", SatNews, 27 April 2021, https://news.satnews.com/2021/04/27/trial-integration-completed-by-norwegian-space-agency-of-thrustmes-propulsion-system-for-norsat-td/
(186) "TrustMe exceeds 100 orders for propulsion systems", TURUSTME web post, 25 May 2023, URL: https://www.thrustme.fr/post/91-thrustme-exceeds-100-orders-for-propulsion-systems
(187) "ThrustMe selected to provide Turion Space with vital propulsion capacities", TURUSTME web post, 4 May, 2023, URL: https://www.thrustme.fr/post/90-thrustme-selected-to-provide-turion-space-with-vital-propulsion-capacities
(188) Malphrus, B. "The Lunar IceCube EM-1 Mission: Prospecting the Moon for Water Ice." IEEE Aerospace and Electronic Systems Magazine, Vol 34, Issue 4, DOI: 10.1109/MAES.2019.2909384.
(189) Clark, P. "Lunar Ice Cube Mission: Determining Lunar Water Dynamics with a First Generation Deep Space CubeSat." 47th Lunar and Planetary Science Conference, 2016.
(190) Foust, J., "Deep space smallsats face big challenges," Spacenews, 17 February 2023, URL: https://spacenews.com/deep-space-smallsats-face-big-challenges/
(191) Hardgrove, C. "The Lunar Polar Hydrogen Mapper CubeSat Mission." IEEE Aerospace and Electronic Systems Magazine, Vol 35, Issue 3, DOI: 10.1109/MAES.2019.2950747.
(192) Morton, Erin, "NASA's LunaH-Map Mission Ends, Validates Science Instrument Performance," NASA blog, 3 August 2023, URL: https://blogs.nasa.gov/lunah-map/2023/08/03/nasas-lunah-map-mission-ends-validates-science-instrument-performance/
(193) Foust, J., "Artemis 1 cubesate nearing end of mission," SpaceNews, 2 May 2023, URL: https://spacenews.com/artemis-1-cubesat-nearing-end-of-mission/
(194) eoPortal, "SMART-1 (Small Mission for Advanced Research in Technology." June, 2020. URL: https://directory.eoportal.org/web/eoportal/satellite-missions/s/smart-1
(195) eoPortal, "TacSat-2 / Roadrunner." June 2020. URL: https://earth.esa.int/web/eoportal/satellite-missions/content/-/article/tacsat2
(196) de Grys, K., Mathers, A., Welander, B., and Khayms, V. "Demonstration of 10,400 Hours of Operation on a 4.5 kW Qualification Model Hall Thruster." 46th AIAA Joint Propulsion Conference, Nashville, TN, July, 2010.
(197) Sheetz, M. and Petrova, M. "Why in the next decade companies will launch thousands more satellites than in all of history." CNBC, 15 December, 2019. URL: https://www.cnbc.com/2019/12/14/spacex-oneweb-and-amazon-to-launch-thousands-more-satellites-in-2020s.html
(198) Gorbunov, A., Khodnenko, V., Khromov, A., Murashko, V., Koryakin, A., Zhosan, V., and Grikhin, G. "Vernier Propulsion System for Small Earth Remote Sensing Satellite "Canopus-V." 32nd IEPC, Wiesbaden, Germany, 2011.
(199) eoPortal, "Kanopus-V 1 (Kanous-Vulkan N1, Environmental Satellite)." June, 2020. URL: https://directory.eoportal.org/web/eoportal/satellite-missions/k/kanopus-v-1
(200) eoPortal, "Kanopus-V-IK 1 (Kanous-Vulkan-Infra-Krasny-1)." June, 2020. URL: https://directory.eoportal.org/web/eoportal/satellite-missions/k/kanopus-v-ik-1
(201) EDB Fakel, "Stationary Plasma Thrusters." Company Website, Datasheet, 2020.
(202) Gnizdor, R., Markov, A., Mitrofanova, O., and Semenenko, D. "The research of the modified SPT-70 thruster parameters and characteristics." 36th IEPC, Vienna, Austria, 2019.

(203) Hruby, P., Demmons, N., Courtney, D., Tsay, M., Szabo, J., and Hruby, V. "Overview of Busek Electric Propulsion." 36th IEPC, Vienna, Austria, 2019.
(204) Busek, "BHT-200 Busek Hall Effect Thruster." Company Website, Datasheet, 2019.
(205) Hruby, V., Monheiser, J., Pote, B., Freeman, C., and Connolly, W. "Low Power, Hall Thruster Propulsion System." IEPC-99-092, Kitakyushu, Japan, 1999.
(206) Kamhawi, H., Haag, T., Benavides, G., Hickman, T., Smith, T., Williams, G., Myers, J., Polzin, K., Dankanich, J., Byrne, L., Szabo, J., and Lee, L. "Overview of Iodine Propellant Hall Thruster Development Activities at NASA Glenn Research Center." 52nd AIAA Joint Propulsion Conference, Salt Lake City, Utah, 2016.
(207) NASA TechPort, "Iodine Satellite (iSAT)." June, 2020. URL: https://techport.nasa.gov/view/91492
(208) Lev, D., Zimmerman, R., Shoor, B., Appel, L., Ben-Ephraim, M., Herscovitz, J., and Epstein, O. "Electric Propulsion Activities at Rafael in 2019." 36th IEPC, Vienna, Austria, 2019.
(209) Herscovitz, J., and Karnieli, A. "VENuS Program: Broad and New Horizons for Super-Spectral Imaging and Electric Propulsion Missions for a Small Satellite." 22nd AIAA/USU Conference on Small Satellites, Logan, Utah, 2008.
(210) Herscovitz, J., Appel, L., Barnett, D., Baron, D., Davidson, A., Gontmacher, P., Kedem, M., Lev, D., Merenstein, A., Rbinovich, L., Reiner, D., Salama, O., Amit-Shapira, Y., Shechter, Y., Shoor, B., Warshavsky, A., and Zhuravel, N. "VENuS – A Novel Technological Mission Using Electric Propulsion." 35th IEPC, Atlanta, Georgia, 2017.
(211) Herscovitz, J., Lev, D., Shoor, B., Katz-Franco, D., Berkman, S., Baron, D, and Adler, S. "VENuS – Updates on Technological Mission using the Israeli Hall Effect Thruster (IHET)." 36th IEPC, Vienna, Austria, 2019.
(212) eoPortal "VENuS (Vegetation and Environment monitoring on a New Microsatellite)." June, 2020. URL: https://directory.eoportal.org/web/eoportal/satellite-missions/v-w-x-y-z/venus
(213) Misuri, T., Ducci, C., Gregucci, S., Pedrini, D., Cannelli, F., Cesari, U., Nania, F., Vicini, A., Pace, G., Magistro, F., Cenni, J., Dignani, D., Farauanu, C., Quaranta, V., Tiseo, B., Orefice, M., and Bruno, G. "SITAEL HT100 Thruster Unit, Full Ground Qualification." 36th IEPC, Vienna, Austria, 2019.
(214) Tambini, A., Antonini, F., Melega, N., Mariotti, G., Centonze, V., and Gabriele, A. "uHETsat small Power System for high demanding payloads." European Space Power Conference, 2019.
(215) SITAEL "Electric Propulsion – HT100 Hall Effect Thruster Product Sheet." Company Website, Datasheet, 2020.
(216) ESA, "Compact electric thruster cleared for space firing," ESA website, Press release, 27 June 2023, URL: https://www.esa.int/Enabling_Support/Space_Engineering_Technology/Compact_electric_thruster_cleared_for_space_firing
(217) SITAEL, "Earth Observation," Company website, Accessed 26 Nov 2023, URL: https://www.sitael.com/space/references-and-customers/earth-observation/
(218) Astra, "Astra Spacecraft Engine." Company Website, Accessed Nov 2023, URL: https://astra.com/space-products/astra-spacecraft-engine/
(219) Spaceflight, "Sherpa Program: New Orbital Transfer Vehicles Launch Smallsats to Custom Orbital Destination," Spaceflight Website, URL: https://spaceflight.com/sherpa/#sherpalte, Accessed October 2022.
(220) Gill, M., Martinez, R., Zannos, A., Bailey, T., Cassidy, M., Cooney, J., Crawford, A., Cuadra, J., Delgado, J., Fuller, J., Haverty, M., Hopkins, M., Keyes, E., Lee, L, Massey, D., McClellan, K., Pittman, J., Schmidt, M., Shiyani, D., and Sorel, T., "On-Orbit Data and

Validation of Astra's ACE Electric Propulsion System," 36th Annual Small Spacecraft Conference, Logan, UT, Paper No.: SSC22-S2-08, 2022.
(221) Astra, "On-Orbit Validation of the Astra Spacecraft Engine," Press Release, 22 Aug 2022, URL: https://astra.com/news/on-orbit-validation-of-the-astra-spacecraft-engine/
(222) Foust, J., "Astra to sell electric thrusters to Airbus OneWeb Satellites," SpaceNews, August 30, 2022.
(223) Foust, J., "Astra wins order for electric thrusters from LeoStella," SpaceNews, April 12, 2022.
(224) Foust, J., "Astra, Exotrail win satellite propulsion orders," SpaceNews, 28 April 2023, URL: https://spacenews.com/astra-exotrail-win-satellite-propulsion-orders/
(225) Astra, "Astra Announces Spacecraft Engine Contract with Astroscale," Press Release, 13 Oct 2022, URL: https://astra.com/news/spacecraft-engine-contract-with-astroscale/
(226) Astra, "Astra Announces Spacecraft Engine Contract with Maxar Technologies," Press Release, 11 Oct 2022, URL: https://astra.com/news/spacecraft-engine-contract-with-maxar-technologies/
(227) Exotrail, "Hall effect thruster for small satellites." Company Website, Online Datasheet, May, 2022.
(228) Exotrail Press Release, "Exotrail delivers its first Hall Effect Propulsion system prototype for In-Orbit Demonstration mission." 02 October, 2019.
(229) Exotrail Press Release, "Exotrail secures contract with AAC Clyde Space to equip their spacecrafts for Eutelsat's ELO 3 and 4." 18 February, 2020.
(230) Gurciullo, A., Jarrige, J., Lascombes, P., and Packan, D. "Experimental performance and plume characterization of a miniaturized 50W Hall thruster." 36th IEPC, Vienna, Austria, 2019.
(231) ExoTrail Press Release, "Exotrail paves the way for new space mobility with first of its kind successful in-orbit demonstration mission," 01 January 2021.
(232) Moriconi, B., Hallouin, T., Gurciullo, A., "Hall Thruster ExoMG™ - micro, ExoMG™ - nano and low current cathode development at Exotrail: cyclic life testing results." 37th IEPC, Cambridge,USA, 2022.
(233) Aerospacelab Press Release, "Press release – Arthur, Aerospacelab launched the first Belgianprivate commercial satellite", 30 June 2021.
(234) NASA, "NASA Selects New Mission to Study Storms, Impacts on Climate Models," NASA website, 5 Nov 2023, URL: https://www.nasa.gov/news-release/nasa-selects-new-mission-to-study-storms-impacts-on-climate-models/
(235) Exotrail, "Exotrail will provide Blue Canyon Technologies with electric propulsion systems for NASA's INCUS mission," Press Release, 22 June 2023, URL: https://www.exotrail.com/blog/exotrail-blue-canyon-technologies
(236) Exotrail, "York Space Systems tackled Exotrail electric propulsion system for its Cislunar mission", Press Release, 08 September 2021.
(237) Exotrail, "OHB LUXSPACE inks contract with exotrail for ExoMG™ electric propulsion system onboard triton-x heavy platform.", Press Release, 23 November 2021.
(238) Exotrail, "Exotrail electric propulsion systems selected by Korean satellite manufacturer Satrec Initiative," Press Release, 26 April 2023, URL: https://www.exotrail.com/blog/satrec-initiative
(239) "Exotrail to provide Astro Digital with electric propulsion systems," Spacewatch, 4 April 2023, URL: https://spacewatch.global/2023/04/exotrail-to-provide-astro-digital-with-electric-propulsion-systems/
(240) Exotrail, "Exotrail's electric propulsion systems to be used in Muon Space's climate monitoring constellation," Press Release, 24 Aug 2023, URL: https://www.exotrail.com/blog/exotrail-muon-contract-for-climate-monitoring

(241) Exotrail Press Release, "Exotrail to debut its SpaceVan™ in-space mobility service on October2023 SpaceX Falcon 9 mission", 12 April 2022.
(242) Exotrail, "Exotrail and Airbus to partner on small Earth Observation satellites propulsion," Press Release, 20 Sept 2023, URL: https://www.exotrail.com/blog/exotrail-and-airbus-to-partner-on-small-earth-observation-satellites-propulsion
(243) Werner, D., "ExoTerra to provide Blackjack Satellite thrusters," SpaceNews, 30 August, 2021.
(244) Werner D., "ExoTerra gains flight heritage for Halo thrusters," SpaceNews, 8 Aug 2023, URL: https://spacenews.com/exoterra-gains-flight-heritage-for-halo-thrusters/
(245) ExoTerra Resource, "Halo Hall-Effect Thruster - Xenon." Company Website, Datasheet.
(246) VanWoerkom, M., Gorokhovsky, V., Pulido, G., Pettigrew, R., and Seidcheck, A. "Test Results of ExoTerra's Halo Micro Electric Propulsion System." 36th IEPC, Vienna, Austria, 2019.
(247) NASA Press Release, "NASA Announces New Tipping Point Partnerships for Moon and Mars Technologies." 27 September, 2019.
(248) Glascock, M. S., Kiefer, E., and VanWoerkom, M.; "Performance and Capability Overview of the Halo Electric Propulsion System", International Electric Propulsion Conference, IEPC-2022-301, 2022.
(249) Glascock, M. S. and VanWoerkom, M., "Channel Erosion Measurements and Predictions in a 400 W Hall Thruster," International Electric Propulsion Conference, IEPC-2022-300, 2022.
(250) NASA, "Space Act Agreement 36647 Between NASA and Exoterra Resource, LLC for Astraeus Flight Opportunity with Boeing," NASA Current Space Agreements, Accessed 26 Nov 2023, URL: https://www.nasa.gov/wp-content/uploads/2023/10/ntaa-domestic.pdf
(251) Exoterra Resource, "Halo12 Electric Propulsion System - Xenon," Company Datasheet, Dated 16 December 2022.
(252) Werner, D., "Orbion to supply propulsion for General Atomics weather satellite," SpaceNews, 24 June, 2022.
(253) Sommerville, J., Fruncek, C., King, L., Makela, J., Terhune, K., Washeleski, R., and Myers, R. "Performance of the Aurora Low-Power Hall-Effect Thruster." 36th IEPC, Vienna, Austria, 2019.
(254) Orbion, "Aurora," Company Website, Datasheet, Accessed 26 Nov 2023.
(255) Busek, "BHT-350 Hall Effect Thruster," Company Datasheet, Copyright 2021.
(256) Busek, "Busek Scales Thruster Production for Airbus OneWeb Satellites," Company Press Release, June 22, 2022.
(257) Werner, D., "Busek ramps up production for OneWeb constellation," SpaceNews, 6 Feb 2023, URL: https://spacenews.com/busek-ramps-up-production-for-oneweb-constellation/
(258) Werner, D., "Busek lauds on-orbit performance of OneWeb's Hall-effect thrusters," SpaceNews, 8 Aug 2023, URL: https://spacenews.com/busek-lauds-on-orbit-performance-of-onewebs-hall-effect-thrusters/
(259) Busek Co Press Release, "Busek Ships First LEO to GEO Capable Electric Propulsion System – Design Demonstrated One Million Newton-Seconds in NASA-Led Test," PR Newswire, 23 February 2021.
(260) Szabo, J., Byrne, L., Strain, M., Paintal, S., Sawyer, S., Yu, T., Kolencik, G., and Hruby, V., "One Million Newton-Second Duration Test of a 600 Watt Hall Effect Thruster Fueled By Xenon," 2020 AIAA Propulsion and Energy Forum, Virtual Event, August 24-28, 2020.
(261) Nikrant, A., Glogowski, M., Cochran, D., Moquin, T., Choi, Y., Benavides, G., Kamhawi, H., Sarver-Verhey, T., Baird, M., Rhodes, C., and Mackey, J., "Overview and Performance Characterization of Northrop Grumman's 1kW Hall Thruster String," 37th IEPC, Boston, MA, Paper No. IEPC-2022-303, 19-23 June, 2022.

(262) Erwin, S., "Northrop Grumman says customers are 'lined up' for on-orbit satellite servicing," SpaceNews, 24 March, 2022.
(263) Jewett, R., "Optus Orders Mission Extension Pod from Northrop Grumman's SpaceLogistics," Via Satellite Article, 22 February, 2022.
(264) NASA, "Small Spacecraft Electric Propulsion Opens New Deep Space Opportunities," NASA Press Release, 19 April, 2022.
(265) Northrop Grumman, "Northrop Grumman's SpaceLogistics Continues Revolutionary Satellite Life-Extension Work with Sale of Third Mission Extension Pod," Press Release, 20 June 2023, URL: https://news.northropgrumman.com/news/releases/northrop-grummans-spacelogistics-continues-revolutionary-satellite-life-extension-work-with-sale-of-third-mission-extension-pod
(266) SETS, "SETS' SPS-25 Propulsion System Proves Successful in Space Testing Despite Challenging Circumstances in Ukraine," Press Release, 21 June 2023, URL: https://www.prnewswire.com/news-releases/sets-sps-25-propulsion-system-proves-successful-in-space-testing-despite-challenging-circumstances-in-ukraine-301855868.html
(267) Werner, D. and Klaczynska, M., "EOS Data Analytics and Space Electric Thruster Systems demonstrate technology in orbit," SpaceNews, 5 June 2023, URL: https://spacenews.com/eos-data-analytics-and-space-electric-thruster-systems-demonstrate-technology-in-orbit/
(268) Aliena Pte Ltd, "MUSIC Electric Propulsion System," Company website, Accessed 4 December 2023, URL: https://www.aliena.sg/music
(269) Aliena and NuSpace Press Release, "Singapore's Aliena & NuSpace to launch demonstration mission for constellation management system," SpaceTechAsia, 28 September 2019, URL: https://www.spacetechasia.com/singapores-aliena-nuspace-to-launch-demonstration-mission-for-constellation-management-system/
(270) Press Release, "Lightweight engine launched into space," Nanyang Technological University – Singapore, 14 January 2022, URL: https://www.ntu.edu.sg/news/detail/ntu-singapore-startup-aliena-deploys-compact-and-fuel-efficient-satellite-engine-into-space
(271) Khoo, K. S., Laterza, M., Lim, J. W. M., and Potrivitu, G.-C., "Novel cathodeless and very low-power Hall thruster: qualification and integration on a 3U platform for an in-orbit demonstration mission," Proceedings of the 37th International Electric Propulsion Conference, IEPC-2022-326, Cambridge, MA, USA, 2022.
(272) Laterza, M., Potrivitu, G-C., Agarwal, D., Khoo, K. S., Pontianus, N., Lim, J. W. M., Eunseo, E. H., Supriyadi, S. D., Li, C., Liau, L., Ramachandran, P., Teo, C., Ng, Z. K., Tsang, S. H., and Teo, H. T. E., "MUlti-Stage Ignition Compact thruster concept and testing," Proceedings of the 37th International Electric Propulsion Conference, IEPC-2022-294, Cambridge, MA, USA, 2022.
(273) Press Release "Aliena signs contract with Orbital Astronautics for ORB-12 STRIDER mission," SatNews, 23 September 2021, URL: https://news.satnews.com/2021/09/23/aliena-signs-contract-with-orbital-astronautics-for-orb-12-strider-mission/
(274) Burton, R. "Pulsed Plasma Thrusters." In Encyclopedia of Aerospace Engineering (eds R. Blockley and W. Shyy). doi:10.1002/9780470686652.eae120.
(275) Kolbeck, J., Anders, A., Beilis, I., and Keidar, M. "Micro-propulsion based on vacuum arcs." Journal of Applied Physics, June, 2019, https://doi.org/10.1063/1.5081096.
(276) Benchmark Space Systems, "Xantus." Datasheet, June 2023, Accessed Sept. 2023, URL: https://www.benchmarkspacesystems.com/xantus-data-sheet
(277) Werner, D., "Benchmark signs contracts for metal plasma thrusters," SpaceNews, 15 March 2023, URL: https://spacenews.com/benchmark-signs-contracts-for-metal-plasma-thrusters/

(278) Comat, "Plasma Jet Pack technology," Datasheet, Accessed Sept. 2023, URL: https://comat.space/en/p/small-sat-en/en-pjp-technology/
(279) Comat, "Great news: Our PJP boards on Vega Mission VV23," Accessed Sept. 2023, URL: https://comat.space/en/actualites/micro-propulsion/
(280) Neumann Space, "Neumann Drive Datasheet," Datasheet, Sept. 2023.
(281) Satnews, "Neumann Drive® propulsion system now integrated onto Australia's Skykraft satellite," News article, 14 March 2023, URL: https://news.satnews.com/2023/03/14/neumann-drive-propulsion-system-now-integrated-onto-australias-skykraft-satellite/
(282) Foust, J., "SpaceX launches eighth dedicated smallsat rideshare mission," SpaceNews, 12 June 2023, URL: https://spacenews.com/spacex-launches-eighth-dedicated-smallsat-rideshare-mission/
(283) Neumann Space, "Australia's world leading propulsion technology achieves first space launch," Press Release, 13 June 2023, URL: https://neumannspace.com/wp-content/uploads/2023/06/2023-06-13_Media-Release-Neumann-Space-First-Launch.pdf
(284) Australian Space Agency, "Ground-breaking nanosatellite showcases Australian technology," 2 March 2023, URL:https://www.industry.gov.au/news/ground-breaking-nanosatellite-showcases-australian-technology
(285) Dalton, Angus, "How do you break Australia's satellite drought? Hitch a ride with Elon Musk," The Sydney Morning Herald, Oct. 9, 2023, Accessed Nov. 2023, URL: https://www.smh.com.au/national/how-do-you-break-australia-s-satellite-drought-hitch-a-ride-with-elon-musk-20231003-p5e9bs.html
(286) University of Melbourne, "SpIRIT of science: University of Melbourne-led satellite mission launch countdown begins," News article, Oct. 9, 2023, Accessed Nov. 2023, URL: https://www.unimelb.edu.au/newsroom/news/2023/october/spirit-of-science-university-of-melbourne-led-satellite-mission-launch-countdown-begins
(287) EX2, "Avalon 2023: Inovor Technologies prepares for busy 2023," 14 March 2023, URL:https://www.ex2.com.au/news/avalon-2023-inovor-technologies-prepares-for-busy-2023/
(288) Garman, L., "Neumann Space to test electric propulsion system in orbit," Space Connect, 6 June 2023, URL:https://www.spaceconnectonline.com.au/satellites/5911-neumann-space-space-inventor-partner-to-trial-propulsion-tech
(289) "SSTL+Neumann Space partner to test Australian made propulsion system on demo mission," Satnews, 8 August 2023, URL:https://news.satnews.com/2023/08/08/sstl-neumann-space-partner-to-test-australian-made-propulsion-system-on-demo-mission/
(290) J. Foust, "Phase Four Launches First Plasma Propulsion Systems", SpaceNews, 25 January 2021, https://spacenews.com/phase-four-launches-first-plasma-propulsion-systems/
(291) Werner, D., "Phase Four Demonstrates Maxwell Block 2 Thruster In Orbit," SpaceNews, 12 September 2023, https://spacenews.com/phase-four-demonstrates-maxwell-block-2-thruster-in-orbit/
(292) Debra Werner "Capella Space and Phase Four reveal Maxwell Engine performance" SpaceNews 30 June 2021.
(293) REGULUS, 6 May 2020, URL: https://www.t4innovation.com/wp-content/uploads/2020/05/T4i-REGULUS-Magazine.pdf
(294) Bellomo, N. "Enhancement of Microsatellites' Mission Capabilities: Integration of REGULUS Electric Propulsion Module into UniSat-7." 70th International Astronautical Congress, Washington, D.C., October, 2019.
(295) Bellomo, N., et al., "Design and In-Orbit Demonstration of REGULUS, an Iodine Electric Propulsion System", CEAS Space Journal, 7 June 2021.

(296) "Cube Quest Challenge NASA Facts." NASA, FS-2019-12-073-MSFC. URL: https://www.nasa.gov/sites/default/files/atoms/files/cubequest_fs_june2020_508.pdf

(297) "Three DIY CubeSats Score Rides on NASA's First Flight of Orion, Space Launch System." NASA Press Release 17-055, 8 June, 2017. URL: https://www.nasa.gov/press-release/three-diy-CubeSats-score-rides-on-nasa-s-first-flight-of-orion-space-launch-system

(298) Alhorn, D., Casas, J., Agasid, E., Adams, C., Laue, G., Kitts, C., and O'Brien, S. "NanoSail-D: The Small Satellite That Could!" SmallSat Conference, 2011.

(299) Mansell, J., Spencer, D., Plante, B., Diaz, A., Fernandez, M., Bellardo, J., Betts, B., and Nye, B. "Orbit and Attitude Performance of the LightSail 2 Solar Sail Spacecraft." AIAA 2020-2177.

(300) "NEA Scout, NASA's solar sail mission to an asteroid," The Planetary Society, Accessed Nov. 2023, URL: https://www.planetary.org/space-missions/nea-scout

(301) Johnson, L, et al. "Near Earth Asteroid Scout: Exploring Asteroid 1991VG Using a SmallSat." *Proceedings of the 70th International Astronautical Congress*, IAC-19/B4/2.

(302) NASA, "Advanced Composite Solar Sail System (ACS3): Using Sunlight to Power Deep Space Exploration," Press Release, October 16, 2023, URL: https://www.nasa.gov/smallspacecraft/what-is-acs3/

(303) "AE's Prox-1 Satellite Launches from Kennedy Space Center." Georgia Institute of Technology, 25 June, 2019. URL: https://ae.gatech.edu/news/2019/06/aes-prox-1-satellite-launches-kennedy-space-center

(304) C. Henry. "Tethers Unlimited Says Early Results of Deorbit Hardware Test Promising." SpaceNews, 23 January, 2020. URL: https://spacenews.com/tethers-unlimited-says-early-results-of-deorbit-hardware-test-promising/

(305) Clark, S. "SmallSat Companies Teaming Up on Deorbit Experiment." Spaceflight Now, 15 August, 2019. URL: https://spaceflightnow.com/2019/08/15/SmallSat-companies-teaming-up-on-deorbit-experiment/

(306) Satnews, "SmallSat Space Debris Remediation Technology Being Demo'd By Millennium Space Systems," 7 December 2020.

(307) Stankey, H. and Hoyt, R., "In-Flight Performance of the Terminator Tape End-of-Life Deorbit Module," 35th Small Satellite Conference, Logan, UT, Paper No: SSC21-XI-03, 2021.

(308) Applied Ion Systems, "Products," Technical Data, Accessed: Sept. 2023. URL: https://appliedionsystems.com/products/

(309) Federal Communications Commission, "FCC Adopts New '5-Year Rule' for Deorbiting Satellites to Address Growing Risk of Orbital Debris", press release, 9 September 2022, accessible at: https://docs.fcc.gov/public/attachments/DOC-387720A1.pdf

(310) Hejmanowski, N. J., Woodruff, C. A., Burton, R. L., Carroll D. L., Palla, A. D., and Cardin J. M. "CubeSat High Impulse Propulsion System (CHIPS) Design and Performance." 63rd JANNAF Propulsion Meeting, Paper Tracking # 4800, Phoenix, AZ, 2016, Dist. A.

(311) Hejmanowski, N. J., Woodruff, C. A., Burton, R. L., and Carroll D. L. "Electrothermal Space Thruster Heater for Decomposable Propellants." U.S. Patent # 9,909,574 B1, 2018.

(312) Hejmanowski, N. J., Woodruff, C., Burton, R.L., Carroll, D.L., and Cardin, J. "CubeSat High Impulse Propulsion System (CHIPS)." 62nd JANNAF Propulsion Meeting, Paper Tracking # 4032, Nashville, TN, 1-5 June, 2015, Dist. A.

(313) CU Aerospace, "CubeSat High Impulse Propulsion System." Company Website, URL: https://cuaerospace.com/products-services/space-propulsion-systems/cubesat-high-impulse-propulsion-chips, Datasheet, 2022.

BODY-155

(314) Electric Propulsion Laboratory, Inc, "Space-based Resistojets," Company website, Accessed 3 December 2023, URL: https://electricpropulsionlab.com/resistojet-propulsion-systems/
(315) Alamalhodaei, Aria, "Atomos Space books launch to demonstrate rendezvous, docking and refueling in-orbit," TechCrunch, 25 July 2023, URL: https://techcrunch.com/2023/07/25/atomos-space-books-launch-to-demonstrate-rendezvous-docking-and-refueling-in-orbit/
(316) Pale Blue, "Propulsion Systems Options," Product Page. Accessed 24 Nov 2023, URL: https://pale-blue.co.jp/product/
(317) Cifali, G., Gregucci, S., Andreussi, T., and Andrenucci, M. "Resistojet Thrusters for Auxiliary Propulsion of Full Electric Platforms." 35th International Electric Propulsion Conference, Atlanta, GA, IEPC-2017-371, 2017.
(318) Sitael, "XR Resistojet Product Family." Datasheet, Company Website, Copyright 2015.
(319) Courtney, D., Wright, P., Lafko, D., Metivier, E., and Demmons, N. "Electrospray Propulsion for Precise Attitude and Position Control." AAS 17-088, Breckenridge, CO., 2017.
(320) D. Courtney et al. "Electrospray Thrusters for Small Spacecraft Control: Pulsed and Steady State Operation." AIAA-2018-4654.
(321) Courtney, D., Wood, Z., and Fedkiw, T. "Reconstructing Electrospray Plume Current Spatial Distributions Using Computed Tomography." 36th IEPC, Vienna, Austria, IEPC-2019-A-787.
(322) D. Courtney et al. "High-Speed Transient Characterization of the Busek BET-300-P Electrospray Thruster." 36th IEPC, Vienna, Austria, IEPC-2019-A-788.
(323) T. Fedkiw, Z. Wood, and N. Demmons, "Environmental and Lifetime Testing of the BET-300-P Electrospray Thruster", AIAA 2020-3614.
(324) Demmons, N., et al., "Qualification of the BET-MAX Electrospray Propulsion System", IEPC-2022-196, 37th International Electric Propulsion Conference, Cambridge, MA, 19-23 June 2022.
(325) Busek, "BET-MAX Electrospray Thruster System", Company Datasheet, Dated August 2021.
(326) Enpulsion, "IFM Nano Thruster." Company Website, Datasheet ENP2018-001.F.1, Accessed 2020.
(327) Schönherr, T., Little, B., Krejci, D., Reissner, A., and Seifert, B. "Development, Production, and Testing of the IFM Nano FEEP Thruster." 36th IEPC, Vienna, Austria, IEPC-2019-362, 2019.
(328) Krejci, D., Hugonnaud, V., Schonherr, T., Little, B., Reissner, A., Seifert, B., Kock, Q, Borras, E., and Amo, J. "Full Performance Mapping of the IFM Nano Thruster, Including Direct Thrust Measurements." *Journal of Small Satellites*, 2019, Vol. 8, No. 2, pp. 881-893.
(329) Enpulsion, "Enpulsion Nano R3." Company Website, Datasheet ENP2019-086.E.1, Accessed 2022.
(330) Enpulsion, "Enpulsion Micro R^3." Company Website, Datasheet ENP2018-002.H.1, Accessed 2021.
(331) Grimaud, L., Krejci, D., and Seifert, B. "The IFM Micro FEEP Thruster: A Modular Design for SmallSat Propulsion." 36th IEPC, Vienna, Austria, IEPC-2019-A675, 2019.
(332) Enpulsion, "Enpulsion Neo." Company Datasheet, Accessed October 2023, URL: https://www.enpulsion.com/order/neo/
(333) Morpheus Space, "nanoFEEP and multiFEEP." Company Website, Datasheet, June, 2020. URL: https://www.morpheus-space.com/static/MSWeb/documents/M-Space%20Products.pdf.

(334) Bock, D., Spethmann, A., Trottenberg, T., Kersten, H., and Tajmar, M. "In-Plume Thrust Measurement of NanoFEEP Thruster with a Force Measuring Probe Using Laser Interferometry." 36th IEPC, Vienna, Austria, IEPC-2017-391, 2017.
(335) Avant Space, "GT-50 RF Ion Thruster." Company Datasheet. URL: http://www.avantspace.com/upload/iblock/dc9/Data-sheet.pdf
(336) Kralkin, E., et al. "Exploratory Testing of a Radio-Frequency Thruster for Small Satellites." IEPC-2017-425.
(337) Busek, "BIT-3 RF Ion Thruster." Company Website, Datasheet, Copyright 2019.
(338) "All-Electric CubeSat Propulsion Technologies for Versatile Mission Application." 32nd International Symposium on Space Technology and Science (ISTS), Fukui, Japan, June, 2019, ISTS-2019-b-002.
(339) Tsay, M., Frongillo, J., Model, J., Zwahlen, J., Barcroft, C., Feng, C. "Neutralization Demo and Thrust Stand Measurement for BIT-3 RF Ion Thruster." 2017 AIAA Propulsion and Energy Forum, Atlanta, GA, July, 2017, AIAA-2017-4890.
(340) Tsay, M., "3,500-Hour Wear Test Result of BIT-3 RF Ion Propulsion System", IEPC-2022-255, 37th International Electric Propulsion Conference, Cambridge, MA, 19-23 June 2022.
(341) Tsay, M., et al., "Volume Production of Gen-2 Iodine BIT-3 Ion Propulsion System", IEPC-2022-267, 37th International Electric Propulsion Conference, Cambridge, MA, 19-23 June 2022.
(342) Tsay, M., Terhaar, R., Emmi, K., and Barcroft, C., "Twin Ion Engine Demonstration for Small Spacecraft Applications," 36th Annual Small Satellite Conference, Logan, UT, 2022, SSC22-X-04.
(343) Pale Blue, "Products – Ion Thruster," Website Datasheet, Accessed October 2022.
(344) Nakagawa, Y., Yaginuma, K., Asakawa, J., and Koizumi, H., "1U+Water Ion and Resistojet Thruster Module for On-Orbit Demonstration," IEPC-2022-276, 37th International Electric Propulsion Conference, Cambridge, MA, 19-23 June 2022.
(345) Navin, J., "JAXA Epsilon fails on sixth flight carrying RAISE-3 and others," NASA Spaceflight.com, 11 October 2022
(346) "Development of Water Gridded Ion Thruster for Small Satellites: Toward On-Orbit Demonstration" Poster, Small Satellite Conference, 02 August 2023, Utah State University, Logan, UT, URL: https://digitalcommons.usu.edu/smallsat/2023/all2023/217/
(347) ThrustMe, "NPT30-I2 1.5U," Company Website, Datasheet, Copyright 2020.
(348) Martinez, J., Rafalskyi, D., and Aanesland, A. "Development and Testing of the NPT30-I2 Iodine Ion Thruster." 36th IEPC, Vienna, Austria, IEPC-2019-811, 2019.
(349) ThrustMe, "NPT30-I2 1U," Company Website, Datasheet, Copyright 2021.
(350) Leiter, H., Lauer, D., Bauer, P., Berger, M., and Rath, M. "The Ariane Group Electric Propulsion Program 2019-2020." 36th IEPC, Vienna, Austria, IEPC-2019-592, 2019.
(351) Leiter, H., Altmann, C., Lauer, D., Rath, M., Becker, F., Feili, D., and Amo, J. "A Nouvelle Neutralization Concept for RIT-µX Miniaturized Radio Frequency Ion Thruster Systems." 36th IEPC, Vienna, Austria, IEPC-2019-806, 2019.
(352) Ariane Group, "Electric Propulsion Systems and Components." Company Website, Datasheet, Accessed 2020.
(353) Leiter, H., Lotz, B., Feili, D., Tartz, M., Neumann, H., and Cara, D. "Design Development and Test of the RIT-µX Mini Ion Engine System." IEPC-2009-179.
(354) Leiter, H., et al., "Recent Development of the RIT-µX Thruster and System", IEPC-2022-252, 37th International Electric Propulsion Conference, Cambridge, MA, 19-23 June 2022.
(355) Holste, K., et al., "Endurance and Coupling Test of the RIT-µX", IEPC-2022-247, 37th International Electric Propulsion Conference, Cambridge, MA, 19-23 June 2022.

(356) Leiter, H., Altmann, C., Porst, J., Lauer, D., Berger, M., and Rath, M. "Six Decades of thrust – The Ariane Group Radiofrequency Ion Thrusters and Systems Family." IEPC-2017-027.

(357) Szabo, J., Tedrake, R., Metivier, E., Paintal, S., and Taillefer, Z. "Characterization of a One Hundred Watt, Long Lifetime Hall Effect Thruster for Small Spacecraft." 53rd AIAA Joint Propulsion Conference, Atlanta, Georgia, 2017.

(358) Busek, "BHT-600 Busek Hall Effect Thruster." Company Website, Datasheet, 2016.

(359) Szabo, J., Pote, B., Tedrake, R., Paintal, S., Byrne, L., Hruby, V., Kamhawi, H., and Smith, T. "High Throughput 600 Watt Hall Effect Thruster for Space Exploration." 52nd AIAA Joint Propulsion Conference, Salt Lake City, Utah, 2016.

(360) Benavides, G., Kamhawi, H., Mackey, J., Haag, T., and Costa, G., "Iodine Hall-Effect Electric Propulsion System Research, Development, and System Durability Demonstration," 2018 Joint Propulsion Conference, Cincinnati, Ohio, July 9-11, 2018.

(361) Saevets, P., Semenenko, D., Albertoni, R., and Scremin, G. "Development of a Long-Life Low-Power Hall Thruster." 35th IEPC, Atlanta, Georgia, 2017.

(362) ExoTerra Resource, "Halo Hall-Effect Thruster - Krypton." Company Website, Datasheet.

(363) Virgin Orbit, "Exoterra, Virgin Orbit Win NASA Phase II SBIR Award for Solar Electric Propulsion Upper Stage," Company Press Release, June 1, 2021.

(364) Exoterra Resource, "Halo12 Electric Propulsion System - Krypton," Company Datasheet, Dated 16 December 2022.

(365) Conversano, R., Goebel, D., Katz, I., and Hofer, R. "Low-Power Hall Thruster with an Internally Mounted Low-Current Hollow Cathode." U.S. Patent No. 10,723,489, 2020.

(366) Conversano, R., Reilly, S., Kerber, T., Brooks, J., and Goebel, D. "Development of and Acceptance Test Preparations for the Thruster Component of the Ascendant Sub-kW Transcelestial Electric Propulsion System (ASTRAEUS)." 36th IEPC, Vienna, Austria, 2019.

(367) Conversano, R., Lobbia, R., Kerber, T., Tilley, K., Goebel, D., and Reilly, S. "Performance characterization of a low-power magnetically shielded Hall thruster with an internally-mounted hollow cathode." Plasma Sources Science and Technology, Vol. 28, No. 10, 2019.

(368) Conversano, R., Barchowsky, A., Lobbia, R., Chaplin, V., Lopez-Ortega, A., Loveland, J., Lui, A., Becatti, G., Reilly, S., Goebel, D., Snyder, J., Hofer, R., Randolph, T., Mikellides, I., Vorperian, V., Carr, G., Rapinchuk, J., Villalpando, C., and Grebow, D. "Overview of the Ascendant Sub-kW Transcelestial Electric Propulsion System (ASTRAEUS)." 36th IEPC, Vienna, Austria, 2019.

(369) Lobbia, R., Conversano, R., Ortega, A., Reilly, S., Mikellides, I. "Pole Erosion Measurements for the Development Model of the Magnetically Shielded Miniature Hall Thruster (MaSMi-DM)." 36th IEPC, Vienna, Austria, 2019.

(370) Ortega, A., Mikellides, I., Conversano, R., Lobbia, R., Chaplin, V. "Plasma Simulations for the Assessment of Pole Erosion in the Magnetically Shielded Miniature Hall Thruster (MaSMi)." 36th IEPC, Vienna, Austria, 2019.

(371) JPL Press Release, "2020 Technology Highlights, Jet Propulsion Laboratory," CL#20-6401.

(372) Conversano, R., Barchowsky, A., Vorperian, V., Chaplin, V., Becatti, G., Carr, G., Stell, C., Loveland, J., and Goebel, D., "Cathode & Electromagnet Qualification Status and Power Processing Unit Development Update for the Ascendant Sub-kW Transcelestial Electric Propulsion System," 34th Small Satellite Conference, 2020, SSC20-VI-10.

(373) Conversano, R., Lobbia, R., Lopez-Ortega, A., Chaplin, V., Reilly, S., Arestie, S., Goebel, D., "Long-Duration Wear Testing of the ASTRAEUS Hall Thruster, Phase 1: 50 kg Xe Total Throughput," Space Propulsion 2020+1, 17-19 March 2021, SP2020-005.

(374) Conversano, R. W., Lobbia, R. B., Arestie, S. M., Lopez-Ortega, A., Chaplin, V. H., Reilly, S. W., and Goebel, D. M.; "Demonstration of 100 kg Xenon Propellant Throughput by a Low-Power Magnetically Shielded Hall Thruster," Journal of Propulsion and Power, 2022.
(375) Conversano, R. W., Becatti, G., Goebel, D. M., and Chaplin, V. H.; "Demonstration of 13,011-hours of Operation of a Proto-Flight Compact Heaterless Lanthanum Hexaboride Hollow Cathode," Acta Astronautica, Vol. 197, 2022, pp. 53-59. DOI: https://doi.org/10.1016/j.actaastro.2022.05.015.
(376) Becatti, G., Conversano, R. W, and Goebel, D. M., "Demonstration of 25,000 ignitions on a proto-flight compact heaterless lanthanum hexaboride hollow cathode," Acta Astronautica, Vol. 178, 2021, pp. 181-191, DOI: https://doi.org/10.1016/j.actaastro.2020.09.013.
(377) Chaplin, V., Simmonds, J., Bryne, M., Goebel, D., Ortega, A., Lobbia, R., Mikellides, I., Hofer, R., Zhu, B., Ratliff, J., and Conversano, R., "Extended Life Qualification of the Magnetically Shielded Miniature (MaSMi) Hall Thruster," 37th Small Satellite Conference, Logan, UT, Aug 2023, SSC23-VI-01.
(378) Lev, D., Franco, D., Zimmerman, R., Tordjman, M., Auslender, B., and Epsteain, O., "96 kN-sec Endurance Test of the R-200 Low Power Hall Thruster," 37th IEPC, Boston, MA, Paper No: IEPC-2022-399, 19-23, June 2022.
(379) Lev, D., Franco, D., Auslender, B., and Epstein, O., "Extension of the Operation Envelope of the R-200 Low Power Hall Thruster," 37th IEPC, Boston, MA, Paper No: IEPC-2022-357, 19-23, June 2022.
(380) Appel, L., Medvinsky, G., Shoor, B., Sirota, A., Zimmerman, R., Lev, D., Epstein, O., "Integration Test of the R-800 Low Power Hall Thruster Electric Propulsion System," 37th IEPC, Boston, MA, Paper No: IEPC-2022-358, 19-23, June 2022.
(381) Safran, "PPS-X00 Stationary Plasma Thruster," Company Datasheet, Dated 11/09/19.
(382) Vaudolon, J., Vial, V., Cornu, N., and Habbassi, I. "PPS-X00 Thruster Development Status at Safran." 36th IEPC, Vienna, Austria, 2019.
(383) Erwin, S., "Safran, Terran Orbital to explore joint production of satellite propulsion systems," SpaceNews, 23 June 2023, URL: https://spacenews.com/safran-and-terran-orbital-to-explore-joint-production-of-satellite-propulsion-systems-in-the-u-s/
(384) SITAEL "Electric Propulsion – HT400 Hall Effect Thruster Product Sheet." Company Website, Datasheet, 2020.
(385) Ducci, C., Raiji, H., Pedrini, D., Misuri, T., and Onida, L. "SITAEL's HT400 Hall Effect Thruster for Constellation Applications." 36th IEPC, Vienna, Austria, 2019.
(386) Pedrini, D., Ducci, C., Misuri, T., Paganucci, F., Andrenucci, M. "Sitael Hollow Cathodes for Low-Power Hall Effect Thrusters." IEEE Transactions on Plasma Science, Vol. 46, No. 2, February, 2018.
(387) Space Electric Thruster Systems "ST25 Hall-Effect Thruster." Company Website, Datasheet, 2020.
(388) Petrenko, O., Tolok, S., Maslov, V., Kulagin, S., Serbin, V., Alekseenko, O., and Shcherbak, D. "Electric propulsion system SPS-25 with Hall Thruster." 70th International Astronautical Congress (IAC),Washington D.C., October 2019.
(389) Space Electric Thruster Systems "ST40 Hall-Effect Thruster." Company Website, Datasheet, 2020.
(390) Krishnan, M., Velas, K., and Leemans, S. "Metal Plasma Thruster for Small Satellites." Journal of Propulsion and Power, March, 2020, DOI: 10.2514/1.B37603.
(391) Jarrige, J., Packan, D., Blanchet, A., and Herrero, L., "Direct Thrust Measurement of a Vacuum Arc Thruster," IEPC-2019-521, Vienna, Austria, 15-20 September 2019.
(392) Hypernova Space Technologies, "NanoThruster A," Datasheet.

(393) Ciaralli et al. "Results of the Qualification Test Campaign of a Pulsed Plasma Thruster for CubeSat Propulsion (PPTCUP)." Acta Astronautica, Volume 121, Pages 314-322, April–May, 2016.
(394) Mars Space Ltd., "Pulsed Plasma Thruster (PPT) projects," Company website, Accessed Sept. 2023, URL: http://mars-space.co.uk/ppt
(395) Siddiqui, M., Cretel, C., Synowiec, J., Hsu, A., Young, J., and Spektor, R. "First Performance Measurements of the Phase Four RF Thruster." 35th IEPC, Atlanta, GA, October, 2017.
(396) Siddiqui, M. and Cretel, C. "Updated Performance Measurements and Analysis of the Phase Four RF Thruster." Joint Propulsion Conference, Cincinnati, OH, AIAA-2018-4817.
(397) Cretel, C., Ajamia, M., Thompson, D., and Siddiqui, M. "Torsional Balance Thrust Measurement Techniques for Small RF Thrusters."
(398) "Meet Maxwell." Phase Four, website, accessed 2020, URL: https://www.phasefour.io/maxwell/
(399) Phase Four, "Maxwell Product Suite," Website Datasheet, Accessed October 2022
(400) "The Maxwell RF Propulsion System", Phase Four Product Brochure, Available at 2023 Small Satellite Conference, Logan, UT.
(401) "REGULUS Electrical Propulsion System for Micro-satellites." T4i, datasheet. URL: https://www.t4innovation.com/wp-content/uploads/2019/10/T4i_REGULUS_datasheet_.pdf
(402) "T4i REGULUS Family", Company Website, Accessed October 2023, https://www.t4innovation.com/regulus-electric-propulsion/
(403) "The M1.4 Thruster with ConstantQ Propulsion", Company Website, Accessed October 2023, URL: https://miles-space.com/thruster/
(404) "Team Miles", NASA Space Science Data Coordinated Archive, Accessed October 2023, https://nssdc.gsfc.nasa.gov/nmc/spacecraft/display.action?id=TEAMMILES/.
(405) "Terminator Tape." Tethers Unlimited, datasheet, 2020. URL: https://www.tethers.com/wp-content/uploads/2020/03/2020-Terminator-Tape.pdf

Chapter Contents

Chapter Glossary

(ADCS)	Attitude Determination and Control System
(CoCom)	Coordinating Committee for Multilateral Export Controls
(COTS)	Commercial-off-the-Shelf
(DOF)	Degrees of Freedom
(DSAC)	Deep Space Atomic Clock
(DSN)	Deep Space Network
(EAR)	Export Administration Regulations
(FOGs)	Fiber Optic Gyros
(GNC)	Guidance, Navigation & Control
(GSO)	Geo-stationary Orbit
(USAF)	U.S. Air Force
(HCI)	Horizon Crossing Indicators
(IMUs)	Inertial Measurement Units
(JPL)	Jet Propulsion Laboratory
(Lidar)	Light Detection and Ranging
(LMRST)	Low Mass Radio Science Transponder
(MarCO)	Mars Cube One
(PMSM)	Permanent-magnet Synchronous Motor
(SDST)	Small Deep Space Transponder
(SWaP)	Size, weight, and power
(TLE)	Two-Line Element
(TRL)	Technology Readiness Level

5.0 Guidance, Navigation & Control

5.1 Introduction

The Guidance, Navigation & Control (GNC) subsystem includes the components used for position determination and the components used by the Attitude Determination and Control System (ADCS). In Earth orbit, onboard position determination can be provided by a Global Positioning System (GPS) receiver. Alternatively, ground-based radar tracking systems can also be used. If onboard knowledge is required, then these radar observations can be uploaded and paired with a suitable propagator. Commonly, the U.S. Air Force (USAF) publishes Two-Line Element sets (TLE) (1), which are paired with a SGP4 propagator (2). In deep space, position determination is performed using the Deep Space Network (DSN) and an onboard radio transponder (3). There are also technologies being developed that use optical detection of celestial bodies such as planets and X-ray pulsars to calculate position data (23).

Using SmallSats in cislunar space and beyond requires a slightly different approach than the GNC subsystem approach in low-Earth orbit. Use of the Earth's magnetic field, for example, is not possible in these missions, and alternate ADCS designs and methods must be carefully considered. Two communication relay CubeSats (Mars Cube One, MarCO) successfully demonstrated such interplanetary capability during the 2018 Insight mission to Mars (4). This interplanetary mission demonstrated both the capability of this class of spacecraft and the GNC fine pointing design for communication in deep space.

ADCS includes sensors to determine attitude and spin rate, such as star trackers, sun sensors, horizon sensors, magnetometers, and gyros. In addition, the ADCS is often used to control the vehicle during trajectory correction maneuvers and, using accelerometers, to terminate maneuvers when the desired velocity change has been achieved. Actuators are designed to change a spacecraft's attitude and to impart velocity change during trajectory correction maneuvers. Common spacecraft actuators include magnetic torquers, reaction wheels, and thrusters. There are many attitude determination and control architectures and algorithms suitable for use in small spacecraft (5).

Miniaturization of existing technologies is a continuing trend in small spacecraft GNC. While three-axis stabilized, GPS-equipped, 100 kg class spacecraft have been flown for decades, it has only been in the past few years that such technologies have become available for micro- and nano-class spacecraft. Table 5-1 summarizes the current state-of-the-art of performance for GNC subsystems in small spacecraft. Performance greatly depends on the size of the spacecraft and values will range for nano- to micro-class spacecraft.

The information described below is not intended to be exhaustive but provides an overview of current state-of-the-art technologies and their development status for a particular small spacecraft subsystem. It should be noted that Technology Readiness Level (TRL) designations may vary with changes specific to payload, mission requirements, reliability considerations, and/or the environment in which performance was demonstrated. Readers are highly encouraged to reach out to companies for further information regarding the performance and TRL of described technology. There is no intention of mentioning certain companies and omitting others based on their technologies or relationship with NASA.

Table 5-1: State-of-the-Art GNC Subsystems		
Component	**Performance**	**TRL**
Reaction Wheels	0.00023 – 0.3 Nm peak torque, 0.0005 – 8 N m s storage	7-9
Magnetic Torquers	0.15 A m^2 – 15 A m^2	7-9
Star Trackers	8 arcsec pointing knowledge	7-9
Sun Sensors	0.1° accuracy	7-9
Earth Sensors	0.25° accuracy	7-9
Inertial Sensors	Gyros: 0.15° h^{-1} bias stability, 0.02° $h^{-1/2}$ ARW Accels: 3 μg bias stability, 0.02 (m s^{-1})/$h^{-1/2}$ VRW	7-9
GPS Receivers	1.5 m position accuracy	7-9
Integrated Units	0.002-5° pointing capability	7-9
Atomic Clocks	10 – 150 Frequency Range (MHz)	5-6
Deep Space Navigation	Bands: X, Ka, S, and UHF	7-9
Altimeters	~15 meters altitude, ~3 cm accuracy	7

5.2 State-of-the-Art – GNC Subsystems

5.2.1 Integrated Units

Integrated units combine multiple different attitude and navigation components to provide a simple, single-component solution to a spacecraft's GNC requirements. Typical components included are reaction wheels, magnetometer, magnetic torquers, and star trackers. The systems often include processors and software with attitude determination and control capabilities. Table 5-2 describes some of the integrated systems currently available that are associated with a TRL value of 7-9. Blue Canyon Technologies' XACT (figure 5.1) flew on the NASA-led missions MarCO and ASTERIA, both of which were 6U platforms, and have also flown on 3U missions (MinXSS was deployed from NanoRacks in February 2016).

Figure 5.1: BCT XACT Integrated ADCS Unit. Credit: Blue Canyon Technologies.

Table 5-2: Currently Available Integrated Systems						
Manufacturer	**Model**	**Mass (kg)**	**Actuators**	**Sensors**	**Processor**	**Pointing Accuracy**
Arcsec	Arcus ADC	0.715	3 reaction wheels 3 magnetic torquers	1 star tracker 3 gyros 6 photodiodes 3 magnetometers	Yes	0.1°
Berlin Space Technologies	IADCS-100	0.4	3 reaction wheels 3 magnetic torquers	1 star tracker 3 gyros, 1 magnetometer, 1 accelerometer	Yes	<<1 deg
AAC Clyde Space	iADCS-200	0.470	3 reaction wheels 3 magnetic torquers	1 star tracker 1 IMU, Optionally high precision magnetometer and sun sensors	Yes	<1°
AAC Clyde Space	iADCS-400	1.7	3 reaction wheels 3 magnetorquers	1 star tracker, 1 IMU, Optionally high precision magnetometer and sun sensors	Yes	<1°
Blue Canyon Technologies	XACT-15	0.885	3 reaction wheels 3 magnetorquers	1 star tracker 3-axis magnetometer	Yes	0.003/0.007°
Blue Canyon Technologies	XACT-50	1.230	3 reaction wheels 3 magnetorquers	1 star tracker 3-axis magnetometer	Yes	0.003/0.007°
Blue Canyon Technologies	XACT-100	1.813	3 reaction wheels 3 magnetorquers	1 star tracker 3-axis magnetometer	Yes	0.003/0.007°

Blue Canyon Technologies	Flexcore	configuration dependent	3 – 4 reaction wheels 3 magnetorquers	2 star trackers 3-axis magnetometer	Yes	0.002°
CubeSpace Satellite Systems	CubeADCS 3-Axis Small	0.55	3 reaction wheels 3 magnetorquers	10 coarse sun sensors 2 fine sun/earth sensors 1 magnetometer	Yes	<1°
CubeSpace Satellite Systems	CubeADCS 3-Axis Small with Star Tracker	0.61	3 reaction wheels 3 magnetorquers	10 coarse sun sensors 2 fine sun/earth sensors 1 magnetometer 1 star tracker	Yes	<0.1°
CubeSpace Satellite Systems	CubeADCS 3-Axis Medium	0.79	3 reaction wheels 3 magnetorquers	10 coarse sun sensors 2 fine sun/earth sensors 1 magnetometer	Yes	<1°
CubeSpace Satellite Systems	CubeADCS 3-Axis Medium with Star Tracker	0.84	3 reaction wheels 3 magnetorquers	10 coarse sun sensors 2 fine sun/earth sensors 1 magnetometer 1 star tracker	Yes	<0.1°
CubeSpace Satellite Systems	CubeADCS 3-Axis Large	1.1	3 reaction wheels 3 magnetorquers	10 coarse sun sensors 2 fine sun/earth sensors 1 magnetometer	Yes	<1°
CubeSpace Satellite Systems	CubeADCS 3-Axis Large with Star Tracker	1.15	3 reaction wheels 3 magnetorquers	10 coarse sun sensors 2 fine sun/earth sensors 1 magnetometer 1 star tracker	Yes	<0.1°
CubeSpace Satellite Systems	CubeADCS Y-Momentum	0.3	1 momentum wheel 3 magnetic torquers	10 coarse sun sensors 1 magnetometer	Yes	<5°

5.2.2 Reaction Wheels

Miniaturized reaction wheels provide small spacecraft with a three-axis precision pointing capability. They must be carefully selected based on several factors including the mass of the spacecraft and the required rotation performance rates. Reaction wheels provide torque and momentum storage along the wheel spin axis which results in the spacecraft counter-rotating around the spacecraft center of mass due to conservation of angular momentum from the wheel spin direction. Table 5-3 lists a selection of high-heritage miniature reaction wheels. Except for three units, all the reaction wheels listed have spaceflight heritage. For full three-axis control, a spacecraft requires three wheels mounted orthogonally. However, a four-wheel configuration is often used to provide fault tolerance (6). Reaction wheels need to be periodically desaturated using an actuator that provides an external torque, such as thrusters or magnetic torquers (7).

In addition, the multiple reaction wheels are often assembled in a "skewed" or angled configuration such that there exists a cross-coupling of torques with two or more reaction wheels. While this reduces the torque performance in any single axis, it allows a redundant, albeit reduced, torque capability in more than one axis. The result is that should any single reaction wheel fail, one or more reaction wheels are available as a reduced-capability backup option.

Table 5-3: High Heritage Miniature Reaction Wheels

Manufacturer	Model	Mass (kg)	Peak Power (W)	Peak Torque (Nm)	Momentum Capacity (Nms)	# Wheels	Radiation Tolerance (krad)
Berlin Space Technologies	RWA05	1.700	0.5	0.016	0.5	1	30
Blue Canyon Technologies	RWP015	0.130	1	0.004	0.015	1	Unk
Blue Canyon Technologies	RWP050	0.240	1	0.007	0.050	1	Unk
Blue Canyon Technologies	RWP100	0.330	1	0.007	0.100	1	Unk
Blue Canyon Technologies	RWP500	0.750	6	0.025	0.500	1	Unk
Blue Canyon Technologies	RW1	0.950	10	0.07	1.000	1	Unk
Blue Canyon Technologies	RW4	3.200	10	0.250	4.000	1	Unk
Blue Canyon Technologies	RW8	4.400	10	0.250	8.000	1	Unk
CubeSpace Satellite Systems	CubeWheel Small	0.060	0.65	0.00023	0.00177	1	24
CubeSpace Satellite Systems	CubeWheel Small+	0.090	2.3	0.0023	0.0036	1	24
CubeSpace Satellite Systems	CubeWheel Medium	0.150	2.3	0.001	0.01082	1	24

CubeSpace Satellite Systems	CubeWheel Large	0.225	4.5	0.0023	0.03061	1	24
GomSpace	NanoTorque GSW-600	0.940	0.3	0.0015	0.019	1	Unk
Comat	RW20	0.180	1	0.002	0.02	1	Up to 20Krad*
Comat	RW40	0.230	1	0.004	0.04	1	Up to 20Krad*
Comat	RW60	0.275	1	0.006	0.06	1	Up to 20Krad*
AAC Clyde Space	RW210	0.48	0.8	0.0001	0.006	1	36
AAC Clyde Space	RW400	0.375	15	0.008	0.050	1	36
AAC Clyde Space	Trillian-1	1.5	24	47.1	1.2	1	Unk
NanoAvionics	RWO	0.137	3.25	0.0032	0.020	1	20
NanoAvionics	4RWO	0.665	6	0.0059	0.037	4	20
NewSpace Systems	NRWA-T6	<5	136	0.3	0.00783	1	20
NewSpace Systems	NRWA-T065	1.55	1.7	0.02	0.00094	1	10
NewSpace Systems	NRWA-T2	2.8	0.4	0.09	0.00163	1	10
Rocket Lab	RW-0.03	0.185	1.8	0.002	0.040	1	20
Rocket Lab	RW-0.003	0.048	Unk	0.001	0.005	1	10
Rocket Lab	RW-0.01	0.122	1.05	0.001	0.018	1	20
Rocket Lab	RW3-0.06	0.235	23.4	0.020	0.180	1	20
Rocket Lab	RW4-0.2	0.6	Unk	0.1	0.2	1	60
Rocket Lab	RW4-0.4	0.77	Unk	0.1	0.4	1	60
Rocket Lab	RW4-1.0	1.38	43	0.1	1	1	60
Vectronic Aerospace	VRW-A-1	1.90	110	0.090	6.000	1	20
Vectronic Aerospace	VRW-B-2	1.00	45	0.020	0.200	1	20

Vectronic Aerospace	VRW-C-1	2.3	45	0.020	1.20	1	20
Vectronic Aerospace	VRW-D-2	2	65	0.05	2.0	1	20
Vectronic Aerospace	VRW-D-6	3	110	0.09	6	1	20

*Printed Circuit Board (PCB) level

5.2.3 Magnetic Torquers

Magnetic torquers provide control torques perpendicular to the local external magnetic field. Table 5-4 lists a selection of high heritage magnetic torquers and figure 5.3 illustrates some of ZARM Technik's product offerings. Magnetic torquers are often used to remove excess momentum from reaction wheels. As control torques can only be provided in the plane perpendicular to the local magnetic field, magnetic torquers alone cannot provide three-axis stabilization.

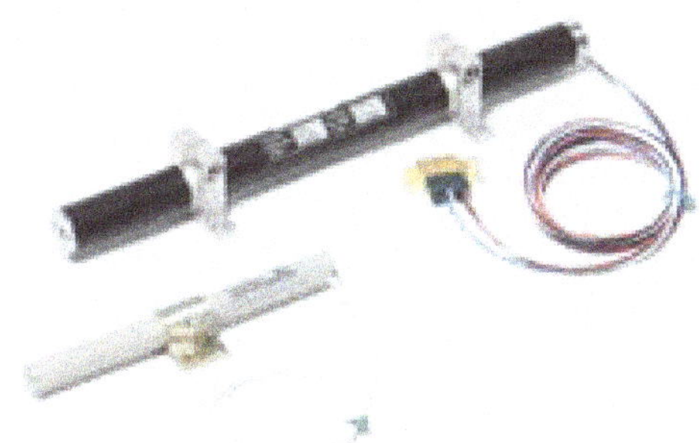

Figure 5.3: Magnetorquers for micro satellites. Credit: ZARM Technik.

Use of magnetic torquers beyond low-Earth orbit and in interplanetary applications need to be carefully investigated since their successful operation is relying on a significant local external magnetic field. This magnetic field may or may not be available in the location and environment for that mission and additional control methods may be required.

Table 5-4: High Heritage Magnetic Torquers

Manufacturer	Model	Mass (kg)	Power (W)	Peak Dipole (A m^2)	# Axes	Radiation Tolerance (krad)
CubeSpace Satellite Systems	CubeTorquer Small	0.028	0.42	0.24	1	24
CubeSpace Satellite Systems	CubeTorquer Medium	0.036	0.37	0.66	1	24
CubeSpace Satellite Systems	CubeTorquer Large	0.072	0.37	1.90	1	24
CubeSpace Satellite Systemse	CubeTorquer Coil(Single)	0.046	0.31	0.13	1	24
CubeSpace Satellite Systems	CubeTorquer Coil(Double)	0.074	0.64	0.27	1	24
GomSpace	Nano Torque GST-600	0.156	Unk	0.31 – 0.34	3	Unk

BODY-169

GomSpace	NanoTorque Z-axis Internal	0.106	Unk	0.139	1	Unk
ISISPACE	Magnetorquer Board	0.196	1.2	0.20	3	Unk
MEISEI	Magnetic Torque Actuator for Spacecraft	0.5	1	12	1	Unk
AAC Clyde Space	MTQ800	0.395	3	15	1	Unk
NanoAvionics	MTQ3X	0.205	0.4	0.30	3	20
NewSpace Systems	NCTR-M003	0.030	0.25	0.29	1	Unk
NewSpace Systems	NCTR-M012	0.053	0.8	1.19	1	Unk
NewSpace Systems	NCTR-M016	0.053	1.2	1.6	1	Unk
Rocket Lab	TQ-40	0.825	Unk	48.00	1	Unk
Rocket Lab	TQ-15	0.400	Unk	19.00	1	Unk
ZARM Technik**	MT0.2-1	0.012-0.014	0.135-0.25	0.2	1	NA*
ZARM Technik	MT0.5-1	0.009	0.275	0.5	1	NA*
ZARM Technik	MT0.7-1-01	0.035	0.5	0.7	1	NA*
ZARM Technik	MT1-1-01	0.065	0.23	1	1	NA*
ZARM Technik	MT1.5-1-01	0.097	0.4	1.5	1	NA*
ZARM Technik	MT2-1-02	0.1	0.5	2	1	NA*
ZARM Technik	MT3-1-D22042701	0.15	0.7	3	1	NA*
ZARM Technik	MT4-1	0.15	0.6	4	1	NA*
ZARM Technik	MT5-1	0.19-0.3	0.73-0.75	5	1	NA*
ZARM Technik	MT5-2	0.31	0.77	5	1	NA*
ZARM Technik	MT6-2	0.25-0.3	0.48-1.1	6	1	NA*
ZARM Technik	MT7-2	0.4	0.9	7	1	NA*
ZARM Technik	MT10-1	0.35-0.4	0.53-0.8	10	1	NA*

ZARM Technik	MT10-2	0.37-0.48	0.7-1	10	1	NA*
ZARM Technik	MT15-1	0.4-0.55	1.0-1.55	15	1	NA*
ZARM Technik	MT15-2	0.5-0.55	0.9-1.5	15	1	NA*

* Only EEE parts are connector and wires. Magnetorquer is not sensitive to ionizing radiation.

** ZARM Technik: Over 200 models available with design to mass/power optimization

5.2.4 Thrusters

Thrusters used for attitude control are described in Chapter 4: In-Space Propulsion. Pointing accuracy is determined by minimum impulse bit, and control authority by thruster force.

5.2.5 Star Trackers

A star tracker can provide an accurate estimate of the absolute three-axis attitude by comparing a digital image to an onboard star catalog (8). Star trackers identify and track multiple stars and provide three-axis attitude several times a second. Table 5-5 lists some models suitable for use on small spacecraft. For example, Arcsec's Sagitta Star Tracker was launched on the SIMBA cubesat in 2020.

Table 5-5: Star Trackers Suitable for Small Spacecraft								
Manufacturer	**Model**	**Mass (kg)**	**Power (W)**	**FOV**	**Cross axis accuracy (3s)**	**Twist accuracy (3s)**	**Radiation Tolerance (krad)**	**TRL**
Redwire Space	Star Tracker	0.475	2.5	14x19	10/27"	51"	75	7-9
Arcsec	Sagitta	0.275	1.4	25.4°	6	30	20	7-9
Arcsec	Twinkle	0.04	0.6	10.4°	30	180	Unk	7-9
Ball Aerospace	CT-2020	3.000	8	Unk	1.5"	1"	Unk	5-6
Berlin Space Technologies / AAC Clyde Space	ST200	0.040	0.65	22°	30"	200"	11	7-9
Berlin Space Technologies / AAC Clyde Space	ST400	0.250	0.75	15°	15"	150"	11	7-9
Blue Canyon Technologies	Standard NST	0.350	1.5	10° x 12°	6"	40"	Unk	7-9
Blue Canyon Technologies	Extended NST	1.300	1.5	10° x 12°	6"	40"	Unk	7-9
Creare	UST	0.840	Unk	Unk	7"	15"	Unk	5-6
CubeSpace Satellite Systems	CubeStar	0.055	0.264	58-47°	55.44" 0.02°	77.4	19	7-9
Danish Technical University	MicroASC	0.425	1.9	Unk	2"	Unk	Unk	7-9
Leonardo	Spacestar	1.600	6	20° x 20°	7.7"	10.6"	Unk	7-9
NanoAvionics	ST-1	0.108	1.2	21° full-cone	8"	50"	20	7-9
Rocket Lab	ST-16RT2	0.185	1	8° half-cone	5"	55"	Unk	7-9
Sodern	Auriga-CP	0.205	1.1	Unk	2"	11"	Unk	7-9
Sodern	Hydra-M	2.75	7	Unk	Unk	Unk	Unk	5-6
Sodern	Hydra-TC	5.3	8	Unk	Unk	Unk	Unk	5-6
Solar MEMS Technologies	STNS	0.14	1	12°	40"	70"	20	7-9
Space Micro	MIST	0.520	3	14.5°	15"	105"	30	7-9

Space Micro	μSTAR-100M	1.800	5	Unk	15"	105"	100	Unk
Space Micro	μSTAR-200M	2.100	8-10	Unk	15"	105"	100	Unk
Space Micro	μSTAR-200H	2.700	10	Unk	3"	21"	100	Unk
Space Micro	μSTAR-400M	3.300	18	Unk	15"	105"	100	Unk
Terma	T1	0.76	0.8	20° circular	2.2"	9"	100	5-6
Terma	T3	0.35	2	20° circular	2.6"	10"	8	5-6
Vectronic Aerospace	VST-41MN	0.7 - 0.9	2.5	14° x 14°	27"	183"	20	7-9
Vectronic Aerospace	VST-68M	0.470	3	14° x 14°	7.5"	45"	20	Unk

5.2.6 Magnetometers

Magnetometers provide a measurement of the local magnetic field which can be used to estimate 2-axis information about the attitude (9). Table 5-6 provides a summary of some three-axis magnetometers available for small spacecraft, one of which is illustrated in figure 5.4.

Figure 5.4: NSS Magnetometer. Credit: NewSpace Systems.

Table 5-6: Three-axis Magnetometers for Small Spacecraft							
Manufacturer	**Model**	**Mass (kg)**	**Power (W)**	**Resolution (nT)**	**Orth ogon ality**	**Radiation Tolerance (krad)**	**T R L**
GomSpace	NanoSense M315	0.008	Unk	Unk	Unk	Unk	7-9
AAC Clyde Space	MM200	0.012	0.01	1.18	Unk	30	7-9
MEISEI	3-Axis Magnetometer for Small Satellite	0.220	1.5	Unk	1°	Unk	7-9
NewSpace Systems	NMRM-Bn25o485	0.085	0.75	8	1°	10	7-9
AAC Clyde Space	MAG-3	0.100	Voltage Dependent	Unk	1°	10	7-9
ZARM Technik	Analogue High-Rel Fluxgate Magnetometer FGM-A-75	0.33	0.75 W	±75000	1°	50	9
ZARM Technik	Digital AMR Magnetometer AMR-D-100-EFRS485	0.18	0.3 W	±100000	1°	unk	6-7

5.2.7 Sun Sensors

Sun sensors are used to estimate the direction of the Sun in a spacecraft body frame. Sun direction estimates can be used for attitude estimation, though to obtain a three-axis attitude estimate at least one additional independent source of attitude information is required (e.g., the Earth nadir vector or the direction to a star). Because the Sun is easily identifiable and extremely bright, Sun sensors are often used for fault detection and recovery. However, care must be taken to ensure the Moon or Earth's albedo is not inadvertently perturbing the measurement.

Figure 5.5: Redwire Coarse Sun Sensor Detector (Cosine Type). Credit: Redwire Space.

There are several types of Sun sensors which operate on different principles.

Cosine detectors are photocells. Their output is the current generated by the cell, which is (roughly) proportional to the cosine of the angle between the sensor boresight and the Sun. Typically several cosine detectors (pointing in different directions) are used on a spacecraft for full sky coverage. Cosine detectors (e.g., figure 5.5) are inexpensive, low-mass, simple and reliable devices, but their accuracy is typically limited to a few degrees, and they do require analog-to-digital converters.

Quadrant detectors. Quadrant sun sensors typically operate by shining sunlight through a square window onto a 2 x 2 array of photodiodes. The current generated by each photodiode is a function of the direction of the Sun relative to the sensor boresight. The measured currents from all four cells are then combined mathematically to produce the angles to the Sun.

Digital Sun Sensor. The Sun illuminates a narrow slit behind which, is located a geometric coded bit mask and a number of photodiodes under the mask. Depending on the angle to the Sun, the photodiodes will be illuminated as per the geometric pattern resulting in correpondingly different photocurrents which are then amplified and thresholded against an average value. Given the known slit geometries, this digital bit output can be then converted to a sun angle.

Sun Camera. Some sun sensors are build as a small camera imaging the Sun. Since the Sun is so bright, the optics will include elements to decrease the thoughput. A computer will identify the image of the Sun and calculate the centroid. Sun sensors can be made very accurate this way. Sometimes, multiple apertures are included to increase accuracy.

Examples of small spacecraft sun sensors are described in table 5-7.

Table 5-7: Small Spacecraft Sun Sensors										
Manufacturer	**Model**	**Sensor Type**	**Mass (kg)**	**Peak Power (W)**	**Analog or Digital**	**FOV**	**Accuracy (3s)**	**# Measurement Angles**	**Radiation Tolerance (krad)**	**TRL**
Redwire Space	Coarse Analog Sun Sensor	Coarse Analog Sun Sensor	0.045	0	Analog	±40° (Can be modified to meet specific FOV requirements)	±1°	1	>100	7-9
Redwire Space	Coarse Sun Sensor (Cosine Type)	Coarse Sun Sensor (Cosine Type)	0.010	0	Analog	APPROXIMATE COSINE, CONICAL SYMMETRY	±2° to ±5°	Depends on configuration	>100	7-9
Redwire Space	Coarse Sun Sensor Pyramid	Coarse Sun Sensor Pyramid	0.13	0	Analog	2π STERADIAN PLUS	±1° to ±3°	2	>100	7-9
Redwire Space	DIGITAL SUN SENSOR (±32°)	DIGITAL SUN SENSOR (±32°)	Sensor 0.3 kg Electronics ~1	1	Digital	±32° x ±32° (each sensor)	±0.125°	2	100	7-9
Redwire Space	Digital Sun Sensor (±64°)	Digital Sun Sensor (±64°)	Sensor0.25 Electronics 0.29 - 1.1	0.5	Digital	128° X 128° (EACH SENSOR) NOTE: 4π STERADIANS ACHIEVED WITH 5 SENSORS	±0.25°	2	100	7-9
Redwire Space	Fine Pointing Sun Sensor	Fine Pointing Sun Sensor	Sensor .95 Electronics 1.08	< 3	Digital	±4.25° x ±4.25° (Typical)	Better than ±0.01°	2	100	7-9
Redwire Space	Fine Spinning Sun Sensor (±64°)	Fine Spinning Sun Sensor (±64°)	Sensor 0.109 Electronics 0.475 – 0.725	0.5	Analog and Digital	±64° FAN SHAPED (each sensor)	±0.1°	1 plus Sun Pulse	100	7-9

Redwire Space	Micro Sun Sensor	Micro Sun Sensor	< 0.002	< 0.02	Analog	± 85° MINIMUM	±5°	2	Approx. 10	5-6
Redwire Space	Miniature Spinning Sun Sensor (±87.5°)	Miniature Spinning Sun Sensor (±87.5°)	< 0.25	0.5	Digital	±87.5° (FROM NORMAL TO SPIN AXIS)	±0.1°	1 plus Sun Pulse	100	7-9
Redwire Space	FINE SUN SENSOR (±50°)	FINE SUN SENSOR (±50°)	Unk	Unk	Digital	100 X 100 Each Sensor	±0.01° TO ±0.05°	2	100, 150, or 300	7-9
Bradford Space	CoSS	Cosine	0.024	0	Analog	160° full cone	3°	1	40000	7-9
Bradford Space	CoSS-R	Cosine	0.015	0	Analog	180° full cone	3°	1	120000	7-9
Bradford Space	CSS-01, CSS-02 Only shows one CSS	Cosine	0.215	0	Analog	180° full cone	1.5°	2	70000	7-9
Bradford Space	FSS	Quadrant	0.375	0.25	Analog	128° x 128°	0.3°	2	100	7-9
Bradford Space	Mini-FSS	Quadrant	0.050	0	Analog	128° x 128°	0.2° With on-board implementation	2	20000	7-9
CubeSpace Satellite Systems	CubeSense	Camera	0.030	<0.2	Digital	180°	0.2°	2	24	7-9
GomSpace	NanoSense FSS	Quadrant	0.002	Unk	Digital	{45°, 60°}	{±0.5°, ±2°}	2	Unk	Unk
AAC Clyde Space	SS200	Unk	.003	0.04	Digital	110°	<1°	Unk	>36	7-9
Lens R&D	BiSon64-ET	Quadrant	0.023	0	Analog	±58° per axis	0.5°	2	9200	7-9
Lens R&D	BiSon64-ET-B	Quadrant	0.033	0	Analog	±58° per axis	0.5°	2	9200	7-9
Lens R&D	MAUS	Quadrant	0.014	0	Analog	±57° per axis	0.5°	2	9200	7-9

NewSpace Systems	NFSS-411	Unk	0.035	0.150	Digital	140°	0.1°	TBD	20	7-9
NewSpace Systems	NCSS-SA05	Unk	0.005	0.05	Analog	114°	0.5°	TBD	Unk	7-9
Solar MEMS Technologies	nanoSSOC-A60	Orthogonal	0.004	0.007	Analog	±60° per axis	0.5°	2	100	7-9
Solar MEMS Technologies	nanoSSOC-D60	Orthogonal	0.007	0.076	Digital	±60° per axis	0.5°	2	30	7-9
Solar MEMS Technologies	SSOC-A60	Orthogonal	0.025	0.01	Analog	±60° per axis	0.5°	2	100	7-9
Solar MEMS Technologies	SSOC-D60	Orthogonal	0.035	0.315	Digital	±60° per axis	0.5°	2	30	7-9
Solar MEMS Technologies	ACSS	Quadrant & Redundant	0.035	0.072	Analog	±60° per axis	0.5°	2	200	7-9
Space Micro	CSS-01, CSS-02	Cosine	0.010	0	Analog	120° full cone	5°	1	100	7-9
Space Micro	MSS-01	Quadrant	0.036	0	Analog	48° full cone	1°	2	100	7-9

5.2.8 Horizon Sensors

Horizon sensors can be simple infrared horizon crossing indicators (HCI), or more advanced thermopile sensors that can detect temperature differences between the poles and equator. For terrestrial applications, these sensors are referred to as Earth Sensors, but can be used for other planets. Examples of such technologies are described in table 5-8 and illustrated in figure 5.6.

Figure 5.6: MAI-SES. Credit: Redwire Space.

In addition to the commercially-available sensors listed in table 5-8, there has been some recent academic interest in horizon sensors for CubeSats with promising results (24) (10) (11).

Table 5-8: Commercially Available Horizon Sensors

Manufacturer	Model	Sensor Type	Mass (kg)	Peak Power (W)	Analog or Digital	Accuracy	# Measurement Angles	Rad Tolerance (krad)	TRL
CubeSpace Satellite Systems	CubeSense	Camera	0.030	0.200	Digital	0.2°	2	24	7-9
Servo	Mini Digital HCI	Pyroelectric	0.050	Voltage Dependent	Digital	0.75°	Unk	Unk	7-9
Servo	RH 310 HCI	Pyroelectric	1.5	1	Unk	0.015°	Unk	20	Unk
SITAEL	Digital Earth Sensor	Microbolometer	0.4	<2	Digital	<1°	Unk	Unk	Unk
Solar MEMS Technologies	HSNS	Infrared	0.120	0.150	Digital	1°	2	30	7-9

5.2.9 Inertial Sensing

Inertial sensors include gyroscopes for measuring angular change and accelerometers for measuring velocity change. They are packaged in different ways that range from single-axis devices (i.e., a single gyroscope or accelerometer), to packages which include 3 orthogonal axes of gyroscopes (Inertial Reference Unit (IRU)) to units containing 3 orthogonal gyros and 3 orthogonal accelerometers (Inertial Measurement Unit (IMU)). These sensors are frequently used to propagate the vehicle state between measurement updates of a non-inertial sensor. For example, star trackers typically provide attitude updates at a few Hertz. If the control system requires accurate knowledge between star tracker updates, then an IMU may be used for attitude propagation between star tracker updates.

Gyroscope technologies typically used in modern small spacecraft are fiber optic gyros (FOGs) and MEMS gyros, with FOGs usually offering superior performance at a mass and cost penalty

(12). Other gyroscope types exist (e.g., resonator gyros, ring laser gyros), but these are not common in the SmallSat/CubeSat world due to size, weight, and power (SWaP) and cost considerations.

Gyro behavior is a complex topic (13) and gyro performance is typically characterized by a multitude of parameters. Table 5-9 only includes bias stability and angle random walk for gyros, and bias stability and velocity random walk for accelerometers, as these are often the driving performance parameters. That said, when selecting inertial sensors, it is important to consider other factors such as dynamic range, output resolution, bias, sample rate, etc.

Table 5-9: Gyros Available for Small Spacecraft													
						Gyros				Accelerometers			
							Bias Stability		ARW		Bias Stability		VRW
Manufacturer	Model	Sensor Type	Technology	Mass (kg)	Power (W)	# Axes	(°/hr)	stat	(°/rt(hr))	# Axes	(µg)	stat	(m/sec)/rt(hr)
Emcore	QRS11	Gyro	MEMS	≤0.06	0.8	1	6	Typical	N/A	N/A	N/A	N/A	N/A
Emcore	QRS28	Gyro	MEMS	≤0.025	0.5	2	N/A	N/A	N/A	N/A	N/A	N/A	N/A
Honeywell	MIMU	IMU	RLG	4	34	3	0.05	Unk	0.01	Unk	100	Unk	Unk
Honeywell	HG1700	IMU	RLG	0.9	5.000	3	1.000	1σ	0.125	3	1000	1σ	0.65
L3	CIRUS	Gyros	FOG	15.400	40.000	3	0.000	1σ	0.100	0	N/A	Unk	N/A
NewSpace Systems	NSGY-001	IRU	Image-based rotation estimate	0.055	0.200	3	N/A		N/A	0	N/A	Unk	N/A
Northrop Grumman	LN-200S	IMU	FOG, SiAc	0.748	12	3	1.000	1σ	0.070	3	300	1σ	Unk
NovAtel	OEM-IMU-STIM300	IMU	MEMS	0.055	1.50	3	0.500	TBD	0.150	3	50	TBD	0.060
Safran	STIM202	IRU	MEMS	0.055	1.500	3	0.400	TBD	0.170	0	N/A	TBD	N/A
Safran	STIM210	IRU	MEMS	0.052	1.500	3	0.300	TBD	0.150	0	N/A	TBD	N/A
Safran	STIM300	IMU	MEMS	0.055	2.000	3	0.300	TBD	0.150	3	50	TBD	0.07

BODY-181

Safran	STIM318	IMU	MEMS	0.057	2.500	3	0.300	TBD	0.150	3	3	TBD	0.015
Safran	STIM320	IMU	MEMS	0.057	2.500	3	0.300	TBD	0.100	3	3	TBD	0.015
Safran	STIM277H	IRU	MEMS	0.052	1.500	3	0.300	TBD	0.150	0	N/A	TBD	N/A
Safran	STIM377H	IMU	MEMS	0.055	2.000	3	0.300	TBD	0.150	3	50	TBD	0.07
Silicon Sensing Systems	CRH03	Gyro	MEMS	0.42	0.2W	1	CRH03-010 – 0.03 CRH03-025 – 0.04 CRH03-100 – 0.04 CRH03-200 – 0.05 CRH03-400 – 0.1		CRH03-010 – 0.005 CRH03-025 – 0.006 CRH03-100 – 0.006 CRH03-200 – 0.008 CRH03-400 – 0.010	0	N/A	-	N/A
Silicon Sensing Systems	CRH03 (OEM)	Gyro	MEMS	0.18	0.2W	1	CRH03-010 – 0.03 CRH03-025 – 0.04 CRH03-100 – 0.04 CRH03-200 – 0.05 CRH03-400 – 0.1		CRH03-010 – 0.005 CRH03-025 – 0.006 CRH03-100 – 0.006 CRH03-200 – 0.008 CRH03-	0	N/A	-	N/A

									400 – 0.010				
Silicon Sensing Systems	RPU30	Gyro	MEMS	1.35	<0.8W	3	0.06		0.006	0	N/A	-	N/A
Silicon Sensing Systems	DMU41	9 DoF IMU	MEMS	<2	<1.5W	3	0.1		0.015	3	15	-	0.05
Silicon Sensing Systems	CAS	Acc	MEMS	0.004	Unk	0	N/A		N/A	2	CAS2X1S - 7.5 CAS2X2S - 7.5 CAS2X3S - 7.5 CAS2X4S - 25 CAS2X5S - 75		CAS2X1S - TBC CAS2X2S - TBC CAS2X3S - TBC CAS2X4S - TBC CAS2X5S - TBC
VectorNav	VN-100*	IMU + magnetometers +barometer	MEMS	0.015	0.220	3	10.000	max	0.210	3	40	max	0.082
VectorNav	VN-110*	IMU + magnetometers	MEMS	0.125	2.500	3	1.000	max	0.0833	3	10	max	0.024

*Small form-factor versions of these products available.

5.2.10 GPS Receivers

For low-Earth orbit spacecraft, GPS receivers are now the primary method for performing orbit determination, replacing ground-based tracking methods. Onboard GPS receivers are now considered a mature technology for small spacecraft, and some examples are described in table 5-10. There are also next-generation chip-size COTS GPS solutions, for example the NovaTel OEM 719 board has replaced the ubiquitous OEMV1.

GPS accuracy is limited by propagation variance through the exosphere and the underlying precision of the civilian use C/A code (14). GPS units are controlled under the Export Administration Regulations (EAR) and must be licensed to remove Coordinating Committee for Multilateral Export Control (COCOM) limits (15).

Although the usability of GPS is limited to LEO missions, past experiments have demonstrated the ability of using a weak GPS signal at GSO, and potentially soon to cislunar distances (16) (17). Development and testing in this fast-growing area of research and development may soon make onboard GPS receivers more commonly available.

Table 5-10: GPS Receivers for Small Spacecraft

Manufacturer	Model	Mass (kg)	Power (W)	Accuracy (m)	Radiation Tolerance (krad)	TRL
AAC Clyde Space	GNSS-701	0.16	Unk	<5	10	7-9
APL	Frontier Radio Lite	0.4	1.4	15	20	5-6
General Dynamics	Explorer	1.2	8	15	Unk	7-9
General Dynamics	Viceroy-4	1.1	8	15	Unk	7-9
GomSpace	GPS-kit	0.031	1.3	1.5	18	7-9
SkyFox Labs	piNAV-NG	0.024	0.124	10	30	7-9
Spacemanic	Celeste_gnss_rx	0.025	~0.1	1.5	40	7-9
Surrey Satellite Technology	SGR-Ligo	0.09	0.5	5	5	7-9
Syrlinks	GPS (L1/L5) GALILEO (E1/E5/E6) BeiDou (B1/B3)	0.435	Unk	<0.1	15	5-6

5.2.11 Deep Space Navigation

Figure 5.7: General Dynamics SDST. Credit: General Dynamics.

In deep space, navigation is performed using radio transponders in conjunction with the Deep Space Network (DSN). As of 2020, the only deep space transponder with flight heritage suitable for small spacecraft was the JPL-designed and General Dynamics-manufactured Small Deep Space Transponder (SDST). JPL has also designed IRIS V2, which is a deep space transponder that is more suitable for the CubeSat form factor. Table 5-11 details these two radios, and the SDST is illustrated in figure 5.7. IRIS V2, derived from the Low Mass Radio Science Transponder (LMRST), has flown on the MarCO CubeSats in 2018, LICIACube that performed an asteroid flyby in September 2022, 12U lunar CAPSTONE spacecraft that entered a lunar orbit November 13, 2022, and was on six Artemis 1 secondary CubeSat payloads (Lunar Flashlight, LunaH-Map, ArgoMoon, CubeSat for Solar Particles, Biosentinel, and NEA Scout). It is also scheduled to fly on INSPIRE (18).

Table 5-11: Deep Space Transponders for Small Spacecraft

Manufacturer	Model	Mass (kg)	Rx Power (W)	Bands	Radiation Tolerance (krad)	TRL
General Dynamics	SDST	3.2	12.5	X, Ka	50	7-9
Space Dynamics Laboratory	IRIS V2.1	1.1	10.3	X, Ka, S, UHF	15	7-9

5.2.12 Atomic Clocks

Atomic clocks have been used on larger spacecraft in low-Earth orbit for several years now, however integrating them on small spacecraft is relatively new. Table 5-12 provides examples of commercially available atomic clocks and oscillators for SmallSats. The conventional method for spacecraft navigation is a two-way tracking system of ground-based antennas and atomic clocks. The time difference from a ground station sending a signal and the spacecraft receiving the response can be used to determine the spacecraft's location, velocity, and (using multiple signals) the flight path. This is not a very efficient process, as the spacecraft must wait for navigation commands from the ground station instead of making real-time decisions, and the ground station can only track one spacecraft at a time, as it must wait for the spacecraft to return a signal (19). In deep space navigation, the distances are much greater from the ground station to spacecraft, and the accuracy of the radio signals needs to be measured within a few nanoseconds.

More small spacecraft designers are developing their own version of atomic clocks and oscillators that are stable and properly synchronized for use in space. They are designed to fit small spacecraft, for missions that are power- and volume-limited or require multiple radios.

Manufacturer	Model	Dimensions (mm)	Mass (kg)	Power (W)	Frequency Range (MHz)	Rad Tolerance (krad)	TRL
AccuBeat	Ultra Stable Oscillator	131 x 120 x 105	2	6.5	57.51852	50	7-9
Bliley Technologies	Iris Series 1"x1" OCXO for LEO	19 x 11 x 19	0.016	1.5	10 -100	39	7-9
	Aether Series TCVCXO for LEO	21 x 14 x 8	Unk	0.056	10 - 150	37	Unk
Microsemi	Space Chip Scale Atomic Clock (CSAC)	41 x 36 x 12	0.035	0.12	10	20	5-6
Safran Timing Technologies SA	MO	44 x 54 x 57	0.22	3.5 Nom 5.5 Max	10	100	Unk
	Space Qualified mRO-50	51 x 51 x 20	0.080	0.4 Nom	10	25 Min	Unk
	miniRAFS	108 x 53 x 68	0.45	< 12 Max	60 and 10	TBD	Unk
	LNMO	50 x 50 x 30	0.1	1.5 Nom 2.5 Max	5 – 40	100	Unk

Table 5-12: Atomic Clocks and Oscillators for Small Spacecraft

5.2.13 LiDAR

Light Detection and Ranging (LiDAR) is new type of sensor that is emerging. The technology has matured in terrestrial applications (such as automotive applications) over the last decade and is used in larger spacecraft that are capable of proximity operations, like Orion. This sensor type has applications for small spacecraft altimetry and relative navigation (e.g., a Mars helicopter, rendezvous and docking, and formation flying). Table 5-13 lists examples of flown LiDARs.

Table 5-13: Lidar for Small Spacecraft

Manufacturer	Model	Mass (kg)	Power (W)	Max Range (m)	Radiation Tolerance (krad)	TRL
Garmin	Lidar Lite V3	0.022	0.7	40	Unk	5-6*
ASC	GSFL-4K (3D)	3	30	>1 km in altimeter mode	Unk	7-9

*Specific units were qualified for Mars Ingenuity helicopter. Product line in general is not space qualified.

5.3 On the Horizon

Figure 5.8: High-speed magnetically levitated reaction wheel. Credit: Celeroton AG.

In general, technological progress in guidance, navigation, and control is advancing quickly in automotive research areas but is lagging slightly in the aerospace industry. Given the high maturity of existing GNC components, future developments in GNC are mostly focused on incremental or evolutionary improvements, such as decreases in mass and power, and increases in longevity and/or accuracy. This is especially true for GNC components designed for deep space missions that have only very recently been considered for small spacecraft. However, in a collaborative effort between the Swiss Federal Institute of Technology and Celeroton, there is progress being made on a high-speed magnetically levitated reaction wheel for small satellites (figure 5.8). The idea is to eliminate mechanical wear and stiction by using magnetic bearings rather than ball bearings. The reaction wheel implements a dual hetero/homopolar, slotless, self-bearing, permanent-magnet synchronous motor (PMSM). The fully active, Lorentz-type magnetic bearing consists of a heteropolar self-bearing motor that applies motor torque and radial forces on one side of the rotor's axis, and a homopolar machine that exerts axial and radial forces to allow active control of all six degrees of freedom. It can store 0.01 Nm of momentum at a maximum of 30,000 rpm, applying a maximum torque of 0.01 Nm (21).

Several projects funded via NASA's Small Spacecraft Technology (SST) program through the University Smallsat Technology Partnerships (USTP) initiative have begun advancing GNC systems. Listed below in table 5-14 are projects that focused on GNC advancement, and further information can be found at the USTP website:

https://www.nasa.gov/smallspacecraft/university-smallsat-technology-partnership-initiative/

Each presentation is from the USTP Technology Exposition that was held in May 2021 and June 2022.

Table 5-14: STP Initiative GNC Projects

Project	University	Current Status	Reference
On-Orbit Demonstration of Surface Feature-Based Navigation and Timing	University of Texas, Austin	Still in development	USTP Technology Expo presentation
Autonomous Nanosatellite Swarming (ANS) using Radio Frequency and Optical Navigation	Stanford University	Onboard Starling mission (Launched in 2023)	USTP Technology Expo presentation
Distributed multi-GNSS Timing and Localization (DiGiTaL)	Stanford University	Leveraged technology used in Starling mission	USTP Technology Expo presentation

Mems Reaction Control and Maneuvering for Picosat beyond LEO	Purdue University	Awarded a suborbital flight test through NASA's Flight Opportunities program	(29)
A Small Satellite Lunar Communications and Navigation System	University of Boulder, Colorado	Still in development	USTP Technology Expo presentation
A high-precision continuous-time PNT compact module for the LunaNet small spacecraft	University of California, Los Angeles	Still in development	USTP Technology Expo presentation

5.4 Summary

Conventional small spacecraft GNC technology is a mature area, with many high TRL components previously flown around Earth offered by several different vendors. These GNC techniques are generally semi/non-autonomous as on-board observations are collected with the assistance of ground-based intervention. As the interest for deep space exploration with small spacecraft grows, semi-to-fully autonomous navigation methods must advance. It is likely that future deep space navigation will rely solely on fully autonomous GNC methods that require zero ground-based intervention to collect/provide navigation data. This is a desirable capability as the spacecraft's dependence on Earth-based tracking resources (such as DSN) is reduced and the demand for navigation accuracy increases at large distances from Earth. However, current methods advancing deep space navigation involve both ground- and space-based tracking in conjunction with optical navigation techniques. To support this maturity, the small spacecraft industry has seen a spike in position, navigation, and timing (PNT) technology progression in inertial sensors and atomic clocks, and magnetic navigation for near-Earth environments.

Other GNC advances involve research on SmallSats performing on-orbit proximity operations. Several research papers have discussed ways to accomplish this, and previous extravehicular free flyers have demonstrated this innovative capability in the past few decades. The CubeSat Proximity Operations Demonstration (CPOD) project is the most recent CubeSat mission to attempt the characterization of low-power proximity operations technologies, however its mission ended June 2023 and was unable to demonstrate rendezvous, proximity operations and docking maneuvers as planned. Seeker, a 3U CubeSat that was deployed September 2019, was built to demonstrate safe operations around a target spacecraft with core inspection capabilities. While Seeker was unable to perform its underlining goal, there were still several benefits for improving future missions (29).

The rising popularity of SmallSats in general, and CubeSats in particular, means there is a high demand for components, and engineers are often faced with prohibitive prices. The Space Systems Design Studio at Cornell University is tackling this issue for GNC with their PAN nanosatellites. A paper by Choueiri et al. outlines an inexpensive and easy-to-assemble solution for keeping the ADCS system below $2,500 (22). Lowering the cost of components holds exciting implications for the future and will likely lead to a burgeoning of the SmallSat industry.

For feedback solicitation, please email: arc-sst-soa@mail.nasa.gov. Please include a business email so someone may contact you further.

References

(1) L.C.G Shepherd. and A.F.S.C Shepherd. "Space Surveillance Network." Shared Space Situational Awareness Conference. Colorado Springs. 2006.

(2) D. Vallado, P. Crawford, R. Hujsak and T.S. Kelso. "Revisiting Spacetrack Report #3," AIAA 2006-6753. AIAA/AAS Astrodynamics Specialist Conference and Exhibit. August 2006. Thornton, C L and Border, J S: "Radiometric Tracking Techniques for Deep-Space Navigation." s.l: John Wiley & Sons, 2003.

(3) JPL. "MarCO: Mission Overview. CubeSat." [Online] 2020. Available at: https://www.jpl.nasa.gov/CubeSat/missions/marco.php

(4) J.R. Wertz. "Spacecraft attitude determination and control." Springer Science & Business Media. Vol. 73. 2012,

(5) Jin, J, Ko, S and Ryoo, C K: "Fault Tolerant Control for Satellites with Four Reaction Wheels." vol. 16, no. 10, pp. 1250–1258. 2008.

(6) R. Kulczycki and P. Wisniewski. "Slew Maneuver Control for Spacecraft Equipped with Star Camera And Reaction Wheels." Vol. 13, no. 3, pp. 349–356. 2005.

(7) C.C. Liebe, "Star Trackers for Attitude Determination," IEEE Aerospace and Electronic Systems Magazine, vol. 10, no. 6, pp. 10-16, June 1995, doi: 10.1109/62.387971

(8) M.L. Psiaki, F. Martel, F and P.K. Pal. "Three-Axis Attitude Determination Via Kalman Filtering of Magnetometer Data." Vol. 13, no. 3, pp. 506–514. 1990.

(9) J.H. Wessels. "Infrared Horizon Sensor for CubeSat Implementation." Master's Thesis, Stellenbosch University. March 2018.

(10) A. Pelemeshko et al. "High-Precision Cubesat Sun Sensor Coupled with Infrared Earth Horizon Detector." IOP Conf. Ser.: Mater. Sci. Eng. Vol. 734. pp. 0121-8. 2020,

(11) D. Greenheck et al. "Design and Testing of a Low-Cost MEMS IMU Cluster for SmallSat Applications." 28th Annual AIAA/USU Conference on Small Satellites, 2014.

(12) S. Merhav. "Aerospace Sensor Systems and Applications". Springer New York, 1998.

(13) O. Montenbruck et al. "Precision Spacecraft Navigation Using a Low-Cost GPS Receiver." Vol. 16, no. 4, pp. 519–529. 2014.

(14) "Foreign Availability Determination Procedures and Criteria." Office of the Federal Register. Title 15 Part 768.7. 2015.

(15) J. Foust: "GPS in Space." MIT Technology Review. [Online] January 2002. https://www.technologyreview.com/2002/01/01/275613/gps-in-space/

(16) A. Hadhazy. "Cosmic GPS." Aerospace America. [Online] May 2020. Available at: https://aerospaceamerica.aiaa.org/features/cosmic-gps/

(17) F.H. Aguirre. "X-Band Electronics for The INSPIRE CubeSat Deep Space Radio." IEEE Aerospace Conference. 2015.

(18) D. Baird. "NASA Tests Atomic Clock for Deep Space Navigation." [Online] 2018. Available at: https://www.jpl.nasa.gov/news/news.php?feature=7053

(19) D. Cornwell. "NASA's Deep Space Atomic Clock and Optical Communications Program for PNT Applications." [Online] 2016. Accessed June 19, 2018. Available at: https://www.gps.gov/governance/advisory/meetings/2016-12/cornwell.pdf

(20) J.W Kolar et al. "High-Speed Magnetically Levitated Reaction Wheels for Small Satellites." Anacapri, Capri: 23rd International Symposium on Power Electronics, Electrical Drives, Automation and Motion (SPEEDAM 2016), 2016.

(21) M.N. Choueiri, M. Bell, and M.A. Peck. "Cost-Effective and Readily Manufactured Attitude Determination and Control System for NanoSatellites." AIAA/ASCE/AHS/ASC Structures, Structural Dynamics, and Materials Conference, AIAA SciTech Forum. 2018.

(22) D. Dickinson. "NICER and SEXTANT demonstrate XNAV pulsar navigation system that may be used on Artemis." [Online] 2020. Available at: https://skyandtelescope.org/astronomy-news/nasa-to-use-pulsar-navigation-for-deep-space-missions/

(23) K. Kapás, T. Bozóki, G. Dálya et al. "Attitude determination for nano-satellites – I. Spherical projections for large field of view infrasensors." Exp Astron 51, 515–527, 2021.

(24) M. M. Kobayashi et al. "The Iris Deep-Space Transponder for the SLS EM-1 Secondary Payloads," in IEEE Aerospace and Electronic Systems Magazine, vol. 34, no. 9, pp. 34-44, 1 Sept. 2019.

(25) NASA: "Deep Space Atomic Clock." [Online] 2021. Available at: https://www.jpl.nasa.gov/missions/deep-space-atomic-clock-dsac

(26) M. M. Kobayashi et al., "The Iris Deep-Space Transponder for the SLS EM-1 Secondary Payloads," in IEEE Aerospace and Electronic Systems Magazine, vol. 34, no. 9, pp. 34-44, 1 Sept. 2019, doi: 10.1109/MAES.2019.2905923.

(27) NASA. "CubeSat Proximity Operations Demonstration (CPOD)." [Online] July 1, 2022. Available at: https://www.nasa.gov/directorates/spacetech/small_spacecraft/cpod_project.html

(28) S. M. Pedrotty, "Seeker Overview and Mission 1 Review-- A New Development Approach for In-Space Inspectors," [Online] 2021. Available at: https://www.nasa.gov/smallsat-institute/seeker-overview-and-mission-1-review--a-new-development-approach-for-in-space-inspectors

(29) A.G. Cofer. "FILM EVAPORATION MEMS THRUSTER ARRAY FOR MICROPROPULSION." Open Access Dissertations. 1106. 2014. Available at: https://core.ac.uk/download/pdf/220145833.pdf

Table of Contents

Glossary

(ABS)	Acrylonitrile Butadiene Styrene
(ACS3)	Advanced Composite Solar Sail System
(AE)	Aerospace Corporation Electron
(AM)	Additive manufacturing
(AMODS)	Autonomous On-orbit Diagnostic System
(AP)	Aerospace Corporation Proton
(CAM)	Computer Aided Manufacturing
(CFRP)	Carbon Fiber Reinforced Polymers
(CNC)	Computerized Numerical Control
(COBRA)	Compact On-Board Robotic Articulator
(COTS)	Commercial-off-the-shelf
(CSLI)	CubeSat Launch Initiative
(CTD)	Composite Technology Deployment
(CTE)	Coefficient of Thermal Expansion
(DCB)	Deployable Composite Boom
(DDD)	Displacement Damage Dose
(DLP)	Digital Light Projection
(DOF)	Degrees of Freedom
(EEE)	Electrical, Electronic and Electro-mechanical
(EELV)	Evolved Expendable Launch Vehicle
(ESD)	Electrostatic Discharge
(ESPA)	EELV Secondary Payload Adapter
(FDM)	Fused Deposition Modeling
(FFF)	Fused Filament Fabrication
(FPGAs)	Field Programmable Gate Arrays
(FST)	Flame, Smoke, and Toxicity
(GCD)	Game Changing Development
(GEVS)	General Environmental Verification Standard
(HDT)	Heat Deflection Temperature
(ISS)	International Space Station
(MOSFETs)	Metal Oxide Semiconductor Field Effect Transistors
(PAEK)	Polyaryletherketone
(PC)	Polycarbonate

(PCB)	Printed Circuit Board
(PEEK)	Polyetheretherketone
(PEI)	Polyetherimide
(PEKK)	Polyetherketoneketone
(PLA)	Polylactic Acid
(PLEO)	Polar Low-Earth Orbit
(PSC)	Planetary Systems Corporation
(RECS)	Robotic Experimental Construction Satellite
(ROC)	Roll Out Composite
(SADA)	Solar Array Drive Actuator
(SEUs)	Single Event Upsets
(SLA)	Stereolithography
(SLS)	Selective Laser Sintering
(SPEs)	Solar Particle Events
(STELOC)	Stable Tubular Extendable Lock-Out Composite
(TID)	Total Ionizing Dose
(TRAC)	Triangle Rollable and Collapsible
(TRL)	Technology Readiness Level
(ULA)	United Launch Alliance

6.0 Structure, Mechanisms, and Materials

6.1 Introduction

Material selection is of primary importance when considering small spacecraft structures. Requirements for both physical properties (density, thermal expansion, and radiation resistance) and mechanical properties (modulus, strength, and toughness) must be satisfied. The manufacture of a typical structure involves both metallic and non-metallic materials, each offering advantages and disadvantages. Metals tend to be more homogeneous and isotropic, meaning properties are similar at every point and in every direction. Non-metals, such as composites, are inhomogeneous and anisotropic by design, meaning properties can be tailored to directional loads. Recently, resin or photopolymer-based AM has advanced sufficiently to create isotropic parts. In general, the choice of structural materials is governed by the operating environment of the spacecraft, while ensuring adequate margin for launch and operational loading. Deliberations must include more specific issues, such as thermal balance and thermal stress management. Payload or instrument sensitivity to outgassing and thermal displacements must also be considered.

Additive manufacturing (AM) has increased custom structural solutions for SmallSats and demonstrated high throughput of complex structures. Materials that were once out of reach of AM are now readily available in higher end systems. Once only for secondary structures, AM has seen an expansion in primary structures – especially in small CubeSat or PocketQube buses.

However, for larger CubeSats and Evolved Expendable Launch Vehicle (EELV) Secondary Payload Adapter (ESPA) SmallSats, conventionally machined assemblies constructed from aluminum alloys still have their place for primary structures. Secondary structures, such as solar panels, thermal blankets, and subsystems, are attached to primary structures. They stand on their own and transmit little to no critical structural loads. When a primary structure fails, catastrophic failure of the mission occurs, and while failure of a secondary structure typically does not affect the integrity of the spacecraft, it can have a significant impact on the overall mission. These structural categories serve as a good reference but can be hard to distinguish for small spacecraft that are particularly constrained by volume. This is especially true for SmallSats, as the capabilities of these spacecraft may be similar to full size buses, but the volume afforded by dispensers or deployment rings becomes the constraining factor. Therefore, it is imperative that structural components are as volume efficient as possible. The primary structural components need to serve multiple functions to maximize volume efficiency. Such functions may include thermal management, radiation shielding, pressure containment, and even strain actuation. These are often assigned to secondary structural components in larger spacecraft.

Structural design is not only affected by different subsystems and launch environments, but also the spacecraft application and intended environment. There are different configurations for spin-stabilized and 3-axis stabilized systems, and the instrumentation used places requirements on the structure. Some instruments require mechanisms, such as deployable booms, to create enough distance between a magnetometer and the spacecraft to minimize structural effects on the measurement. The spacecraft exterior and interior material and electronic subsystems need to be understood in the specific mission environment (e.g., in-space charging effects). Mitigation for charge build-up is provided in section 6.3.2 Thermoplastics and Photopolymers.

Highly configurable or modular systems may be desirable in quick-turn products, as prototyping and firmware and software development can be extended further into the spacecraft design cycle with flight hardware in the loop. Card slot systems not only provide those benefits, but when paired with certain standards, they can still fulfill the same structural, mechanical, and thermal requirements as the current CubeSat method of "stacking" electronics and payloads.

Small satellite mechanisms have advanced with deployable structures, actuators, and switches. Deployable structures enable large structural applications with minimal volume requirements. Actuator and switch mechanisms expand the capabilities of small satellites with motion and deployment applications. These mechanisms enable increased small satellite capabilities beyond original structural volume constraints.

An overview of radiation effects and some mitigation strategies is included in this chapter because radiation exposure can impact the structural design of small spacecraft. For SmallSats operating out of low-Earth orbit with increased radiation exposure, mission planners may also want to consider risk mitigation strategies associated with specific radiation environments. This includes both interplanetary missions, where solar radiation dominates, and polar low-Earth orbit (PLEO) missions, where solar radiation risk increases over the poles. In addition, as solar maximum approaches in 2025 (1) with an increased number of solar particle events (SPEs), mission planners will need to consider many orbital environments.

The information described below is not intended to be exhaustive but provides an overview of current state-of-the-art technologies and their development status for a particular small spacecraft subsystem. It should be noted that Technology Readiness Level (TRL) designations may vary with changes specific to payload, mission requirements, reliability considerations, and/or the environment in which performance was demonstrated. Readers are highly encouraged to reach out to companies for further information regarding the performance and TRL of described technology. There is no intention of mentioning certain companies and omitting others based on their technologies or relationship with NASA.

6.2 State-of-the-Art – Primary Structures

6.2.1 CubeSat Standard

Two general approaches are common for primary structures, often called frames or chassis, in the small spacecraft market: commercial-off-the-shelf (COTS) structures and custom machined or printed components. It is not surprising that most COTS offerings are for the CubeSat market. Often COTS structures can simplify development, but only when the complexity of the mission, subsystems, and payload requirements fall within the design intent of a particular COTS structure. Custom machined structures enable greater flexibility in mission specific system and payload design. The typical commercially available structure has been designed for low-Earth orbit applications and limited mission durations, where shielding requirements are confined to limited radiation protection from the Van Allen Belts.

Table 6-1: CubeSat Standard Structure Dimensions

Type	Dimension (mm)	Average Weight (kg)
1U	100 x 100 x 113.5	0.118
1.5U	100 x 100 x 170.2	0.142
2U	100 x 100 x 227	0.220
3U	100 x 100 x 340.5	0.352
6U	100 x 226.3 x 366	0.916
12U	226.3 x 226.3 x 366	1.84

The CubeSat standard structure has evolved with increasing use over many years. The CubeSat standard structures, also referred to as canisterized satellites, include 1U, 1.5U, 2U, 3U, 6U, and 12U. Table 6-1 shows the nominal weight limits and dimensions of each CubeSat structure from the CubeSat Design Specification document. There is an extra volume (XL) option available for

3U, 6U, and 12U CubeSats; this additional volume, commonly referred to as the “tuna can” volume, is associated with an individual dispenser type. This cylindrical XL additional space allows for structural extensions of the CubeSat that can be used for various components. Steamjet Space has developed Steam Thruster, a tuna can-sized electrochemical thruster specifically designed for CubeSats. The 3U CubeSat Elfin mission used this tuna can space for antenna deployment. Shields mission also fit a radiator within its tuna can volume. Figure 6.1 shows this optional volume and location on the CubeSat.

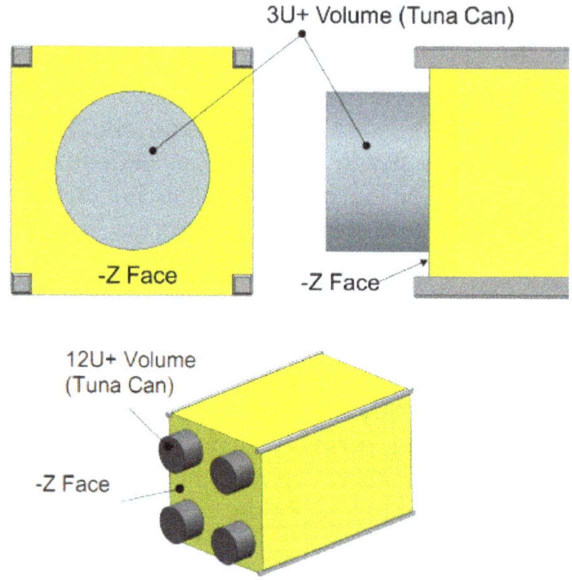

Figure 6.1: Optional Extra Volume shown on 3U and 12U –Z Face (also known as a "Tuna Can"). Credit: Cal Poly CubeSat Laboratory.

There are several companies that provide CubeSat primary structures. Most are machined from aluminum alloy 6061 or 7075 and are designed with several mounting locations for components to allow flexibility in spacecraft configuration. The SmallSat community has witnessed an increase in CubeSat standard configuration over the last 10 years from 1U to 3U, to include 6U and 12U. This was due to a higher demand for more science on a smaller platform, and by the need for more volume to design more complex CubeSats that can handle greater responsibility. Table 6-2 lists several commercial primary CubeSat structures. Of the offerings included here, 1U, 3U and 6U frames are most prevalent, however 12U frames are becoming more widely available as there are now more dispensers for the 12U CubeSat structure. Figure 6.2 shows some commercial examples of 3U, 6U and 12U CubeSat structures.

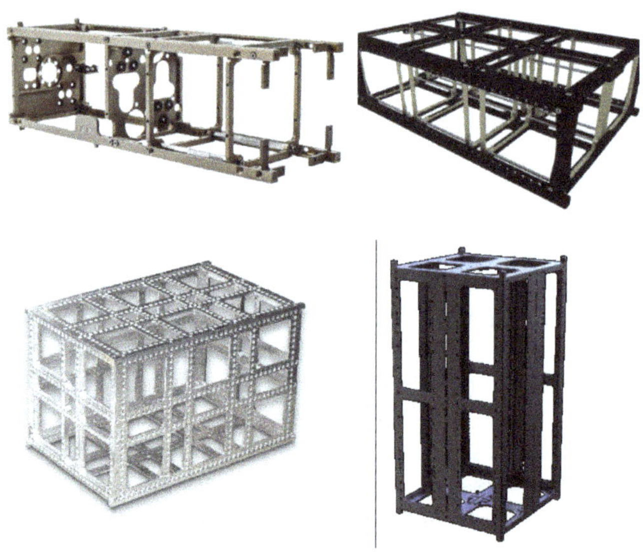

Figure 6.2: Various commercial CubeSat structures. Top Left: NanoAvionics 3U Structure. Credit: NanoAvionics. Top Right: 6U nanosatellite structure. Credit: GomSpace. Lower Left: 12U Structure. Credit: C3S Electronics Development, LLC. Lower Right: 16U structure. Credit: EnduroSat.

8U and 16U CubeSat Structure

Following the trend of larger CubeSat structures that is driven by the needs of the SmallSat market, several companies are now offering CubeSat structures not officially recognized by the CubeSat standard such as the 8U and 16U. Customized dispensers are available that will host these larger volumes.

Table 6-2: Commercial Primary CubeSat Structures	
Manufacturer	**Structure (U)**
AAC Clyde Space	ZAPHOD 1U, 2U, 3U, 6U, 12U
C3S Electronics Development LLC	3U/3U Plus, 6U, 12U, 16U
Cervos Space	1U, 2U, 3U, 6U
Cosats Satellite Technology	COSTR 1U, 1.5U, 3U, 6U, 8U,12U, 16U
EnduroSat	1U, 1.5U, 3U, 6UXL, 8U 12UXL, 16U
German Orbital Systems	1U, 2U, 3U, 6U, 12U
GomSpace	3U, 6U, 8U, 12U, 16U
Gran Systems	1U, 1.5U, 2U, 3U, 6U, 6UXL
Gumush	n-ART 1U, 2U, 3U
ISISPACE	1U, 2U, 2UXL, 3U, 6U, 8U, 12U, 16U
Ishitoshi Machining	MBF-1U, 3U
NanoAvionics	1U, 2U, 3U, 6U, 12U, 16U
Pumpkin Space Systems	Supernova 1U, 3U, 6U, 12U
NPC Spacemind	SM 1U, 1.5U, 2U, 3U, 6U, 12U
Nara Space Technology	12U, 16U

6.2.2 Custom CubeSat Primary Structures

A growing development in building custom small satellites is the use of detailed interface requirement guidelines. These focus on payload designs with the understanding of rideshare safety considerations for mission readiness and deployment methods. Safety considerations include safety switches, such as the "remove before flight" pins and foot switch, and requirements that the spacecraft remain powered-off while stowed in the deployment dispensers. Other safety requirements often entail anodized aluminum rails and specific weight, center of gravity, and external dimensions for a successful canister or dispenser deployment.

DiskSat Structure

The Aerospace Corporation is developing a DiskSat (figure 6.3) demonstration flight with support from NASA's Space Technology Mission Directorate (STMD). The DiskSat is a 1-m circular disk, 2.5 cm thick, graphite-epoxy composite sandwich, with a structural mass less than 3 Kg/m^2. The volume is close to 20 liters, which is equivalent to a hypothetical '20U' spacecraft. While the entire volume will not be filled, the increased surface area is useful for power, aperture, thermal management, and for manufacturing simplification. First launch for the demonstration mission is planned for 2024 (2).

Figure 6.3: DiskSat structure. Courtesy of and reprinted by permission of The Aerospace Corporation.

6.2.3 Primary Structure Standard Dispenser

The box that houses the CubeSats in the launch vehicle is called a dispenser (or deployer), and they dispense (or deploy) the CubeSat into the desired orbit. The CubeSat uses the entire volume of the dispenser to make use of its full capacity. Since the CubeSat adopts a standard size and form factor, CubeSat dispensers have also been standardized with two constraint systems: rail- or tab-type. This allows spacecraft designers and launch service providers to minimize launch integration cost, increase access to space, and sustain frequent launches (3). The CubeSat Design Specification document by the CubeSat Program at Cal Poly was created to provide CubeSat developers baseline requirements that are compatible with as many CubeSat dispensers and launch opportunities as possible to eliminate launch interface failures (4). To view the most updated versions of the CubeSat Design Specification, please visit: http://www.cubesat.org. The CubeSat Design Specification document includes rail systems. The Canisterized Satellite Dispensers (CSD) tab system created by Planetary Systems Corporation (now Rocket Lab) is the most widely available tab dispenser that offers design flexibility for structures that do not require the use of rails. See CSD datasheet for detailed information on tab dispenser (5).

A tab-style canister deployment system uses tabs that are loaded to hold the CubeSat to a wall of the canister which are released upon deployment. The vibrational load during launch passes from the launch vehicle to the canister structure with the pre-loaded CubeSat. A CubeSat using a rail dispenser is lightly loaded on the z-axis. On the x and y axis a thin gap exists between the rail of the dispenser and rails on the CubeSat which can cause vibrational chatter. The vibrational chatter adds to the mechanical load of the CubeSat during testing and launch. For more CubeSat rail vs tab dispensers, please refer to Chapter 10: Launch, Deployment, Integration, and Orbital Services.

The required interface documents originate with the rideshare integrator for the specific dispenser being used with the launch vehicle. The launch vehicle provider typically provides the launch vibrational conditions. The NASA CubeSat Launch Initiative (CSLI) requires CubeSat or SmallSat systems be able to withstand the General Environmental Verification Standard (GEVS) vibration environment of approximately 10 G_{rms} over a 2-minute period (6). The NASA CSLI rideshare provides electrical safety recommendations for spacecraft power-off requirements during launch and initial deployment. The detailed dispenser or canister dimensional requirements provide enough information, including CAD drawings in many cases, to enable a custom structural application.

Table 6-3 lists some dispenser and canister companies that provide spacecraft physical and material requirements for integration. In response to the demand for larger CubeSats, dispensers for 12U CubeSats are now available through several launch service providers like NanoRacks and United Launch Alliance (ULA) through the Atlas series. There are several European companies providing deployment for 16U platforms that expand the limits of the CubeSat Design Specification. The DSOD, EXOpod, and the Quad Pack are all dispensers that can fit a single 16-unit CubeSat platform or several smaller CubeSats.

Table 6-3: Spacecraft Physical Dimension and Weight Requirements from Deployers			
Manufacturer	**U**	**Requirements**	**Available Documents**
P-POD by Cal Poly	1U, 3U	Dimensions, Weight, Rail	Follows CubeSat Standard (4)
CSD by Planetary Systems Corp.	1U, 3U, 6U, 12U	Dimensions, Weight, Tab	
Tyvak Railpod III, 6U NLAS, 12U Deployer	3U, 6U, 12U	Dimensions, Weight, Rail	Interface Control Documentation (8)
PSC by Rocket Lab	3U, 6U, 12U	Dimensions, Weight, Tabs	Interface Guide, CAD Drawings (5)
ISIPOD ISISPACE	1U, 2U, 3U, 4U, 6U, 8U, 12U	Dimensions, Weight, Rail	Follows CubeSat Standard (7)
Gran Systems MyPOD Deployer and Test PODs	3U, 6U	Dimensions, Weight, Rail	Website (9)
Dhruva Space CubeSat Deployers DSOD	1U, 3U, 6U, 12U, 16U	Dimensions, Weight, Rail	Website (10)
Exolaunch EXOpod CubeSat Deployer	1U, 2U, 3U, 6U, 8U, 12U, 16U	Dimensions, Weight, Rail	User Manual (11)

6.2.4 CubeSat Structures Construction Methods

Monocoque Construction

Monocoque structures are load-bearing skins that have significant heritage on aircraft. On small spacecraft, the intent of this design is several-fold – it maximizes internal volume, it provides more thermal mass for heat sinks or sources, it allows for more mounting points, and it has more surface area to potentially reduce total ionizing dose (TID). Monocoque construction is common, and "extruded" designs are relatively easy to fabricate through computerized numerical control (CNC) machining, waterjet, or laser cutting.

Modular Frame Designs

Modular frames allow for a flexible internal design for quick-turn missions, while still ensuring strict adherence to external dimensions of the CubeSat standard, especially when deployment from a standardized, reusable dispenser is required. Open frames are suitable for low-Earth orbit, as radiation shielding is not provided by the structure. Care must also be taken to design for thermal mass requirements, as modular frames are inherently light.

6.3 State-of-the-Art – Mechanisms

Spacecraft commonly contain onboard devices whose function are based on mechanical movement (i.e.: slide, roll, rotate, separate, unfold, or spin) to either modify part of the spacecraft's geometry or to ensure operational function of a component or instrument. These devices are known as mechanisms, and as spacecraft become more sophisticated with the advances in miniaturization of electronics and systems, their reliance of mechanisms greatly increases.

The domain of spacecraft mechanisms is quite broad as there are many different types in the design and life of a spacecraft that include the moving parts associated in each phase:

- Deployment: dispensing spacecraft into orbit

- Beginning of mission life: deployments of solar arrays, booms, antennas, instrumentation, etc.
- Mission maintenance: sun tracking, pointing antennas and instruments, active doors or shields, gyroscopes and reaction wheels, thrusters, etc.
- End-of-life: deorbiting methods

The technology within the mechanism to perform the movement is accomplished with an actuator. Depending on the actuation method, spacecraft mechanisms are either passively or actively driven. Passive mechanisms do not consume electric energy and provide driving power via spring load, and active mechanisms are motorized to produce driving power for mechanism operation. Most mechanisms can use both passive and active capabilities depending on the application. Table 6-4 provides an overview of common spacecraft mechanisms and examples of technologies used.

The state-of-the-art of small spacecraft mechanisms is quantified on their high reliability, low power, and light weight characteristics, and the common mechanisms listed below are considered state of the art for small spacecraft use. For the purposes of this chapter, the mechanisms focus on deployable extensions, robotic manipulations, release actuation, component pointing, and gimbal mechanisms. Reliability considerations are provided for optimal operational capabilities, as well as a brief explanation of the factors that affect spacecraft mechanisms.

Table 6-4: Type of Spacecraft Mechanisms

Type of Mechanism	Description	Technology Examples
Separation and Release	Reliable stowage and release of spacecraft and deployable components upon an external command (active) or spring-loaded (passive).	Clamp band systems, Frangibolts, release nuts, pin pullers, bolts, burn wire, hinges, and passive spring-loaded switches
Motorized	Allows for rotatory motion of spacecraft components.	Solar Array Drive Assembly, directional antennas, combination of dampeners and absorbers
Attitude Control	Provides pointing accuracy and stability for spacecraft and components.	Reaction (momentum) wheel assembly, gimbals, component pointing, passive methods

6.3.1 Actuators

By classical definition, actuators are devices that convert electrical, thermal, hydraulic, and/or pneumatic energy into mechanical motion when said energy is allowed to flow. Active, or commanded, actuators use onboard data links and electrical transistors to determine the transfer of energy; whereas passive, or reactive, actuators allow the spacecraft environment (including external launch systems) to dictate actuator energy transfer. Table 6-5 provides some commercial actuators.

Specifically, spacecraft actuators are used for a variety of purposes, including:

- Attitude control and gimbaling: to control the orientation of either part (gimbaling), or all (attitude control), of a spacecraft in space. This is important for pointing sensors, instruments, and/or communications antennas in a direction required for their use.
    - Attitude control general types: reaction control thrusters, momentum wheels, control moment gyros, magnetic torquers, aerodynamic control surfaces, solar sails, and gravity gradient stabilizers.

- Gimbal general types: single-axis, dual-axis, and triple-axis system.
- Propulsion: supporting attitude control system operations, maneuvering to a new orbit, or reducing orbital velocity to begin atmospheric reentry.
  - General types: chemical rocket engines (which can be the same as the upper stage launch vehicle engines), reaction control thrusters, and electric propulsion systems. These systems typically require actuated valves to operate.
- Deployment, docking and separation: extend and unfold solar panels, antennas, and other spacecraft components requiring unpacking to function.
  - Deployment general types: hinge-&-spring based, linear-actuator-&-scissor-frame based, roll-out systems, and inflatable structures.
  - Docking general types: probe-and-drogue, peripheral, and soft-capture systems.
  - Separation general types: spring-powered or gas-powered systems.
- Thermal control: manage all or part of the spacecraft's temperature. This is important for protecting internal components from extreme temperatures.
  - General types: louvers, heat pipes, thermoelectric/Peltier devices, and pumped thermal fluid systems.

Mechanical actuation methods/techniques that are found in many of the above systems include:

- Electric & electromagnetic: AC/DC motor, piezoelectric ceramics, and push/pull & rotary solenoids (including solenoid valves), and microelectromechanical systems (MEMS).
- Thermal & thermoelectric: Shape memory alloys (SMA), phase-change liquids/solids (paraffin wax, liquid metals), thermofluidic gas systems, thermal bimorph structures, harmonic drive micro actuators (HMAs), thermal knife cutters, and magnesium alloy band systems.

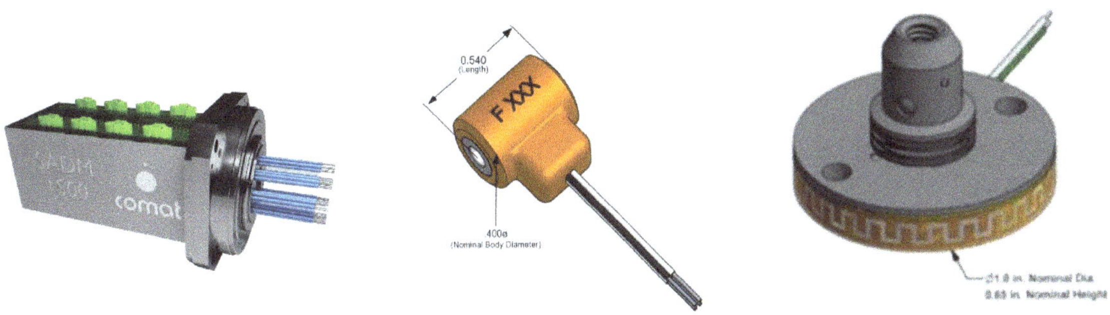

Figure 6.4: (top left) SADM 1500. Credit: Comat. (right) TiNi Aerospace Frangibolt Actuator and (right) ML50 microlatch. Credit: Ensign-Bickford Aerospace & Defense.

Table 6-5: Commercial Actuators						
Manufacturer	**Product**	**Mass (Kg)**	**Size (mm)**	**Power Consumption**	**Actuation method**	**Ref**
Ensign-Bickford Aerospace & Defense Company	TiNi™ FD04 Frangibolt	0.007	13.72x10.16	15 W @ 9 VD	SMA	(13)
Ensign-Bickford Aerospace & Defense Company	TiNi™ ML50	0.015	-	-	SMA	(14)
Moog	Type 2 Side-Drive Solar Array Drive Mechanism (SADM)	5	234x278.6	15	-	(15)
Honeybee and MMA Design	Solar Array Drive Actuator (SADA)	3.1	127x210	-	Stepper Motor	(16)
Comat Space	Solar Array Drive Mechanism - 400	0.465	83x62x46	4	Geared motor	(17)
Comat Space	Solar Array Drive Mechanism - 1500	3.5	201x132	13	Geared motor	(18)
DHV Technology	MicroSADA-10	<0.25	100x100x100	-	Stepper motor	(19)
DHV Technology	MicroSADA-18	<0.95	226x80x18	-	Stepper motor	(19)
DCUBED	Micro Pin Puller (uD3PP)	0.08	25.5x25.5 x 25.5	-	SMA	(20)
DCUBED	Nano Pin Puller (nD3PP)	0.025	17x17x17	-	SMA	(20)
DCUBED	Micro Release Nut (uD3RN)	0.078	25x25x25	-	SMA	(21)
Beyond Gravity	Separation Nut PSM 3/8B	0.23	58.5x36x56	-	-	(22)
Revolv Space	Solar Array Rotary Actuator (SARA)	<0.35	97x97x23	1 W (average)	-	(23)

Data unknown is represented by -

6.3.2 Deployable Structures

Space deployable mechanisms are structures folded into a compact configuration and deployed into a larger predetermined shape. The development of deployable structures on spacecraft is appealing to enable greater mission performance. Once deployed, the structures reconfigure, changing shape and size from folding and unfolding. Common spacecraft deployables include antennas, radiators, solar panels, gravity assists, and other science instruments. Small spacecraft are great candidates for using deployable structures to raise the functionality of a smaller platform. However, there are limited designs for compact, lightweight, low power deployable structures that can be folded or rolled up for launch and then self-deployed in space to support these kinds of systems on small satellites.

There are different types of deployment mechanisms to ensure the deployed structure effectively expands to the desired configuration in-orbit: folding, sleeve, truss, and inflatable. Deployable solar arrays are a common folded-type of passive deployment mechanism achieved by connecting the spring and hinge to increase solar energy for the spacecraft. Please refer to the *Power* chapter for deployable solar panels and arrays. The sleeve-type deployment mechanism is implemented using a rolling or sliding screw conveyor and is commonly seen on SmallSats for various antennas (24). Inflatable deployment structures are light-weight film material typically used for larger deployed structures, like solar sails. Please refer to the *Deorbit Systems* chapter for deployable mechanisms used for deorbit devices.

Figure 6.5: GPX-2 CAD image with gravity gradient boom deployed. Credit: NASA.

For SmallSat applications, it is common that deployable components are on a boom – a cantilever arm ejected from the spacecraft – that can perform various tasks once deployed. See figure 6.5 for NASA's GPX-2 CubeSat mission with a Redwire Space deployable boom to create gravity gradient stabilization as an example. SmallSat deployable structures are common and are associated with high reliability. Engineers have started developing deployables with different materials to decrease the stowage area, mass, and power. Table 6-6 lists a selection of commercially available deployable booms.

Figure 6.6: NASA Deployable Composite Boom (DCB) Technology. Credit: NASA.

NASA Langley Research Center (LaRC) has developed Deployable Composite Booms (DCB) through the Space Technology Mission Directorate (STMD) Game Changing Development (GCD) program and a joint effort with the German Aerospace Center, see figure 6.6. DCBs have high bending and torsional stiffness, packaging efficiency, thermal stability, and 25% less weight than metallic booms (25). The Advanced Composite Solar Sail System (ACS3) project will demonstrate DCB technology for solar sailing applications with an anticipated 2024 launch. The DCB/ACS3 7-m boom technology is extensible to 16.5 m deployable boom lengths (26).

Engineers have started using origami – the art of paper folding – as a strategy of deployable structure design. Origami structures are flexible in their deployment direction so that they can be easily collapsed along the same path they are deployed. One advantage of origami-inspired mechanisms is potentially faster and cheaper prototyping. Instead of relying on laser cutting or 3D-printing, prototyping of origami-inspired mechanisms can be accomplished using inexpensive materials like paper before moving to other more expensive materials. Many resources and patterns already exist that detail how designs can be created and modified or adapted for engineering purposes (27). Solar panels and arrays, solar sails, and sunshades are examples of ongoing origami engineered SmallSat components.

Table 6-6: Commercial Deployable Booms

Manufacturer	Product	Reference
Composite Technology Development	Stable Tubular Extendable Lock-Out Composite (STELOC)	(28)
Oxford Space Systems	AstroTube deployable boom	(29)
Redwire Space	Roll Out Composite (ROC) booms	(30)
Redwire Space	CubeSat ROC Boom Deployer	(31)
Redwire Space	ROC-FALL system	(31)
Magellan Aerospace	Deployable Boom	(32)
Rolatube Technology	Deployable Composite Booms	(33)

6.3.3 Robotic Manipulator

The need for in-space servicing is receiving more attention from the SmallSat community with the increasing demand of more complex SmallSat with greater capability and longer mission life. These types of challenges are being solved with robotic manipulations that can perform intricate actions in space. Tasks such as repairing defunct satellites, in-orbit assembly, satellite servicing, debris capture, spacecraft system up-keep, construction, and repair are important advances for future space operations; these challenges are currently expensive and risky to perform. Current robotic solutions for in-space construction and repair involve humans and use very large, expensive, custom-built robotic arms with limited capabilities, such as the Canadian Arm. As NASA's Artemis program prepares for astronaut presence in lunar and deep space missions on the Lunar Gateway, there is a greater need for more advanced and maneuverable space robotic systems. The use of these sophisticated robotic systems on a SmallSat are more alluring than traditional larger platforms as SmallSats present a more cost-effective and agile solution. A more agile robotic system can be stowed in small space and deployed to perform several tasks automatically or semi-automatically.

This section provides an overview of the continuous work occurring to further develop robotic systems on SmallSats. Table 6-7 lists a non-exhaustive list of the ongoing work. This type of SmallSat mechanism is maturing with research and development at government, academia, and commercial entities (34). For example, the Naval Academy Satellite Team for Autonomous Robotics (NSTAR) has developed an autonomous 3U CubeSat robotic arm system called the Robotic Experimental Construction Satellite (RECS) to be tested on the ISS. RECS was launched November 2022 (35).

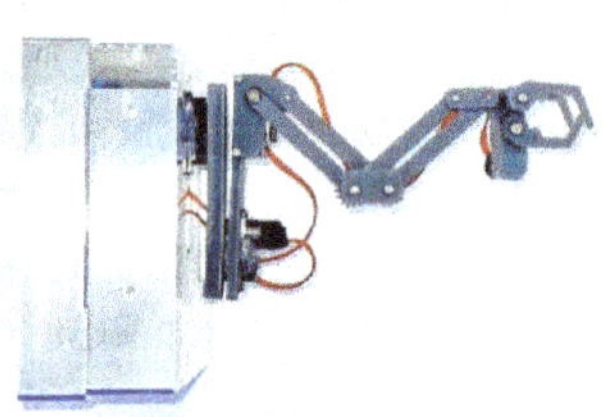

Figure 6.7: RECS robotic arm. Credits: US Naval Academy.

Table 6-7: Robotic Arms for Small Spacecraft								
Manufacturer	**Product**	**Mass (kg)**	**Extendable length (mm)**	**Stowed Envelope (mm)**	**DOF**	**Power Consumption (W)**	**Actuator method**	**Form Factor**
U.S. Naval Academy	3U CubeSat with two robotic arms	4	600	300x100x100	6 plus "claw" end-effector actuation	-	Stepper motor	RECS 3U (Nov 2022)
Redwire Space	-	-	1 to 4 m reach	-	5 to 7	8 to 65	-	ESPA class satellites
Sierra Lobo	Sierra Lobo Arm (SLAC 1)	-	100x100x100	30x50x65	3	1.5	-	-

Data unknown is represented by -

6.3.4 Reliability Considerations

Mechanisms add capabilities and complexities to small satellite design. Additional integration and testing are required. NASA Reliability and Maintainability Standard (36) describes maintainability, and "test as you fly," in addition to multiple other mitigation strategies and considerations. For mechanisms, it is important to test the full sub-system and system integration for power consumption and sub-system dependencies. Mechanisms have lifetimes, so it is important to have a maintainable mechanism and to understand the lifetime of the mechanism from test to flight. Because mechanisms add complexity and a single point failure risk in some instances, such as attitude control or solar panel pointing, directional antenna control, or one-time sub-system deployment switches, it is important to focus on reliability strategies. Mechanisms have contributed to over 10% of reported small satellite failures (37). Adding a mechanism to enable a mission increases risk.

The space environment adds to reliability considerations for operational considerations in vacuum, plasma, and/or thermal environments. For the reliability of mechanisms, there are multiple steps that contribute to risk mitigation. Sarafin et al. describes a multi-step approach for a reliable mechanism from design simplicity, margin, supplier selection, to test (38). The steps include guidance for torque margin for rotating parts, such as solar panel and antenna pointing motors (38)(39)(40). During ground testing of mechanisms, it is important to understand the mechanism lifetime, so that the component performs throughout the planned mission duration. Materials considerations contribute to mechanism reliability in the space environment, such as lubricant use and material coatings to avoid corrosion and welding of dissimilar materials (39)(40)(41). Because mechanisms are critical for advanced spacecraft capabilities for power, communications, and science/research instruments, it is important to add mechanism margin and tests to the spacecraft development and/or sub-system and system integration.

6.4 State-of-the-Art – Additive Manufacturing

Additive manufacturing (AM) processes for primary spacecraft structures have long been proposed but only recently have such methodologies been adopted for flight. AM has been common for SmallSat secondary structural elements for many years. Typically, the advantage of AM is to free the designer from constraints imposed by standard manufacturing processes and allow for monolithic structural elements with complex geometry. In practice, additive manufacturing has a separate design space and design process, which has seen tighter integration into computer-aided design, computer-aided manufacturing, and modal and structural analysis packages in the past few years. Such tools can enable quicker turnaround times for SmallSat development, and have been instrumental in mass optimization, using AM materials in radiation shielding, and enabling high-throughput, high-quality manufacturing. As the AM field is rapidly evolving, this section makes a best attempt to cover as many materials and printers as possible that are potentially applicable to SmallSat development.

6.4.1 Applicability of TRL to Polymer AM

While AM systems and platforms might be considered mature and of high TRL, the TRL of AM parts configured for spaceflight depends on the material, the configuration of the actual part, the manufacturing process of the material, the postprocessing of the manufactured part, the testing and qualification process, and many other factors. For example, nylon fabricated with a fused filament fabrication (FFF) system will have different bulk structural properties from nylon fabricated with a selective laser sintering system.

In other words, a TRL might be assignable to a component created through a particular manufacturing process with a specific material. If a particular component manufactured with nylon on an FFF system was flown to LEO successfully, the TRL for this component would be 7. If this component was subsequently flown on another mission manufactured with Antero 840 PEEK also on an FFF system, the TRL would still be 7. Documentation of the manufacturing process is important to properly account for TRL. This section focuses on polymer AM and does not address metal AM for SmallSats.

6.4.2 Inspection and Testing

When new materials and/or processes are used, testing must be performed to minimize risk and bridge the gap between TRL levels. In particular, the only way to validate a tailored structure, component, or material is through testing, especially if more freedom is allocated to research and development. For new material types, if there is latitude afforded in upfront research and development, mechanical, modal, and thermal tests should be performed to compare against a known, proven structural design.

6.4.3 Thermoplastics and Photopolymers

With the expansion of available open-source AM platforms in the last decade, thermoplastics and photopolymer materials have rapidly gained traction and acceptance in many applications ranging from mechanical validation and fit-checking to engineering-grade, low-rate production products. Photopolymer or “thermoset” resins and associated manufacturing processes have improved to the point where microfluidics experiments may be additively manufactured, with the microfluidics channels and growth chambers directly manufactured as one piece, as opposed to the more traditional microfluidics approach of machining a plastic block.

As of publication, there are three primary methods of conducting AM for plastics: FFF, which uses thermoplastics in either a spool or pellet form; stereolithography (SLA), which uses photopolymer resin; and selective laser sintering (SLS), which uses a fine powder. Within SLA, there are two methods of curing resin: digital light projection (DLP), which uses a very high-resolution LED matrix – a monochrome display – to cure the entire layer nearly instantly; and polyjets which deposit resin from a line array of jets, much like an inkjet printer with a large print head.

Certain thermoplastics are quickly gaining acceptance for high-reliability parts and applications on Earth, although, as of this writing, they have yet to gain widespread acceptance for space applications. One reason for this is AM methods cannot yet produce surfaces as smooth as machined metals, which is often a requirement for parts with tight tolerances. However, some thermoplastics are machinable, such as Nylon or polyetherimide (PEI). Like the manufacture of cast iron parts, machining to a final, high tolerance specification may allow these thermoplastics to gain further acceptance.

Except for some large-format AM centers, almost all thermoplastics are manufactured in spools, and may or may not be packaged for proprietary solutions. For SLA, almost all resins are used specifically for commercial solutions and AM centers. Additionally, some manufacturers may mix in additives to enhance material properties or ease the printing process. Because of this, the following sections on each material include a table of materials for both open-source and commercial solutions, and selected properties of interest. Availability of recommended nozzle and bed temperature is indicative of the ability to be printed on an open-source machine, except otherwise noted in the material description. Materials are not picked according to preference but through availability of technical specifications and potential applicability. For various types of AM solutions, readers are encouraged to use these sections as a rough guide for currently available commercial filaments. Additionally, the material tables will be expanded as more data is obtained on the following materials.

Surface discharge, or electrostatic discharge (ESD), is a result of in-space charging effects and is caused by interactions between the in-flight plasma environment and spacecraft materials and electronic subsystems (42). The field buildup and ESD can negatively affect the spacecraft and there are design precautions which must be considered depending on the spacecraft's operational environment. Per ESD guidelines from NASA Spacecraft Charging Handbook 4002A, dielectric materials above 10^{12} Ohm (Ω) cm should be avoided because charge accumulation occurs regardless. Please refer to the NASA Handbook 4002A, 5.2.1.5 Material Selection for more information. Historically, ESD due to faulty grounding has been a leading cause of spacecraft or subsystem failures (42).

Polylactic Acid (PLA)

PLA is the most common filament used in AM and table 6-7 lists several PLA filaments. It exhibits very low shrinkage and is extremely easy to print because it does not require a heated bed or build chamber and requires a relatively low extruder (nozzle) temperature. It also has low off gassing during printing, important in open-frame AM systems in rapid prototyping environments such as lab settings. Unless the application has a very short-term exposure to harsh conditions, and if the conditions are well characterized and controlled, it is not recommended to use PLA for an application beyond TRL 3-4. For laboratory settings in controlled environments not subject to excessive mechanical forces, ESD-compatible filaments are available.

Table 6-7: Polylactic Acid Filaments

Filament Name (Citation)	ISO 75/ASTM D648 Deflection Temp (°C)	ISO 179-1 Hardness (kJ/m^2) or Izod D256-10A (J/m)	ISO 527-1/ASTM D638 ZX Tensile strength (MPa)	ASTM D790/ISO 178 Flexural strength (MPa)	Nozzle Temp (°C)	Bed Temp (°C)	Density (g/cc)	ESD Risk* (Ω-cm)
Prusament PLA	55	12 kJ/m^2	57	N/A	215	50-60	1.24	No
Verbatim PLA	50	16 kJ/m^2	63	N/A	210	50-60	1.24	No
ColorFabb PLA-PHA (43)	N/A	30 kJ/m^2	61	89	210	50-60	1.24	No
Stratasys PLA (44)	51	27 kJ/m^2	26	84	N/A	N/A	1.264	No, 10^{15}
3DXSTAT™ ESD-PLA	55	N/A	55	95	210	23-60	1.26	Yes, 10^6-10^9

Acrylonitrile Butadiene Styrene (ABS)

ABS has traditionally been the choice for higher strength, lightweight prints from the Fused Deposition Modeling (FDM) process in the open-source community. It is generally temperature resistant and UV resistant but turns yellow and eventually becomes more brittle over time when exposed to sunlight. It is a marginally difficult filament to print, especially in open-frame systems. High temperature gradients during printing may cause warping as parts get larger. Enclosed AM systems with heated chambers print ABS well. Additionally, ABS shrinks 1 to 2 percent of its printed size upon cooling – the shrinkage varies from manufacturer to manufacturer. ABS has flown as the complete structure for KickSat-2, a FemtoSat deployer for chip-scale satellites (45). The single-use, short mission duration, and intricate dispenser frame made a conventionally machined deployer mass- and cost-prohibitive. Table 6-8 lists some examples of ABS filaments.

Table 6-8: ABS Filaments

Filament Name	ISO 75/ASTM D648 Deflection Temp (°C)	ISO 179-1 Hardness (kJ/m²) or Izod D256-10A (J/m)	ISO 527-1/ASTM D638 Tensile strength (MPa)	ASTM D790/ISO 178 Flexural strength (MPa)	Nozzle Temp (°C)	Bed Temp (°C)	Density (g/cc)	ESD Risk (Ω-cm)
Stratasys ABS-CF10	100	20-51 J/m	21	29-69	N/A	N/A	1.0972	Marginal 10^4-10^9
Stratasys ABS-ESD7	105	36.2 J/m	35	44	N/A	N/A	1.07	Marginal 10^4-10^9
3DXSTAT™ ESD-ABS	97	N/A	58	80	230	110	1.09	Yes, 10^6-10^9
Verbatim ABS	106 (ISO 306)	21 J/m	47	78	240-260	90	1.05	No

Nylon

Versatile and tough, there are multiple formulations for nylon that allow for a very wide range of applications and material properties. In general, nylon is more difficult to manufacture than ABS on open-source FFF systems because it requires an enclosure for thermal stability and additional bed preparation for higher adhesion. Secondary structural pieces have been flown through the TechEdSat program using Markforged Onyx carbon fiber filaments. Table 6-9 lists some examples of nylon filaments.

Table 6-9: Nylon Filaments

Filament Name (Citation)	ISO 75/ASTM D648 Deflection Temp (°C)	ISO 179-1 Hardness (kJ/m²) or Izod D256-10A (J/m)	ISO 527-1/ASTM D638 ZX Tensile strength (MPa)	ASTM D790/ISO 178 Flexural strength (MPa)	Nozzle Temp (°C)	Bed Temp (°C)	Density (g/cc)	ESD Risk (Ω-cm)
Taulman3D Alloy 910 (46)	82	N/A	56	N/A	250-255	30-65	N/A	Unk
Taulman3D Alloy 910 HDT (46)	112	N/A	56	N/A	285-300	55	N/A	Unk
Taulman3D Nylon 680 Food Grade (47)	N/A	N/A	47	N/A	250-255	30-65	N/A	No
Markforged Onyx ESD (48)	138	44 J/m	52	83	N/A	N/A	1.2	Yes, 10^5-10^7
3DXTECH CARBONX™ HTN+CF (49)	240	N/A	87	95	295	130	1.24	Marginal 10^9
Stratasys Nylon 12 (50)	92-95	71-138 J/m	33-42	55-57	N/A	N/A	1.01	No, 10^{13}

Polycarbonate (PC)

Also known as Lexan™, this thermoplastic has some of the highest impact resistance, tensile strength, and temperature resistance available for most open source-based AM systems. After manufacturing, it is dimensionally stable and very stiff. However, it is difficult to print on open-frame, open-source AM systems due to very high warping especially when printing large components. Very high bed and nozzle temperatures are required, and poor adhesion to the bed is a typical issue. It is also highly hygroscopic; if possible, the filament should be baked out before printing, or should be kept in a dedicated dry box while printing. Certain filaments, like the Prusament PC Blend, have additives to mitigate some of the difficulties of printing PC. If PC is desired for a SmallSat structure, it should be printed on a commercial AM system. Table 6-10 lists some polycarbonate filaments.

Table 6-10: Polycarbonate Filaments

Filament Name (Citation)	ISO 75/ASTM D648 Deflection Temp (°C)	ISO 179-1 Hardness (kJ/m²) or Izod D256-10A (J/m)	ISO 527-1/ASTM D638 ZX Tensile strength (MPa)	ASTM D790/ISO 178 Flexural strength (MPa)	Nozzle Temp (°C)	Bed Temp (°C)	Density (g/cc)	ESD Risk (Ω-cm)
Prusament PC Blend (51)	113	No break for ISO 179	63	88-94	275	110	1.22	No
Prusament PC Blend Carbon Fiber (51)	114	35 kJ/m²	55-65	85-106	285	110	1.16	No
Stratasys PC (52)	143	27-77 J/m	60	75	N/A	N/A	1.20	No

Windform

Manufactured by CRP Technology, these proprietary materials are classified as a carbon fiber reinforced polymer originally designed for the automotive racing industry. They are unique in that these composites are manufactured through SLS (53). This results in higher dimensional stability and more isotropic properties than FFF. Windform XT 1.0 and 2.0 have been used on CubeSat and PocketQube platforms and have flight heritage through KySat-2 launched on ELaNa IV, and TANCREDO-1, launched through the ISS via JEM in 2017 (54). Table 6-11 lists CRP Windform filaments. The NASA GPX-2 Windform XT 2.0 structure launched in July 2022 and is operational.

Table 6-11: CRP Windform

Filament Name (Citation)	ISO 75/ASTM D648 Deflection Temp (°C)	ISO 179-1 Hardness (kJ/m²) or Izod D256-10A (J/m)	ISO 527-1/ASTM D638 ZX Tensile strength (MPa)	ASTM D790/ISO 178 Flexural strength (MPa)	Manufacturing process	Bed Temp (°C)	Density (g/cc)	ESD Risk (Ω-cm)
Windform XT 2.0	173	4.72 kJ/m²	84	133	N/A, SLS	N/A, SLS	1.097	Yes, 10^8
Windform RS (56)	181	10.8 kJ/m²	48-85	139	SLS	SLS	1.10	Yes, 10^8

Polyetherimide

Polyetherimide (PEI), also known by the Saudia SABIC trade name Ultem™, is a very tough thermoplastic resin with high thermal and chemical stability. It is inherently flame-resistant and can be machined. Some formulations of PEI are FAA-approved for flame, smoke, and toxicity (FST), and may also have ESD formulations. PEI is also known for extremely low off gassing, crucial for optical components and sensitive scientific packages. PEI is a common bed material for higher end open-source FFF systems due to its adhesive properties with other thermoplastics at higher temperatures. PEI has similar characteristics to polyetheretherketone (PEEK). Due to these similarities, PEI is only practically printable on commercial FFF systems. Table 6-12 lists some PEI filaments.

Table 6-12: PEI Filaments

Filament Name (Citation)	ISO 75/ASTM D648 Deflection Temp (°C)	ISO 179-1 Hardness (kJ/m²) or Izod D256-10A (J/m)	ISO 527-1/ASTM D638 ZX Tensile strength (MPa)	ASTM D790/ISO 178 Flexural strength (MPa)	Nozzle Temp (°C)	Bed Temp (°C)	Density (g/cc)	ESD Risk
THERMAX™ Ultem™ 9085	158	N/A	63	90	275	115	1.34	No
3DXSTAT™ Ultem™ 1010 CF-ESD (57)	205	N/A	62	115	395	150	1.34	Yes, 10^7-10^9
Stratasys Ultem™ 1010 CG(58)	212	22-27 J/m	81	82-128	N/A	N/A	1.29	No, 10^{14}
Stratasys Ultem™ 9085 (59)	153	39-88 J/m	69	80-98	N/A	N/A	1.27	No, 10^{15}
Zortrax Z-PEI 9085 (60)	186	N/A	54	90	N/A	N/A	1.34	No

PAEK

Polyetheretherketone (PEEK) and polyetherketoneketone (PEKK) – in the polyaryletherketone (PAEK) family – are the highest performing thermoplastics developed as of this writing. With certain additives and matrix materials, they can rival the strength of stainless steel and withstand over 200°C continuously in some formulations, after annealing. PEEK/PEKK are naturally flame-retardant; they are accepted for use in aviation ducting. They also achieve extremely low off gassing in operation, which makes these thermoplastics good candidates for compatibility with optical components in space. Due to the extreme conditions required for manufacturing and the very high filament cost, these materials are only practically available for printing in extremely robust commercial FFF systems with sealed and heated chambers. PEEK has heritage on long-term, external ISS experiments, and structural elements on the Juno spacecraft, making it suitable for extreme radiation environments (61). Table 6-13 lists some PAEK-based filaments.

Table 6-13: PAEK-based Filaments

Filament Name (Citation)	ISO 75/ASTM D648 Deflection Temp (°C)	ISO 179-1 Hardness (kJ/m²) or Izod D256-10A (J/m)	ISO 527-1/ASTM D638 ZX Tensile strength (MPa)	ASTM D790/ISO 178 Flexural strength (MPa)	Nozzle Temp (°C)	Bed Temp (°C)	Density (g/cc)	ESD Risk (Ω-cm)
3DXSTAT™ ESD-PEEK (62)	140	N/A	105	141	380-400	150	1.32	Yes, 10^7-10^9
3DXSTAT™ ESD-PEKK	185	N/A	109	135	375	140	1.34	Yes, 10^7-10^9
CarbonX™ CF PEKK-Aerospace	285	N/A	126	178	390	140	1.33	Yes, 10^7
Stratasys Antero 840 (63)	150	28-43 J/m	95	87-139	N/A	N/A	1.27	Yes, 10^4-10^9
Zortrax Z-PEEK (64)	160	N/A	100	130	N/A	N/A	1.30	N/A

Photopolymers

Otherwise known as "thermosets," these materials are liquid polymers cured by an optical and thermal process. Compared to other AM processes, photopolymers and their manufacturing processes allow for superior isotropic material properties, very high resolution, and the ability to manufacture optical quality parts. Some formulations, especially from 3D Systems and Stratasys, are designed for extreme temperature resistance and strength, desirable in aerospace applications. In some cases, the listed heat deflection temperature (HDT) may be superior to those of PAEK. As previously discussed, there are three major methods of curing photopolymers, one of which is proprietary. Many photopolymers are specifically paired for commercial systems. As a result, the table 6-14 includes the commercial system associated with the photopolymer.

Some of the photopolymers listed below have several additional characteristics not listable in this table, including, but not limited to, elasticity, tear strength, optical clarity, water absorption, and medical grade certifications. Such characteristics may be useful for biological experiments in future SmallSats. Please consult the products' specific websites and datasheets for additional information. Additionally, photopolymers have the advantage of being able to be mixed, in-situ, as the object is being manufactured. This allows for continuously varying material properties throughout the object. Table 6-14 lists some photopolymers.

Table 6-14: Photopolymers

Photopolymer Name (Citation)	ISO 75/ASTM D648 HDT (°C)	ISO 179-1/ASTM D256-10A (J/m)	ISO 527-1/ASTM D638 Tensile (MPa)	ASTM D790 Flexural (MPa)	Density (g/cc) at 25°C	ESD Risk (Ω-cm)	Manufacturing and/or Machine Type
Accura Bluestone (65)	267-284	13-17	66-68	124-154	1.78	ND	3D Systems ProX 800
VisiJet M2S-HT250 (66)	250	10	51	83	1.15	ND	3DS MJP 2500 Plus
DSM Somos® Watershed XC	50	25	50	69	1.12	ND	Stratasys V650 Flex SL
Henkel LOCTITE® IND402 A70 Flex (67)	N/A	N/A	5.5	N/A	1.068	ND	Several
Henkel LOCTITE® 3D 3843 (68)	80	54	60	81	N/A	ND	DLP SLA types only

6.4.4 AM Design Optimization

Design optimization is an integral part of manufacturing validation and testing. As previously discussed for AM, validation, testing, and optimization encompass all materials and manufacturing processes. Software platforms help speed up this process, especially those that integrate toolpathing generation, computer aided manufacturing (CAM), load analysis, and fill generation. The inherent advantage of AM to allow monolithic structural elements implies a much-expanded design space compared to subtractive manufacturing. Software has kept up with the pace of manufacturing advances and incorporates tools to assist with AM designs.

The manufacturing ecosystem includes software ranging from simple CAM solutions generating toolpaths (G-code) to complete, structural analysis and high-fidelity manufacturing simulations. As of this writing, AM has gained significant traction and value in low-TRL demonstrations and physical validation, partly due to the ease of fabrication in typical AM ecosystems. It is beginning to displace traditional machining – "subtractive" manufacturing – as AM systems have matured enough to print advanced thermoplastics, resins, and metals.

Infill Patterns

Due to the flexibility that AM offers, new methods of lightweighting are now possible. "Lightweighting" refers to the reduction of mass of structural elements, without compromising structural integrity. The best examples of well-proven heritage methods of lightweighting are "honeycomb" sandwiched aluminum panels, subtractive machining, and truss structures. However, such methods have certain limitations. Honeycomb panels for example, do not have uniform, or isotropic, properties – they do not exhibit the same stiffness in all directions.

Lightweighting in AM encompasses what is called "infill," or the internal structure of a hollow body or panel. With a minimal increase in mass, an internal structure manufactured with AM can vastly increase the strength of a body. Very recently, the AM community has renewed interest in the use of the gyroid pattern, discovered by NASA researcher Alan Schoen in 1970, due to the ease of generation in AM toolpath programs. Aside from honeycomb and gyroids, several options for infill exist. Different options are offered with different AM-focused software packages.

Digital Materials

Both honeycomb panels and AM parts with infill have a common repetitive unit cell. By repeating this unit cell throughout the interior of a part, or as a structure on its own, a larger structure can be made. Further, by defining properties into this unit cell, information can effectively be encoded into the design, allowing for differing behavior of different parts of the structure. Digital materials can dramatically expand the design space of a structure, allowing for targeted optimization of various properties such as mass to strength ratios, structural lightweighting, and others. As previously discussed, with certain resin polyjet AM centers, resins can be mixed in real time to form an object that has continuously varying properties.

6.5 Radiation Effects and Mitigation Strategies

6.5.1 Shielding from the Space Environment

Radiation Shielding has been described as a cost-effective way of mitigating the risk of mission failure due to total ionizing dose (TID) and internal charging effects on electronic devices. In space mission analysis and design, the average historical cost for adding shielding to a mission is below 10% of the total cost of the spacecraft (69). The benefits include reducing the risk of early total ionizing dose electronics failures (70). Some of the key CubeSat and SmallSat commercial electronic semiconductor parts include processors, voltage regulators, and memory devices, which are key components in delivering science and technology demonstration data (71).

Shielding the spacecraft is often the simplest method to reduce both a spacecraft's ratio of total ionizing dose to displacement damage dose (TID/DDD) accumulation, and the rate at which single event upsets (SEUs) occur if used appropriately. Shielding involves two basic methods: shielding with the spacecraft's pre-existing mass (including the external skin or chassis, which exists in every case whether desired or not), and spot/sector shielding. This type of shielding, known as passive shielding, is only very effective against lower energy radiation, and is best used against high particle flux environments, including the densest portions of the Van Allen belts, the Jovian magnetosphere, and short-lived solar particle events. In some cases, increased shielding is more detrimental than if none was used, owing to the secondary particles generated by highly penetrating energetic particles. Therefore, it is important to analyze both the thickness and type of materials used to shield all critical parts of the spacecraft. Due to the strong omni-directionality of most forms of particle radiation, spacecraft need to be shielded from the full 4π steradian celestial sphere. This brings the notion of "shielding-per-unit-solid-angle" into the design space, where small holes or gaps in shielding are often only detrimental proportionally to the hole's solid angle as viewed by the concerned electrical, electronic and electro-mechanical (EEE) components. Essentially, completely enclosing critical components should not be considered a firm design constraint when other structural considerations exist.

6.5.2 Inherent Mass Shielding

Inherent mass shielding consists of using the entirety of the pre-existing spacecraft's mass to shield sensitive electronic components that are not heavily dependent on location within the spacecraft. This often includes the main spacecraft bus processors, power switches, etc. Again, the notion of "shielding-per-unit-solid-angle" is invoked here, where a component could be well shielded from its “backside” (2π steradian hemisphere) and weakly shielded from the “front” due to its location near the spacecraft surface. It would only then require additional shielding from its front to meet operational requirements. The classic method employed here is to increase the spacecraft's structural skin thickness to account for the additional shielding required. This is the classic method largely due to its simplicity, where merely a thicker extrusion of material is used for construction. The disadvantage to this method is the material used, very often aluminum, is mass optimized for structural and surface charging concerns and not for shielding either protons/ions or electrons. Recent research has gone into optimizing structural materials for both structural and shielding concerns; currently an active area of NASA's Small Business Innovation Research (SBIR) program research and development.

The process to determine exactly how much inherent shielding exists involves using a reverse ray tracing program on the spacecraft solid model from the specific point(s) of interest. After generating the "shielding-per-unit-solid-angle" map of the critical area(s) of the spacecraft, a trade study can be performed on what and where best to involve further additional shielding.

Numerous CubeSat and SmallSat systems use commercial processors, radios, regulators, memory, and SD cards. Many of these products rely on silicon diodes and metal oxide semiconductor field effect transistors (MOSFETs) in these missions. A comprehensive NASA guidance document on the use of commercial electronic parts was published for the ISS orbit, which is a low-Earth orbit where the predominant radiation source is the South Atlantic anomaly. The hardness of commercial parts was noted as having a range from 2 – 10 kRad (72). For typical thin CubeSat shielding of 0.20 cm (0.080 in) aluminum, yearly trapped dose is 1383 Rad; with an additional estimated 750 Rad from solar particle events, the total dose increases to 2133 Rad for the ELaNaXIX Mission environment at 85 degrees inclination and 500 km circular orbit (table 6-16) (73). Adding a two-fold increase for the trapped belt radiation uncertainty brings the total radiation near the TID lifetime of many commercial parts (72), even before estimating a SPE TID contribution. The uncertainty of radiation model results of low-Earth orbit below 840 km has been estimated as at least two-fold; Van Allen Belt models are empirical and rely on data in the orbital environment (74). The NASA Preferred Reliability Series "Radiation Design Margin Requirements" also recommends a radiation design margin of 2 for reliability (75). Currently, The Aerospace Corporation proton (AP) (63) and The Aerospace Corporation electron (AE) (77) Models do not have radiation data below 840 km, and radiation estimates are extrapolated for the lower orbits (74). For spacecraft interplanetary trajectories near the Sun or Earth, the radiation contributions from SPEs will be higher than low-Earth orbit, where there is some limited SPE radiation protection by the magnetosphere. By reducing the total ionizing dose on commercial parts, the mission lifetimes can be increased by reducing the risk of electronic failures on sensitive semiconductor parts.

Figure 6.14: Shields-1 Z-shielding structure and final Preship picture, ELaNaXIX Mission. Credit: NASA.

6.5.3 Shields-1 Mission, Radiation Shielding for CubeSat Structural Design

Shields-1 has operated in polar low-Earth orbit and was launched through the ELaNaXIX Mission in December 2018. The Shields-1 mission increased the development level of atomic number (Z) Grade Radiation Shielding with an electronic enclosure (vault) and Z-grade radiation shielding slabs with aluminum baselines experiments (figure 6.14) (78). Preliminary results in table 6-15 show a significant reduction in total ionizing dose in comparison to typical modeled 0.20 cm (0.080 in) aluminum structures sold by commercial CubeSat providers. The 3.02 g cm^{-2} Z-shielding vault has over 18 times reduction in total ionizing dose compared to modeled 0.20 cm aluminum shielding (73).

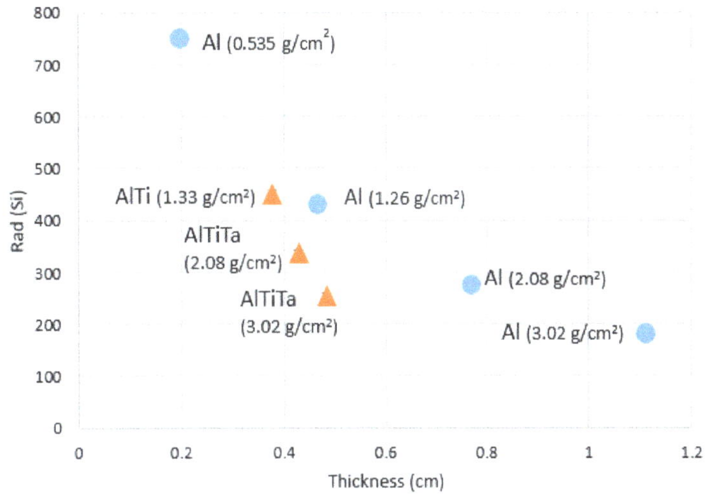

Figure 6.15: SPE Contribution to TID in PLEO, King Sphere Model, ELaNaXIX Shields-1 orbit. Credit: NASA.

Z-shielding enables a low volume shielding solution for CubeSat and SmallSat applications where reduced volume is important. AlTiTa, Z–shielding, at 2.08 g cm^{-2} reduces the dose from a SPE by half when compared to a standard 0.2 cm aluminum structure (figure 6.15). NASA has innovated “Methods of Making Z-Shielding” with patents in preparing different structural shieldings (79)(80)(81)(82), from metals to hybrid metal laminates and thin structural radiation shielding, to enable low-volume integrated solutions with CubeSats and SmallSats (83).

Table 6-15: Shields-1 Experimental Total Ionizing Dose Measurements in PLEO

Shielding	Areal Density (g/cm^2)	Thickness (cm)	Trapped Belts TID Total (Rad (Si)/Year)	SPE King Sphere Model, (Rad (Si))
Al	0.535	0.198	1383+/-47 #	750+/-5
Al	1.26	0.465	**90.9 +/-2.7 (SL)**	432 +/- 7
Al	1.69	0.624	**84.3 +/-2.5 (SL)**	345 +/- 9
Al	3.02	1.11	**73.6 +/-3.2 (SL)**	183 +/- 11
AlTi	1.33	0.378	**89.7 +/-2.7 (SL)**	451 +/- 6
AlTiTa20	2.08	0.429	**84.3 +/-2.5 (SL)**	338 +/- 6
AlTiTa40	3.02	0.483	**81.9 +/-3.4 (SL) 75.6+/-3.2 (Vault)**	253 +/- 6

Shields-1 Experimental total ionizing dose measurements in PLEO in comparison to typical 0.20 cm aluminum shielding commercially available for CubeSats and SPE additional contributions to dose. ***Bold values*** *Shields-1 experimental results. SL = Slab, Vault = Z-Shielding electronics enclosure. # sphere Space Environment Information System (SPENVIS) Multi-layered Shielding Simulation Software (MULASSIS) AP8 Min AE8 Max modeled results. SPE King Sphere Model SPENVIS MULASSIS modeled results.*

6.5.4 Ad Hoc Shielding

There are two types of ad hoc shielding used on spacecraft: spot shielding, where a single board or component is covered in shield material (often conformally), and sector shielding, where only critical areas of the spacecraft have shielding enhancement. These two methods are often used in concert as necessary to further insulate particularly sensitive components without unnecessarily increasing the overall shield mass and/or volume. Ad hoc shielding is more efficient per unit mass than inherent mass shielding because it can be optimized for the spacecraft's intended radiation environment while loosening the structural constraints. The most recent methods include multiple layer shields with layer-unique elemental atomic numbers which are layered advantageously (often in a low-high-low Z scheme), known as "graded-Z" shielding, and advanced low-Z polymer or composite mixtures doped with high-Z, metallic micro-particles. Low-Z elements are particularly capable at shielding protons and ions while generating little secondary radiation, where high Z elements scatter electrons and photons much more efficiently. Neutron shielding is a unique problem, where optimal shield materials often depend on the particle energies involved. Commercial options include most notably Tethers Unlimited's VSRS system for small spacecraft, which was specifically designed to be manufactured under a 3D printed fused filament fabrication process for conformal coating applications (a method which optimizes volume and minimizes shield gaps).

6.5.5 Charge Dissipation Coating

The addition of conformal coatings over finished electronic boards is another method to mitigate electrostatic discharge on sensitive electronic environments. Arathane, polyurethane coating materials (84), and HumiSeal acrylic coatings (85) have been used to mitigate discharge and provide limited moisture protection for electronic boards. This simple protective coating over sensitive electronic boards supports mission assurance and safety efforts. Charge dissipation films have decreased electrical resistances in comparison to standard electronics and have been described by NASA as a coating that has volume resistivities between 10^8 – 10^{12} ohm-cm. In comparison, typical conformal coatings have volume resistivities from 10^{12} – 10^{15} ohm-cm (42).

6.5.6 LUNA Innovations, Inc. XP Charge Dissipation Coating

The XP Charge Dissipation Coating has volume resistivities in the range of $10^8 - 10^{12}$ ohm-cm (table 6-16) and is currently developing space heritage through the NASA MISSE 9 mission and Shields-1 (86). The XP Charge Dissipation Coatings were developed through the NASA SBIR program from 2010 to present for extreme electron radiation environments, such as outer planets, medium-Earth, and geostationary orbits, to mitigate charging effects on electronic boards.

Table 6-16: XP Charge Dissipation Coating and Commercial Conformal Coating Resistivity Comparisons

Material	Volume Resistivity (Ohm-cm)
XP Charge Dissipation Coating	$10^8 - 10^{12}$, 4.7 x 10^9 at 25°C
Arathane 5750 A/B	9.3 X 10^{15} at 25°C, 2.0 X 10^{13} at 95°C
Humiseal 1B73	5.5 x 10^{14} Ohms (Insulation Resistance per MIL-I-46058C)

Figure 6.16: Transparent LUNA XP Charge Dissipation Coating on an electronic board. Credit: LUNA Innovations, Inc.

The LUNA XP Charge Dissipation Coating has reduced resistance compared to typical commercial conformal coatings, which reduces surface charging risk on electronic boards. LUNA XP Coating (figure 6.16) on an electronic board has transparency for visual parts inspection. For extreme radiation environments, a combination of radiation shielding, and charge dissipation coating reduces the ionizing radiation that contributes to charging and provides a surface pathway for removing charge to ground (42).

6.6 Summary

This chapter has been updated with the current status of structures, materials, and mechanisms for small satellite missions. Additions include custom structure references with the dimensional and material requirements of integrating deployment systems, new mechanisms technology to reflect the ongoing growth in SmallSat mechanical devices, and more commercially procured deployable booms and larger CubeSat primary structures (12U and 16U), as well as the upcoming DiskSat structure. The radiation environment section, state-of-the-art radiation shielding, and charge dissipation materials have been updated. Reflecting the fast pace of development in additive manufacturing, a selection of available thermoplastics and resin-based materials suitable for different TRL levels have been detailed.

There has been high focus on deployment mechanisms with respect to light weighting and reliability. Small spacecraft subsystems related to antenna booms, gravity gradients, stabilization, sensors, sails, and solar panels are some examples. These technologies are gaining space heritage through operations and more often are being included in mission planning. The growth of these deployment mechanisms increase the capabilities of SmallSat technology and will be a continued focus in the next edition of this report.

For feedback solicitation, please email: arc-sst-soa@mail.nasa.gov. Please include a business email so someone may contact you further.

References

(1) Space Weather Prediction Center. Solar Cycle 25 Forecast Update. [Online] December 19, 2019. Available at: https://www.swpc.noaa.gov/news/solar-cycle-25-forecast-update

(2) R. Welle, C. Venturini, D. Hinkley, J. Gangestad, S. Grasso, A. Muszynski, R. Hunter, C. Frost, D. Mayer, and C. Baker, "DiskSat: Demonstration Mission for a Two-Dimensional Satellite Architecture," Proceedings of the AIAA/USU Conference on Small Satellites, Out of this World!, SSC22-VIII-01, 2022.

(3) R. Hevner, W. Holemans, J. Puig-Suari, and R. Twiggs. "An Advanced Standard for CubeSat." 2003. 25th Annual AIAA/USU Conference for Small Satellites.

(4) CubeSat Design Specification (1U – 12U), REV 14.1, CP-CDS-R14.1, CubeSat Program, CAL POLY San Luis Obispo, February 2022. (https://static1.squarespace.com/static/5418c831e4b0fa4ecac1bacd/t/62193b7fc9e72e0053f00910/1645820809779/CDS+REV14_1+2022-02-09.pdf

(5) Rocket Lab. "Canisterized Satellite Dispenser datasheet version 2002337G". [Online] Available at: https://www.rocketlabusa.com/assets/Uploads/2002337G-CSD-Data-Sheet-compressed2.pdf

(6) NASA Goddard Technical Standards, "GSFC General Environmental Verification Standard (GEVS) for GSFC Flight Programs and Projects", GSFC-STD-7000B, 28 April 2021.

(7) ISISPACE. "CubeSat Deployers." [Online] Available at: https://www.cubesatshop.com/wp-content/uploads/2016/07/ISIS-CubeSat-deployers-Brochure-v2-compressed.pdf

(8) Tyvak. "Launch Services". [Online] 2021. [Accessed: August 24, 2021. Available at: https://www.tyvak.com/launch-services/

(9) GranSystems. "Products," [Online] Available at: https://www.gransystems.com/Products.html

(10) Dhruva Space. "Launch Services," [Online] Available at: https://www.dhruvaspace.com/launch-services

(11) Exolaunch. "EXOpod User Manual," [Online] Available at: https://exolaunch.com/documents/EXOpod_User_Manual_September_2022.pdf

(12) Planetary Systems Corporation. "Canisterized Satellite Dispenser." [Online] Accessed: August 28, 2021. Available at: https://www.planetarysystemscorp.com/product/canisterized-satellite-dispenser/

(13) Ensign-Bickford Aerospace & Defense. "TiNi™ Mini Frangibolt® Actuator: Small, SImple, Powerful." [Online] 2022. Available at: https://www.ebad.com/tini-mini-frangibolt

(14) Ensign-Bickford Aerospace & Defense. "TiNi™ Micro Latch, Developed Specifically for "New Space" Applications." [Online] 2022. Available at: https://www.ebad.com/tini-micro-latch/

(15) Moog, Inc. "Type 2 Side-Drive Solar Array Drive Mechanism," [Online] Available at: https://www.moog.com/products/space-mechanisms/solar-array-drive-assemblies/type-2-side-drive-solar-array-drive-mechanism.html

(16) Honeybee Robotics. "Solar Array Drive Assemblies," [Online] Available at: https://www.honeybeerobotics.com/products/solar-array-drive-assemblies/#1562267018138-ae149058-3c0374f5-d928

(17) Comat Space. "SADM 200/400200/400: Solar Array Drive Mechanism" datasheet.

(18) Comat Space. "SADM 1500: Solar Array Drive Mechanism" datasheet.

(19) DHV Technologies. "Solar Array Drive Assembly (SADA)," [Online] Available at: https://dhvtechnology.com/products/sada/

(20) Dcubed. "Pin Puller," [Online] Available at: https://dcubed.space/products/pin-pullers/

(21) Dcubed. "Release Nut," [Online] Available at: https://dcubed.space/products/release-nut/

(22) Beyond Gravity. "Separation Nut PSM 3/8B," datasheet.

(23) Revolv Space. "Introducing SARA: [Online] Available at: https://www.revolvspace.com/product

(24) Tianyuan Ye *et al.* "Review of Deployable Mechanism Test Technologies Oriented Towards Deep Space Exploration," 2021 *IOP Conference Ser.: Material Science Engineering.* **1043** 052022

(25) NASA. "Deployable Composite Booms (DCB)." [Online]. August 26, 2020. Accessed May 18, 2021. Available at: https://www.nasa.gov/directorates/spacetech/game_changing_development/projects/dc

(26) NASA. "Advanced Composite Solar Sail System: Using Sunlight to Power Deep Space Exploration," [Online] June 23, 2021. Accessed October 2022. Available at: https://www.nasa.gov/directorates/spacetech/small_spacecraft/ACS3/

(27) Brigham Young University. "Origami-Inspired Mechanisms Explained," [Online] 2022. Accessed October 2022. Available at: https://www.compliantmechanisms.byu.edu/about-origami-mechanisms

(28) Composite Technology Development, Inc. "STELOC Deployable Composite Booms." [Online] 2018. Available at: https://ctd-materials.com/engineered-materials/tembo/

(29) J. Reveles et al. "In-Orbit Performance of AstroTube: AlSat Nano's Low Mass Deployable Composite Boom Payload." Proceedings of the AIAA/USU 31st Annual AIAA/USU Conference of Small Satellites, Logan, UT, 2017.

(30) B.L. Davis, W.H. Francis, J.A. Goff, M.C. Cross, and D.J. Copel. "Big Deployables in Small Satellites", Proceedings of the AIAA/USU Conference on Small Satellites, The Commerce of Small Satellites, SSC12-XII-1, 2014.

(31) Redwire Space. "Open Lattice Mast Boom (OLM)," [Online] 2022. Accessed October 2022. Available at: https://redwirespace.com/products/open-lattice-mast-boom/?rdws=nnn.xffxcv.tfd&rdwj=57342-41850-43938-1-44045-35348-41850-41622\

(32) Magellan Aerospace. "Deployable Booms," [Online] Available at: https://magellan.aero/wp-content/uploads/Deployable%20Booms%20-%20Digital%20Version.pdf

(33) Rolatube. "Space Satellite Technology: Deployable Composite Booms," [Online] Available at: https://www.rolatube.com/sector/spacetechnology/

(34) NASA. "In-space Servicing, Assembly, and Manufacturing (ISAM) State of Play," 2023 Edition. Available at: https://www.nasa.gov/wp-content/uploads/2023/10/isam-state-of-play-2023.pdf

(35) A. Nance and M. Thibault. "CubeSat-scale Robotic Arms in Space," US Naval Academy. CubeSat Developer's Workshop Conference 2023. Available at: http://mstl.atl.calpoly.edu/~workshop/archive/2023/presentations/2023_Day2_Session6_ThibaultNance.pdf

(36) NASA Reliability and Maintainability (R&M) Standard for Spaceflight and Support Systems, NASA-STD-8729.1A – 2017-06-13.

(37) Raja Pandi Perumal, Holger Voos, Florio Dalla Vedova, Hubert Moser. "Small Satellite Reliability, A Decade in Review", SSC21-WKIII-02, 35th Annual Small Satellite Conference, Logan, UT, 2021, p.12.

(38) Thomas P. Sarafin Ed, Spacecraft Structures and Mechanisms. Damon D. Phinney and William R. Britton, Ch.19 "Developing Mechanisms", pp.665-734.

(39) NASA Standard 5017B, "Design and Development Requirements for Mechanisms", 12/06/2022, p. 117

(40) MIL-A-83577B (USAF), 2/1/1988, "Assemblies, Moving Mechanical, for Space and Launch Vehicles, General Specification For", p.70.

(41) MIL-PRF-8625F w/amendement 2, 11/23/2020, "Anodic Coatings for Aluminum and Aluminum Alloy", p.20.

(42) NASA. Technical Handbook, "Mitigating In-Space Charging Effects – A Guideline." NASA-HDBK-4002A. October 19, 2017.

(43) ColorFabb. "PLA/PHA." Technical Datasheet, [Online] April 8, 2020. Available at: https://colorfabb.com/media/datasheets/tds/colorfabb/TDS_E_ColorFabb_PLA_PHA.pd

(44) Stratasys. "PLA: Economy Thermoplastic for Stratasys F123 Series Printers." Technical Datasheet 2018. Available at: https://stratasysstorage01.file.core.windows.net/ssys-websites-files-prod/Public1/Materials/FDM/PLA/PLA%20-%20EN%20Data%20Sheet%20FDM%20Material.pdf?sv=2017-04-17&sr=f&sig=colFyWKHlkaQ9GUZsyg0%2FFwD%2Fllpuft2a2SR%2FYKqFEg%3D&st=2022-11-27T23%3A00%3A27Z&se=2023-11-28T23%3A00%3A27Z&sp=rwl

(45) NASA. "What is KickSat-2?" [Online] June 3, 2019. Accessed July 12, 2021. Available at: https://www.nasa.gov/ames/kicksat

(46) Taulman3D. "Alloy 910 Specifications." [Online] Accessed July 12, 2021. Available at: https://taulman3d.com/alloy-910-spec.html

(47) Taulman3D. "Alloy 680 Specifications." [Online] Accessed July 12, 2021. Available at: https://taulman3d.com/nylon-680-spec.html

(48) Markforged. "Onyx ESD™." Technical Datasheet, [Online] Accessed July 12, 2021. Available at: https://markforged.com/materials/plastics/onyx-esd

(49) 3DXTech Additive Manufacturing. "CARBONX HTN+CF [HIGH TEMP CF NYLON]." [Online] Accessed July 12, 2021. Available at: https://www.3dxtech.com/product/carbonx-htn-cf/

(50) Statasys. "Nylon 12." [Online] Accessed July 12, 2021. Available at: https://www.stratasys.com/materials/search/fdm-nylon-12

(51) Prusa Polymers. "Prusament PC Blend." Technical Datasheet, [Online] Available at: https://prusament.com/media/2022/10/PCBlendCF_Prusament_TDS_2022_16_EN.pdf

(52) Stratasys. "PC (Polycarbonate): FDM Thermoplastic Filament." Technical Datasheet, [Online] Accessed July 12, 2021. Available at: https://www.stratasys.com/-/media/files/material-spec-sheets/mds_fdm_pc_0920a.pdf

(53) CRP Technology. "High performance composite materials for SLS." [Online] 2018. Available at: https://www.windform.com/top-line/#strength

(54) CRP Technology. "Windform 3D Printing materials launch into Orbit on KySat-2." [Online]. Accessed July 12, 2021. Available at: https://www.crp-usa.net/windform-3d-printing-materials-launch-orbit kysat-2/

(55) CRP Technology. "Winform XT 2.0" Technical Datasheet, [Online] Accessed July 12, 2021. Available at: http://www.windform.com/windform-xt-2-0.html

(56) CRP Technology. "Winform RS" Technical Datasheet, [Online] Accessed July 12, 2021. Available at: http://www.windform.com/windform-rs.html

(57) 3DXTech Additive Manufacturing. "3DXSTAT™ ESD-Ultem™ 3D Printing Filament." Technical Datasheet, [Online] Available at: https://www.3dxtech.com/wp-content/uploads/2020/11/ESD_Ultem_v3.pdf
(58) Stratasys. "ULTEM 1010 Resin" [Online] Accessed July 12, 2021. Available at: https://www.stratasys.com/en/materials/materials-catalog/fdm-materials/ultem-1010/
(59) Stratasys. "ULTEM 9085 Resin" [Online] Accessed July 12, 2021. Available at: https://www.stratasys.com/en/materials/materials-catalog/fdm-materials/ultem-9085/
(60) Zortrax. "Aerospace-Grade Flame-Retardant Filament." Technical Datasheet, [Online] October 9, 2020, Updated June 22, 2021. Available at: https://cf.zortrax.com/wp-content/uploads/2020/10/Z-PEI_9085_Technical_Data_Sheet_eng.pdf
(61) TechBriefs. "Plastic Components Ride Juno Spacecraft to Jupiter." [Online] August 1, 2012. Available at: https://www.techbriefs.com/component/content/article/tb/pub/features/applications/14486
(62) 3DXTech Additive Manufacturing. "3DXSTAT™ ESD-PEEK 3D Printing Filament." [Online] Available at: https://www.3dxtech.com/product/3dxstat-esd-peek/
(63) Stratasys. "Antero 840CN03 High-Performance PEKK-Based ESD Thermoplastic." [Online] Accessed July 12, 2021. Available at: https://www.stratasys.com/materials/search/antero-840cn03
(64) Zortrax. "Z-PEEL." [Online] Accessed July 12, 2021. Available at: https://zortrax.com/filaments/z-peek/
(65) 3D Systems Inc. "Accura Bluestone Technical Data." [Online] September 2020. Available at: https://www.3dsystems.com/sites/default/files/2020-11/3d-systems-accura-bluestone-datasheet-usen-2020-09-15-a-print.pdf
(66) 3D Systems, Inc. "VisiJet M2S-HT250 (MJP)." [Online] Accessed June 24, 2021. Available at: https://www.3dsystems.com/materials/visijet-m2s-ht250
(67) Henkel Corporation, Loctite. "IND402TM: PhotoElastic A70 High Rebound Black." Technical Datasheet, [Online] March 22, 2022. Available at: https://www.loctiteam.com/wp-content/uploads/2022/03/Loctite-3D-IND402-A70-High-Rebound-2022_03_22.pdf
(68) Stratasys. "LOCTITE ® 3D 3843 High Toughness by Henkel." [Online] Available at: https://www.stratasys.com/en/materials/materials-catalog/p3-materials/loctite-3d-3843-high-toughness/
(69) J.R. Wertz and W.J. Larson. "Space Mission Analysis and Design, Third Edition." Space Technology Library, Space Technology Series, Hawthorne, CA: Microcosm Press and New York: Springer, 1999, p.976.
(70) D. Hastings and H. Garrett, "Spacecraft Environment Interactions." New York, NY: Cambridge University Press, 1996, p.292.
(71) M. Langer and M. Boumeester. "Reliability of CubeSats – Statistical Data, Developers' Beliefs and the Way Forward", 30th Annual AIAA/USU Conference on Small Satellites, August 2016, Logan, UT, SSC16-X-2, 1-12.
(72) NASA Preferred Reliability Series PD 1258, "Space Radiation Effects on Electronic Components in Low Earth Orbit", August 1996.
(73) D.L. Thomsen. "Shields-1 Preliminary Radiation Shielding Dosimetry in Polar Low Earth Orbit", Committee on Space Research (COSPAR), Small Satellites for Sustainable Science and Development, Herzliya, Israel, 7 November 2019.
(74) Vampola, A. L. "The Space Particle Environment", NASA/SDIO Space Environmental Effects on Materials Workshop, NASA Conference Proceedings 3035, part 2, NASA Langley Research Center, Hampton, VA June 28 – July 1, 1988, pg. 367.

(75) NASA Preferred Reliability Series PD 1260, "Radiation Design Margin Requirement", May 1996.

(76) D.M. Sawyer and J.I. Vette, “AP-8 Trapped Proton Environment for Solar Maximum and Solar Minimum.” NSSDC/WDC-A-R&S 76-06, 1976.

(77) J.I. Vette. “The AE-8 Trapped Electron Model Environment.” NSSDC/WDC-A-R&S 91-24, 1991a.

(78) D.L. Thomsen III, W. Kim, and J.W. Cutler. “Shields-1, A SmallSat Radiation Shielding Technology Demonstration”, 29th AIAA/USU Conference on Small Satellites, SSC15-XII-9, August 2015.

(79) D.L. Thomsen III, R.J. Cano, B.J. Jensen, S.J. Hales, and J.A. Alexa. “Methods of Making Z-Shielding.” U.S. Patent No. 8,661,653, 4 March 2014.

(80) U.S. Patent No. 10,039,217, July 31, 2018, “Methods of Making Z-Shielding.” D.L. Thomsen III, R.J. Cano, B.J. Jensen, S.J. Hales, and J.A. Alexa.

(81) D.L. Thomsen III. “Method of Making Thin Atomic (Z) Grade Shields.” U.S. Patent No. 10,600,52212, March 24, 2020.

(82) D.L. Thomsen III, S.N. Sankaran, and J.A. Alexa. “Atomic Number (Z) Grade Shielding Materials and Methods of Making Atomic Number (Z) Grade Shielding.”U.S. Patent No. 10,919,650, February 16, 2021.

(83) D.L. Thomsen III and W.R. Girard. “Method of Making Atomic Number (Z) Grade Small SAT Radiation Shielding Vault.” U.S. Patent No. 11,043,311, June 22, 2021.

(84) Arathane. [Online] Available at: https://apps.huntsmanservice.com/WebFolder/ui/browse.do?pFileName=/opt/TDS/Huntsman%20Advanced%20Materials/English%20US/Long/Arathane%205750%20AB_LV_US_e.pdf

(85) Chase Electronic Coatings. "HumiSeal® 1B73 Acrylic Conformal Coating Technical Data Sheet." [Online] Available at: https://chasecorp.com/humiseal1/wp-content/uploads/sites/12/2018/10/1B73-TDS.pdf

(86) LUNA. "Luna coating provides radiation protection in space." [Online] February 2, 2018. Accessed June 28, 2020. Available at: https://lunainc.com/blog/luna-coating-provides-radiation-protection-space

Chapter Contents

Chapter Glossary

(APG)	Annealed Pyrolytic Graphite
(ARC)	Ames Research Center
(ATA)	Active Thermal Architecture
(BIRD)	Bi-Spectral Infrared Detection
(CSE, USU)	Center for Space Engineering at Utah State University
(ESPA)	EELV Secondary Payload Adapter
(FEP)	Fluorinated Ethylene Propylene
(FETS)	Folding Elastic Thermal Surface
(FOX)	Flat-Plate Heat Pipe On-Orbit Experiment
(GFTS)	Graphite Fiber Thermal Straps
(GSFC)	Goddard Space Flight Center
(HEC)	High Efficiency Cooler
(IR)	Infrared
(ISS)	International Space Station
(KGS)	Kaneka Graphite Sheets
(LPT)	Linear Pulse Tube
(MLI)	Multi-Layer Insulation
(MPFL)	Mechanically Pumped Fluid Loop
(MWIR)	Midwave Infrared
(NLAS)	Nanosatellite Launch Adapter System
(OHP)	Oscillating Heat Pipe
(P-POD)	Poly-Picosatellite Orbital Deployer
(PFL)	Pumped Fluid Loop
(PGF)	Pyrovo Pyrolytic Graphite Film
(PGS)	Pyrolytic Graphite Sheets
(PRISM)	Portable Remote Imaging Spectrometer
(q_{albedo})	Solar heating reflected by the planet
(Q_{gen})	Heat generated by the spacecraft
($Q_{out,rad}$)	Heat emitted via radiation
($q_{planetshine}$)	IR heating from the planet
(q_{solar})	Solar heating
(Q_{stored})	Heat stored by the spacecraft
(SDL)	Space Dynamics Laboratory

(SI)	International System of Units
(SPOT)	Standard Passive Orbital Thermal-Control
(SST)	Small Satellite Technology
(TAFTS)	Two Arm Flexible Thermal Strap
(TEC)	Thermoelectric Coolers
(TMT)	Thermal Management Technologies
(TRL)	Technology Readiness Level
(TSU)	Thermal Storage Unit
(UAM)	Ultrasonic additive manufacturing
(ULP)	Ultra-Low Power
(VDA)	Vacuum Deposited Aluminum

7.0 Thermal Control

7.1 Introduction

All spacecraft components have a range of allowable temperatures that must be maintained to meet survival and operational requirements during all mission phases. Spacecraft temperatures are determined by how much heat is absorbed, stored, generated, and dissipated by the spacecraft. Figure 7.1 shows a simplified overview of heat exchange from a satellite orbiting Earth, but the heating principles apply to any planet or body a spacecraft orbits.

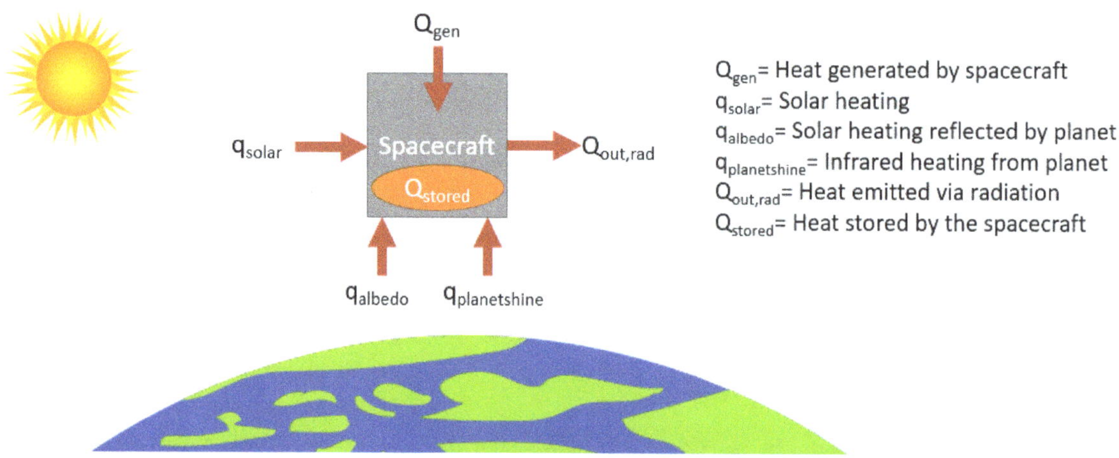

Figure 7.1: Orbiting spacecraft heating simplified overview. Q_{gen}, $Q_{out,rad}$, and Q_{stored} are represented as heat values, Watts per square meter in International System of Units (SI), whereas q_{solar}, q_{albedo}, and $q_{planetshine}$ are represented as heat fluxes. Credit: NASA.

The heat exchange depends on several factors listed below. Solar absorptivity and infrared (IR) emissivity are surface optical properties referenced below and are described further in section 7.2.1: Paints, Coatings, and Tapes. Thermal control of a spacecraft is achieved by balancing the energy as shown in Equation 1.

$$q_{solar} + q_{albedo} + q_{planetshine} + Q_{gen} = Q_{stored} + Q_{out,rad} \quad (1)$$

- Q_{gen} (heat generated by the spacecraft) depends on the power dissipation of spacecraft components.
- The amount of q_{solar} (solar heating) absorbed by the spacecraft depends on the solar flux, which is determined by distance to the sun, the surface area viewing the sun (view factor), and the solar absorptivity of that surface.
- The amount of q_{albedo} (solar heating reflected by the planet) absorbed by the spacecraft depends on the planet, the surface area viewing the planet (view factor), and the solar absorptivity of that surface.
- The amount of $q_{planetshine}$ (IR heating from the planet) absorbed by the spacecraft depends on the planet, the surface area viewing the planet (view factor), and the IR emissivity of that surface.
- $Q_{out,rad}$ (heat emitted via radiation) includes the surface area designated as radiator space, the IR emissivity of the surface, and the difference in temperature between the spacecraft radiator and the heat sink to which it is dissipating, typically and most effectively deep space. $Q_{out,rad}$ also include heat lost through insulation or other surfaces not specifically intended to function as radiators.

- Q_{stored} (heat stored by the spacecraft), is based on the thermal capacitance of the spacecraft.

Temperatures are regulated with passive and/or active thermal management technology and design methods. Many of the same thermal management methods used on larger spacecraft are also applicable to SmallSats and given the increased interest in small spacecraft over the last decade, some spacecraft thermal control technologies have been miniaturized or otherwise adapted to apply to SmallSats. Thermal control methods and technologies as applied to large spacecraft are considered state-of-the-art for the purposes of this review but may have a Technology Readiness Level (TRL) value less than 9 for small spacecraft applications.

Challenges of designing a thermal control system for a SmallSat stem from several intrinsic properties, summarized in table 7-1. Due to the small size and volume limitations inside the deployer or around deployables, there is often no room for multi-layer insulation (MLI) for CubeSats. The thermal solution must be worked out as a coatings problem, exposing the CubeSat to more transient thermal behaviors.

Table 7-1: SmallSat Thermal Control Challenges

SmallSat Property	Challenge
Low thermal mass	The spacecraft is more reactive to changing thermal environments.
Limited external surface area	There is less real estate to be allocated to solar cells, designated radiator area, and/or viewports required for science instruments.
Limited volume	There is less space for electronic components, science instruments, and thermal control hardware. Components can be more thermally coupled and it can be harder to isolate different thermal zones.
Limited power	There is less power available for powered thermal control technology.
Power Density	There is a big challenge to dissipate power as electronics are stacked close to each other, sometimes with no direct path to radiator.
MLI Edge Effects	MLI can "short" along the edges resulting in degraded performance, not specific to SmallSats; more of a general spacecraft issue.

The information described in this section is not exhaustive but provides an overview of current state-of-the-art thermal technologies and their development. TRL designations may vary with changes specific to the payload, mission requirements, reliability considerations, and/or the environment in which performance was demonstrated. Readers are highly encouraged to reach out to companies for further information regarding the performance and TRL of described technology. There is no intention of mentioning certain companies and omitting others based on their technologies or relationship with NASA.

7.2 State-of-the-Art – Passive Systems

Passive thermal control maintains component temperatures without using powered equipment. Passive systems are typically associated with low cost, volume, weight, and risk, and are advantageous to spacecraft with limited, mass, volume, and power, like SmallSats and especially CubeSats. MLI, coatings/surface finishes, interface conductance, heat pipes, sunshades, thermal straps, interface materials, and louvers are some examples of passive thermal control technology.

In addition to passive thermal control technology, structural and electrical design methods also contribute to managing the thermal environment, passively. These design methods include:

- Material selection

BODY-231

  - Structural component materials chosen based on needed heat transfer through the structure. A high or low thermal conductivity may be more advantageous based on the application.
- Spacecraft orientation
  - If orientation is not dictated by science objectives, changing the orientation of the spacecraft can help maintain temperatures.
  - Changing orientation may only be needed during certain mission phases, such as science operation if larger amounts of heat are dissipated.
  - This method is often used in conjunction with other thermal control methods, such as orienting the spacecraft so that the radiator area can face deep space.
- Thermal interfaces:
  - Definition of the thermal contact between components through specific mounting methods can thermally isolate components or allow more heat to be transferred to a structural element (or radiator area) when each is needed. For example:
    - Heat transfer can be reduced by mounting a component through multiple stacked washers with low thermal conductivity.
    - Heat transfer can be increased by mounting components with more fasteners (if applicable) and can be further increased by using thermal interface materials between a component and mounting surface.
- Circuit board design considerations, include:
  - Copper layers within each board can be increased, in number or thickness, to conduct heat away from electrical components through the boards to their structural connection points.
  - Circuit boards can be mounted to increase heat transfer away from the boards to the structure, such as by mounting with wedge locks.

Table 7-2: SmallSat Passive Thermal Technologies

Passive Technology	Description
Sprayable Thermal Control Coatings	Specialized liquid coatings applied to the spacecraft surfaces to manage and regulate the ratio of absorptivity and emissivity for optimal energy balance and thermal performance.
Films, Tapes, and MLI	Materials used to modify spacecraft surface property to manage and regulate temperature.
Thermal Straps	Provide a conductive link between a heat source and thermal sink to conductively transfer heat
Thermal Contact Conductance and Bolted Joint Conductance	Heat transfer by "contact" conductance between two surfaces pressed together by uniform pressure can be varied by using interface filler materials and conductive gaskets.
Sunshields	Deployed material that reduces the amount of incident solar flux on a spacecraft.
Thermal Louvers	Controls heat transfer between spacecraft surfaces.
Deployable Radiators	Dedicated surface for dissipating excess heat via radiative heat transfer.
Phase Change Materials	Energy is absorbed as material within a metal compartment changes phase because of exposure to a heat source.
Thermal Switches	An electromechanical device which controls the flow of electrical current in response to temperature change.

Table 7-2 provides a description of current passive thermal control technology as applied to SmallSats. One key factor to consider when choosing thermal control technology, both passive and active, is the temperature limits of the technology itself. The goal is to use the appropriate technology to maintain the temperatures of spacecraft components within their limits, but the technology used to achieve this also has limits. It is recommended to verify that the technology used is applicable to the given design not only with respect to needed function, but to the environment (temperature limits) as well.

7.2.1 Sprayable Thermal Control Coatings, Tapes, and MLI

In a vacuum, heat is transferred only by radiation and conduction, with no convection. The internal environment of a fully enclosed small satellite is usually dominated by conductive heat transfer, while heat transfer to/from the outside environment is driven via thermal radiation. Many missions with electrical surface resistivity requirements drive the use of coatings with these properties to handle these surface charging concerns (this also applies to MLI). For SmallSat missions where extensive use of MLI is not practical (see MLI paragraph on page 207), a mixed use of several different coatings is needed to achieve optimal energy balance and thermal performance. There are also coatings that better approximate the use of MLI by being relatively low emissivity (such as 0.25) with a lower alpha (0.1) so they don't overheat in the sun. These are colloquially known as tailorable emittance coatings that involve some oxide depositions starting with a vacuum deposited aluminum (VDA) base to drive up the emissivity while keeping the alpha low.

The thermal radiation band of the electromagnetic spectrum is between 0.1 and 100 µm in wavelength, as shown in Figure 7.2. Outside of the thermal radiation waveband, electromagnetic energy generally passes through objects or has very little heat energy under practical conditions. Thermal analyses are typically conducted using a two waveband absorptance model which subdivides the thermal energy spectrum into solar (< 3 µm) and IR (> 3 µm) wavelengths. Thermal radiation heat transfer is controlled by using materials that have specific optical surface properties, namely: solar absorptivity and IR emissivity. Solar absorptivity governs how much incident heating from solar radiation a spacecraft absorbs, while IR emissivity determines how much heat a spacecraft emits to space, relative to a perfect blackbody emitter, and what fraction of thermal radiation from IR sources (e.g., the Earth, Moon, any particularly hot spacecraft components) are absorbed by that spacecraft surface.

The surface properties of a spacecraft can be modified by adding specialized paints, coatings, surface finishes, or adhesive tapes, depending on the needs of the spacecraft. For example,

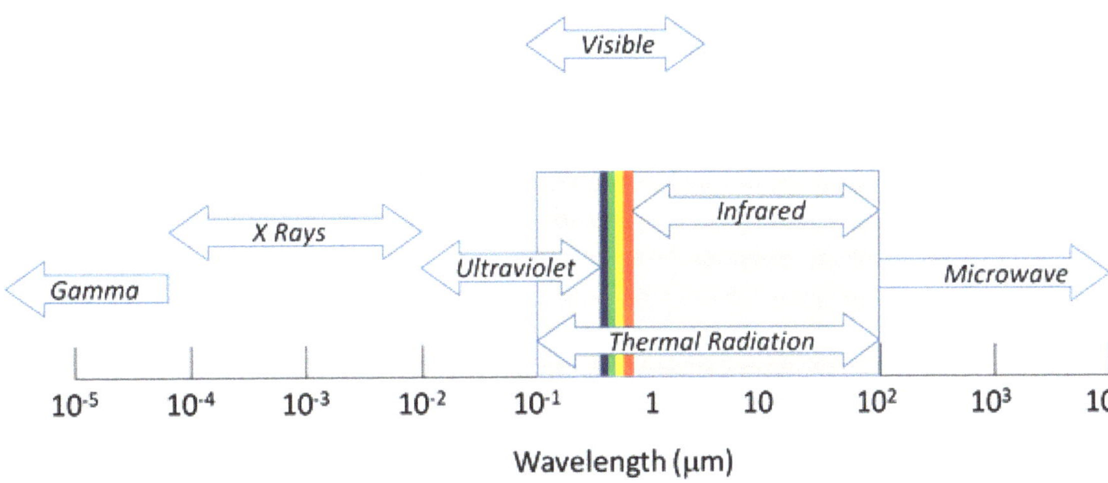

Figure 7.2: Electromagnetic spectrum showing the range of Thermal Radiation. Credit: NASA.

matte black paint has a high solar absorptivity and high IR emissivity for surfaces required to absorb a high percentage of solar heating and emit a high percentage of spacecraft heat. Alternatively, matte white paint has a low solar absorptivity and high IR emissivity (1) for surfaces required to absorb a low percentage of solar heating and emit a high percentage of spacecraft heat (e.g., radiator). Second-surface silver Fluorinated Ethylene Propylene (FEP) tapes offer excellent performance as radiator coatings, reflecting incident solar energy (low solar absorptivity) while simultaneously emitting spacecraft thermal energy efficiently (high IR emissivity). The selection between paints, coatings, and tapes depends on the application. Tape is typically easy to apply and remove, is comparatively inexpensive, and has a longer usable lifetime than paint. Tape can also be added later in the assembly process if changes to thermal control need to be made after the spacecraft has already begun assembly. Some tapes, however, must be handled carefully to maintain optical properties and can be difficult to bond properly to curved surfaces.

While optical films/tapes may have solar absorptivities and IR emissivities very close to the datasheet value as received from the manufacturer, these two values will change throughout the mission. Coatings will darken due to atomic oxygen bombardment (low earth orbit), UV light (high earth orbit, GEO, and beyond), and other cosmic rays. This will increase the solar absorptance values, some quite substantially. The longer the mission duration, the more the optical coating's properties will change. It's important the thermal engineer take this into consideration; thermal performance at the beginning-of-life (BOL) of a tape/film may not be the same at its end-of-life (EOL). Thermal analysis should take this into consideration and bias the optical property values in their thermal models to account for this degradation. BOL measurements from the datasheets (used for cold biased conditions) should not be used for EOL modeling, which are hot biased conditions.

Coatings and paints must often be applied earlier in the assembly process but can cover non-flat surfaces more easily. However some paints, like Parker-Lord's Aeroglaze 306/307, are expensive and require extensive and highly specialized processes to apply. Different options may also have different temperature limits. All these factors must be considered with regard to the needed application when selecting the final solution. Table 7-3 has a list of non-exhaustive sprayable thermal control coatings that include application difficulty and specific notes on some of the products.

One example, BioSentinel, a 6U spacecraft that launched as a secondary payload on the Artemis I mission in 2022, made extensive use of Sheldahl metallized tape coatings and second-surface silvered FEP tapes to control its external thermal radiative properties and overall energy balance (2). Another example, Picard, a 150 kg SmallSat, used white paint on the Sun pointing face to reduce the amount of solar flux absorbed and lower temperatures. For most small spacecraft projects to date, adhesive tapes, such as silver FEP, or other standard surface finishes (e.g., polishing, anodize, alodine) have been the preferred choices.

Table 7-3: Sprayable Thermal Control Coatings					
Manufacturer	**Product**	**Color**	**Cost**	**Difficulty to Apply***	**Notes**
Socomore	Z306 Polyurethane	Black	$	1	
	Z307 Polyurethane	Black	$$	1	
	A276 Polyurethane	White	$	1	Not UV stable
Huntington Ingalls Industries (formerly Alion and IITRI)	Z93P Silicate	White	$$	3	**
	Z93-C55 Silicate	White	$$$	4	Thickness control paramount **
	S13GP:6N/LO-1 Silicone	White	$$$	1	Dedicated equipment/cross contamination
AZ Technology	AZ-93 Silicate	White	$$	3	**
	AZ2000 Silicate	Off-white	$$$	4	May require specialty, epoxy primer **
	AZW/LA-II Silicate	White	$$$$	5	Thickness control paramount **

*1 not difficult, 5 extremely
**All silicates require temperature and RH control

A MLI blanket is typically comprised of multiple inner layers of a thin material with low IR emissivity (usually 10 to 20 layers) and a durable outer layer. The amount of radiative heat transfer allowed is limited by the many layers of reflectors. The low IR emissivity layers are either embossed or alternated with thin netting to limit conduction through the layers. Perforations may be added to allow the MLI to vent trapped gas once arriving on-orbit, although this can also be achieved via edge venting. MLI is used as a thermal radiation barrier to both protect spacecraft from incoming solar and IR flux, and to prevent undesired radiative heat dissipation to space. It is commonly used to maintain temperature ranges for components in-orbit.

MLI is delicate and performance drops drastically if compressed (causing a thermally conductive "short circuit"), so it should be used with caution or avoided altogether on the exterior of small satellites that fit into a deployer (e.g., P-POD, NLAS). MLI blankets can also pose a potential snagging hazard in these tight-fitting, pusher-spring style deployers. Additionally, MLI blankets tend to drop efficiency as size decreases because heat transfer through the blanket increases closer to the blanket edges, and the specific attachment method has a large impact on performance because attachment to the spacecraft creates a heat path. Due to these challenges, MLI generally does not perform as well on small spacecraft (more specifically CubeSat form factors) as on larger spacecraft. Surface coatings are typically less delicate and more appropriate for the exterior of a small spacecraft that will be deployed from a dispenser. Internal MLI blankets that do not receive direct solar thermal radiation can often be replaced by a variety of low emissivity tapes or coatings that perform equally well in that context, using less volume and at a potentially lower cost.

Table 7-4: Films, Tapes and MLI		
Manufacturer	**Film Product**	**Typical Solar Absorptivity BOL**
Astral Technologies Unlimited	StaMet/100XC Kapton	0.5
	StaMet/275XC Kapton	0.56
Astral Technologies Unlimited Multek/Sheldahl	0.010" FEP/Ag/Inconel	0.08
	0.005" FEP/Ag/Inconel	0.07
	ITO/0.010" FEP/Ag/Inconel, perforated	0.09
	ITO/0.005" FEP/Ag/Inconel, perforated	0.08
	Germanium/100XC Kapton	0.55
Dunmore Multek/Sheldahl	VDA/200HN/PSA	0.08
	200HN/VDA/PSA	0.41
	100HN/VDA/PSA	0.38
	VDG/200HN/PSA	0.2
NASA GSFC	GSFC Silver Composite Coating/ 200HN Kapton	0.1
	GSFC Aluminum Composite Coating/200HN Kapton	0.14
	GSFC Aluminum Composite Coating/100XC Kapton	0.2
	GSFC Aluminum Composite Coating/275XC Kapton	0.28
	GSFC Tailorable Emittance Coating/200HN Kapton	0.14
Thales	Optical Solar Reflector CMX 0.005"	0.08

7.2.2 Thermal Straps

Figure 7.3: Flexible Thermal Straps. Credit: Thermal Management Technologies. Q-Strap. Credit: Redwire Space.

A thermal strap is a flexible, thermally conductive link added between a heat source and sink to conductively transfer heat. They are often used between high heat dissipating chips or components and a chassis wall or other radiator surface. Their flexibility prevents the addition of structural loads. Thermal straps can be made of metal, traditionally copper or aluminum, or high conductivity carbon materials, such as graphite. They can be formed of multiple foil sheets or wound cables (also referred to as ropes and braids), with end blocks at each end to hold the sheets/cables in place and to mount or otherwise attach to the needed surfaces. Straps with more than two end blocks and multiple material combinations can also be produced and have been used on large spacecraft. Advances in thermal straps are being developed to further increase heat transfer capability and custom thermal straps are now commonly fabricated and tested using graphite material due to improved thermal conductivity.

Space Dynamics Laboratory (SDL) developed solderless, flexible thermal straps without solder, epoxy, or other filler materials. Figure 7.4 shows a comparison of the as-tested conductance for

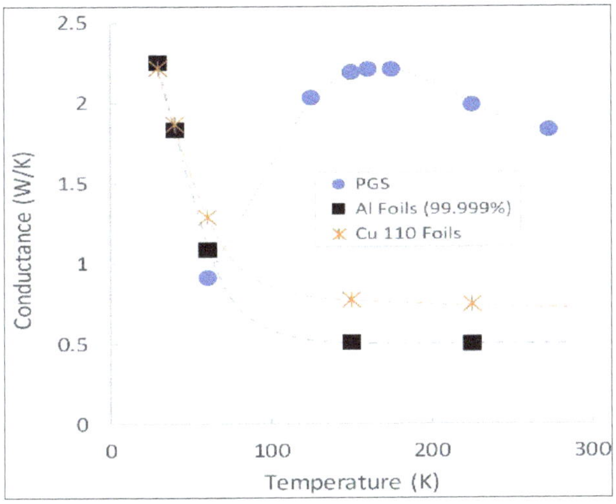

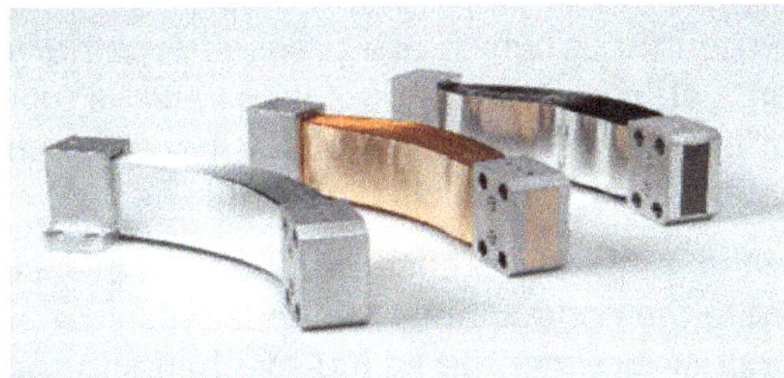

Figure 7.4: Thermal strap design with aluminum foils, copper foils, and PGS in aluminum end blocks (above), and their respective measured thermal conductance (left). The dashed lines connecting data points are based on material thermal conductivity curves. Credit: SDL.

the same strap geometry fabricated with three different foil materials of aluminum, copper, and pyrolytic graphite sheets (PGS) showing the PGS increases the overall thermal conductivity. SDL supplied Utah State University with a PGS strap for the Active Thermal Architecture (ATA) project sponsored by the Small Spacecraft Technology (SST) program. A follow-on to this ATA project is referenced in the cryocooler section. Boyd Corporation has designed thermal straps using their patented k-core technology that has an annealed pyrolytic graphite (APG) core within an encapsulating structure (4). Technology Applications, Inc. has specialized in testing and developing Graphite Fiber Thermal Straps (GFTS), with flight heritage only on larger spacecraft missions (Orion and Spice) (5). The Pyrovo Pyrolytic Graphite Film (PGF) thermal straps developed by Thermotive have already flown in optical cooling applications for high altitude cameras and avionics on larger spacecraft and flew on JPL's ASTERIA CubeSat in 2017 and were used on the Mars 2020 rover mission (6).

Table 7-5: Thermal Straps

Manufacture	Product	Material
Technology Applications, Inc.	Copper Cable	Copper
	Copper Foil	Copper
	Aluminum Foil	Aluminum 1100
	PGL	Pyro Graphite
	GFTS	Graphite Fiber
Thermotive	Copper Foil	Copper
	Aluminum Foil	Aluminum 1100
	Pyrovo	Pyro Graphite Film
Space Dynamics Laboratory	PGS	Pyro Graphite Strips
Redwire Space	Q-Strap	High-k graphite material
Thermal Management Technologies	TMT010-200 Series	Copper Foil
	TMT010-300 Series	Copper Braid (small)
	TMT010-400 Series	Copper Braid (large)
	Customized	Copper braid or foil or using aluminum foil
Boyd Corporation	k-Core Graphite-Encapsulated	Pyrolytic graphite

7.2.3 Thermal Contact Conductance and Bolted Joint Conductance

Two surfaces which are pressed together by uniform pressure will transfer heat via "contact" conductance. This conductance value is a product of the heat transfer coefficient and the contact surface area. The heat transfer between such interfaces can be varied by using interface filler materials and conductive gaskets (7).

Bolted joints experience non-uniform pressure, creating a more complex heat transfer scenario where the conductance depends on screw size, torque, surface properties, and other values. The conductance can be varied by changing torque, surface properties and finishes, and materials. Table 7-6 provides the conductance of various screws (7).

Table 7-6: Bolted Joint Thermal Conductance Design Guideline

	Conductance [W/K]	
Screw Size	**Small Stiff Surface**	**Large Thin Surfaces**
2-56	0.21	0.105
4-40	0.26	0.132
6-32	0.42	0.176
8-32	0.80	0.264
10-32	1.32	0.527
1/4-28	3.51	1.054

7.2.4 Thermal Interface Materials and Conductive Gaskets

Thermal interface materials can be inserted between two components to increase the conductive heat transer between them. They are often made as a sheet or pad of material sandwiched between surfaces, but there are many different types that vary in material, thickness, thermal conductivity, temperature limits, and vacuum-compatibility. Thermal interface materials can also be a grease or paste.

Thinner sheets of materials are commonly used between heat dissipating electronics boxes and mounting surfaces to thermally sink the hot components to a colder surface and reduce the temperature of the electronics. The performance of these types of materials depends on reaching a certain contact pressure between components to ensure the needed heat transfer. Laird Performance Materials has developed many different types of thermal interface materials for a variety of applications. For example, their Tflex series, shown in figure 7.5, is about 1 to 5 mm thick with a thermal conductivity of 6 W mK^{-1} (8), whereas their Tgon series of materials are about 0.13 to 0.5 mm thick with a thermal conductivity of 5 W mK^{-1} (9).

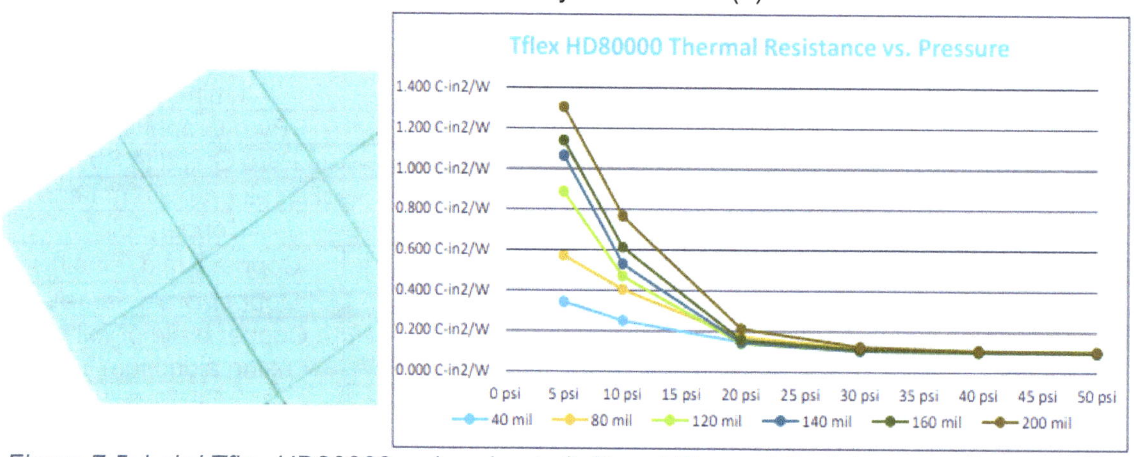

Figure 7.5: Laird Tflex HD80000 series sheets (left) and Thermal Resistance vs. Pressure (right). Credit: Laird Performance Materials.

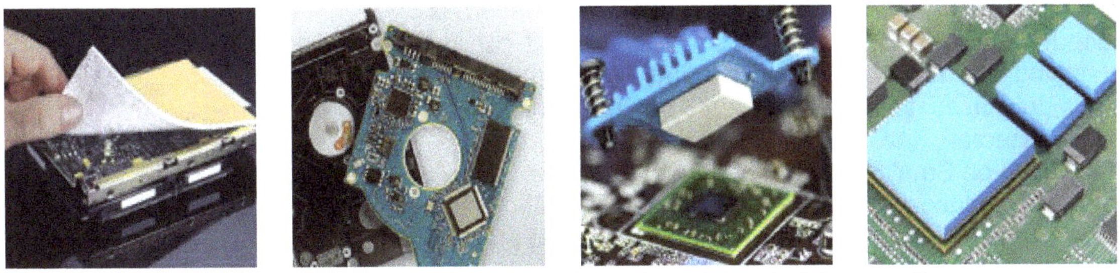

Figure 7.6: A variety of thermal interface materials. Credit: Henkel Corporation.

Thicker pad-like materials, such as Henkel brand GAP PADs®, are often used between high heat dissipating chips on an electronics boards and the electronics enclosure. These are also made to fit a variety of applications, with varying material, thickness, conformability, tear-resistance, electrical isolation, thermal conductivity, and more (10). Several additional thermal interface materials developed by Henkel Corporation are shown in figure 7.6. For conductive gaskets and interface materials, see table 7-7.

Table 7-7: Thermal Interface Materials and Conductive Gaskets

Manufacturer	Product	Recommended Conductance (W/in2-C)
Parker Chomerics	Nusil CV-2946	3.5
NeoGraf	Egraf HT-1210	2
NeoGraf	Egraf HT-C3200	2.5
Parker Chomerics	Chotherm 1671	0.75
Parker Chomerics	Therm a gap 579	0.9
Laird Performance Materials	T-PLI 220	1.5
Henkel Corporation	Gap Pad 300	1.25
Indium Corporation	Indium foil 0.005"	0.9

7.2.5 Sunshields

A sunshield, or sunshade, is an often-deployed device made up of a material with low solar absorptivity that reduces the amount of incident solar flux impinging a spacecraft, by blocking the view to the sun. Sunshields are commonly used for spacecraft thermal control, although only recently designed for small spacecraft.

7.2.6 Thermal Louvers

Thermal louvers are thermally activated shutters that regulate how much heat the louvered surface can dissipate. As the louvers open, the average IR emissivity of the surface changes, changing how much heat the surface dissipates. Full-sized louvers on larger spacecraft have high efficacy for thermal control, however, integration on small spacecraft is challenging. Typical spacecraft louvers are associated with a larger mass and input power, which are both limited on small spacecraft. Although commonly defined as active thermal control, here we consider louvers as a passive thermal control component because the CubeSat-adapted design considered does not require a power input from the spacecraft.

7.2.7 Deployable Radiators

A radiator is a dedicated surface for dissipating excess heat via radiative heat transfer and has a high IR emissivity and low solar absorptivity, an optical property combination typically referred to as "radiator properties." A deployable radiator is stowed during transit or when the radiator is not needed and deployed when excess heat dissipation is required. Deployable radiators on small spacecraft can be challenging due to volumetric constraints. While paint has been widely used to create efficient radiator surfaces on larger spacecraft, the relatively limited available external surface area on SmallSats that already have body-mounted solar cells reduces the potential for creating dedicated radiative surfaces. For a system that requires a large amount of heat dissipation, a passive deployable radiator would greatly enhance thermal performance by increasing the available radiative surface area. Since deployable radiators may be needed because of a lack of radiator surfaces on the spacecraft body due to body-mounted solar cells, an alternate approach (perhaps more common for CubeSats) is to use the chassis body as the radiator area and have a deployable solar array. Also, deployed solar arrays would be able to radiate off a high emissivity/low solar absorptance backside for improved thermal management of the array. Figure 7.7 shows an example of a deployable thermal dissipation technology from Redwire Space.

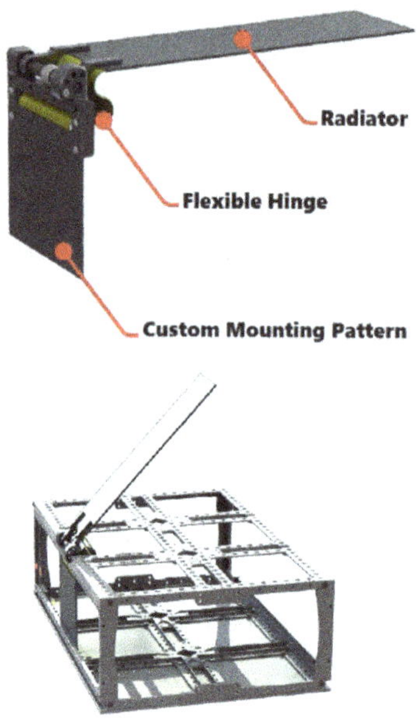

Figure 7.7: Q-Rad Deployable Radiator technology. Credit: Redwire Space.

A novel deployable radiator is being developed by JPL, California Polytechnic San Luis Obispo, and California State Los Angles. At the core of this technology is an Additively Manufactured Deployable Radiator with embedded Oscillating Heat Pipes (AMDROHP) that enables heat to be efficiently transported across moving interfaces. The current AMDROHP radiator design is shown in figure 7.8 and consists of an evaporator and a condenser plate, and a series of flexible joints connecting the two plates. AMDROHP can be stowed within a 3U CubeSat and can be passively deployed without use of an actuator. This AMDROHP technology is currently in the testing phase and further design optimization is ongoing. This project is funded by NASA's SST program in the 2020 cohort of the SmallSat Technology Partnerships initiative.

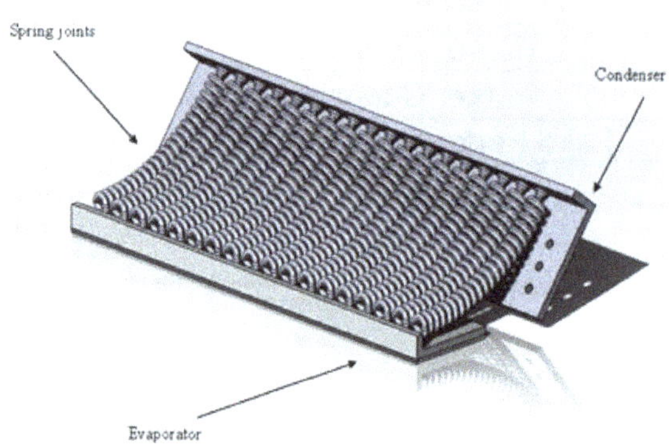

Figure 7.8: Rendering of an AMDROHP radiator design. Credits: California State Los Angles, Jet Propulsion Laboratory, and California Polytechnic San Luis Obispo.

7.2.8 Heat Pipes

A traditional heat pipe is a passive device comprised of a metal container (pipe) that holds a liquid under pressure and has a porous wick-like structure within the container. When heat is applied to one end of the tube, the liquid inside the tube near the hot end vaporizes into a gas that moves through the tube to the cooler end, where it condenses back into a liquid. The wick transports the condensed liquid back to the hot end via capillary action. There are also more complicated and non-passive types of heat pipes such as variable conductance, diode, and loop heat pipes, which are not further explained in this document.

BODY-240

Heat pipes are an efficient passive thermal transfer technology, where a closed-loop system transports excess heat via temperature gradients, typically from electrical devices to a colder surface, which is often either a radiator itself, or a heat sink that is thermally coupled to a radiator. Traditional constant conductance heat pipes are cylindrical in shape with a grooved inner wick, like those used on Bi-Spectral Infrared Detection (BIRD), a 92 kg satellite launched in 2001, to join satellite segments (13). Heat pipes can also be configured as flat plates with tubing sandwiched between two plates and charged with a working fluid inside. SDS-4, a 50 kg small spacecraft launched in 2012, incorporated the Flat-Plate Heat Pipe On-Orbit Experiment (FOX), developed at JAXA (14).

Figure 7.9: FlexCool conformable micro heat pipe before integrating with TechEdSat-10 DVB-S2 radio. Credit: Redwire Space.

The FlexCool heat pipe by Redwire Space is a bent, flat heat pipe developed as a cross between a heat pipe and a thermal strap that can be customized for higher heat fluxes by increasing the thickness. This heat pipe flew on TechEdSat-10, a 6U CubeSat deployed from the ISS in 2020, to thermally manage the radio. An image of this technology in a 1U CubeSat model is shown in figure 7.9.

7.2.9 Phase Change Materials and Thermal Storage Units

A phase change material used as a thermal storage unit is made up of a material (e.g., wax) within a metal housing with a heat source attached so that, as the source conducts heat to the enclosure, the phase change material within absorbs the energy as it changes phase (usually from solid to liquid). Then, as the heat source energy output reduces, the phase change material releases the energy as it changes back to its initial phase (usually from liquid to solid). Owing to the low thermal conductivity of the phase change material, the metal housing must conduct heat into the phase change medium for efficient solidification or melting. See table 7-8 for common phase change materials.

Table 7-8: Phase Change Materials

Material	Melting Point (°C)	Heat of Fusion (kJ/kg)
Saliclic Acid	159	199
Bee Wax	61.8	177
Paraffin	20-60	140-280
Polyethylene glycol 600	20-25	146
Glycerol	18	199
Acetic acid	17	187
Water	0	333
Isopropyl alcohol	-89	88
Butane	-135	76
Ethane	-172	93
Methane	-183	59

Thermal storage units are typically used with components that will experience repeated temperature cycling or to slow down the temperature transient caused by a high heat dissipation event, or a temporary change in the environment such an eclipse. They can be challenging to apply to CubeSats and other small satellites because of the extra mass of the housing needed.

Thermal Management Technologies developed a phase-changing thermal storage unit (TSU) that considers desired phase-change temperatures, interfaces, temperature stability, stored energy, and heat removal methodologies. This device will allow the user to control temperature peaks, stable temperatures and/or energy storage (15).

Redwire Space developed multiple phase change materials (PCM)-based thermal energy storage panels for the CubeSat form factor allowing them to be easily stacked in between critical components (16). Shown in figure 7-10 are two examples of thermal storage technology solutions, Q-Store and Q-Cache.

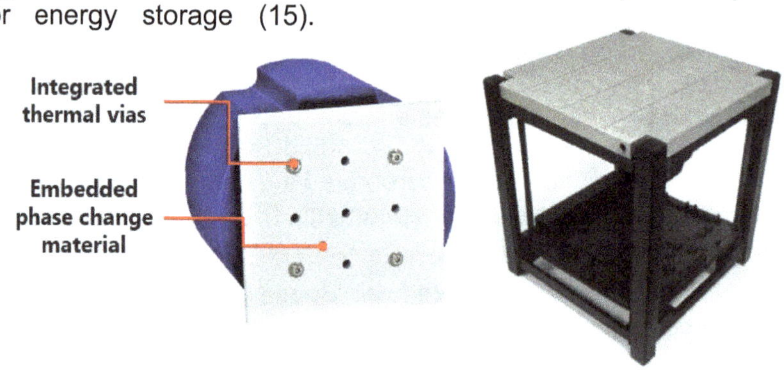

Figure 7.10: Redwire Space's thermal energy storage technologies: (left) Q-Store and (right) Q-Cache. Credit: Redwire Space.

7.2.10 Thermal Switches

A thermal switch is a device that switches a heat conduction path between either a strong thermal coupling or weak thermal coupling (thermal isolation) as needed to control the temperature of heat producing components. A switch typically connects a heat producing component and a low temperature sink, such as a radiator. Heat switches differ from thermostats in that they passively modulate a thermal coupling while thermostats modulate heater circuits (17). Part of the challenge with integration of a thermal switch in SmallSats is that they take up additional space between a component and heat sink. Typical, heat switches may provide a conduction ratio of 10:1 with a technology goal of 100:1 (18). See table 7-9 for example commerical thermal switches.

Table 7-9: Thermal Switches

Manufacture	Product	Type	Max Conductance	Min Conductance	Mass	Max Q	Notes:
ACT	(Study)	VCHP	20 W/K	0.01 to 0.04 W/K	0.3 to 0.5 kg		VCHP Variable Cond. Heat Pipe)
Sierra Nevada	Passive Thermal Control Heat Switch	Mechanical	1 W/K	0.015 W/K	~0.110 kg	6 W	Originally designed for Martian surface ops.
Sierra Nevada	HP Thin Plate	Mechanical	607 W/m^2/°C	7.8 W/m^2/°C	2.72 g/cm^3	12 W	Based on single "cell" design (25.4 mm x 25.4 mm)
Sierra Nevada	Diaphragm Thin Plate	Mechanical	607 W/m^2/°C	7.1 W/m^2/°C	2.72 g/cm^3	12 W	Based on single "cell" design (41.3 mm x 41.3 mm)

7.2.11 Multifunctional Thermal Structures

A newer development in passive thermal control for small spacecraft are multi-functional thermal structures. These integrate thermal control capabilities directly into the structure. This is particularly advantageous for small spacecraft due to strict mass and volume constraints. Currently, Thermal Management Technologies has adapted its multifunction heat spreading structure technology, scaled it to smaller satellite configurations, and called it Standard Passive Orbital Thermal-control (SPOT) Structures. SPOT Structures come in four standard configurations: 6U, 12U, Launch U, and ESPA (19). Each incorporates heat-spreading technology that improves the ability to radiate waste heat. They incorporate features such as low mass, high stiffness/strength, and integrated heat pipes. This new technology is at TRL 4.

7.3 State-of-the-Art – Active Systems

Active thermal control methods rely on input power for operation and have been shown to be more effective in maintaining tighter temperature control for components with stricter temperature requirements or higher heat loads (20). Typical active thermal devices used on large-scale spacecraft include electrical resistance heaters, cryocoolers, thermoelectric coolers, and fluid loops. Electrical heaters are usually easily integrated into SmallSat architectures as they do not typically use much mass or volume. Heaters are frequently used in all space applications, including small and large satellites, so they are often included as passive thermal control technology. Other active systems are challenging to integrate into CubeSats and other small satellites because of the power, mass, and volume needs associated with each given technology. Until spacecraft designers can miniaturize existing actively controlled thermal techniques and reduce power requirements or increase available spacecraft power, most of active thermal systems in small spacecraft are limited and have a TRL range of 3-6. Descriptions of SmallSat active thermal technologies are shown in table 7-10.

Table 7-10: SmallSat Active Thermal Technologies

Active Technology	(Description)
Electrical Heaters	Space heaters use an electrical-resistance element in between two sheets of flexible electrically insulating material.
Cryocoolers	Miniature refrigeration system designed to cool components to extremely low temperatures.
Thermoelectric Coolers (TEC)	miniature solid-state heat pumps which provide localized cooling via the Peltier effect, which is cooling resulting from passing electric current through a junction formed by two dissimilar metals
Fluid Loops	System that circulates a working fluid through the spacecraft to a heat sink.

7.3.1 Heaters

Electrical resistance heaters used on small spacecraft are most often Kapton heaters, which consist of a polyimide film with etched foil circuits that produce heat when a current is applied. Kapton heaters also often have a pressure sensitive adhesive on one side for easy application. Heaters are typically controlled by a thermostat or temperature sensor and used in cold environments to maintain battery temperature, typically the component with the narrowest temperature limits. The low mass of SmallSats requires little additional heater power to maintain temperature limits, and so heaters do not typically need to be very high power to effectively manage temperatures. The TRL values for electrical heaters on SmallSats are 7-9 in LEO environments.

The 1U CubeSats Compass-1, MASAT-1, and OUTFI-1 each required an electrical heater attached to the battery in addition to passive control for the entire spacecraft system to maintain thermal regulation in eclipses (21). Additionally, as biological payloads become more common on small spacecraft, their temperature limits must be considered and maintained as well. NASA ARC biological nanosats (GeneSat, PharmaSat, O/OREOS, SporeSat, EcAMSat, and BioSentinel) all used actively-controlled heaters for precise temperature maintenance for their biological payloads, with closed-loop temperature feedback to maintain temperatures. See table 7-11 for examples of electrical heaters for active thermal systems.

Table 7-11: Active Thermal Systems - Electrical Heaters					
Manufacturer	**Product**	**Power (W/cm^2)**	**Temp Limits**	**Material**	**Notes**
Omega	KHLVA, PLM-Series	0.4 to 1.56	-40°C to 149°C (w/PSA) -57°C to 232°C (w/o PSA)	Kapton (polyimide)	Various sizes, rectangular/square/round, w & w/o pressure sensitive adhesive (PSA), Low-Outgassing
Omega	Polyimide Heater Kit	0.4 to 1.56	-200°C to 200°C (w/PSA)	Kapton (polyimide)	Various shapes. Temp limit max for short excursions. 149°C for continuous ops due to PSA limits.
Tempco	SHK Series	0.8 to 6.2	-35°C to 150°C (w/ Al Foil)	Kapton (polyimide) Some w/ Al Foil	Rectangular/square/round, w/ (PSA). Some model available w/ AL foil for extended temp limits.
Minco	Polyimide Thermofoil HK Series	0.8 to 5.1	-200°C to 200°C -200°C to 150°C (w/ Al Foil)	Kapton (polyimide) Some w/ Al Foil	Mountings: #12 & arcrylic PSA, epoxy, shrink band, stretch tape, clamping. NASA rated for materials & vacuum.
Birk Manufacturing	Polyimide-Traditional	0.02 to 3.1	See Notes	Kapton (polyimide)	Max Temps: FEP-Bonded & All Polyimide(260°C); Acrylic Bonded(120°C).
Birk Manufacturing	Polyimide-High Temp	0.02 to 7.75	See Notes	Kapton (polyimide)	Max Temp: up to 300°C
Chromalox	KPH, KPM Series	0.4 to 1.56	-200°C to 200°C* -200°C to 121°C**	Kapton (polyimide)	* FEP Construction, Acrylic construction
All Flex Solutions	Custom Designs				See website for design guide, options.

7.3.2 Cryocoolers

Cryocoolers are refrigeration devices designed to cool around 100K and below. A summary of cryocooler systems is given in figure 7.11 and a detailed review of the basic types of cryocoolers and their applications is given by Radebaugh (22). The first two systems (a) and (b) are recuperative cycles, and (c), (d), and (e) are regenerative cycles. Cryocoolers are used on instruments or subsystems requiring cryogenic cooling, such as high precision IR sensors. Instruments such as imaging spectrometers, interferometers and midwave infrared (MWIR) sensors require cryocoolers to function at extremely low temperatures. The low temperature improves the dynamic range and extends the wavelength coverage. The use of cryocoolers is also associated with longer instrument lifetimes, low vibration, high thermodynamic efficiency, low mass, and supply cooling temperatures less than 50K (23). Cryocoolers on small spacecraft are still a new concept, however there have been two CubeSats with cryocooling on board. Lunar IceCube, a 6U secondary payload launched from Artemis I in 2022 and developed by Morehead State University, used a 600 mW cryocooler for its BIRCHES point spectrometer (24). For more examples of cryocooler solutions, see table 7-12.

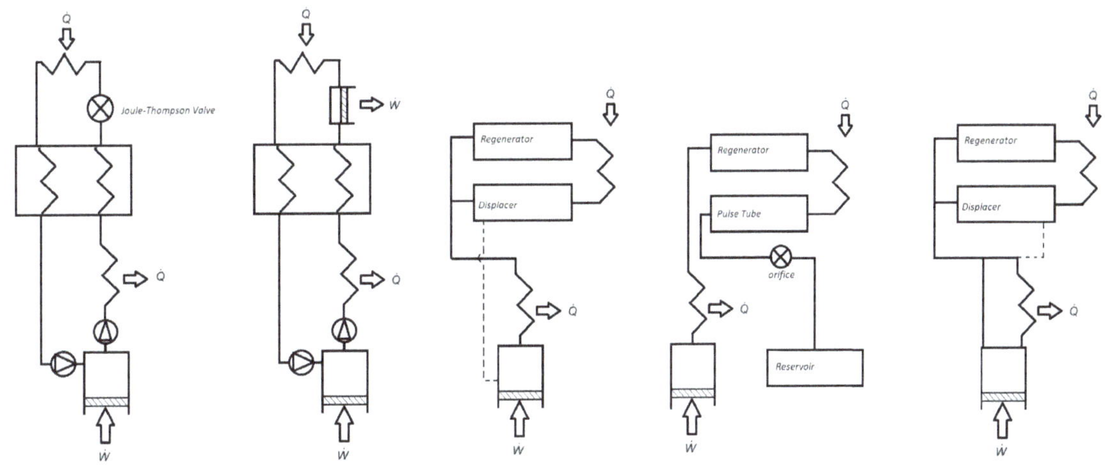

(a) Joule-Thomson (b) Brayton (c) Stirling (d) Pulse Tube (e) Gifford-McMahon

Figure 7.11: A comparison of cryocooler types. Credit: NASA.

Table 7-12: Active Thermal Systems - Cryocoolers							
Manufacturer	**Product**	**Form factor**	**Life (yrs)**	**Mass (kg)**	**Power (W)**	**TRL**	**Notes:**
Lockheed Martin Space	MICRO 1-1	~1U	10	0.350	15	7 - 9	1 W cooling @150K cold tip. Integrated on LunIR CubeSat (launcher: Artemis I).
	MICRO 1-2	~1U	10	0.475	25	6	2 W cooling @105K cold tip. Tested for MISE. Tentatively slated for integration into Europa Clipper.
	MICRO 1-3	~1U	10		40	5	Original design point: 1.4 W cooling @105K (40 W). Max Cooling: 2.9 W @200K (30 W).
AIM Infrarot-Module GmbH	SF070	~1U	>3.5 (MTTF)	0.85	24	6 - 9	0.6 W @80K cold tip. Slated from HyTi: Hyperspectral Thermal Imager (6U CubeSat).
SunPower - Ametek	CryoTel DS Mini	<2U	>14	1.20	45	6	1.8 W @77K (23°C reject). Op to 40K.
	CryoTel MT	<3U	>23	2.10	80	6	5 W @77K (23°C reject). Op to 40K.
Ricor	K508N	1U+	15	0.475	5.5	4 - 6	0.2 W @80K (23°C reject). Demo for 'Active CryoCubesat (ACCS)' project.

TRL values provided are specific to low-Earth environments for SmallSats <180 kg

7.3.3 Thermoelectric Coolers (TECs)

TECs are miniature solid-state heat pumps which provide localized cooling via the Peltier effect, which is cooling resulting from passing electric current through a junction formed by two dissimilar metals. TECs have been used to cool star trackers, IR sensors and low noise amplifiers on large spacecraft. Advantages of TECs are that they have no moving parts, are reliable, noiseless, lightweight, and compact. Their use is limited by low efficiency below temperatures of 130K and low performance with large temperature differences. Furthermore, the TECs are fragile to mount and highly sensitive to thermal expansion stresses. External stresses can be mitigated by adding a conductive strap on the cold side (25). Table 7-13 provides an overview of some commercial thermoelectric coolers.

Table 7-13: Thermoelectric Coolers					
Manufacturer	**Product**	**Cooling Capacity**	**ΔT @ 27°C**	**Operating Temp**	**Notes:**
Laird Technologies	*Examples:* CP Series PolarTec PT Series UltraTEC UTX Series HighTemp ETX Series Multistage MS Series	1.8 - 118 W 18 - 71 W 69 - 299 W 7.7 - 322 W 0.3 - 38 W	68 - 74°C 68 - 74°C 68 - 74°C 68 - 74°C 81 - 129°C	< 80°C < 80°C < 80°C < 150°C < 80°C	
Custom Thermoelectric	04801-9A30-18RB	6.1 W	70°C (Max)	< 200°C	
FerroTec	9502/031/018 M	4.1 W	70°C (Max)		Values @ 50°C

II-IV Coherent	MT30-0.9A-21AN	2.2 W	70°C (Max)		Values @27°C, Designed for zero/high G
II-IV Coherent	CM23-1.9-08AC	3.4 W	71°C (Max)		Values @27°C, Designed for zero/high G

7.3.4 Fluid Loops

A pumped fluid loop (PFL) consists of a circulating pump that moves a liquid through tubing connected to a heat exchanger and heat sink. A heat source is mounted to the heat exchanger and the pumped fluid carries the heat from the source to a heat sink, typically a radiator, and then the cooled fluid is returned to the heat source to continue providing cooling. A PFL is capable of cooling multiple locations via forced fluid convective cooling. Mechanically pumped fluid loops (MPFL) are not typically used on SmallSats because they are associated with high power consumption and mass.

The Active Thermal Architecture (ATA) system is an advanced, active thermal control technology for small satellites in support of advanced missions in deep space, helio-physics, earth science, and communications. The ATA technology is capable of high-power thermal rejection and zonal temperature control of satellite busses, payloads, and high-energy density components. The ATA system includes integrated heaters and a PID-based control algorithm to dynamically tune the satellite's thermal rejection and zonal temperature control as a function of payload and mission parameters. The ATA project was developed by the Center for Space Engineering at Utah State University (CSE, USU) and funded by the NASA SST program in partnership with JPL. The ATA is a sub 1U two-stage active thermal control system targeted at 6U CubeSat form factors and larger. The first stage consists of a MPFL with a micro-pump that circulates a single-phase heat transfer fluid between an internal heat exchanger and a deployed tracking radiator. The second stage is composed of a miniature tactical cryocooler, which directly provides cryogenic cooling to payload instrumentation. The conceptual operation of the ATA system is shown in figure 7.13. The ATA system is scheduled to fly on the upcoming ACMES mission, a technology demonstration flight funded by NASA ESTO through an InVEST grant.

Ultrasonic additive manufacturing (Fabrisonics UAM) techniques were used to simplify and miniaturize the ATA system by embedding the MPFL fluid channels directly into the integrated HX, CubeSat chassis, and the external radiator, to create integrated multi-function structures. By leveraging UAM 3D printing, the ATA system can improve baseline thermal performance by

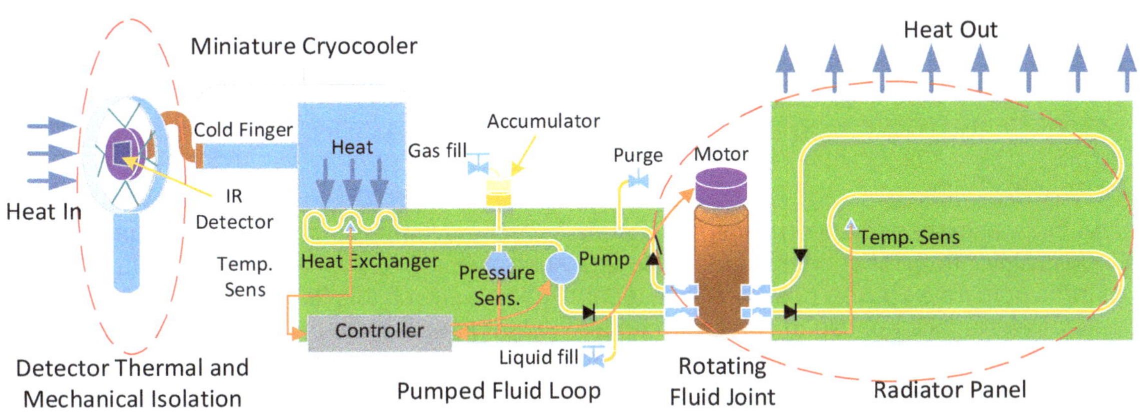

Figure 7.13: Conceptual operation of the ATA thermal control system. Credits: CSE/USU/NASA/JPL.

>200%, a mass savings of more than >30%, and a savings on cost and schedule of better than

>50%. The ATA system also features flexible, multi-axis rotary fluid unions and an integrated geared micro-motor, which allows for the two-stage deployment and solar tracking of the ATA radiator. The ATA features passive vibration isolation and jitter cancellation technologies such as a floating wire-rope isolator design, particle damping, flexible PGS thermal links and a custom Kevlar isolated cryogenic electro-optical detector mount. Figure 7.14 shows some of the technologies developed for ATA as well as the ground-based prototype CubeSat.

Figure 7.14: From top left: ATA CubeSat prototype, ATA subsystem testing, ATA prototypes, UAM radiator with copper backing, UAM heat exchanger, Kevlar isolated Cryogenic Electro-optical prototype mount. Credits: CSE/USU/NASA/JPL.

Table 7-14: Active Thermal Systems - Fluid Loops						
Manufacturer	**Product**	**Cooling Capacity**	**Mass**	**Volume**	**TRL**	**Notes**
Netherlands Aerospace Centre	Mini-MPL	>20 W		< 1U	4 - 6	Scalable to ≤ 200W
Lockheed Martin	Joule-Thomson Microcryocooler		0.2 kg	< 1U (Est.)	4 - 5	

TRL values provided are specific to low-Earth environments for SmallSats <180 kg.

7.4 Summary

As thermal management on small spacecraft is limited by mass, surface area, volume, and power constraints, traditional passive technologies such as paints, coatings, tapes, MLI, and thermal straps dominate thermal design. Active technologies, such as thin flexible resistance heaters have also seen significant use in small spacecraft, including some with advanced closed-loop control. Many technologies that have to date only been integrated on larger spacecraft are being designed, evaluated, and tested for small spacecraft to meet the growing needs of SmallSat developers as small satellites become more and more advanced. Deployable solar panels that have been used by many other SmallSats are paving the way for thermal deployable components, while advanced deployable radiators and thermal storage units are still undergoing testing for small spacecraft.

More active thermal control system technologies are being developed to accommodate volume and power restrictions of a smaller spacecraft; cryocoolers are being designed to fit within 0.5U volume that will allow small spacecraft to use optical sensors and imaging spectrometers.

For feedback solicitation, please email: arc-sst-soa@mail.nasa.gov. Please include a business email.

References

(1) A. Anvari, F. Farhani, and K. S. Niaki. "Comparative Study on Space Qualified Paints Used for Thermal Control of a Small Satellite." 2009.

(2) Sheldahl, A Flex Company. "The Red Book". Jan 24, 2020 REV E. Available at: https://www.sheldahl.com/sites/default/files/2020-02/RedBook_0.pdf

(3) Thermal Management Technologies. “Products, Thermal Components: Flexible Thermal Straps.” [Online] Accessed 2018. Available at: https://www.tmt-ipe.com/thermal-components

(4) Boyd Corporation. “Flexible Thermal Conductors -- Thermal Straps & Strap Assemblies.” Datasheet. [Online] 2020. Accessed 2022. Available at: https://info.boydcorp.com/hubfs/Thermal/Conduction-Cooling/Boyd-Thermal-Straps-Technical-Datasheet.pdf

(5) Technology Applications, Inc. “Graphite Fiber Thermal Straps (GFTS).” [Online] 2022. Available at: https://www.techapps.com/graphite-fiber-thermal-straps

(6) Thermotive. “Products: Thermal Straps.” [Online] 2019. Available at: http://www.thermotive.com/thermalstraps.html

(7) D., Gilmore. “*Spacecraft Thermal Control Handbook, Volume I, Fundamental Technologies, Chapter 8: Mountings and Interfaces."* 2002.

(8) Laird Technologies. Tflex HD80000 Series Thermal Gap Filler. Laird Technologies. Datasheet. [Online] Available at: https://www.laird.com/sites/default/files/tflex-hd80000-datasheet.pdf

(9) Laird Technolgies. “Tgon 800 Series Electrically and Thermally Conductive Interface Pad.” Datasheet. [Online] Available at: https://www.laird.com/sites/default/files/tgon-800-datasheet.pdf

(10) Henkel Corporation, Bergquist. “Thermal Interface Materials Selection Guide. Bergquist.” [Online] 2019. Available at: https://dm.henkel-dam.com/is/content/henkel/lt-8116-brochure-thermal-interface-materials-selection-guidepdf

(11) Thermal Management Technologies. “Thermally Efficient Deployable Radiators.” Datasheet. [Online] 2021. Available at: https://www.tmt-ipe.com/product-sheets

(12) E. Urquiza, B.X. Zhang, M.P. Thelen, J.I. Rodriguez, and S. Pellegrino. “Folding Elastic Thermal Surface - FETS.” [Online] June 1, 2013. Available at: https://www.techbriefs.com/component/content/article/tb/pub/techbriefs/mechanics-and-machinery/16598

(13) H.J. Kramer. “BIRD (Bi-Spectral Infrared Detection).” [Online] May 25, 2012. Available at: https://directory.eoportal.org/web/eoportal/satellite-missions/b/bird

(14) Y. Nakamura et al. "Small Demonstration Satellite-4 (SDS-4): Development, Flight Results, and Lessons Learned in JAXA’s Microsatellite Project." 27th Annual AIAA/USU Conference on Small Satellites, Logan, UT. 2013.

(15) Thermal Management Technologies. “Thermal Storage Units.” Datasheet. [Online] 2021. Available at: https://www.tmt-ipe.com/product-sheets

(16) Redwire Space. “Thermal Energy Storage Panel: Q Store,” [Online] 2022. Available at: https://redwirespace.com/products/qstore/?rdws=nnn.szex.tfd&rdwj=43870

(17) D. Gilmore. Spacecraft Thermal Control Handbook, Volume I, Fundamental Technologies, Chapter 10: Heat Switches. 2002.

(18) NASA. NASA Technology Taxonomy and the NASA Strategic Technology Integration Framework, TX 14: Thermal Management Systems. [Online] July 2020. Available at: https://www.nasa.gov/sites/default/files/atoms/files/2020_nasa_technology_taxonomy_lowres.pdf

(19) Thermal Management Technologies. "Products: Small Spacecraft Structures. Thermal Management Technologies." [Online] 2019. Available at: https://www.tmt-ipe.com/small-spacecraft-structures

(20) K. Hogstrom. "State-of-the-Art Thermal Analysis Methods and Validations for Small Spacecraft." Ae241 Literature Survey. March 4, 2013.

(21) D. Hengeveld, J. Braun, E. Groll, and A. Williams. "Review of Modern Spacecraft Thermal Control Technologies and Their Application to Next-Generation Buildings." 2010. International High Performance Buildings Conference. Paper 40.

(22) R. Radebaugh. "Refrigeration for superconductors." 2004, IEEE, pp. 1719 - 1734 .

(23) R. Hon, C. Kesler, and D. Sigurdson. "Integrated Testing of a Complete Low Cost Space Cryocooler System." 2009.

(24) B. K. Malphrus et al. "The Lunar IceCube EM-1 Mission: Prospecting the Moon for Water Ice." Institute of Electrical and Electronics Engineers A&E Systems Magazine. April 2019

(25) D., Gilmore. Spacecraft Thermal Control Handbook, Volume I, Fundamental Technologies, Chapter 13: Thermoelectric Coolers. 2002

(26) P. Champagne, J.R. Olsen, T. Nast, E. Roth, A. Collaco, G. Kaldas, E. Saito, V. Loung. "*Development of a J-T Micro Compressor.*" IOP Conf. Series: Materials Science and Engineering 101 (2015) 012009.

Table of Contents

Glossary

(AI/ML)	Artificial Intelligence/Machine Learning
(ASICs)	Application-specific Integrated Circuits
(CDH)	Command and Data Handling
(ConOps)	Concept of Operations
(COTS)	Commercial-off-the-shelf
(CRAM)	Chalcogenide RAM
(DDD)	Displacement Damage Dose
(DRAM)	Dynamic RAM
(EPS)	Electrical Power System
(FERAM)	Ferro-Electric RAM
(FPGAs)	Field Programmable Gate Arrays
(FSW)	Flight Software
(I/O)	Input & Output
(LEO)	Low-Earth Orbit
(MRAM)	Magnetoresistive RAM
(OBC)	Onboard Computer
(PCM)	Phase Change Memory
(Rad-hard)	Radiation-hardened
(SDRs)	Software-defined Radios
(SEEs)	Single-event Effects
(SEL)	Single-event Latch-up
(SEUs)	Single-event Upsets
(SoCs)	System-on-chip
(SRAM)	Static Random-Access Memory
(SSA)	Small Spacecraft Avionics
(SWaP)	Size, Weight, and Power
(SWaP-C)	Size, Weight, Power, and Cost
(TID)	Total Ionizing Dose

8.0 Small Spacecraft Avionics

8.1 Introduction

Small Spacecraft Avionics (SSA) consist of all the electronic subsystems, components, instruments, and functional elements of the spacecraft platform, including the primary flight sub-elements Command and Data Handling (CDH) and Flight Software (FSW), as well as other critical flight subsystems such as Payload and Subsystems Avionics (PSA). All must be configurable into specific mission platforms, architectures, and protocols, and be governed by appropriate operations concepts, development environments, standards, and tools. The CDH and FSW are the brain and nervous system of the integrated avionics system, and generally provide command, control, communication, and data management interfaces with all other subsystems in some manner, whether in a direct point-to-point, distributed, integrated, or hybrid computing mode. The avionics system is essentially the foundation for all components and their functions integrated on the spacecraft. As the nature of the mission influences the avionics architecture design, there is a large degree of variability in avionics systems.

There are two major factors to consider for SmallSat avionics:

1. Spacecraft scale: a traditional spacecraft is a high-size, weight, power, and cost (SWaP-C), flagship system, so it'll have a high-SWaP-C avionics system, typically to reduce risk and address higher reliability requirements. A SmallSat is a low-SWaP-C, miniature system, so it'll have a low-SWaP-C avionics system. Typically, due to low cost, more risk is often tolerable, but nonetheless, enhancements can be applied to increase reliability. Individually, the avionics system scales with the spacecraft, however constellations of SmallSats can "match" the capabilities of a traditional spacecraft (using multiple cheap units versus one expensive unit).

2. Architecture design: the architecture design is not necessarily dependent on the scale of the spacecraft. In both traditional spacecraft and SmallSats, the avionics system can be either centralized or decentralized, simplex or fault-tolerant, and modular or monolithic. Traditional spacecraft are very expensive, and to reduce risk, the avionics may employ redundancy such that if one element fails the entire architecture is able to continue, but SmallSat avionics designs are more centralized, whereby if one element fails, the entire system fails. Figure 8.1 illustrates an architectural block diagram of a centralized small spacecraft system. In anticipation of extended durations in low-Earth orbit (LEO) and deep space missions, designers are now incorporating radiation-hardened (rad-hard) or radiation-tolerant architecture designs in their SSA packages to further increase their overall reliability.

While a significant focus of this chapter is on commercial products and developments, vendors are not the only ones developing avionics platforms; there are also numerous government and academic efforts worth considering, with a few examples below:

- SpaceCube and MUSTANG, by NASA GSFC (government)
- Sabertooth by JPL
- CHREC/SHREC Space Processor, by NSF SHREC (academic)
- RadPC by Montana State University (academic)

Given the distributed and integrated nature of modern SSA, this chapter organizes the state-of-the-art in SSA into CDH (8.3) and FSW (8.4). On-the-Horizon activities (TRL <5) for CDH and FSW (8.5 and 8.6, respectively) highlight recent developments in next-generation SSA systems. Avionics Systems Platform and Mission Development Considerations (8.2) discusses how these considerations are being addressed and/or mitigated by state-of-the-art advances in CDH, FSW, and PSA products. A summary of future SSA systems is provided in (8.7).

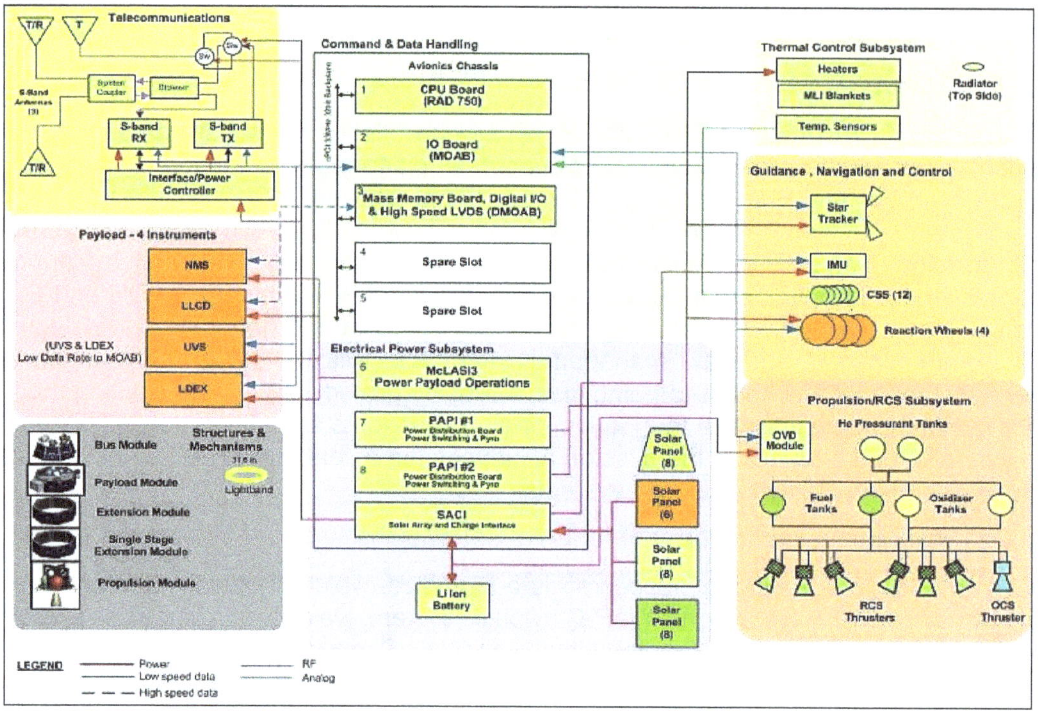

Figure 8.1: Functional block diagram of the LADEE spacecraft. Credit: NASA.

The information described below is not intended to be exhaustive but provides an overview of current state-of-the-art technologies and their development status. It should be noted that Technology Readiness Level (TRL) designations may vary with changes specific to payload, mission requirements, reliability considerations, and/or the environment in which performance was demonstrated. Readers are highly encouraged to reach out to companies for further information regarding the performance and TRL of described technology. There is no intention of mentioning certain companies and omitting others based on their technologies or relationship with NASA.

8.2 Avionics Systems Platform and Mission Development Considerations

There are many factors to be considered in selecting the optimum configuration and implementation of avionics subsystems, components, and elements for small spacecraft missions. Overall spacecraft concerns of Size, Weight, and Power (SWaP) always need to be considered. Some of the more pertinent issues and concerns that all small spacecraft missions must address include:

- Mission applicability and tailoring
- Element, module, and component modularity and interoperability
- Manufacturing and production efficiency, complexity, and scaling
- Mission environment, especially radiation and long-duration space exposure
- Standards and regulatory concerns
- SWaP-C constraints

In addition to CDH and FSW, state-of-the-art SSA systems should consider the following subsystem/payload specific electronic systems:

- Small spacecraft platform size ranges and configurations

- Integrated avionics platform architectures
- Mission avionics configurations
- Spacecraft and mission autonomy

Flight payload and subsystems avionics elements include:

- Subsystem integrated onboard computer (OBC) controllers
- Integrated systems health avionics
- Onboard payload processors
- Cloud-based processors

Modular avionics architectures for small spacecraft can be characterized as either federated or integrated. In a federated avionics architecture, each subsystem of the spacecraft is considered an independent, dedicated autonomous element, with the avionic components performing all functions independently and exchanging data over standardized communications protocols and interfaces. An integrated avionics architecture is a shared, distributed functionality, that can be configured with distributed, heterogeneous and/or mixed criticality elements. In either case, modular avionics architectures can be configured with smart subsystem capabilities, redundant fault tolerant radiation, and anomaly mitigation procedures.

Constellation networks and swarms, synchronized formations, and other multi-satellite cluster formations are creating new opportunities for SSA. The increased need for synchronization, intersatellite communications, controlled positioning for integrated CDH functionality, coordination and conduct, Concept of Operations (ConOps), and autonomous operations impose new constraints on the avionics system. This is true not only for single satellites, but now also for multi-satellite configurations, whereby overall mission performance is dependent on all the platform elements acting in a co-dependent fashion.

8.3 State-of-the-Art (TRL 5-9): Command and Data Handling

Current trends in small spacecraft CDH generally appear to be following those of previous, larger scale CDH subsystems. The current generation of microprocessors can easily handle the processing requirements of most CDH subsystems and will likely be sufficient for use in spacecraft bus designs for the foreseeable future. Cost and availability are likely primary factors for selecting a CDH subsystem design from a given manufacturer, but many groups develop their own custom platforms. The ability to spread nonrecurring engineering costs over multiple missions and reduce software development through reuse are both desirable factors in a competitive market. Heritage designs work well for customers looking to select components with proven reliability for their mission. SmallSat CDH should consider the following:

1. Avionics and onboard computing form factors
2. Highly integrated onboard computing products
3. Rad-hard processors and FPGAs
4. Memory, electronic function blocks, and components
5. Bus electrical CDH interfaces
6. Radiation mitigation and tolerance schemes

As small satellites move from the early CubeSat designs with short-term mission lifetimes to potentially longer missions, radiation tolerance becomes a significant factor when selecting parts. These distinguishing features, spaceflight heritage and radiation tolerance, are the primary differentiators in the parts selection process for long-term missions, verses those which rely heavily on commercial-off-the-shelf (COTS) parts. Experimental missions typically focus on low-cost, easy-to-develop systems that take advantage of open-source software and hardware to provide an easy entry into space systems development, especially for hobbyists or those who lack specific spacecraft expertise.

Small spacecraft CDH technologies and capabilities have been continuously evolving, enabling new opportunities for developing and deploying next-generation SSA. When small spacecraft were first introduced, a primary purpose was to observe and send information back to Earth. As awareness and utility have expanded, there is a need to improve the overall capability of data collection for specific mission environments beyond LEO. Small spacecraft, including nanosatellites and CubeSats, currently perform a wide variety of science in LEO, and these smaller platforms are emerging as candidates for more formidable missions beyond LEO.

The adoption of CubeSat and SmallSat technology is enabled by the miniaturization of electronics, sensors, and instruments. As spacecraft manufacturers begin to use more space-qualified parts, they find that those devices can often lag their COTS counterparts by several generations in performance but may be the only means to meet the radiation requirements placed on the system. Presently, there are several commercial vendors who offer highly integrated systems that contain the onboard computer, memory, electrical power system (EPS), and the ability to support a variety of Input & Output (I/O) for the CubeSat class of small spacecraft. Several CDH developments for CubeSats have resulted from in-house development, the rise of new companies that specialize in CubeSat avionics, and the use of parts from established companies who provide spacecraft avionics for the space industry in general. While parallel developments are impacting the growth of CubeSats, vendors with ties to the more traditional spacecraft bus market are increasing CDH processing capabilities within their product lines.

In-house designs for CDH units are being developed by some spacecraft bus vendors to better accommodate small vehicle concepts. While these items generally exceed CubeSat form factors in size, they can achieve similar environmental performance and may be useful in small satellite systems that replicate more traditional spacecraft subsystem distribution.

8.3.1 Avionics and Onboard Computing Form Factors

The CompactPCI and PC/104 form factors continue generally to be the industry standard for CubeSat CDH bus systems, with multiple vendors offering components that can be readily integrated into space-rated systems. Overall, form factors should fit within the standard CubeSat dimension of less than $10 \times 10\ cm^2$. Spacecraft avionics components are performance-driven and not necessarily dependent on spacecraft platform sizes, but some noncontainerized spacecraft platforms may need to consider the availability and utility of higher TRL avionics products. The PC/104 form factor was the original inspiration to define standard architecture and interface configurations for CubeSat processors, but with space at a premium, many vendors have been using all available space exceeding the formal PC/104 board size. Although the PC/104 board dimension continues to inspire CubeSat configurations, some vendors have made modifications to stackable interface connectors to address reliability and throughput concerns. Many vendors have adopted the use of stackable "daughter" or "mezzanine" boards to simplify connections between subsystem elements and payloads, and to accommodate advances in technologies that maintain compatibility with existing designs. A few vendors provide a modular package which allows users to select from a variety of computational processors.

8.3.2 Highly Integrated Onboard Computing Products

A variety of vendors are producing highly integrated, modular, onboard computing systems for small spacecraft. These CDH packages combine processors and/or Field Programmable Gate Arrays (FPGAs) with various memory banks, and with a variety of standard interfaces for other subsystems onboard. FPGAs and software-defined architectures also give designers a level of flexibility to integrate uploadable software modifications to adapt to new requirements and interfaces. Table 8-1 summarizes the current state-of-the-art for some of these components. It should be noted that while some products have achieved TRL 9 by virtue of a space-based demonstration, what is relevant in one application may not be relevant to another, and different space environments and/or reliability considerations may result in lower TRL assessments. Some larger, more sophisticated computing systems have significantly more processing capability than what is traditionally used in SmallSat CDH systems, however the increase in processing power may be a useful tradeoff if payload processing and CDH functions can be combined (note that overall throughput should be analyzed to assure proper functionality under the most stressful operating conditions).

System developers are gravitating towards ready-to-use hardware and software development platforms that can provide seamless migration to higher performance architectures. As with non-space applications, there is a reluctance to change controller architectures due to the cost of retraining and code migration. Following the lead of microprocessor and FPGA vendors, CubeSat avionics vendors are now providing simplified tool sets and basic, cost-effective evaluation boards.

Table 8-1: Sample of Highly Integrated Onboard Computing Systems							
Manufacturer	**Product**	**Processor/ SoC/ FPGA**	**Radiation Hardness Assurance (RHA)**	**Board Dimension (cm)**	**Power Consumption (W)**	**Orbits Flown**	**Ref**
AAC Clyde Space	Kryten-M3	Smart Fusion 2 SoC including an ARM Cortex-M3 processor delivering 62.5 DMIPS	TID 20 krad (system level)	9.589x9.017x0.551	0.4	LEO	(1)
AAC Clyde Space	Sirius OBC & TCM	32-bit LEON3FT (IEEE-1754 SPARC v8) fault-tolerant processor	TID 20 krad (system level)	9.589x9.017x1.720	1.3	LEO, lunar lander	(2)
Alén Space	OBC+TTC	ARM 32-bit Cortex-M7 with FPU	n/a	8.93x9.33x1.26	OBC_max: 5.865 TTC_max: 5.865	LEO, 2024	(3)
Aitech Systems Inc.	SP0-S	Power PC 1020, Alcatel RTAX	100 krad TID, SEL/LET 40 Mev-cm2/mg (without enclosure)	3U cPCI: 10.06x16x20.3	12	LEO, MEO, GEO, Lunar, Deep Space*	(4)
Aitech Systems Inc.	SP1-S	64-bit Arm® Cortex®-A72 cores @ 1.8 GHz	100 krad TID (Target)	3U Space VPX: 10.06x16x20.3	<25	TBD	(4)
Aitech Systems Inc.	S-A1760	Pascal™ Architecture GPU w/256 CUDA® cores NVIDIA Denver 2 Dual-Core ARM® CPU + Cortex® A57 Quad-Core ARM® CPU	1.05 krad TID, < 1 type 2 SEFI update in 158 days (without enclosure)	12.7x12.7x5.2	Configurable ≤ 5 Idle, 8-10 under typical CUDA load, <20 when System on Module is fully utilized	LEO*	(5) (6)
Aitech Systems Inc.	S-C8780	Intel® Pentium-D or Xeon-D, discrete 2D GPU, Xilinx UltraScale+ FPGA w/Dual-core ARM® CPU	TBD	3U Open VPX: 10.06x16x20.3	Configurable 25-55	TBD	(7)
Aitech Systems Inc.	S-C8500	Intel® Tiger Lake UP3 SoC, 4 cores/8 threads, integrated GPGPU with 96 Execution Units, Xilinx UltraScale+ FPGA w/Dual-Core ARM® CPU is Optional	1 krad TID, SEL/LET TBD (without enclosure)	3U Open VPX: 10.06x16x20.3	CPU Configurable from 12-28, Total system 22-42 configurations.	TBD	(8)

Table 8-1: Sample of Highly Integrated Onboard Computing Systems (*Continued*)							
Manufacturer	**Product**	**Processor/ SoC/ FPGA**	**Radiation Hardness Assurance (RHA)**	**Board Dimension (cm)**	**Power Consumption (W)**	**Orbits Flown**	**Ref**
Argotec	OBC FERMI	Dual-Core LEON3FT SPARC V8 Processor with fault-tolerant memory controller +Microchip RTG4	Rad-hard	10.24x10x4.49	5 (depending on load)	Deep Space, Lunar Orbit, Earth Orbi	(9)
Argotec	OBC HACK	Quad-Core SPARC V8 +Xilinx Kintex7	Rad-hard + MIL + Automotive	10x12.3x4.6	10 (depending on load)		(10)
Bradford Space	Binary OBC	Dual ARM Cortex R5	30 krad (Si) + functional SEE resilience through SW EDAC and internal cold redundancy	13.3x11.5	3 Peak 1 Idle	TBD	(11)
C3S Electronics Development LLC	OBC	32-bit ARM Cortex-M7	Continuous integrity check on the program memory, multi-level watchdog system, MRAM storage	0.92x0.895x0.123 without daughterboard, 0.92x0.895x0.136 with daughterboard	0.46 (measured in test mode, using eMMC as mass storage)	LEO	(12)
Cesium Astro	SBC-1461	1.4 GHz ARM Cortex	LEO	5x8.4x1.3	8	LEO (2023)	(13)
EnduroSat	OBC	ARM Cortex M7	Tested at 40 krad TID	8.9x9.4x2.3 (with integrated GNSS)	1.5 Peak 0.2 Idle	400-600 km SSO, ISS-like, equatorial	(14)
EnduroSat	Payload Controller	Xilinx UltraScale+ SoC	20 krad TID (SoC testing)	9.8x9.8x7.4	<50 Peak, 20 Idle	TBD	
GOMspace	NanoMind HP MK3	Xilinx Zynq 7030/7045	>20 krad	9.5x9.5x3.15	Dependent on customer application	LEO	(15)
Innoflight	CFC-400XS	Defense Grade Xilinx UltraScale+ MPSoC	TID: 30 krad	8.2x8.2x2.7	6-25	LEO (2021) & GEO (2022)	(16)
Innoflight	CFC-400XP	Defense Grade Xilinx UltraScale+ MPSoC	TID: 30 krad	17.2x10x2.5	4-30	LEO in 2024	(17)

BODY-259

Table 8-1: Sample of Highly Integrated Onboard Computing Systems (*Continued*)							
Manufacturer	**Product**	**Processor/ SoC/ FPGA**	**Radiation Hardness Assurance (RHA)**	**Board Dimension (cm)**	**Power Consumption (W)**	**Orbits Flown**	**Ref**
Innoflight	CFC-510P	AMD Ryzen GPGPU	TID: 30 krad	17.2x10x2.5	12-40	LEO in 2024	
KP Labs	Antelope OBC and DPU	OBC – RM57 Herkules microcontroller (Dual 300 MHz ARM Cortex-R5F with FPU in lock-step) DPU – AMD Xilinx Zynq UltraScale+ MPSoC (ZU2EG, ZU3EG, ZU4EG, ZU5EG), Quad ARM Cortex-A53 CPU, Dual ARM Cortex-R5 in lock-step	COTS with SEE mitigation	Motherboard – 1x1x1 Daughterboard - 7x4x2	From less than 0.5 (DPU is off) to 6 (DPU processes the data)	LEO	(18)
KP Labs	Leopard D PU	AMD Xilinx Zynq UltraScale+ MPSoC (ZU6EG, ZU9EG, ZU15EG); Quad ARM Cortex-A53 CPU; Dual ARM Cortex-R5 in lock-step	COTS with SEE mitigation	Non-redundant: 9x9.5x5 Redundant: 9x9.5x7.8	7 static power consumption; up to 10 for deep learning inference.	LEO	(19)
KP Labs	Lion DPU	AMD Xilinx Kintex Ultrascale FPGA (KU035, KU060, KU095)	COTS with SEE mitigation	16x10x5	<15 for KU035 version	TBD	(20)
Laboratory for Atmospheric and Space Physics	LASP Generic Small-sat FPGA Board	Kintex-7	25 krad	8.763x8.763	1	LEO	(21)
Micro Aerospace Solutions	MAS-SBC-107	Arm® Cortex®-M7	Total Ionizing Dose of 30 krad (Si)	9×9.6	<30	LEO	
Nara Space Technology	NSTOBC	Atmel ARM 9 based Microprocessor Unit	<24 krad	9x9x1.6	0.429 (Idle)	LEO qualified	

Table 8-1: Sample of Highly Integrated Onboard Computing Systems (*Continued*)

Manufacturer	Product	Processor/ SoC/ FPGA	Radiation Hardness Assurance (RHA)	Board Dimension (cm)	Power Consumption (W)	Orbits Flown	Ref
Novo Space	SBC002AV	Zynq Ultrascale+	TID: 30 krad; Automotive parts with Rad-tolerant in high criticality parts; Rich telemetry with local event response; Board sectorization and power control; ECC in Volatile memories; CRC / Reed-Solomon / Interleaving in Non-Volatile memories; FPGA and fabric use selective TRM on critical functionality; Scrubbing on soft FPGAs.	16x10	application dependent active mode max: 15	TBD	(22)
Novo Space	SBC003AV	SmartFusion2		16x10	application dependent active mode max: 8	TBD	(23)
Novo Space	SBC004AF	Versal ACAP		16x10	Under development	TBD	(24)
Novo Space	SBC005AF	Polarfire		16x10	application dependent active mode max: 8	TBD	(25)
Novo Space	GPU001AF	Jetson TX2i		8.7x5	active mode min:9 active mode max:15	TBD	(26)
Novo Space	GPU002AF	Jetson Orin Nano	TID: 30 krad	8.7x5	active mode min:6 active mode max:11	TBD	(27)
Pumpkin Space Systems	PPM A1	MCU+DSP	N/A	90x96	0.05	LEO	
Pumpkin Space Systems	PPM A2	MCU+DSP	N/A	90x96	0.05	LEO	
Pumpkin Space Systems	PPM A3	MCU+DSP	N/A	90x96	0.05	LEO	
Pumpkin Space Systems	PPM B1	MCU+DSP	N/A	90x96	0.05	LEO	
Pumpkin Space Systems	PPM D1	MCU	N/A	90x96	0.05	LEO	
Pumpkin Space Systems	PPM D2	MCU+DSP	N/A	90x96	0.05	LEO	
Pumpkin Space Systems	PPM E1	MCU	N/A	90x96	0.05	LEO	
Pumpkin Space Systems	PPM B1	MCU+DSP	N/A	90x96	0.05	LEO	
Pumpkin Space Systems	PPM D1	MCU	N/A	90x96	0.05	LEO	
Pumpkin Space Systems	MBM2 w/BBB	MCU	~5 krad	90x96	2	LEO	

Table 8-1: Sample of Highly Integrated Onboard Computing Systems (*Continued*)							
Manufacturer	**Product**	**Processor/ SoC/ FPGA**	**Radiation Hardness Assurance (RHA)**	**Board Dimension (cm)**	**Power Consumption (W)**	**Orbits Flown**	**Ref**
Resilient Computing	RadPC-SBC-001	RISC-V 32-Bit (AMD Artix 7)	COTS with SEE mitigation	10x10	1.5	LEO, Lunar (2024)	(28)
SkyLabs	NANOhpc-obc	4x RISC-V 64-bit / PolarFire / SoC	COTS w/ SEE mitigation	9.5x9.1x1.3	~10 Peak		(29)
SkyLabs	NANOhpm-obc	32-bit RISC-V core / PolarFire	COTS w/ SEE mitigation	9.5x9.1x1.3	<5	LEO, MEO	(30) (31)
SkyLabs	NANOobc-2	PicoSkyFT / IGLOO 2	COTS w/ SEE mitigation	9.5x9.1x1.1	<1 Peak, <0.9 Idle	LEO, MEO	(32) (33)
Space Dynamics Laboratory	Pearl Avionics	LEON3 / RTP3	10 krad	13x9	2-6	LEO, GEO	
SPiN USA	MA61C CubeSat	GR712RC dual-core 32-bit LEON3 fault-tolerant, SPARC V8 processor	Processor is 300 krad	9.599x9.027	1-1.2	TBD	(34)
SPiN USA	MA61C smallsat	GR712RC dual-core 32-bit LEON3 fault-tolerant SPARC V8 processor	Processor is 300 krad	10.5x10.5	1-1.2	TBD	(35)
SPiN USA	MA61C cPCI serial space	GR712RC dual-core 32-bit LEON3 fault-tolerant SPARC V8 processor	Processor is 300 krad	16x10	1-1.2	TBD	(36)
Spiral Blue	Space Edge One	Nvidia Jetson Xavier NX	N/A	10x10	7	LEO	(37)
Southwest Research Institute® (SwRI®)	Centaur SBC	GR712RC dual core LEON3-FT CPU Xilinx Ultrascale or Microchip RT-ProAsic or RTG4 FPGA	TID: Up to 60 krad (Si), more with shielding SEL: Immune to >67eV/mg/cm2 SEU: < 1 error per 24 hour period	Available in 3U/6U cPCI form factors	Low power < 4 Operation	LEO	(38)
Southwest Research Institute® (SwRI®)	HP-SBC	PowerPC based MPC8548e CPU Microsemi RT-ProASIC FPGA Optional: Microsemi RTG4 FPGA Optional: Xilinx Ultrascale FPGA	TID: Up to 60 krad (Si), more with shielding SEL: Immune to >67eV/mg/cm2 SEU: < 1 error per 24-hour period	Available in 3U/6U cPCI form factors	12-20 depending on clock frequency and FPGA pairing	Leo and GEO	(38)

Table 8-1: Sample of Highly Integrated Onboard Computing Systems (*Continued*)							
Manufacturer	**Product**	**Processor/ SoC/ FPGA**	**Radiation Hardness Assurance (RHA)**	**Board Dimension (cm)**	**Power Consumption (W)**	**Orbits Flown**	**Ref**
Spacemanic	DeepThought_OBC	SAMV71	-	6.7x4.2x0.7	0.1	GRBAlpha BDSat1 & 2 Planetum Veronika CroCube	(39)
Spacemanic	Eddie_OBC	MSP430	-	6.7x4.2x0.7	0.1	LEO (500-550 SSO), Planetum Veronika CroCube	(40)
Trident Space Electronic Systems	VDRT	Versal VC1902	30 krad / 50MeV LU	10x14.6x2.54	<60	LEO/MEO capable	(41)
Trident Space Electronic Systems	UDRT	MPSoC ZU19	30 krad / 40MeV LU	10x14.6x2.54	<50	Multiple LEO 400-1200km missions	(42)
Unibap	iX10-100A	AMD Ryzen V1000 (CPU and GPU) Intel Myriad-X (VPU)	Radiation Tolerant COTS with SEE mitigation	10x10x6	25-40	LEO	(43)
Unibap	iX5-106	AMD Steppe Eagle (CPU and GPU) Intel Myriad-X (VPU) Microchip SmartFusion2 (FPGA)	Radiation Tolerant COTS with SEE mitigation	10x10x5	15-25		
Xiphos	Q7S	AMD-Xilinx Zynq-7020 Dual-core ARM Cortex-A9	25 krad	7.8x4.3x0.9	2	LEO	
Xiphos	Q8S	AMD-Xilinx Zynq Ultrascale+ MPSOC Quad-core ARM Cortex-A53	30 krad	8x8x1.12	>5	LEO	
Xiphos	Q8JS	AMD-Xilinx Zynq Ultrascale+ MPSOC Quad-core ARM Cortex-A53	30 krad	8x8x1.12	>5	LEO	

*Orbits flown are on larger spacecraft

8.3.3 Radiation-Hardened Processors

Several radiation-hardened embedded processors have recently become available. These are being used as the core processors for a variety of purposes including CDH. Some of these are the Vorago VA10820 (ARM M0) and the VA41620 and VA41630 (ARM M4); Cobham GR740 (quad core LEON4 SPARC V8); BAE 5545 quad core processor; and LS1043 quad processor. These have all been radiation tested to at least 50 kRad total ionizing dose.

8.3.4 Memory, Electronic Function Blocks, and Components

The range of onboard memory for small spacecraft is wide, typically starting around 32 kB and increasing with available technology. For CDH functions, onboard memory requires high reliability. A variety of different memory technologies have been developed for specific traits, including volatile memory, such as Static Random-Access Memory (SRAM) and Dynamic RAM (DRAM), Magnetoresistive RAM (MRAM), Ferro-Electric RAM (FERAM), Chalcogenide RAM (CRAM) and Phase Change Memory (PCM). SRAM is typically used due to price and availability, with numerous SRAM choices (up to 4M x 39 [20 MB]). There are many manufacturers that provide a variety of electronic components that are space-rated with high reliability. A chart comparing the various memory types and their performance is shown in table 8-2.

Table 8-2: Comparison of Memory Types

Feature	SRAM	DRAM	Flash	MRAM	FERAM	CRAM/ PCM
Non-volatile	No	No	Yes	Yes	Yes	Yes
Operating Volt age, ±10%	2.5 – 5 V	1.35-3.3 V	3.3 & 5 V	3.3 V	3.3 V	3.3 V
Organization (bits/die)	512 k × 8 4M × 39	128 M × 8 1Gb × 8	16 M × 8; 4G × 8	2M × 8	16 k × 8	Unk
Data Retention (70°C)	N/A	N/A	10 years	10 years	10 years	10 years
Endurance (Erase/Write cycles)	Unlimited	Unlimited	1E5	1E13	1E13	1E13
Access Time	10-25 ns	25 ns	50 ns after page ready; 200 us write; 2 ms erase	300 ns	300 ns	100 ns
Radiation (TID)	50K - 1 Mrad	50 krad	30 krad	1 Mrad	1 Mrad	1 Mrad
Temperature Range	MIL-STD	Industrial	Commercial	MIL-STD	MIL-STD	MIL-STD
Power	500 mW	300 mW	30 mW	900 mW	270 mW	Unk
Package	4 MB-20 MB	128 MB 1GB	128MB – 4 GB	2 MB	1.5 MB (12 chip package)	Unk

8.3.5 Bus Electrical Interfaces

CubeSat class spacecraft continue to use interfaces that are common in the microcontroller or embedded systems world. Highly integrated systems, especially systems-on-chip (SoCs), FPGAs, and application-specific integrated circuits (ASICs), will typically provide several interfaces to accommodate a wide range of users and to ease the task of interfacing with peripheral devices and other controllers. FPGAs are commonly used for these interfaces because of their flexibility and ability to change interfaces as needed. Some of the most common bus electrical interfaces are listed below with applicable interface standards:

- Serial Communication Interfaces (SCI): RS-232, RS-422, RS-485 etc.
- Synchronous Serial Communication Interface: I2C, SPI, SSC and Enhanced Synchronous Serial Interface (ESSI)
- Multimedia Cards (SD Cards, Compact Flash, etc.)
- Networks: Ethernet, LonWorks, etc.
- Fieldbuses: CAN Bus, LIN-Bus, PROFIBUS, etc.
- Timers: PLL(s), Capture/Compare and Time Processing Units
- Discrete IO: General Purpose Input/Output (GPIO)
- Analog to Digital/Digital to Analog (ADC/DAC)
- Debugging: JTAG, ISP, ICSP, BDM Port, BITP, and DB9 ports
- SpaceWire: a standard for high-speed serial links and networks
- High-speed data: RapidIO, XAUI, SerDes and MGT protocols are common in routing large quantities of mission data in the gigabit per second speeds

8.3.6 Radiation Mitigation and Tolerance Schemes

Deep space and long-duration LEO missions compel developers to consider reliability requirements and possibly incorporate radiation-mitigation strategies into their respective spacecraft designs. CubeSats are often either composed of only COTS components or a hybrid combination of COTS and rad-hard and radiation-tolerant components. COTS components typically offer superior performance, energy efficiency, and affordability compared to their rad-hard alternatives; however, COTS devices tend to be highly susceptible to radiation. The advantages of COTS components have enabled low-cost CDH development, while also allowing developers to leverage start-of-the-art technologies in their designs. A hybrid design combines COTS and rad-hard components, such as COTS processor and memory with rad-hardened supporting electronics (e.g., EPS, watchdog, etc.), to maximize the benefits of both technologies. These designs may also incorporate radiation-mitigation techniques to further enhance overall system reliability.

For space applications, the effects of radiation on electronic devices can vary broadly (44). Radiation effects are often categorized into long-term cumulative effects and transient single-event effects (SEEs). Long-term effects include total ionizing dose (TID) and displacement damage dose (DDD). TID, measured in krad, is the ionizing radiation absorbed by the device material over time causing parametric or functional degradation of the device. DDD is the nonionizing damage caused by particle collisions with the device structure over time. SEEs occur when a single radiation particle strike deposits enough charge to cause an effect. SEEs can be destructive or nondestructive. Single-event upsets (SEUs) are nondestructive SEEs that can affect the logic state of a memory cell. Single-event latch-up (SEL) are destructive SEEs that manifest as parasitic structures in CMOS logic or bipolar transistor structures, potentially causing a high-current state.

Other areas of consideration for CDH elements include memory, imaging, protection circuits (watchdog timers, communications watchdog timers, overcurrent protection, and power control), memory protection (error-correction code memory and software error detection and correction), communication protection (several components), and parallel processing and voting.

8.4 State-of-the-Art (TRL 5-9): Flight Software

The FSW, at a fundamental level, communicates the instructions for the spacecraft to perform all operations necessary for the mission. These include all the science objectives as well as regular tasks (commands) to keep the spacecraft functioning and ensure the storage and communication of data (telemetry). The FSW is usually thought of as all the programs that run on the CDH avionics but should also include all software running on the various subsystems and payload(s).

There are many factors in selecting a development environment and/or operating system for a space mission. A major factor is the amount of memory and computational resources. There are always financial and schedule concerns. Another factor is what past software an organization may have used and their experiences with that software. The maturity of the software and its availability for the target subsystem or payload are additional factors to be considered in the final selection.

FSW complexity can refer to the architecture design (e.g., the interactions between subsystems, especially for spacecraft autonomy) as well as the number of operations to be performed. The more software is required to do, the bigger the task and cost. This complexity (and the associated verification effort) is what primarily drives the cost and schedule for a program or mission. Required reliability and fault management can also increase complexity and cost, regardless of the size of the spacecraft. Changing requirements is also a huge factor, which may be mitigated by involving the software team early in the planning process.

With the increase in processing capability with CDH and other processors, more capable FSW has been enabled. Traditionally, larger spacecraft require rad-hard processors which have poor performance, while CubeSats and SmallSats can take more risks with COTS processors that offer substantially more performance. Several advances have increased the processing capabilities available for CubeSats. Low-power ARM-based processors and embedded COTS SoCs, as well as advances in radiation hardened processors, have brought similar processing capabilities down to the small size of CubeSats. All of this has resulted in increased demands and requirements for FSW.

Generally, CDH and other subsystems need to be able to supervise several inputs and outputs as well as process and store data within a fixed time-period. These all need to be performed in a reliable and predictable fashion throughout the lifetime of the mission. The needs of each mission can vary greatly, but basic deterministic and reliable processing is a fundamental requirement. The following are important when considering FSW design:

- Implication of CDH processors on FSW
- Frameworks
- Operating systems
- Software languages
- Mission operations and ground support suites
- Development environment, standards, and tools

8.4.1 Implication of CDH Processors on FSW

The processor and memory available on the CDH can put significant limitations on the FSW. For some of the smaller jobs, or to reduce electronic complexity, smaller processors are used (distributed processing). These have typically been thought of as embedded processors, with many of them containing dedicated memory. Modern integrated space avionics, including heterogeneous and mixed criticality architectures, also impact operational constructs and can contribute to advanced configurations (such as multiple modular redundant systems architectures) which can allow advanced paradigms for radiation tolerance and system redundancies in critical small spacecraft missions.

8.4.2 Frameworks

In the context of SSA, a FSW framework can be described as a hierarchal architecture, sometimes referred to as a set of lego-like building block constructs, partitions, and functions. This emerging system-of-systems concept describes the large-scale integration of many independent, self-contained systems that work together to satisfy a global need. Examples of commonly used frameworks include:

- cFS (https://cfs.gsfc.nasa.gov)
- F' (https://github.com/nasa/fprime)
- NanoSat Mission Operations Framework (https://nanosat-mo-framework.github.io/)
- Spacecloud (https://space-cloud.io/)
- ROS (https://www.ros.org/)

8.4.3 Operating Systems

Operating systems manage computer hardware, software resources, and provide common services for computer programs. Examples of commonly used operating systems include:

- VxWorks
- RTEMS
- FreeRTOS
- Linux

8.4.4 Software Languages

System programming involves designing and writing computer programs with software languages that allow the computer hardware to interface with the programmer and the user, leading to the effective execution of application software on the computer system. State-of-the-art small spacecraft have used C, C++, Python, Arduino and other software languages.

8.4.5 Mission Operations and Ground Support Suites

Although not directly used on the spacecraft, mission operations and ground support suites must also use software and systems for testing, and to monitor, command, control, and communicate with the spacecraft, as well as display status and disseminate data across all aspects of a space mission (including spacecraft performance and procedures, systems health, science and technology data handling and management, and telemetry tracking and control). For smaller spacecraft and missions, it is usually best to use the same ground support software for mission operations, integration and testing, and development and testing. There are numerous open-source and proprietary tools and programs available for these activities. A small set of tools that have been used at NASA are described below. For more information, please refer to the *Ground Data System and Mission Operations* chapter.

8.4.6 Development Environment, Standards, and Tools

Development environment, standards, and tools are used to design, develop, validate, and operate small spacecraft missions, with adherence to accepted software and space mission standards. Examples of commonly used development tools include:

- Version control tools

- Auto-generation of software
- Simulations and simulators
- Software best practices and NPR7150

8.5 On the Horizon (TRL 1-4): Command and Data Handling

Many CDH systems will continue to follow trends set for embedded systems. Short-duration missions in LEO will continue to take advantage of advances made by industry leaders who provide embedded systems, technologies, and components. In keeping with the low-cost, rapid development theme of CubeSat-based missions, many COTS solutions are available for spacecraft developers.

While traditional CDH processing needs are relatively stagnant, as small satellites are being targeted for flying increasingly data-heavy payloads (i.e., imaging systems) there is new interest in advanced onboard processing for mission data. Typically, these higher performance functions would be added as a separate payload processing element outside of the CDH function.

Next-generation SSA/PSA distributed avionics applications are integrating FPGA-based software-defined radios (SDRs) on small spacecraft (45). A SDR can transmit and receive in widely different radio protocols based on a modifiable, reconfigurable architecture, and is a flexible technology that can enable the design of an adaptive communications system. This can increase data throughput and enable software updates on-orbit, also known as re-programmability. Additional FPGA-based functional elements include imagers, Artificial Intelligence and Machine Learning (AI/ML) processors, and subsystem-integrated edge and cloud processors. The ability to reprogram sensors or instruments while on-orbit have benefited several CubeSat missions when instruments do not perform as anticipated, or when entering an extended mission phase that requires subsystems or instruments to be reprogrammed.

In keeping with trends seen in other disciplines and industries, the Industry 4.0 and "digitally managed everything" is absolutely of critical importance for technological and programmatic efficiencies in SSA systems development. Following are some modern tools, technologies, and approaches that should be considered when developing and deploying next-generation small spacecraft avionic systems:

- Artificial intelligence, machine learning, and machine vision
- Robotics and automation
- Model-based systems engineering
- Embedded systems / edge computing
- Internet-of-space-things
- Cloud computing
- Augmented reality/ virtual reality / mixed reality
- Software-defined-everything
- Advanced manufacturing
- Digital twin

8.6 On the Horizon (TRL 1-4): Flight Software

FSW is key to mission success. The field of software is a very dynamic environment that is continuously evolving. The challenges with flight software usually remain the same regardless of the size of the spacecraft (CubeSat to SmallSat) and are related to the size and complexity of the endeavor. Overall, FSW can be known to cause scheduling and implementation issues, especially during integration and test. There is usually a temptation to add additional features, and all these factors can drive up overall complexity of the FSW and increase risk to the mission as a whole.

It is essential that FSW be as simple as possible. It is critical to survey options and plan early in any FSW effort. Wherever possible, early development and testing should be performed. Efforts to add additional features should be looked at very critically with a strong effort to stick to the existing plan. With good planning and careful execution, a favorable outcome can be achieved. It is becoming more common to update software after the hardware is delivered (or even launched), and there are now software frameworks such as cFS that have features to enable software updates after deployment.

On the horizon FSW will soon include multicore processor operating systems and programming, as learning how to harness multicore processors differently than Microsoft Windows does will enable true real-time multiprocessing. On the horizon FSW will also include artificial intelligence (e.g., Nvidia); FSW for multicore, multiprocessor, and heterogenous platforms (e.g., AMD-Xilinx Versal); and FSW (middleware) for constellations of SmallSats with resource management, scheduling and task assignment, and fault tolerance.

Spacecraft autonomy is an emerging capability and SmallSat designers have particular interest in the following characteristics for autonomous systems:

- Situational and self-awareness
- Reasoning and acting
- Collaboration and interaction
- Engineering and integrity

Spacecraft autonomy can be considered part of management, direction, and control for all subsystems and functions in a spacecraft. CDH takes input from, and provides direction to, all subsystems (ADCS, Power, Propulsion, Comm, vehicle health, etc.). Those subsystems may also have a degree of autonomy depending on the complexity of its local "smart subsystems" processor. The NASA 2020 Technology Roadmap defines autonomous systems as a cross-domain capability that enables the system to operate in a dynamic environment independent of external control (46).

Some autonomous systems now implement a heterogeneous architecture, meaning they contain multiple processors with varying levels of performance and capabilities. For instance, higher performance modules and components can be used for sophisticated data processing, AI and onboard computing for both spacecraft and mission performance optimization—as well as real-time adaptive analysis of science data—while lower performance onboard processors and FPGAs conduct the routine spacecraft operations functions and interact with the subsystems which also may include distributed performance cascades.

8.7 Summary

Space applications now require considerable autonomy, precision, and robustness, and are refining technologies for such operations as on-orbit servicing, relative and absolute navigation, inter-satellite communication, and formation flying. An exciting trend is that small spacecraft missions are becoming more complex as these platforms are now being used for lunar and deep space science and exploration missions. Small spacecraft technology is expanding to meet the needs of increasing small spacecraft mission complexity. This has accelerated over the past few years to achieve the next gen goals of using small spacecraft to collect important science in deep space, and mitigate risk for larger, more complex mission-critical situations. In parallel, spacecraft electronics have matured with higher performance and reliability, and with miniaturized components that meet the growing needs of these now very capable spacecraft.

BODY-269

The 2022 Small Spacecraft Avionics chapter has been updated with a broader, interrelated framework, where CDH, FSW, and smart payloads are not just independent space platform subsystems but are part of an integrated avionics ecosystem which includes all electronic elements of a space platform, now primarily digitally based and or managed. Also, SSA should not be considered as an isolated spaceflight technology component, but rather as a core digital engineering technology emphasis area, capable of taking advantage of and integrating products, processes, and technologies from other disciplines. To continue to be relevant and efficient, the SSA communities must remain cognizant and receptive of the continuously evolving nature of the digital based Industry 4.0 technology revolution now being evidenced in other related and/or associated vertical disciplines and solutions.

For feedback solicitation, please email: arc-sst-soa@mail.nasa.gov. Please include a business email for further contact.

References

(1) AAC Clyde Space. "Kryten-M3" datasheet. Available online at: https://www.aac-clyde.space/wp-content/uploads/2021/10/AAC_DataSheet_Kryten.pdf. Accessed: August 1, 2023.

(2) AAC Clyde Space. "Sirius OBC and TCM" datasheet. Available online at: https://www.aac-clyde.space/wp-content/uploads/2021/10/AAC_DataSheet_Sirius-OBC.pdf. Accessed: August 1, 2023.

(3) Alen Space. "TRISKEL OBC+TTC+OBSW Solution" datasheet. Available at: https://products.alen.space/products/triskel/. Accessed: December 1, 2023.

(4) Aitech Systems Inc. [Online] "SP0-S: Radiation Tolerant 3U CompactPCI SBC". Accessed: December 1, 2023. Available at: https://aitechsystems.com/product/sp0-s-rad-tolerant-3u-compactpci-sbc/

(5) Aitech Systems Inc. [Online] "S-A1760 Venus: Radiation-characterized Space AI GPGPU". Accessed: December 1, 2023. Available at: https://aitechsystems.com/product/s-a1760-space-gpgpu/

(6) Aitech Systems Inc. [Online] "Aitech's Space-rated AI Supercomputer On NASA Atmospheric Re-entry Demonstration". May 2023. Accessed: December 1, 2023. Available at: https://aitechsystems.com/aitechs-space-rated-ai-supercomputer-on-nasa-atmospheric-re-entry-demonstration/

(7) Aitech Systems Inc. [Online] "C878:3U VPX SBC". Accessed: December 1, 2023. Available at: https://aitechsystems.com/product/c878-3u-vpx-sbc/

(8) Aitech Systems Inc. [Online] "U-C850x Series: Innovative SWaP-C-optimized Secure 3U". Accessed: December 1, 2023. Available at: https://aitechsystems.com/product/u-c850x-series-3u-vpx-secure-sbc/

(9) Argotec. [Online] "FERMI: Deep Space On-Board Computer," Technical Datasheet. Available at: https://www.argotecgroup.com/wp-content/uploads/2022/03/Argotec_FERMI_scheda_prodotto.pdf

(10) Argotec. "Hack: High-Performance Modular On-Board Computer," Technical Datasheet. [Online] Available at: https://www.argotecgroup.com/wp-content/uploads/2022/03/Argotec_HACK_scheda_prodotto.pdf

(11) Bradford Space Deep Space Industries. "Deep Space Avionics" Preliminary Datasheet.

(12) C3S Electronics Development LLC. "OBC" Datasheet. Available at: https://c3s.hu/subsystems/#CHP-OBC

(13) Cesium Astro. [Online] "SBC-1461: Single-Board Computer". Accessed: December 1, 2023. Available at: https://www.cesiumastro.com/products/sbc-1461#specifications

(14) EnduroSat. [Online] "Onboard Computer (OBC). Accessed: December 1, 2023. Available at: https://www.endurosat.com/cubesat-store/cubesat-obc/onboard-computer-obc/

(15) GOMspace. [Online] "NanoMind HP MK3". Available at: https://gomspace.com/shop/subsystems/command-and-data-handling/nanomind-hp-mk3-(1).aspx

(16) Innoflight. [Online] "CFC-400XS". 2023. Accessed: December 1, 2023. Available at: https://www.innoflight.com/product/cfc-400xs-2/

(17) Innoflight. [Online] "CFC-400XP". 2023. Accessed: December 1, 2023. Available at: https://www.innoflight.com/product/cfc-400xp/

(18) KP Labs. [Online] "Antelope" datasheet. Available at: https://kplabs.space/wp-content/uploads/Antelope-technical-sheet.pdf

(19) KP Labs. [Online] "Leopard" datasheet. Available at: https://kplabs.space/wp-content/uploads/Leopard-technical-sheet.pdf

(20) KP Labs. [Online] "Lion" datasheet. Available at: https://kplabs.space/wp-content/uploads/Lion-technical-sheet.pdf

(21) Laboratory for Atmospheric and Space Physics, University of Colorado, Boulder. [Online] "State-of-the-art small satellite technology". Accessed December 1, 2023. Available at: https://lasp.colorado.edu/our-expertise/engineering/smallsats/

(22) Novo Space. [Online] "SBC002AV". Accessed: December 1, 2023. Available at: https://www.novo.space/products/sbc002av

(23) Novo Space. [Online] "SBC003AV". Accessed: December 1, 2023. Available at: https://www.novo.space/products/sbc003av

(24) Novo Space. [Online] "SBC004AV". Accessed: December 1, 2023. Available at: https://www.novo.space/products/sbc004av

(25) Novo Space. [Online] "SBC005AV". Accessed: December 1, 2023. Available at: https://www.novo.space/products/sbc005av

(26) Novo Space. [Online] "GPU001AF - General purpose GPU". Accessed: December 1, 2023. Available at: https://www.novo.space/products/gpu001af

(27) Novo Space. [Online] "GPU002AF - General purpose GPU". Accessed: December 1, 2023. Available at: https://www.novo.space/products/gpu002af

(28) Resilient Computing. [Online] "Technology: RadPC - Radiation Tolerant Computing." Accessed: December 1, 2023. Available at: https://resilient-computing.com/technology

(29) SkyLabs. [Online] "NANOhpc-obc". Accessed: December 1, 2023. Available at: https://www.skylabs.si/products/nanohpc-obc/

(30) SkyLabs. [Online] "NANOhpm-obc". Accessed: December 1, 2023. Available at: https://www.skylabs.si/products/nanohpm-obc/

(31) SkyLabs. [Online] "TRISAT-R SUCCESSFULLY REACHED THE MEDIUM EARTH ORBIT". July 14, 2022. Accessed: December 1, 2023. Available at: https://www.skylabs.si/trisat-r-successfully-reached-the-medium-earth-orbit/

(32) SkyLabs. [Online] "NANOobc-2". Accessed: December 1, 2023. Available at: https://www.skylabs.si/products/nanoobc-2/

(33) D. Gačnik. "NANOsky I 2nd-Generation satellite platform for multi-mission constellation projects". Presented at 5th ESA CubeSat Industry Space Days, June 1-3,

2021. Available at: https://www.hermes-sp.eu/wp-content/uploads/2021/07/CID-SkyLabs-NANOsky-I-Platform_v7.pdf

(34) Space Products and Innovation. "ULTIPURPOSE ADAPTER GENERIC INTERFACE CONNECTOR (MA61C) Datasheet". Available at: https://www.spinintech.com/_files/ugd/0ac116_9b444733c51f401898e790bb26040144.pdf

(35) Space Products and Innovation. "MULTIPURPOSE ADAPTER GENERIC INTERFACE CONNECTOR (MA61C) Datasheet". Available at: https://534279dc-dca3-4d3e-b394-5d35d472455d.filesusr.com/ugd/0ac116_8d41f30b0a264087adf66848f2247b69.pdf

(36) Space Products and Innovation. "MULTIPURPOSE ADAPTER GENERIC INTERFACE CONNECTOR (MA61C) cPCI serial space version Datasheet". Available at: https://www.spinintech.com/_files/ugd/0ac116_bad3b7ec01ce4e62ba5b06a9ad3e7c40.pdf

(37) Spiral Blue. "Space Edge Computer Datasheet". April 2023. Available at: https://www.spiralblue.space/_files/ugd/8ebd97_1dee6e72951345a784a2974782a6164a.pdf

(38) Southwest Research Institute (SwRI). "SINGLE BOARD COMPUTERS". 2023. Accessed December 1, 2023. Available at: https://www.swri.org/industry/space-engineering/single-board-computers

(39) Spacemanic. "Deep Thought On-board Computer". 2023. Accessed December 1, 2023. Available at: https://www.spacemanic.com/deep-thought-onboard-computer/

(40) Spacemanic. "Eddie Onboard Computer ". 2023. Accessed December 1, 2023. Available at: https://www.spacemanic.com/eddie-onboard-computer/

(41) Trident Space Electronic Systems. "VERSAL DIGITAL RF TRANSCEIVER" datasheet. September 12, 2023. Available at: https://tridsys.com/wp-content/uploads/2023/10/Trident-VDRT-Cut-Sheet.pdf

(42) Trident Space Electronic Systems. "ZYNQ ULTRASCALE+ DIGITAL RF TRANSCEIVER" datasheet. September 12, 2023. Available at: https://tridsys.com/wp-content/uploads/2023/10/Trident-VDRT-Cut-Sheet.pdf

(43) Unibap. [online] "SpaceCloud's 22 apps aboard the Wild Ride mission verified in orbit". June 9, 2021. Accessed: December 1, 2023. Available online: https://unibap.com/news/spaceclouds-22-apps-aboard-the-wild-ride-mission-verified-in-orbit/

(44) National Academies of Sciences, Engineering, and Medicine. 2018. Testing at the Speed of Light: The State of U.S. Electronic Parts Space Radiation Testing Infrastructure. Washington, DC: The National Academies Press.

(45) M. Wirthlin, "High-Reliability FPGA-Based Systems: Space, High-Energy Physics, and Beyond," in Proceedings of the IEEE, vol. 103, no. 3, pp. 379-389, March 2015.

(46) NASA. 2020 Technology Taxonomy. [Online] Available at: https://www.nasa.gov/sites/default/files/atoms/files/2020_nasa_technology_taxonomy.pdf

Chapter Contents

Chapter Glossary

(ADCS)	Attitude Determination and Control System
(BPF)	BandPass Filters
(CDH)	Command and Data Handling
(COTS)	Commercial-off-the-Shelf
(DORA)	Deployable Optical Receiver Aperture
(DLR)	German Aerospace Center
(DSN)	Deep Space Network
(DSP)	Digital Signal Processing
(DVB-S2)	Digital Video Broadcast Satellite Second Generation
(FCC)	Federal Communications Commission
(FIPS)	Federal Information Processing Standard
(FPGAs)	Field Programmable Gate Arrays
(FSM)	Fine-steering Mirror
(FSO)	Free Space Optical
(IARU)	International Amateur Radio Union
(IEEE)	Institute of Electrical and Electronics Engineers
(ISARA)	Integrated Solar Array and Reflectarray Antenna
(ISM)	Industrial, Scientific, and Medical
(ISOC)	Inter-spacecraft Optical Communicator
(ISS)	International Space Station
(JAXA)	Japanese Aerospace Exploration Agency
(JPL)	Jet Propulsion Laboratory
(LADEE)	Lunar Atmosphere and Dust Environment Explorer
(Lasercom)	Laser Communications
(LCH)	Laser ClearingHouse
(LCT)	Laser Communication Terminals
(LDPC)	Low-Density Parity-check Code
(LLCD)	Lunar Laser Communications Demonstration
(LNA)	Low Noise Amplifier
(LSRB)	Laser Safety Review Board
(MA)	Multiple Access
(MarCO)	Mars Cube One
(MEMS)	Micro-Electro-Mechanical Systems

(MRR)	Modulating Retro-Reflector
(NEN)	Near Earth Network
(NICT)	National Institute of Information and Communications Technology
(NOAA)	National Oceanic and Atmospheric Administration
(NPR)	NASA Procedural Requirements
(NTIA)	National Telecommunications and Information Administration
(OCSD)	Optical Communication and Sensor Demonstration
(OCTL)	Optical Communication Telescope Laboratory
(OSIRIS)	Optical Space Infrared Downlink System
(PAT)	Pointing, Acquisition, and Tracking
(RF)	Radio Frequency
(SBIR)	Small Business Innovative Research
(SCaN)	Space Communications and Navigation
(SDR)	Software Defined Radios
(SME)	Subject Matter Expert
(SNR)	Signal-to-Noise Ratio
(SOTA)	Small Optical Transponder
(SWaP)	Size, Weight, and Power
(TDRS)	Tracking and Data Relay Satellite
(TMA)	Technology Maturity Assessments
(TRL)	Technology Readiness Levels
(TT&C)	Tracking, Telemetry & Command
(VSOTA)	Very Small Optical Transponder
(WFF)	Wallops Flight Facility

9.0 Communications

9.1 Introduction

For most missions the communication system enables the spacecraft to transmit data and telemetry to Earth, receive commands from Earth, and relay information from one spacecraft to another. A communications system consists of the ground segment: one or more ground stations located on Earth, and the space segment: one or more spacecraft and their respective communication payloads. The three functions of a communications system are receiving commands from Earth (uplink), transmitting data down to Earth (downlink) and transmitting or receiving information from another satellite (crosslink or inter-satellite link) (figure 9.1). There are two types of communication systems: radio frequency (RF) and free space optical (FSO), FSO is also referred to as laser communications (lasercom).

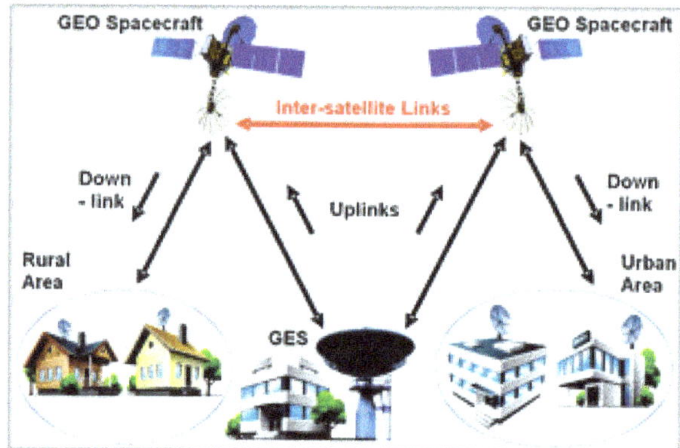

Figure 9.1: Satellite uplink, downlink, and crosslink. Credit: D. Stojce (2019).

Most spacecraft communications systems are radio frequency based. They typically operate within the designated Institute of Electrical and Electronics Engineers (IEEE) radio bands of 300 MHz to 40 GHz. A RF system communicates by sending data using electromagnetic waves to and from antennas. Information is modulated onto radio frequency electromagnetic waves and sent over a channel, through the atmosphere or space, to the receiving system where it is demodulated (figure 9.2).

Although RF systems are typically used for low-rate space communication, recent developments in FSO communications have made it a compelling alternative to RF systems, particularly for high-rate communication. FSO systems consist of a transmitting terminal and receiving terminal. Like an RF system, information is modulated onto electromagnetic waves (at optical frequencies) and sent over a channel to the receiving system. FSO links operate at a much higher frequency than RF links, generally at near-infrared bands (e.g., 1064 nm or 1550 nm). Visible light is often not

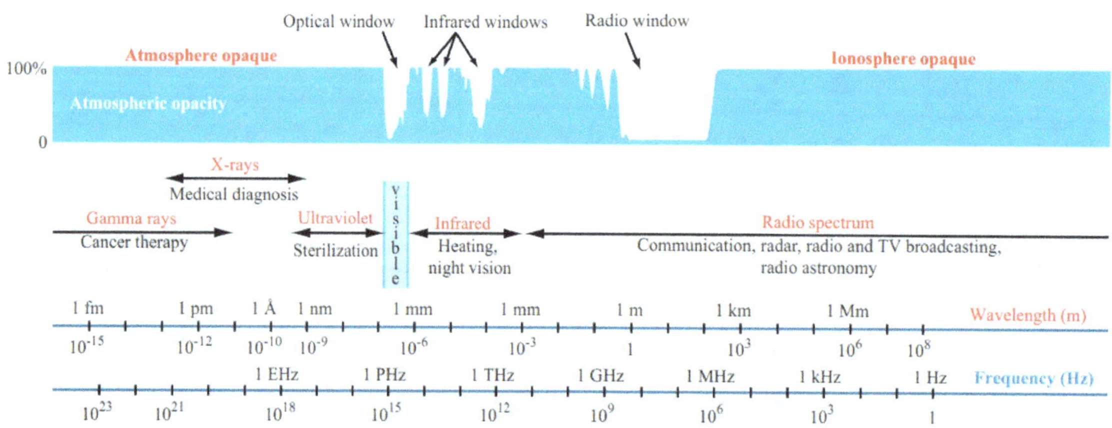

Figure 9.2: Atmospheric opacity of the electromagnetic wave spectrum with the infrared and radio windows used by spacecraft for communication. Credit: Microwave Radar and Radiometric Remote Sensing by Ulaby and Long.

used due to eye safety concerns for technicians at the terminals. The use of higher frequencies and wider bandwidths can support higher data rates, but the shorter wavelengths also result in narrower beamwidths that require more accurate pointing towards the communication terminal both more accurately and precisely.

This chapter organizes the state-of-the-art in small spacecraft communications technologies into two main categories: RF and FSO. Tables at the end of each section list hardware options for RF and developing FSO technologies for mission designers to consider.

This chapter is a survey of small spacecraft communications technologies as discussed in open literature and does not endeavor to be an original source. This chapter only considers literature in the public domain to identify and classify devices. Commonly used sources for data include manufacturer datasheets, press releases, conference papers, journal papers, public filings with government agencies, and news articles. There is no intention of mentioning certain companies and omitting others based on their technologies or relationship with NASA.

9.2 Radio Frequency Communications

A radio communication system includes a radio transmitter, a free space communication channel, and a radio receiver. At the top level, a radio transmitter system consists of a data interface, modulator, power amplifier, and an antenna. The transmitter system uses the modulator to encode digital data onto a high frequency electromagnetic wave. The power amplifier then increases the output RF power of the transmitted signal to be sent through free space to the receiver using the transmit antenna.

The radio receiver system uses a receiving antenna, low noise amplifier, and demodulator to produce digital data output from the received signal. The receiving antenna collects the electromagnetic waves and routes the signal to the receiver, which then demodulates the wave and converts the electrical signals back into the original digital message. Low noise amplifiers are sometimes employed to minimize thermal noise in certain frequency bands and/or increase the received signal strength. In many cases, the functions of the modulator and demodulator are combined into a radio transceiver that can both send and receive RF signals.

Radio frequency communications for spacecraft are conducted between 30 MHz and 60 GHz. The lower frequency bands (up to S-band) are typically more mature for SmallSat use, however extensive use of these bands has led to crowding and challenges acquiring licensing. Higher frequencies offer a better ratio of gain-to-aperture-size, but this is offset by the increased atmospheric attenuation at those frequencies and the higher free space loss that is directly proportional to the square of the frequency.

Table 9-1: Radio Frequency Bands

Band	Frequency
VHF	30 to 300 MHz
UHF	300 to 1000 MHz
L	1 to 2 GHz
S	2 to 4 GHz
C	4 to 8 GHz
X	8 to 12 GHz
Ku	12 to 18 GHz
K	18 to 27 GHz
Ka	27 to 40 GHz
V	40 to 75 GHz

9.2.1 Frequency Bands

Satellite communications are conducted over a wide range of frequency bands. The typical bands considered for small satellites are UHF, S, X, and Ka. The most mature bands used for CubeSat communication are VHF and UHF frequencies. There has been a shift in recent years towards S and X, with Ka-band also being used for recent & future small satellite communications. The move to higher frequency bands has been driven by a need for higher data rates. At the higher frequencies, there is generally greater atmospheric and rain attenuation adding to increased free space loss. This needs to be compensated for with higher power transmission and/or high gain antennas with narrower beamwidths. Moving to higher-gain antennas increases the pointing accuracy required for closing the link. See table 9-1 for a list of RF bands.

NASA spacecraft, which use the government bands of S-band, X-band and Ka-band, may use the NASA Near Space Network (NSN). The primary frequency bands of S, X, and Ka are more advantageous than using the UHF band, which has a higher probability of local interference. Satellite Tracking, Telemetry & Command (TT&C) is typically conducted over S-band. Non-NASA spacecraft have access to a wide variety of ground system options ranging from do-it-yourself to pay-per-pass services.

In L-band, CubeSats can take advantage of legacy communications networks such as Globalstar and Iridium by using network-specific transponders to relay information to and from Earth. These networks remove dependence on dedicated ground station equipment. However, they can only be used at orbital altitudes below the communication constellation and require experimental frequency authorization.

Ku-, K-, and Ka-band communication systems are the state-of-the-art for large spacecraft, especially in spacecraft-to-spacecraft communications, but they are still young technologies in the CubeSat world. They are becoming more attractive to SmallSat designers as the lower frequencies become more congested. At the higher frequencies, rain fade becomes a significant problem for communications between a spacecraft and Earth (1). Nonetheless, the benefits of operating at higher frequencies have justified further research by both industry and government alike. At JPL, the Integrated Solar Array and Reflectarray Antenna (ISARA) mission demonstrated high bandwidth Ka-band CubeSat communications with over 100 Mbps downlink rate (2). The back of the 3U CubeSat was fitted with a high gain reflectarray antenna integrated into an existing solar array. The successful demonstration of the reflectarray on ISARA became the basis for the Mars Cube One (MarCO) mission to Mars. The MarCO mission uses two twin CubeSats for a communications relay between the InSight lander and Earth. Using a X-band reflectarray they were able to successfully complete their mission (3). Another mission to use Ka-band for DTE communications was the Kepler telescope, launched in 2009. With future missions being increasingly data hungry, we are likely to see a shift towards Ka-band and, possibly, even higher frequencies.

CubeSats have also used the unlicensed Industrial, Scientific, and Medical (ISM) bands for communications. The Ames TechEdSat team has successfully demonstrated WiFi to downlink data at 1 Mbps. Notably, a group at Singapore's Nanyang Technological University used a 2.4 GHz ZigBee radio on its VELOX-I mission to demonstrate commercial-off-the-shelf (COTS) land-based wireless systems for inter-satellite communication (4). Similarly, current investigations are looking at using wireless COTS products, such as Bluetooth-compatible hardware, for inter-satellite communications (5).

9.2.2 System Architecture

A small satellite RF communications system consists of a transceiver comprised of a radio, an amplifier, and an antenna. Radios receive a message from the Command and Data Handling (CDH) subsystem, then produce and modulate an electromagnetic wave to create a signal. They are responsible for generating the signal and modulating or demodulating it. The radio is also where coding may be added to the signal. Channel coding is added to provide data error detection and correction capabilities, which ensures reliable communication under the conditions imposed by the satellite transmission path. From Shannon's Equation (6), it is known that the information capacity of a channel is related to its bandwidth and signal-to-noise ratio (SNR). The channel capacity (information flow) can be increased by increasing the SNR or the bandwidth, and many modulation and coding schemes make effective use of this tradeoff.

Radios offer some power amplification, but often the signals from small satellites require a greater boost. The power amplifier will take the signal from the radio and increase the RF output power before sending it to the transmit antenna. On the receive side, a low noise amplifier will take the weak signal from the receive antenna and amplify it while minimizing thermal noise. A bandpass filter might be used before the LNA to reject undesired frequencies. The radio will then be able to process the stronger signal with higher accuracy. In RF communications the role of the antenna is to increase and focus the strength of the signal in a specific direction. The digital message encoded on the RF carrier signal will be sent to and from the antennas of each system. See figure 9.3 for an example transmit and receive block diagram.

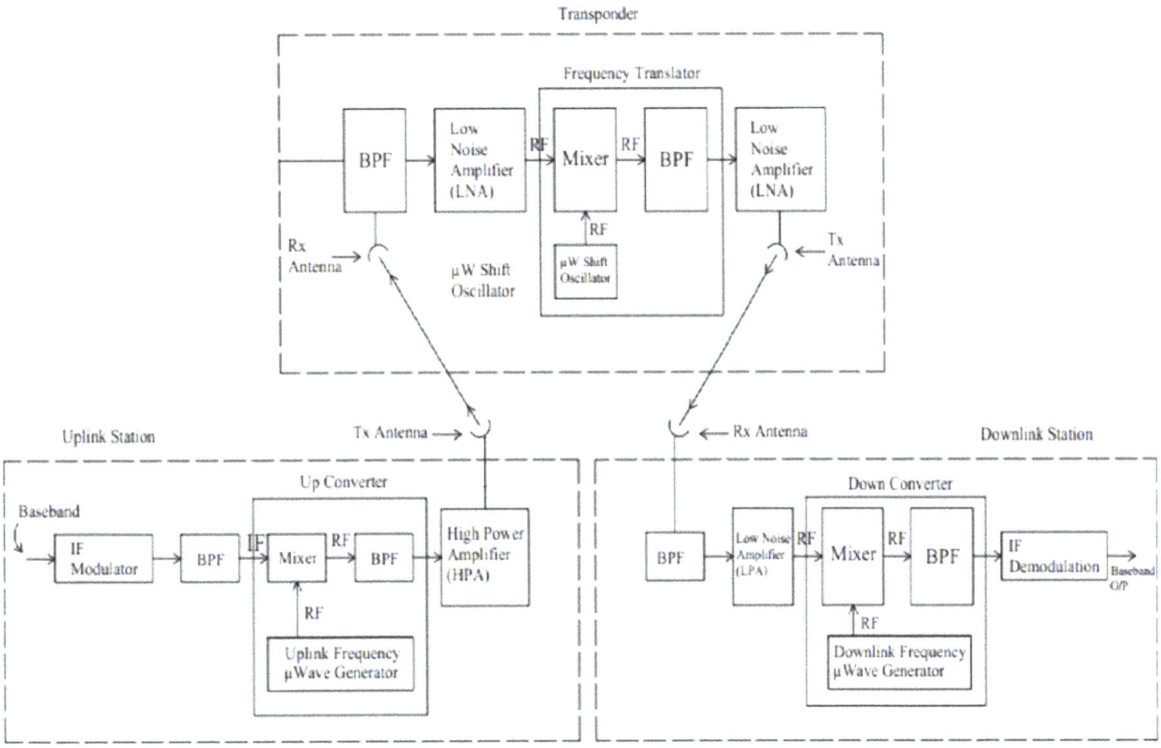

Figure 9.3: Transmit and receive block diagram. Credit: Karim et al. (2018). http://creativecommons.org/licenses/by/4.0/

9.2.3 Major Components in SmallSat Communication Systems

- Radio or Modulator/Demodulator: on the transmit side it produces, modulates, codes, and amplifies an electromagnetic wave to create a signal. Adds modulation and coding as needed. As a receiver it decodes and demodulates received signals.
- Mixers: RF mixers are used in communications systems to change the frequency of the signal. If the frequency generated by the radio is not the desired transmit frequency, then an upconverter will convert the signal to a higher frequency for transmit. Similarly, the downconverter will down convert a receive frequency to a lower one for processing.
- Filters: bandpass filters are used to reject undesired frequencies, typically before the LNA or downconverter.

- Amplifier: a power or gain amplifier is required for a transmit system. A low noise amplifier (LNA) is required for a receive system. LNAs, in addition to amplifying the (low power) received signal, serve to minimize the system noise temperature.
- Antenna: increases the strength of a signal in a specific direction, relative to the same signal strength without directionality. Transmits signals fed to it by a transmitter and receives signals propagated across free space. Antennas can be low-gain & omni-directional with a broad beam, or high-gain & directional with a narrow beam.
- Encryption: a cryptographic unit is an integrated encryptor/decryptor device that provides secure uplink, downlink, or crosslink for satellite communication links. Most small satellite designers will not require a cryptographic payload unit based on their threat level and may be able to use the communications radio for simple encryption schemes.
- Spread-spectrum communication applies a known frequency spreading function to the signal, which helps reduce interference from other transmitters, and provides more secure communications; as such, it is often used for multi-way communication networks. For example, the NASA Tracking and Data Relay Satellite (TDRS) multiple-access mode requires spread spectrum signals to support multiple simultaneous communication links.

9.2.4 Design Considerations

As with all spacecraft subsystems, there are power and mass constraints placed on the comm system. Based on these restrictions several trade studies need to be performed to choose the optimal design.

When designing an RF communications system, the first trades performed are for data rate, power consumption, and total mass. For example, a mission with high data rate needs would select a high frequency such as X-band for downlink and a directional high-gain antenna. Based on the ground station locations available, engineers would perform link budget analyses to determine the minimum power needed for a specific ground station antenna. This analysis would factor in rain and atmospheric attenuation, as well as modulation and coding. A few different link budget trades will be run, varying antenna size, RF output power and data rate. Each link will return a different margin of decibels, representing the reliability of the system. The engineers will proceed to calculate the final mass and power for each configuration. The mission designer will have a limit on mass and power constraints for the communications subsystem. Each configuration traded will compare data rate, power, and mass. A high data rate downlink may cost a high amount of mass for the antenna and power for the amplifier and radio. Conversely, a low-power, low-mass system may have a lower data rate.

Another factor that is considered in the design phase is pointing. Depending on the orbit of the satellite and whether the link is Uplink/Downlink, or Crosslink, the system may have a specific pointing requirement. Large satellites frequently use gimbals--platforms that can pivot to point their antennas. The addition of a gimbal will increase the overall mass and power draws of the system. CubeSats frequently trade high-gain antennas for low-gain, omni-directional ones to maintain the link regardless of directionality. CubeSats may also change their attitude to point a body-mounted antenna, rather than use a gimbal.

9.2.5 Policies and Licensing

Any non-Federal US spacecraft with a transmitter must be licensed by the Federal Communications Commission (FCC). The types of RF licenses used by small satellites are: Amateur (FCC Part 97) and Experimental (FCC Part 5) (7). An amateur license type of authorization is limited to hobbyists and non-profit use and comes with many FCC restrictions. Experimental Part 5 licenses are commonly used for university CubeSats and can be granted for a CubeSat operating in the amateur band (A SmallSat or SmallSat constellation can also apply under provisions of Part 25). A spacecraft with any sort of remote sensing capability must contact the National Oceanic and Atmospheric Administration (NOAA) to find out if a NOAA license is required. A NOAA license is not an RF license and conveys no authority for the radiation of RF energy for communication. For government missions the National Telecommunications and Information Administration (NTIA) is the licensing authority.

For amateur licensing, there must be an FCC licensed amateur radio control operator. Downlink telemetry and communications cannot be obscured (encrypted). Use of science gathered via amateur radio downlink for profit ("pecuniary interest") is prohibited. Frequency "assignment" in the amateur-satellite allocations requires coordination, a process administered by the International Amateur Radio Union (IARU) (8).

In 2018, the FCC adopted a Notice of Proposed Rulemaking to develop a new authorization process tailored specifically to small satellite operations, keeping in mind efficient use of spectrum and mitigation of orbital debris. Small satellites that would qualify for the new rules include those with 10 or lesser number of satellites under a single license. All individual satellites will have to be 10 cm or larger in the smallest dimension and weigh less than 180 kg. The maximum in-orbit lifetime of each individual satellite will be six years, including de-orbiting time, and they would have to be deployed under 600 km altitude. Each satellite will have a unique telemetry marker for tracking and will not release any debris (9).

9.2.6 Encryption

Encryption is the process of encoding information to conceal it from outside actors. Small satellites can use a cryptographic unit to encrypt or decrypt data prior to transmission. When data is being prepared for transmission, it is broken up into packets. These packets are then scrambled according to the encryption scheme being used. An encryption scheme uses an encryption key generated by an algorithm to encode the data. The authorized receiver of the encrypted data will be able to decrypt the message using the appropriate key. Without the authorized key, decrypting the data will be extremely difficult.

With the increased proliferation of small satellites in low-Earth orbit comes an increase in vulnerabilities. Many SmallSats are comprised of COTS hardware and/or open-source software. While this strategy allows for a more flexible design approach, adversaries can gain insight into the design. Encryption of data in transit prevents other actors from commanding satellites or intercepting transmissions.

NASA requires any of its propulsive spacecraft within 2 million kilometers of Earth to protect their command uplink with encryption that is compliant with Level 1 of the Federal Information Processing Standard (FIPS) 140-3 (10). The FCC has also considered requiring encryption on the telemetry, tracking, and command communications as well as mission data for propulsive spacecraft, but decided not to incorporate a specific requirement at this time. A satellite with an amateur license cannot encrypt transmissions in any way and must consist of open information. The eligibility rules are listed in 47 CFR Part 97 (11).

9.2.7 Antennas

Antennas are used for propagating data through free space using electromagnetic waves. RF antennas are typically sized for their respective frequencies. This means that antennas are often chosen or designed specifically for their mission. COTS antennas are available for SmallSats and can be built to order. For missions that don't have high data rate requirements, a simple patch or monopole antenna with low gain and efficiency will suffice. Due to their low directionality, these antennas can generally maintain a communication link even when the spacecraft is tumbling, which is advantageous for CubeSats lacking good attitude and accurate pointing control. New developments in antenna design have put technologies like the deployable reflector antenna, reflectarray, and passive or active array antennas on the horizon for small satellites. Please see table 9-3 for information on commercially available antennas for SmallSats/CubeSats.

Figure 9.4: (from left to right) CubeSat-compatible S-band patch antenna (IQ Wireless), X-band high-gain antenna and pointing mechanism (Surrey Satellite Technology, Ltd.), and Ka-band transmitter with a horn antenna (Astro Digital).

There are two primary classifications of antenna: fixed or deployable. Fixed antennas do not require any power or triggering mechanisms. They remain stationary in the position that they are attached to the spacecraft. This includes patch antennas, array antennas, monopole antennas, omni-directional antennas, and horn antennas (see figure 9.4). Deployable antennas require power to deploy and use mechanisms to configure into their final position. This includes whip antennas, parabolic reflectors, reflectarrays, helical and turnstile antennas (see figure 9.5).

A communications link is often characterized by the frequency and data rate. The antenna is a key design decision for meeting data rate objectives by increasing link margin. Increasing the aperture or diameter of an antenna inceases the link margin, which can allow designers to increase the data rate of the system or reduce the necessary transmit power.

Figure 9.5: (from left to right) Example of deployable quadrifilar helical antenna (Helical Communication Technologies), SNaP spacecraft with Haigh-Farr's deployable UHF Crossed Dipole antenna (Space Missile and Defense Command), and EnduroSat UHF antenna with EnduroSat solar panels (EnduroSat).

9.2.8 Radios

Radios for SmallSat downlink are transceivers (transmitter and receiver in one). Transceivers convert digital information into an analog RF signal using a variety of modulation and coding schemes. Radios for TT&C are designed for low data rates, with high reliability and only need to transmit health data and receive commands. Traditional radios may be locked to a single frequency band and modulation/coding scheme based on their design and build. Software defined radios (SDR) have some or all of the radio's functions implemented in Digital Signal Processing (DSP) software rather than hardware, figure 9.6. Furthermore, spacecraft teams can change such characteristics in-flight by uploading new settings from the ground. By using Field Programmable Gate Arrays (FPGAs), SDRs have great flexibility that allows them to be used with multiple bands, filtering, adaptive modulation, and coding schemes, without much (if any) change to hardware (12). SDRs are especially attractive for use on CubeSats, as they are becoming increasingly small and efficient as electronics become smaller and require less power. NASA has been operating the Space Communications and Navigation (SCaN) Testbed on the International Space Station since 2012 for the purpose of SDR TRL advancement, among other things (13). Many radios can provide RF output power to the antenna directly. For higher power applications, an external RF amplifier or high gain antenna may be used. The reader is encouraged to refer to the SmallSat Avionics chapter for further information on FPGAs and SDRs. Please see table 9-4 for information on commercially available radios for SmallSat/CubeSats.

Figure 9.6: Example of software defined radio, tunable in the range 70 MHz to 6 GHz. Credit: GomSpace.

This report recommends efficient modulation and coding schemes for spacecraft power and bandwidth to increase the data rate and meet bandwidth constraints with the limited power and mass for CubeSat spacecraft. Advanced coding, such as the CCSDS low-density parity-check code (LDPC) family, with various code rates is a powerful technique to provide bandwidth and power with high-order modulation to achieve high data rate requirements for CubeSat missions. Digital Video Broadcast Satellite Second Generation (DVB-S2), a significant satellite communications standard, is a family of modulations and codes for maximizing data rates and minimizing bandwidth use. DVB-S2 uses power and bandwidth efficient modulation and coding techniques to deliver performance approaching theoretical limits of RF channels. NASA's NSN has conducted testing at NASA Wallops Flight Facility (WFF) to successfully demonstrate DVB-S2 over a S-band 5 MHz channel achieving 15 Mbps with 16 APSK LDPC 9/10 code (14).

9.2.9 On the Horizon RF Communications

A CubeSat constellation may involve numerous CubeSats in the constellation (e.g., tens or hundreds). Each CubeSat is typically identical from a communication perspective. One or more CubeSats may be mother ship-capable while the others may be subordinate (e.g., daughterships). CubeSat constellations optimize coverage over specific areas or improve global revisit times to fulfill mission objectives. There is growing interest among the NASA science community in using constellations of CubeSats to enhance observations for Earth and space science. NASA GSFC has conducted research on future CubeSat constellations, including CubeSat swarms, daughter ship/mother ship constellations, NEN S- and X-band direct-to-ground links, TDRS Multiple Access (MA) arrays, and Single Access modes. The MA array requires the use of spread-spectrum to support multiple simultaneous communications links to increase coverage and link availability

As CubeSat missions employ more automation, constellations could exchange information to maintain precise positions without input from the ground. Radiometric ranging is a function recently incorporated into CubeSat transceivers. A timing signal is embedded into the radio signal and is used to determine the range to the spacecraft. Using this method along with directional vectors obtained from ground antennas allows for trajectory determination of satellites beyond low-Earth orbit. Spacecraft may relay data to increase the coverage from limited ground stations. Inter-CubeSat transponders may very well become a vital element of eventual deep space missions, since CubeSats are typically limited in broadcasting power due to their small size, and may be better suited to relay information to Earth via a larger, more powerful mothership.

Spacecraft routinely use transponders, however, networked swarms of CubeSats that pass information to each other and then eventually to ground, have not flown. Developing networked swarms is less of a hardware engineering problem than a systems and software engineering problem in that one must manage multiple dynamic communication links. As of 2023, only the two MarCO SmallSats and BioSentinel have operated beyond low-Earth orbit and both use the IRIS CubeSat Deep Space transponder. Both the MarCO satellites used a deployable reflectarray panel at X-band and were equipped with a full-duplex radio providing both UHF and X-band coverage (15). This allowed for near real-time updates of the InSight rover's landing. After this success, more SmallSats may be deployed beyond low-Earth orbit. The ability to provide crosslink relay hops for large spacecraft will prove to be critical for deep space missions.

Several projects funded via NASA's Small Spacecraft Technology (SST) program through the University Smallsat Technology Partnerships (USTP) initiative have began advancing RF Communication systems. Listed below in table 9-2 are projects that focused on RF technology advancement, and further information can be found at the STP website:

https://www.nasa.gov/smallspacecraft/university-smallsat-technology-partnership-initiative/

Each presentation is from the USTP Technology Exposition that was held in June 2022.

Table 9-2: STP Initiative Communication Projects

Project	University	Current Status	Reference
FIGARO, 5G arrays for lunar relay operations	San Diego State	Still in development	USTP Technology Expo presentation
A Small Satellite Lunar Communications and Navigation System	University of Colorado, Boulder	Still in development	USTP Technology Expo presentation

Table 9-3: Antennas									
Manufacturer	**Product**	**Type**	**Min Frequency**	**Frequency Band**	**Gain**	**Polarization**	**Mass**	**Dimensions**	**Flight Heritage**
---	---	---	[MHz]	--	[dBi]	--	[g]	[cm]	---
MMA Design	LAMBDA	Deployable Sinuous Antenna	100	UHF	>-1	Dual CP	4000	20x15x10 (stowed) 100x100 (deployed)	N
Oxford Space Systems	Yagi antenna	Deployable	156.5-162.5	VHF	6.5	Dual Linear	<1kg	100 x 70	Y
Spacemanic	SAM_Antenna_Module	Dipole Cross Dipole	144 399 420	VHF, UHF	2.1	Linear/RHCP	120	9.8X9.8X0.56	Y
Haigh-Farr, Inc.	Part Number: 17100	Crossed Dipole	307	VHF, UHF	--	RHCP	267	32x8x1	Y
GomSpace	NanoCom ANT430	Omni Canted Turnstile	400-435	VHF, UHF	1.5	Circular	30	10x10	Y
Helical Communications Technologies	Helios Deployable Antenna	Helical	400-3000	VHF, S	3	Circular	180	10x10x3.5	Y
NanoAvionics	CubeSat UHF Antenna System 1x1U	Turnstile	400-500	UHF	1.37	--	33	10x10x0.7	Y
NanoAvionics	CubeSat UHF Antenna System 1x2U	Turnstile	400-500	UHF	2.31	--	50	20x10x0.7	Y
NanoAvionics	CubeSat UHF Antenna System 2x2U	Turnstile	400-500	UHF	3.4	--	65	20x20x0.7	Y
EnduroSat	UHF Antenna III	Whip/Burn-wire	435-438 or 400-403	VHF, UHF	> 0	RHCP	85	10x10	Y
ISISPACE	CubeSat Antenna System for 1U/3U	Tape	--	VHF, UHF	0	Circular, Linear	89	10x10x0.7	Y
Flexitech Aerospace	600MHz - 10GHz Spiral Antenna	Spiral	600-10000	UHF, L, S, C, X	3	Circular	1283	17x17x8.5	N

Table 9-3: Antennas (*Continued*)									
Manufacturer	**Product**	**Type**	**Min Frequency**	**Frequency Band**	**Gain**	**Polarization**	**Mass**	**Dimensions**	**Flight Heritage**
---	---	---	[MHz]	--	[dBi]	--	[g]	[cm]	---
Oxford Space Systems	Helical antenna	Deployable	862–928	UHF	6.5-7.5	RHCP	~300	33	Y
SkyFox Labs	piPATCH-L1E1	Patch	1575.42	GPS-L1 GALILEO E1	--	--	50	9.8x9.8x1.3	Y
NAL Research Corporation	Antenna SYN7391-A/B/C (Iridium)	Flat Mount	1610-1626.5	L	4.9	RHCP	31	4.6x.4.3x1.0	Y
IQ Spacecom	S-Band Single Patch Antenna	Patch	1980-2500	S	6	Circular	49	7x7x0.34	Y
IQ Spacecom	S-Band Dual Patch Antenna	Patch	1980-2500	S	6	Circular	62	8x10x0.34	Y
IQ Spacecom	S-Band High Gain Patch Antenna	Patch	1980-2500	S	11.5	Circular	179	16x16x0.34	Y
Flexitech Aerospace	2-2.5GHz Turnstile Antenna	Turnstile	2000-2500	S	5	Circular	173	--	N
SkyLabs	S-band Patch Antenna	Patch	2025-2110	S	6	LHPC/RHPC	70	8.2x8.2x1.1	Y
Vulcan Wireless	ANT-S/S Unified S-Band Antenna	Patch	2025-2300	S	6.5	Circular	76	8x8x1	Y
EnduroSat	S-band Commercial Patch Antenna	Patch	2025-2110	S	7	Selectable Circular	64	9.8x9.8x0.6	Y
EnduroSat	S-band Wideband Patch Antenna	Patch	2025-2110 2200-2290	S	5	RHCP	115	9.8x9.8x0.7	Y
ANYWAVES	S-Band TT&C Antenna	Patch	2025-2290	S	6.5	RHCP/LHCP	132	8x8x1.2	Y
Haigh-Farr, Inc.	P/N 21060	Waveguide	2020-2120	S	25	LHCP	667	10x10x4.1	N
ISISPACE	S-Band Patch Antenna	Patch	2200-2290	S	6.5	RHCP	50	8x8x1	N

Table 9-3: Antennas (*Continued*)									
Manufacturer	**Product**	**Type**	**Min Frequency**	**Frequency Band**	**Gain**	**Polarization**	**Mass**	**Dimensions**	**Flight Heritage**
---	---	---	[MHz]	--	[dBi]	--	[g]	[cm]	---
Haigh-Farr, Inc.	S-band Patch Antenna	Patch	2245-2245	S	--	RHCP	48	4.8x6.5x6.5	Y
EnduroSat	S-band ISM Patch Antenna	Patch	2400-2450	S	8	LHCP	64	9.8x9.8x0.6	Y
Oxford Space Systems	Deployable Parabolic Offset Reflector	Deployable	5310-5510	C	42	Linear	12kg to 21kg	350	N
IQ Spacecom	X-Band Single Patch Antenna	Patch	7145-7250 8025-8400	X	6	Circular	10	3.5x3.5x0.18	Y
IQ Spacecom	X-Band High Gain Antenna	Patch	7145-7250 8025-8400	X	10	Circular	12	4x6x0.18	Y
Oxford Space Systems	(Single) Hinged Rib Metal Mesh	Deployable	7200-8500	X	30	RHCP/LHCP	5200	90	N
EnduroSat	X-band Patch Antenna	Patch	8025-8400	X	6	RHCP	2.2	2.4x2.4x0.2	Y
EnduroSat	X-band 2x2 Patch Antenna	Patch	8025-8400	X	12	RHCP	23.2	6.0x6.0x0.3	Y
EnduroSat	X-band 4x4 Patch Antenna	Patch	8025-8400	X	16	RHCP	53	9.8x8.3x0.3	Y
ANYWAVES	X-band Payload Telemetry Antenna	Patch	8025-8400	X	11.5	Circular	65	7.3x7.3x11	Y
MMA Design	T-DaHGR	Deployable Reflectarray	8400 to 10000	X	29 – 42.5	Configurable	1300 to 11000	10 x 10 x 10 - 20 x 20 x 20 (stowed) Ø70-Ø200 (deployed)	N
Oxford Space Systems	Deployable Cassegrain Wrapped Rib Antenna	Deployable	9200-10400	X	46 - 49	Linear	25kg to 38kg	300 - 500	N

Table 9-3: Antennas (*Continued)*									
Manufacturer	**Product**	**Type**	**Min Frequency**	**Frequency Band**	**Gain**	**Polarization**	**Mass**	**Dimensions**	**Flight Heritage**
---	---	---	**[MHz]**	--	**[dBi]**	--	**[g]**	**[cm]**	---
MMA Design	NeuSAR	Deployable Reflectarray	10000	X	>45.5	V+H Linear	16550	52 x 52 x 25 (stowed) Ø300 deployed	Y
EnduroSat	K-band Patch Antenna	Patch	17700-20200	K	18	RHCP	90	4.8x4.8x0.5	N
Oxford Space Systems	Deployable Hinged Rib Metal Mesh	Deployable	17700-20200 27700-30000	K,Ka	38/41	Linear	~2-3kg	60	N
Cesium Astro	Nightingale	Phased Array	27000-40000	Ka	30	Circular	1200	18x18x2	N
EnduroSat	W-band Patch Antenna	Patch	71000-75000	W	23-29	RHCP	37	8.7x8.1x2.0	N

Table 9-4: Radios									
Manufacturer	**Product**	**Type**	**Min Frequency**	**Frequency Band**	**Data Rate**	**Tx Power**	**Mass**	**Dimensions**	**Flight Heritage**
---	---	---	**[MHz]**	--	**[kbps]**	--	**[g]**	**[cm]**	---
Space Micro	MicroSDR-C	SDR	70-3000	VHF, UHF, L, S, C	42,000	0	750	10x10x8	Y
Rincon Research	ASTROSDR	SDR	70-6000	VHF, UHF, L, S, C	--	5 dBm	95	9.0x9.0x1.613	Y
GomSpace	NanoCom SDR	SDR	70-6000	VHF, UHF, L, S, X	--	--	271	9x9x6.6	Y
NI Ettus Research	B205mini	SDR	70-6000	VHF, UHF, L, S, X	--	10 dBm	24	8.3x5.1x8	Y
Alén Space	TOTEM	SDR	70 - 6000	VHF, UHF, L, S	–	7 dBm	130	9.33x8.93x1.36	Y

Table 9-4: Radios *Continued*									
Manufacturer	**Product**	**Type**	**Min Frequency**	**Frequency Band**	**Data Rate**	**Tx Power**	**Mass**	**Dimensions**	**Flight Heritage**
---	---	---	[MHz]	--	[kbps]	--	[g]	[cm]	---
Alén Space	TREVO	SDR	70 - 6000	VHF, UHF, L, S	–	–	up to 714	9.33x8.93x27.88	N
Alén Space	TRISKEL	TTC	395 - 410, 430 - 440	UHF	2.2 - 1.9.2	30 dBm	200	9.33x8.93x126	N
SkyLabs	NANOcomm-2	Transceiver	130-470	VHF, UHF	Up to 25	31 dBm	110	9.5x9.1x1.2	Y
AstroDev	Helium-100	Transceiver	120-150, 400-450	VHF, UHF	38.4	3 W	78	9.6x9x1.6	Y
AstroDev	Lithium-1	Transceiver	130-450	VHF, UHF	9.6	0.25-4 W	48	1.0x3.3x6.5	Y
AstroDev	Beryllium-2	Transceiver	130-450	VHF, UHF	9.6	0.25-4 W	52	1x3.3x6.5	Y
GomSpace	NanoCom AX100	Transceiver	143-150, 430-440	VHF, UHF	0.1-38.4	30 dBm	24.5	6.5x4x7	Y
Spacemanic	Murgas_trx_VHF	Transceiver	144	VHF, UHF	9.6	+30dBm	25	6.7x4.2x0.7	Y
LY3H	SatCOM TP0	FM Repeater	144-146, 430-440	VHF, UHF	--	217 mW	59	--	Y
ISISPACE	TRXVU	Transceiver	145.8-150.05, 400.15-440	VHF, UHF	9.6	27 dBm	75	9x9.5x1.5	Y
CeisumAstro	SDR-1001	SDR	300 – 6000	UHF, L, S, C	up to 62,500	-	100	5 x 8.4 x 1.3	N
AAC Clyde Space	TRX-U	Transceiver	390-450	UHF	19.2	2	140	8.3x5.7x1.6	Y
NanoAvionics	SatCOM UHF	Transceiver	395-440	VHF, UHF	2.4-38.4	3 W	7.5	5.6x3.3x6.6	Y
Spacemanic	Murgas_trx_UHFlow	Transceiver	399	UHF	9.6	+30dBm	25	6.7x4.2x0.7	N
EnduroSat	UHF Transceiver Type II	Transceiver	400-403, 430-440	VHF, UHF	9.6	2 W	94	10x10x2	Y

Table 9-4: Radios (*Continued*)									
Manufacturer	**Product**	**Type**	**Min Frequency**	**Frequency Band**	**Data Rate**	**Tx Power**	**Mass**	**Dimensions**	**Flight Heritage**
---	---	---	[MHz]	--	[kbps]	--	[g]	[cm]	---
Spacemanic	Murgas_trx_UHF	Transceiver	420 MHz	UHF	9.6	+30dBm	25	6.7x4.2x0.7	Y
L3 Communications, Inc. /SDL	Cadet	SDR	450	VHF, UHF	3,000	--	200	6.9x7.4x1.34	Y
NearSpace Launch	EyeStar-D2	Transceiver	1610-1625, 2484-2499	L	10,000	0.8 W	138	6.1x11.9x2.2	Y
sci_Zone, Inc.	LinkStar-STX3	Transmitter	1610-1625	L	0.009	--	48	8.6x5.3x2.9	Y
Qualcomm	GSP-1720	Transmitter	1610-1626.5, 2483.5-2500	L, S	9.6	31 dBm	60	11.9x6.5x1.5	Y
NAL Research Corporation	NAL Iridium 9602-LP	Iridium Satellite Tracker	1616-1626.5	L	--	1 W	136	6.9x5.5x2.4	Y
NearSpace Launch	EyeStar-S3	Transmitter	1616.25	L	600	20 dBm	22	1.5x2.6x5.5	Y
L3Harris	CXS-1000	Transponder	1700-2100	L, S	20,000	1-5 W	1360	10x10x11	Y
Tethers Unlimited	SWIFT-SLX	SDR	1700-2500	S	6,000	33 dBm	300	9x9.8x3.6	Y
Tethers Unlimited	SWIFT-XTS S Transceiver X Transmitter	SDR	1700-2500, 7000-8500	S, X	6,000-25,000	34 dBm	800	9x9.8x6	Y
AAC Clyde Space	TX-2400	Transmitter	2000-2300	S	6,000	2.5	70	6.8x3.5x1.5	Y
Syrlinks	EWC27 + OPT27-SRX S/X Transceiver	Transceiver	2025-2110	S	100,000	27-33 dBm	400	9x9.6x3.9	Y

Table 9-4: Radios (*Continued*)									
Manufacturer	**Product**	**Type**	**Min Frequency**	**Frequency Band**	**Data Rate**	**Tx Power**	**Mass**	**Dimensions**	**Flight Heritage**
---	---	---	**[MHz]**	--	**[kbps]**	--	**[g]**	**[cm]**	---
Innoflight, Inc.	SCR-104	SDR with AES-256 Encryption	Tx: 2200-2290 Rx: 1760-1840 Rx: 2025-2110	Tx: S Rx: L/S	Tx: >4,500 Rx: 1000		250 - 290	8.2x8.2x2.5 9.8x8.2x3.3	Y
IQ Technologies for Earth and Space GmbH	HISPICO	Transmitter	2100-2500	S	1,000	27 dBm	100	9.5x4.6x1.5	Y
Emhiser Research, Inc.	ETT-01EBA102-00	Transmitter	2200-2400	S	--	1 W	57	3x8.6x0.8	Y
Quasonix	NanoTX	Transmitter	2200.5-2394.5	S	50	1-10 W	Request	3.3x8.6x0.8	Y
IQ Technologies for Earth and Space GmbH	SLINK-PHY	Transceiver	2200-2290, 2025-2110	S	64-4000	30 dBm	275	6.5x6.5x13.7	Y
ISISPACE	TXS	Transceiver	2200-2290	S	4.3	27-33 dBm	132	9.8x9.3x1.4	Y
Syrlinks	S-band Transponder	Transponder	2200–2290	S	8-2000	27-33 dBm	--	--	Y
EnduroSat	S-band Transmitter	Transmitter	2200-2290, 2400-2450	S	20000	0.5-2 W	250	--	Y
General Dynamics	S-Band TDRSS/DSN	Transponder	Tx: 2200-2300 Rx: 2025-2220	S	12,000	0.03 W	4900	19x23x15	Y
Microhard	Nano N2420	Modem	2400-2483.5	S	230	0.1-1 W	210	5x3x0.6	Y
Laboratory for Atmospheric and Space Physics (LASP)/Blue Canyon Technologies (BCT)	X-band Radio	SDR	Tx: 2200-2500, 8000-8500, 21000-33000 Rx: 1760-1840, 2000-2110, 21000-23000	Downlink: S, X, Ka Uplink: L, S, Ka	100000	30 dBm	--	4.5x4.35x1.25	Y

Table 9-4: Radios (*Continued*)									
Manufacturer	**Product**	**Type**	**Min Frequency**	**Frequency Band**	**Data Rate**	**Tx Power**	**Mass**	**Dimensions**	**Flight Heritage**
---	---	---	**[MHz]**	--	**[kbps]**	--	**[g]**	**[cm]**	---
SkyLabs	NANOlink-base Gen2	SDR	2200-2300	S	Up to 4000	30 dBm	110	9.5x9.1x1.2	Y
SkyLabs	NANOlink-boost Gen2	SDR	2200-2300	S	Up to 4000	37 dBm	250	9.5x9.1x2.2	Y
SkyLabs	NANOlink-boost-dp Gen2	SDR	2200-2300	S	Up to 4000	31.5 dBm	385	9.5x9.1x3.2	Y
Tethers Unlimited	SWIFT-XTX X Transmitter	SDR	7000-8500	X	25,000	33 dBm	300	9x9.8x6	N
General Dynamics	X-Band Small Deep Space	Transponder	7145-7230, 8400-8500	X	100,000	0.06	3200	18x17x11	Y
Space Dynamics Laboratory	IRIS	Transponder	7145, 8400	X, Ka, S, UHF	0.1 - 5313	3.8	875	10.1x10.1x5.6	Y
Innoflight, Inc.	SCR-106	SDR with AES-256 Encryption	Tx: 7800-8500 Rx: 1760-1840 Rx: 2025-2110	Tx: X Rx: L/S	Tx: 20000 Rx: 1000	0.02-2.5 W	250 - 290	8.2 x 8.2 x 2.5 9.8 x 8.2 x 2.8	Y
Innoflight, Inc.	SCR-108	SDR with AES-256 Encryption	Tx: 19200-21200 Rx: 29000- 31000	Tx: Ka Rx: Ka	Tx: 100000 Rx: 20000	0.02-3 W	404	9.8x8.7x3.9	Y
EnduroSat	X-band Transmitter	Transmitter	7900-8400	X	150000	27-33 dBm	270	9x9.6x2.6	Y
EnduroSat	UHF/VHF Transceiver	Transceiver	137-150, 400-403, 430-44	VHF, UHF	19.2	0.5-1 W	< 150	9.6x9.0x1.9	N
EnduroSat	S-band Transceiver I	Transceiver	2200-2290 (downlink)	S	100	0.5-2 W	195	9.6x9.0x1.9	Y

Table 9-4: Radios (*Continued*)									
Manufacturer	**Product**	**Type**	**Min Frequency**	**Frequency Band**	**Data Rate**	**Tx Power**	**Mass**	**Dimensions**	**Flight Heritage**
---	---	---	[MHz]	--	[kbps]	--	[g]	[cm]	---
IQ Technologies for Earth and Space GmbH	XLINK	Transceiver	8025-8500 7145-7250	X	64-25,000	30 dBm	--	<1 U	Y
Syrlinks	EWC27	Transmitter	8025-8500	X	140,000	27-33 dBm	235	9x9.6x2.6	Y
Syrlinks	EWC27 + OPT27-SRX	Transceiver	Rx: 2025-2110 Tx: 8025-8500	X/S	RX: 256 TX:100000	33 dBm	320	9.6x9x3.9	Y
Syrlinks	EWC31	Transceiver	Rx: 2025-2110 Tx: 2200-2290	S	RX: 256 TX: 2,000	33 dBm	405	9.5x9.5x5.3	Y
Syrlinks	EWC31-NG	Transceiver	Rx: 2025-2110 Tx : 2200-2290	S	RX: 512 TX: 2,000	33 dBm	360	9.5x9x3.2	N
Syrlinks	N-XONOS	Transmitter	Rx: 2025-2110 Tx: 8025-8400	X/S	RX: 256 TX:400000	33 dBm	385	9.5x9x3.1	Y
Syrlinks	EWC15-NG	Transceiver	Rx: 2025-2110 Tx: 2200-2290	S	RX: 512 TX: 2,000	33 dBm	1280	17.2x12x6.7	N
Syrlinks	XONOS	Transmitter	Rx: 2025-2110 Tx: 8025-8500	X/S	RX: 256 TX: 628,000	40 dBm	2400	20,6x15,2x6,9	N
Tethers Unlimited	SWIFT-KTX Ka Transmitter	SDR	20200-21200 24000-27000	Ka	25,000	33 dBm	300	9x9.8x4	N
Tethers Unlimited	SWIFT-KTRX Ka Transmitter	SDR	24000-27000	Ka	---	35 dBm	1,000	16x9.6x6	N
SpaceMicro	microKaTx-300	Transmitter	25250-27250	K	1,000,000	2	1000	10x10x8	Y
AAC Clyde Space	PULSAR-DATA XTX X-Band Transmitter	SDR	--	X	50,000	2 W	130	9.6x9x1.1	Y

Table 9-4: Radios (*Continued*)

Manufacturer	Product	Type	Min Frequency	Frequency Band	Data Rate	Tx Power	Mass	Dimensions	Flight Heritage
---	---	---	[MHz]	--	[kbps]	--	[g]	[cm]	---
AAC Clyde Space	PULSAR-DATA STX S-Band Transmitter	SDR	--	S	7,500	1 W	100	9.6x9x1.7	Y
AAC Clyde Space	PULSAR-VUTRX	SDR	--	VHF, UHF	9.6	1.5 W	100	9.6x9x1.6	Y
Honeywell	STC-MS03	Transceiver	--	S	6,250	3.16 W	1000	16x11x4.4	Y
Innoflight, Inc.	SCR-106HDR	SDR with AES-256 Encryption	Tx: 7800-8500 Rx: 1760-1840 Rx: 2025-2110	Tx: X Rx: L/S	Tx: 100000 Rx: 20000		250 - 290	8.2x8.2x2.5 9.8x8.2x2.8	Y
EnduroSat	Versatile Wideband SDR (VW-SDR)	Transceiver	75 - 6000	VHF, UHF, L, S, C	982,000	-10 dBm	<1500	9.8x9.8x7.5	N
Trident	RDRT	SDR – RF System on Chip (RFSoC)	100	L/S	8-channels 250MHz – 6554MHz TX, 8-channels 250MHz – 4096MHz RX	-1 dBm Full-scale	571	10 x 14.6 x 2.54	N
Trident	ADCM	SDR – MPSoC Basecard and -SP converters - RX only	100	L/S/C	6.4GSPS single channel, 3.2 GSPS dual-channel	2.8dBm Full scale input	690	10 x 14.6 x 2.54	N

9.3 Free Space Optical Communications

Free space optical communications, or lasercom, uses optical wavelengths of electromagnetic radiation to transmit messages wirelessly between user terminals. While few small satellite optical communications terminals have flown, availability is rapidly changing, and optical communication is becoming a more common wireless communication technology for small satellites.

Due to the higher frequencies used in lasercom, the amount of bandwidth available for communicating is much larger compared to RF. This increase in bandwidth over RF enables much higher data rates. The beam width of a lasercom link is also typically much narrower than a RF link (figure 9.7). The amount that a transmitted beam spreads as a function of its propagation distance is called its divergence. The divergence of a beam is proportional to the wavelength of the electromagnetic wave transmitted divided by the transmitted beam diameter. The high frequencies used in lasercom mean the wavelength of the transmitted energy is orders of magnitude smaller than RF systems. These small wavelengths mean the transmitter diameters and beam divergence of lasercom systems can also be much smaller, which enables the size, weight, and power (SWaP) of lasercom systems to be lower than similar performing RF systems. Laser communications have a low probability of intercept, are difficult to jam, and encounter very little interference because of the narrow beamwidth. At present, optical frequencies are unregulated, unlike RF systems which require a licensing process to be able to communicate with a spacecraft. Lasercom is not without its disadvantages, which include the required pointing of the beam and the impact weather has on the signal. The small beam divergence of lasercom systems means that the acceptable pointing error is much smaller. The frequencies used in lasercom systems are also susceptible to large amounts of attenuation due to moisture in clouds. This attenuation prohibits communication while there is cloud cover and incentivizes operators to build their optical ground stations in areas that have infrequent cloud cover.

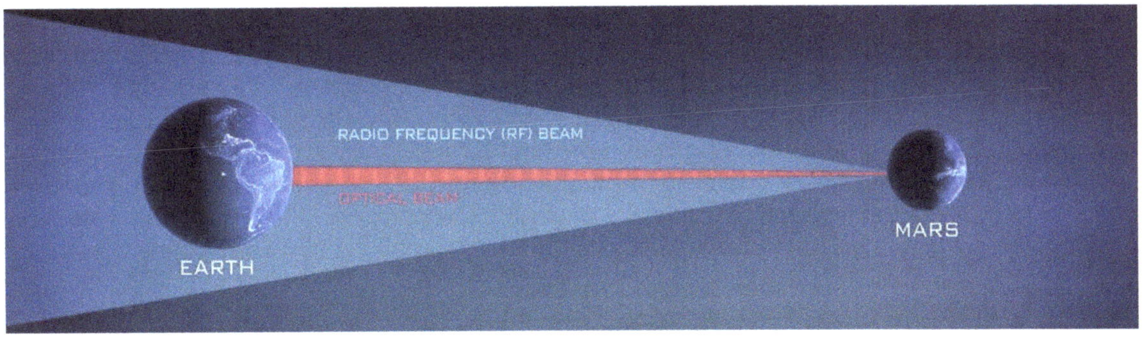

Figure 9.7: Laser vs RF link and data downlink. Credit: NASA.

While larger mission such as Geosynchronous Lightweight Technology Experiment (GeoLITE), Near Field Infrared Experiment (NFIRE), and Lunar Laser Communication Demonstration (LLCD) have demonstrated laser communications downlinks and crosslinks for over a decade, small satellites and CubeSats have also now successfully demonstrated laser communication downlinks from space. For example, the Aerospace Corporation, in cooperation with NASA ARC, launched three CubeSats in its AeroCube Optical Communication and Sensor Demonstration (OCSD) series (figure 9.8). OCSD-B & C demonstrated a 200 Mbps

Figure 9.8: An artist rendering of laser communications for the OCSD. Credit: NASA.

downlink from a 1.5U CubeSat satellite to a 40 cm ground station (16). The Aerospace Corporation transmitter has also successfully flown on follow-on missions that were able to use lasercom systems to downlink science data (17).

9.3.1 System Architecture

An optical modem, optical amplifier, and optical head typically comprise a lasercom terminal (LCT) (see figure 9.9 for an example laser terminal system diagram). As with radio terminals, component locations in optical terminals can vary; for example, the modulator may not be located proximal to the optical front end. Also, the pointing mechanism might differ from the one shown in figure 9.9.

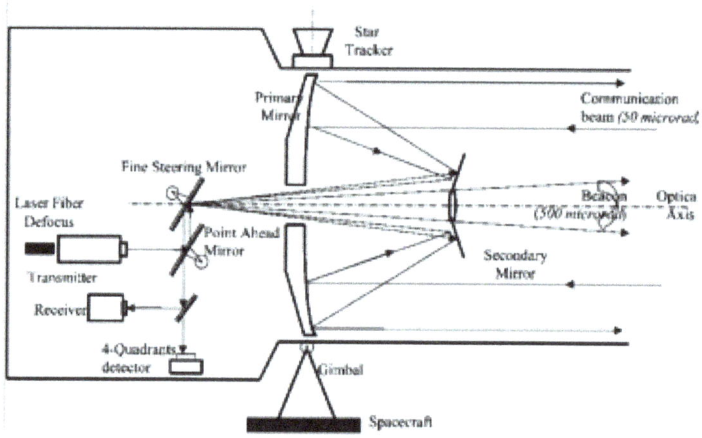

Figure 9.9: Laser terminal architecture diagram. Credit: M. Guelman et al. (2004).

The key parameters of an optical communication system are frequency, modulation, aperture size, and range. Successful optical communications links typically require high pointing accuracy. The optical communication terminal on a spacecraft typically has a two-stage pointing system, with a coarse-pointing stage and a fine-pointing stage. The optical communication system often relies heavily on the spacecraft attitude determination and control system (ADCS) for coarse-pointing, and may use a second pointing mechanism such as a gimbal as additional support for coarse pointing. Fine pointing is often implemented with additional mirrors in the payload. However, pointing that is solely dependent on spacecraft attitude control has also been demonstrated. On transmit, energy passing through the optical aperture forms a very narrow beam. The larger the aperture, the narrower the beam; this creates higher power density at a receiver for a given range. In order for two communication terminals to locate each other, they may shine higher power and broader-beam "beacon" lasers to find each other before engaging the narrower and higher data rate link. The beacon itself may also be modulated. Optical modems may be software defined and can support multiple modulation and coding schemes, similar to RF.

9.3.2 Optical Ground Stations

The ground stations for optical communications understandably differ significantly from RF ground stations due to the need to have the receiving aperture (typically a mirrored telescope) maintain an optical-quality surface to focus the collected optical energy onto a receiver. Optical ground stations are often located at or near astronomical telescope sites located in favorable environments. Optical ground stations are typically mounted inside protected domes or other structures to cover them during bad weather. These structures need to be opened for clear access to the sky. Since optical ground stations often have beacons, it is important to consider laser safety and proximity to airports. Typical ground-to-space beacons are tens of watts of optical power for low-Earth orbit missions. Most optical ground stations are experimental facilities used for campaigns with specific research missions, although there has been recent development in commercial optical ground stations. For a more detailed outline of existing optical ground stations, refer to the *Ground Data Systems and Mission Operations* chapter.

9.3.3 Design Considerations

Lasercom terminals can offer a smaller footprint and power draw compared to an RF terminal. However, lasercom pointing requirements are significantly tighter. One of the largest challenges to mainstream implementation is the required pointing for the LCT. To manage this, each system architecture will describe the specific system of pointing used. The LCTs that have been designed, built, and operated on small satellite and CubeSat platforms have some significant differences from LCTs designed for larger spacecraft. Given the size, weight, and power constraints, SmallSat LCTs usually do not use mechanical gimbals. SmallSat lasercom systems solely or largely rely on the body pointing of the satellite to point the LCT at the ground station and may use an internal fine pointing mechanism to achieve the required pointing performance.

On SmallSat platforms, the limited volume and tight packaging is often a major challenge in the design of low-SWaP LCTs. There are thermal management challenges during operation, as it is difficult to radiate enough heat with limited surface area for radiators. There are also power constraints, due to limited surface area for solar arrays and secondary battery systems. In addition, not all SmallSat platforms can achieve the pointing requirements necessary for laser communications. Typically, precise three-axis reaction wheels and attitude determination from one or more star trackers is necessary.

While RF bands with high frequency and bandwidth are affected by clouds and rain, cloud cover can prove difficult or even insurmountable for optical communications due to the high levels of attenuation by water vapor. If the cloud coverage is too great at a specific ground station, the transmission may be held for a later time or passed off to a different ground station. With advances in intersatellite networking and the development of extensive networks of optical communication ground stations, routing data around weather may become more feasible.

The atmosphere is also a source of aberrations for optical communication systems. For example, some high-rate optical downlink terminals that require coupling the received light into fiber receivers must use adaptive optics to correct atmospheric effects on the incident wavefront. The correction of the wavefront is required because of the lack of power that would couple into the receive optical fiber due to the perturbed wavefront of the received light. Adaptive optics systems take a sample of the incident wavefront and measure the aberration to feed into the control of the adaptive optics system acting on the received light.

Lasercom crosslinks can provide a high bandwidth connection between two satellites, as well as perform ranging between the satellites, potentially with high ranging precision. Connecting two satellites across different orbit planes helps with data routing and can reduce how long it takes to route data to the end use. Lasercom crosslink system are now in use for both commercial and government missions. Lasercom crosslink demonstrations have been performed from GEO-LEO, LEO-GEO, and LEO-LEO, and are operational as part of the European Data Relay Service (37)(38), but these LCTs were developed for much larger spacecraft (19)(20).Crosslinks also have the challenge of both terminals being resource-constrained onboard a spacecraft. Space-to-ground links have an advantage in that the ground station apertures can be large with essentially unconstrained resources. The challenges facing inter-satellite optical communications also centers on pointing, acquisition, and tracking (PAT) requirements. Satellites in different orbital planes can have high relative velocities and performing pointing, acquisition, and tracking of the terminal can be a challenge. An advanced opto-mechanical system may be needed to surmount this challenge, and modifications to the receive optics may be required to manage high Doppler shift.

9.3.4 Policies and Licensing

Given the early stages of development for optical communication systems, both policy and regulatory approaches are still evolving. In the policy realm, there is an initial draft CCSDS Pink Book in process (CCSDS 141.0-P-1.1) with a goal to facilitate interoperability and cross-support between different communication systems. There is also an optical communication working group with NASA and ESA participation.

Regarding licensing and regulation, the situation is very different from the radio frequency domain. Currently there are no licensing requirements for laser communications. In the radio frequency spectrum, the main goal for licensing is to prevent interference between transmitters.

Lasercom interference is not currently coordinated by a regulatory body (like the ITU or NTIA in RF) for two major reasons:

1) Laser communications is highly directional, which makes interference unlikely, due to the narrow divergence of the transmitting beam and corresponding small beam footprint at the receiver.
2) The small number of laser communications systems currently deployed doesn't warrant a complex coordination body like the ITU.

However, in the US there are three regulatory entities that are concerned with aspects of outdoor laser operations: The FAA, DoD Laser Clearing House (for DoD missions) and the NASA Laser Safety Review Board (for NASA missions).

FAA coordination is required if potentially harmful laser irradiance is transmitted through navigable airspace. This includes prevention of injury as well as potential distraction of pilots by visible lasers. The FAA will most likely only be concerned about transmitters at ground stations because transmitters on spacecraft are hundreds of miles above the highest-flying aircraft and beam dispersion is large enough that there are usually no safety implications. Missions should coordinate with their local FAA service center to get approval, documented with a "letter of non-objection."

The DoD Laser Clearinghouse (LCH) works to ensure that DoD and DoD-sponsored outdoor laser use does not impact orbiting spacecraft or their sensors. That includes both US DoD and foreign assets. LCH and mission operators might enter close cooperation where LCH permits specific laser engagements. The process of coordinating with LCH to get to that point can take many months and should be started as early as possible. However, currently LCH will only engage DoD and DOD-sponsored missions.

NASA's Laser Safety Review Board (LSRB) is focused on personnel safety for all outdoor laser operations. NASA missions prepare safety documentation and submit to LSRB for review before launch. LSRB will also verify FAA concurrence. Further information on regulations can be found in ANSI Z136.6 "American National Standard for Safe Use of Lasers Outdoors" and in (39).

9.3.5 Mission Examples

Missions demonstrating lasercom terminals on small satellite and CubeSat platforms have shown viable pathways for overcoming the challenges associated with lasercom to enable high bandwidth communications. Please refer to table 9-5 for more information on lasercom missions.

The Small Optical Transponder (SOTA) was developed by the National Institute of Information and Communications Technology (NICT) in Japan and launched in 2014. This LCT is capable of up to 10 Mbps and has successfully demonstrated a laser space-ground link from a 50 kg microsatellite (21). The Very Small Optical Transponder (VSOTA) LCT, also developed by NICT, is capable of 1 Mbps. VSOTA was integrated into the Rapid International Scientific Experiment Satellite (RISESAT) from Tohoku University and launched in 2019 (22).

The German Aerospace Center (DLR) has been developing LCTs as part of its Optical Space Infrared Downlink System (OSIRIS) program to support lasercom from small satellites. The first, OSIRISv1 is capable of 200 Mbps downlinks and is integrated into the University of Stuttgart's Flying Laptop satellite. This LCT uses a body pointing-only approach. The OSIRISv1 LCT, launched in 2017, has completed commissioning and is being used by DLR to test their optical ground stations. The OSIRISv2 LCT, launched in 2016, is capable of 1 Gbps and is integrated into the BiROS satellite from DLR Berlin. This LCT uses closed-loop body pointing with a beacon reference. The OSIRISv2 LCT has been undergoing commissioning with parts of the terminal having been commissioned (23)(24)(25).

The Aerospace Corporation completed the first demonstration of optical communication from a CubeSat platform with the NASA-sponsored Optical Communication and Sensor Demonstration (OCSD) mission. These terminals were integrated into a 1.5U CubeSat and rely only on body pointing. The use of body pointing-only comes from using high optical power amplifiers with a larger beam divergence tuned to the pointing performance capability of their spacecraft. The terminals achieved a 200 Mbps downlink data rate to a 40 cm ground station and do not use a beacon for a pointing reference (16). This transmitter has been flown since on multiple missions such as R3 (17) and the Rogue Alpha and Beta CubeSats (18).

As part of NASA's CLICK mission, MIT developed the 1.2U CLICK-A terminal. The first phase of the mission is flying the CLICK-A downlink terminal on a 3U CubeSat to demonstrate an optical design that uses a secondary fine pointing micro-electromechanical systems (MEMS) fine-steering mirror (FSM) to achieve the necessary pointing requirements for optical communication without imposing those requirements on the spacecraft pointing or needing large gimbals. This LCT uses closed-loop fine pointing with a beacon reference and is designed to close its link with a 28 cm ground station. The terminal is integrated into a Blue Canyon Technology's XB1 spacecraft bus and was launched to and deployed from the ISS in 2022. CLICK-A ultimately serves as a risk-reduction phase for the CLICK-B/C phases of the mission described later in this section (26).

DLR has also been developing their OSIRIS4 CubeSat transmitter. This optical communication terminal is designed to demonstrate an optical downlink in a 0.3U package. This transmitter also uses a MEMs FSM fine pointing mirror and was launched on the PIXL-1 mission in 2021. A beacon is used for fine pointing reference with this terminal. This terminal is designed to be used with a 60 cm optical receiver and has been commercialized through TESAT with the product name CubeLCT (27).

Sony Computer Science Lab and the Japanese Aerospace Exploration Agency (JAXA) jointly developed a LCT called Small Optical Link for ISS (SOLISS). This LCT is capable of bidirectional 100 Mbps links and was launched to and mounted on the ISS in 2019. This LCT has been successfully demonstrated with NICT's ground station (28)(29).

MIT Lincoln Laboratory developed the TBIRD terminal, which supports 200 Gbps downlinks. The transmitter uses commercial fiber telecommunication components to support very high data rates. This project is planned to downlink to NASA JPL's Optical Communication Telescope Laboratory (OCTL), which hosts a 1 m telescope with the adaptive optics necessary to couple the received light back into a fiber transceiver card. This terminal development was sponsored by NASA and was launched on the PDT-3 6U CubeSat mission in June of 2022 (30).

Future mission launches include the CLICK-B/C terminals. The CLICK-B/C phase of the CLICK mission is developing a 1.5U crosslink LCT. The CLICK-B/C crosslink LCT is designed to establish a 20 Mbps link at separations from 25 to 580 km. CLICK-B & C will each be integrated into its own 3U Blue Canyon Technologies XB1 spacecraft. The LCTs are designed to be capable of precision ranging up to a precision of 50 cm relative to each other. The spacecraft will be launched to and deployed from the ISS in 2024 and fly in the same orbital plane (26).

While results have not been shown on a flown mission, the CubeCat LCT is a commercial product by AAC Clyde Space that offers a bidirectional space-to-ground communication link between a CubeSat and an optical ground station. This LCT offers downlink speeds of up to 1 Gbps and an uplink data rate of 200 Kbps (31).

9.3.6 Future Technologies

While free space optical communication technology development has been making strides towards fielding operational systems, other avenues of research have also been explored. Quantum key distribution is a protocol that shares a secret cryptographic key through entangled photons. Sources and optical front ends have been development for transmitting these keys from small satellite spaceborne platforms (32)(33). The Deployable Optical Receiver Aperture (DORA) project, which is developing a 1 Gbps crosslink LCT (34), is a novel approach to deploying large apertures in space. The inter-spacecraft optical communicator (ISOC), which includes arrays of fast photodetectors and transmit telescopes to provide full-sky coverage, gigabit data rates and multiple simultaneous links, was initially developed at NASA's Jet Propulsion Laboratory with funding from NASA's Small Spacecraft Technology (SST) program from 2018 to 2020. An advanced version of the ISOC is currently being developed by Chascii Inc. with funding from NASA Small Business Innovation Research Program for cislunar applications. There are currently several ISOC versions for short-, mid-, and long-range applications that use appropriate levels of power and aperture size, respectively, to achieve gigabit connectivity (35). Another approach to expanding the communication windows for small satellites in low-Earth orbit is to form an intersatellite link to geosynchronous orbit. Major programs, such as the previously mentioned European Data Relay System use this type of link. NICT is looking to establish this type of link with a CubeSat through the CubeSOTA program (36). In addition to CubeSat terminals, larger terminals for larger SmallSats are under development by Tesat, Mynaric (26), SpaceMicro (27), and SA Photonics. DARPA has funded the Space-BACN program that seeks to develop a reconfigurable and multi-protocol inter-satellite LCT that can be supported on small satellites.

BODY-300

Table 9-5: LCT Technologies									
Vendor/Developer	Terminal	Platform	Data Rate	Mass	Power	Wavelength	Modulation	Launch Date	Reference
---	---	---	[Mbps]	[kg]	[W]	[nm]	---	---	---
NICT	SOTA	SOCRATES	10	5.9	16	976/800/1549	OOK	5.2014	21
DLR	OSIRISv2	BiROS	1000	1.65	37	1550	OOK	6.2016	24
DLR	OSIRISv1	Flying Laptop	200	1.3	26	1550	OOK	7.2017	23, 24, 25
Aerospace Corporation	OCSD-B&C	AeroCube-7	200	<2.3	20	1064	OOK	12.2017	16
NICT	VSOTA	RISESAT	1	<1	4.33	980/1550	OOK/PPM	1.2019	22
Sony/JAXA	SOLISS	ISS	100	9.8	36	1550	OOK	7.2019	28, 29
DLR	OSIRIS4CubeSat	PIXL-1	100	0.4	10	1550	OOK	1.2021	27
MIT Lincoln Labs	TBIRD	PDT-3	200,000	<3	100	1550	QPSK	5.2022	30
MIT	CLICK-A	CLICK	10	1.2	15	1550	PPM	7.2022	26
MIT	CLICK-B/C	CLICK	20	1.5	30	1537/1563	PPM	Est. 2024	26
AAC Clyde Space	CubeCat	---	1000	<1.33	15	1550	OOK	---	31

9.4 Summary

There is already strong flight heritage for many UHF/VHF and S-band communication systems for CubeSats. Less common, but with growing flight heritage, are X-band systems. Higher RF frequencies and laser communication already have CubeSat flight heritage, but with limited (or yet to be demonstrated) performance. Although there are limited Ka-band systems for CubeSats today, high-rate transmitters such as the Astro Digital AS-10075 demonstrated 320 Mbps in the Landmapper-BC 3 v2 mission. While laser communication has been demonstrated on a CubeSat platform, this is still not yet considered to be established technology on spacecraft. More demonstrations are in development, with some already launched and operating, to show higher data rates and increased pointing performance. Since optical communications uplink and downlink can be blocked by clouds, RF is considered complementary to maintain contact under all conditions. There is growing interest among the NASA science community in using constellations of CubeSats to enhance observations for Earth and space science.

For feedback solicitation, please email: arc-sst-soa@mail.nasa.gov. Please include a business email for further contact.

References

(1) J.N Pelton. "The Basics of Satellite Communications." s.l.: Professional Education International, Inc., 2006.

(2) NASA “Integrated Solar Array and Reflectarray Antenna (ISARA) for High Bandwidth CubeSats.” [Online] July 26, 2016. Available at: https://www.nasa.gov/sites/default/files/atoms/files/isara_fact_sheet-26july2016.pdf

(3) JPL. Mars Cube One (MarCO). Jet Propulsion Laboratory California Institute of Technology | CubeSat. [Online] Available at: https://www.jpl.nasa.gov/cubesat/missions/marco.php

(4) S. Xie et al. "Wireless Sensor Network for Satellite Applications: A Survey and Case Study." Unmanned Systems, Vol. 2, pp. 261-277, 2014.

(5) R. Schoemaker and J. Bouwmeester. "Evaluation of Bluetooth Low Energy Wireless Internal Data Communication for Nanosatellites." Small Satellites Systems and Services Symposium. pp. 1-15, 2014.

(6) P.G. Farrell and A.P. Clark. "Modulation and coding." International Journal of Satellite Communications, 1984.

(7) A. Datta. “New FCC rules make small satellite licensing easier, faster, and cheaper. Geospatial World | Advancing Knowledge for Sustainability.” [Online] August 22, 2020. Available at: https://www.geospatialworld.net/blogs/new-fcc-rules-make-small-satellite-licensing-easier-faster-and-cheaper/

(8) J. Buxton and D. Glasbrenner. “AMATEUR RADIO AND EXPERIMENTAL LICENSING FOR CUBESATS.”

(9) CalPoly. “CubeSat Licensing Support.” [Online] Available at: https://www.cubesat.org/cubesatlicensingsupport

(10) National Institute of Standards and Technology. "Security Requirements for Cryptographic Modules, Federal Information Processing Standard (FIPS). ISO/IEC 19790:2012; ISO/IEC 24759:2014

(11) 47 CFR Part 97 - AMATEUR RADIO SERVICE. [Online] Available at: https://www.law.cornell.edu/cfr/text/47/part-97

(12) J.R. Wertz, D.F. Everett, and J.J. Puschell. "Space Mission Engineering: The New SMAD." Microcosm Press, Hawthorne, CA, 2015.

(13) S.K. Johnson, R.C. Reinhart, and T.J. Kacpura. "CoNNeCT's approach for the development of three Software Defined Radios for space application." IEEE Aerospace Conference. pp. 1-13, Big Sky, MT, 2012.

(14) Y. Wong et al. "NASA Near Earth Network (NEN) DVB-S2 Demonstration Testing for Enhancing Data Rates for CubeSats." 31st Annual AIAA/USU Conference on Small Satellites, Logan, UT, 2019.

(15) JPL. "Iris V2 CubeSat Deep-Space Transponder (IRIS)." [Online] Available at: https://www.jpl.nasa.gov/cubesat/missions/iris.php

(16) [101] T.S. Rose, D.W. Rowen, S.D. LaLumondiere, N.I. Werner, R. Linares, A.C. Faler, J.M. Wicker, C.M. Coffman, G.A. Maul, D.H. Chien, A.C. Utter, R.P. Welle, and S.W. Janson. "Optical communications downlink from a low-earth orbiting 1.5U CubeSat," Opt. Express 27, 24382-24392, 2019.

(17) D.W. Pack, G. Kinum, P.D. Johnson, T.S. Wilkinson, C.M Coffman, J.C. Mauerhan, B.S. Hardy, R. Russell, and K. Mercy. "Landsat Imagery from a CubeSat: Results and Operational Lessons from the R3 Satellite's First 18 Months in Space." Small Satellite Conference, Logan UT, 2020.

(18) D.W Pack, B.S Hardy, J.R. Santiago, D. Pietrowski, J.C. Mauerhan, P.F. Zittel, D.W. Rowen, C.R. Purcell, P. Thiyanaratnam, L.J. Gelinas, P.K. Su, J. Gussy, and J.M. Santiago. "Flight Operations of Two Rapidly Assembled CubeSats with Commercial Infrared Cameras: The Rogue-Alpha,Beta Program." Small Satellite Conference, Logan UT, 2021.

(19) T. Jono, Y. Takayama, N. Kura, K. Ohinata, Y. Koyama, K. Shiratama, Z. Sodnik, B. Demelenne, A. Bird, and K. Arai. "OICETS on-orbit laser communication experiments. Page 610503, San Jose, CA, February 2006.

(20) R. Fields, D. Kozlowski, H. Yura, R Wong, J Wicker, C. Lunde, M. Gregory, B Wandernoth, and F. Heine. "5.625 Gbps bidirectional laser communications measurements between the NFIRE satellite and an Optical Ground Station." International Conference on Space Optical Systems and Applications (ICSOS), pages 44–53, Santa Monica, CA, USA, May 2011. IEEE.

(21) H. Takenaka, Y. Koyama, M Akioka, D. Kolev, N. Iwakiri, H. Kunimori, A. Carrasco-Casado, Y. Munemasa, E. Okamoto, M. Toyoshima. "In-orbit verification of small optical transponder (SOTA): evaluation of satellite-to-ground laser communication links." Proc. SPIE 9739, Free-Space Laser Communication and Atmospheric Propagation XXVIII, 973903, March 15, 2016.

(22) H. Tomio, T. Kuwahara, S. Fujita, Y. Sato, M. Sakal, H. Kunimori, T. Kubooka, H. Takenaka, Y. Saito, M. Toyoshima. "Assembly and integration of optical downlink terminal VSOTA on microsatellite RISESAT." Proc. SPIE 11180, International Conference on Space Optics — ICSO 2018, 111805Z, July 12, 2019.

(23) C. Schmidt, C. Fuchs, ""The OSIRIS program at DLR,"" Proc. SPIE 10524, Free-Space Laser Communication and Atmospheric Propagation XXX, 105240R, 5 February 15, 2018.

(24) C. Fuchs, C. Schmidt, J. Keim, F. Moll, B. Rödiger, M. Lengowski, S. Gaißer, and D. Giggenbach. "Update on DLR"s OSIRIS program and first results of OSIRISv1 on Flying Laptop." Proc. SPIE 10910, Free-Space Laser Communications XXXI, 109100S, March 4, 2019.

(25) D. Giggenbach, C. Fuchs, C. Schmidt, B. Rödiger, S. Gaißer, S. Klinkner, D. Phung, J. Chabé, C. Courde, N. Maurice, H. Mariey, E. Samain, and G. Artaud. "Downlink communication experiments with OSIRISv1 laser terminal onboard Flying Laptop satellite." Appl. Opt. 61, 1938-1946, 2022.

(26) W. Kammerer, P. Grenfell, L. Hyest, P. Serra, H. Tomio, N. Belsten, C. Lindsay, O. Čierny, K. Cahoy, M. Clark, D. Coogan, J. Conklin, D. Mayer, J. Stupl, J. Hanson. "CLICK Mission Flight Terminal Optomechanical Integration and Testing." International Conference on Space Optics — ICSO, 2022.

(27) Tesat. " SMALLEST LASER COMMUNICATION TRANSMITTER WORLDWIDE." Technical Datasheet. [Online] 2022. Available at: https://www.tesat.de/images/tesat/products/220607_DataSheet_CubeLCT100M_A4.pdf

(28) K. Iwamoto, H. Komatsu, S. Ohta, Y. Kubo, T. Nakao, H. Yamazoe, T. Kamata, Y. Munemasa, H. Kunimori, M. Toyoshima, D. Koda, H. Sawada, T. Ikeda. "Experimental results on in-orbit technology demonstration of SOLISS." Proc. SPIE 11678, Free-Space Laser Communications XXXIII, 116780D, March 5, 2021.

(29) H. Komatsu, S. Ohta, H. Yamazoe, Y. Kubo, T. Nakao, T. Ito, D. Koda, H. Sawada, T. Ikeda, Y. Munemasa, H. Kunimori, T. Kubooka, M. Toyoshima, K. Iwamoto. "The pointing performance of the optical communication terminal, SOLISS in the experimentation of bidirectional laser communication with an optical ground station." Proc. SPIE 11678, Free-Space Laser Communications XXXIII, 116780F, March 5, 2021.

(30) Curt M. Schieler, Kathleen M. Riesing, Andrew J. Horvath, Bryan C. Bilyeu, Jesse S. Chang, Ajay S. Garg, Jade P. Wang, Bryan S. Robinson, ""200 Gbps TBIRD CubeSat downlink: pre-flight test results,"" Proc. SPIE 11993, Free-Space Laser Communications XXXIV, 119930P, March 4, 2022.

(31) AAC Clyde Space. "LASER COMMUNICATIONS CUBECAT." Technical Datasheet. [Online] 2022. Available at: https://www.aac-clyde.space/wp-content/uploads/2021/10/CUBECAT.pdf

(32) R. Bedington, J.M. Arrazola, and A. Ling. "Progress in satellite quantum key distribution." Npj Quantum Information Vol. 3, 30, 2017.

(33) P. Serra, O. Čierny, W. Kammerer, E.S. Douglas, D.W. Kim, J.N. Ashcraft, G. Smith, C. Guthery, T. Vergoossen, A. Lohrmann, R. Bedington, C. Perumangatt, A. Ling, K. Cahoy. "Optical front-end for a quantum key distribution cubesat." Proc. SPIE 11852, International Conference on Space Optics, 118523C, June 11, 2021.

(34) D. Jacobs, J. Bowman, M. Patterson, M. Horn, C. McCormick and M. Adkins. "Plan for On-Orbit Demonstration of the Deployable Optical Receiver Array." IEEE Aerospace Conference (AERO), pp. 1-8, 2022.

(35) J.E. Velazco and J.S. de la Vega. "Q4 – A CubeSat Mission to Demonstrate Omnidirectional Optical Communications." IEEE Aerospace Conference, pp. 1-6, 2020.

(36) A. Carrasco-Casado et al., ""Intersatellite-Link Demonstration Mission between CubeSOTA (LEO CubeSat) and ETS9-HICALI (GEO Satellite)." IEEE International Conference on Space Optical Systems and Applications (ICSOS), pp. 1-5, 2019.

(37) European Space Agency. "European Data Relay Service." [Online] Available at: https://www.esa.int/Applications/Telecommunications_Integrated_Applications/EDRS

(38) F. Heine, P. Martin-Pimentel, N. Höpcke, D. Hasler, C. Rochow, and H. Zech. "Status of tesat lasercomms activities." Free-Space Laser Communications XXXIII, vol. 11678, pp. 45-50. SPIE, 2021.

(39) R. Lafon, J. Wu, B. Edwards. "Regulatory Considerations: Laser Safety and the Emerging Technology of Laser Communication." Commercial Laser Communications Interoperability and Regulatory Workshop, Washington, DC, June 12, 2017.

Chapter Contents

Chapter Glossary

(CBOD)	Clamp Band Opening Device
(CDS)	CubeSat Design Specification
(CSLI)	CubeSat Launch Initiative
(DPAF)	Dual Payload Attach Fittings
(EAGLE)	ESPA Augmented Geostationary Laboratory Experiment
(EELV)	Evolved Expendable Launch Vehicle
(ENRCSD)	Nanoracks External CubeSat Deployer
(ESA)	European Space Agency
(ESPA)	EELV Secondary Payload Adapter
(GEO)	Geostationary Equatorial Orbit
(HEO)	Highly Elliptical Orbit
(ISS)	International Space Station
(J-SSOD)	JEM Small Satellite Orbital Deployer
(JAXA)	Japan Aerospace Exploration Agency
(JEM)	Japanese Experimental Module
(JEMRMS)	Japanese Experimental Module Remote Manipulator System
(M-OMV)	Minotaur Orbital Maneuvering Vehicle
(MEO)	Medium Earth Orbit
(MLB)	Motorized Light Bands
(MPAF)	Multi Payload Attach Fittings
(MPEP)	Multi-Purpose Experiment Platform
(NICL)	Nanoracks Interchangeable CubeSat Launcher
(NOAA)	National Oceanic and Atmospheric Administration
(NRCSD)	Nanoracks ISS CubeSat Deployer
(OMV)	Orbital Maneuvering Vehicle
(PCBM)	Cygnus Passive Common Berthing Mechanism
(RUG)	Rideshare User Guide
(SL-OMV)	Small Launch Orbital Maneuvering Vehicle
(SSMS)	Small Spacecraft Mission Service
(SSOD)	Small Satellite Orbital Deployer
(TRL)	Technology Readiness Level

10.0 Integration, Launch, Deployment, and Orbital Transport

10.1 Introduction

Of the 2,510 total spacecraft launched in 2022, 96% were SmallSats with a mass less than 600kg and 25% were SmallSats with mass under 200 kg. Roughly 40% of all spacecraft with a mass under 200 kg were launched in the past ten years. The significant increase in SmallSats launched in the past few years is due to an expansion in SmallSat and CubeSat constellations by operators like SpaceX, OneWeb, and Planet. With more SmallSat missions currently being planned, the demand for SmallSat launches is expected to continually rise (1).

Since launch vehicle capability usually exceeds primary spacecraft requirements, there is typically enough mass, volume, and other performance margins to include secondary small spacecraft. This surplus capacity can be used by SmallSats as a cost-effective ride to space. A large market of adapters and dispensers has been created to compactly house multiple small spacecraft on existing launchers. These technologies provide a structural attachment to the launcher and deployment mechanisms. This method, known as "rideshare," is still the main way of putting small spacecraft into orbit. The terms 'rideshare' and 'hosted payload' are sometimes used interchangeably, however there are distinct and subtle differences; hosted payload services offer space for a payload on a shared platform to a predetermined orbit, while rideshare services provide space for a dedicated spacecraft integrated onto the launch vehicle or separation system. For more information on hosted payloads, readers are encouraged to review the Complete Spacecraft Platforms chapter of this report.

As both SmallSat and CubeSat adapters and dispensers have become more developed, rideshares have become a more popular way to access space. Additionally, nanosatellite form factors are increasing in dimensions and mass, requiring larger dispensers to accommodate these larger CubeSat sizes. Orbital transport vehicles (OTVs), along with generally more capable orbital maneuvering vehicles (OMVs) can offer "last mile" delivery services to intended orbits for small satellites. These vehicles are becoming a more common paradigm for SmallSat and CubeSat deployment and operational logistics. While historically referred to as "Space Tugs," this term is in the process of being phased out by many in the commercial industry. Several companies are developing OTVs/OMVs with on-board propulsion systems that can be launched to an approximate orbit and then propel themselves to one or more target orbits, where they can either deploy the small spacecraft or serve as an integral part of the hosted payload. Some OTVs are based on a traditional rocket kick stage and are intended to work with specific launch vehicles. As of 2023, several commercial companies have successfully flown OTVs/OMVs and are booking future launch manifests.

Expanding future capabilities of small satellites will demand dedicated launchers. Flying a spacecraft as a dedicated payload may be the best method of ascent for missions that need a very specific orbit, near complete capability of available launcher performance, interplanetary trajectories, precisely timed rendezvous, or special environmental considerations. Technology developers and hard sciences can take advantage of the quick iteration time and low capital cost of small spacecraft to yield new and exciting advances in space capabilities and scientific understanding. The emergence of very small launch vehicles has altered the landscape by providing dedicated rides for small spacecraft to specific destinations on more flexible timelines.

NASA's Launch Services program developed a new Indefinite Delivery/Indefinite Quantity (IDIQ) mechanism in Q1 2022: the Venture Class Acquisition of Dedicated and Rideshare (VADR) launch services. The principal purpose of the VADR IDIQ contract is to embrace a commercial approach that provides NASA a new class of launches. VADR-procured launch services enable unique launch capabilities for Class D or higher risk tolerant payloads and provide FAA licensed launch services capable of delivering payloads to a variety of orbits. This contract mechanism offers a broad range of commercial launch services for traditional and dedicated rideshare options. The commercial approach uses a lower level of mission assurance for payloads with higher risk tolerance and contributes to NASA's science and research development efforts as an ideal platform for technical development. The 2022 Heliophysics Small Explorers Announcement of Opportunity and Mission of Opportunity were the first NASA AO's to use this contract structure. The VADR IDIQ contract provides a new mechanism for traditional and dedicated rideshare launches for risk-tolerant payloads. While 13 companies were initially selected, an on-ramp provision allows new launch services and capabilities to be proposed (2).

In 2022, NASA's Flight Opportunities program's TechFlights solicitation, in cooperation with the agency's Small Spacecraft Technology program, included opportunities for flight tests on commercial orbital platforms capable of hosting payloads. This capability was also included in the 2023 NASA solicitation for Suborbital/Hosted Orbital Flight and Payload Integration Services. IDIQ contracts for these services are expected to be in place with commercial providers in early 2024.

The information described below is not intended to be exhaustive but provides an overview of current state-of-the-art technologies and their development status for particular small spacecraft launch, integration, deployment, and logistics systems. It should be noted that Technology Readiness Level (TRL) designations may vary with changes to mission requirements, reliability considerations, and/or the environment in which performance was demonstrated. Readers are highly encouraged to reach out to companies for further information regarding the performance and TRL of described technology. There is no intention of mentioning certain companies and omitting others based on their technologies or relationship with NASA.

10.2 State-of-the-Art – Launch Integration Role

Launch options for a SmallSat include dedicated launch, traditional rideshare launch, or multi-mission launch, as described in the launch section below. Regardless of the approach, however, integration with the launch vehicle is a complex and critical portion of the mission. The launch integration effort for a primary spacecraft typically includes the launch service provider, the spacecraft manufacturer, the spacecraft customer, the launch range operator, and sometimes a launch service integration contractor (3). When launching on either a multi-mission or rideshare launch, the launch integration becomes even more complex.

When flying as a rideshare payload, it is generally the primary spacecraft customer who decides whether secondary spacecraft will share a ride with the primary spacecraft and, if so, how, and when the secondary spacecraft are dispensed. This is not always the case, however, as there are occasions where the launch vehicle contractor or a third-party integration company determines rideshare possibilities. More flexibility may be available to secondary spacecraft through such a program, although the mission schedule is normally still determined by the primary spacecraft.

There are several options for identifying and booking a ride for a SmallSat. For rideshare and multi-mission launches, the spacecraft customer may choose to use a launch broker, or aggregator to facilitate the launch manifest, or work directly with the launch service provider. A launch broker matches a spacecraft with a launch opportunity, whereas an aggregator provides additional services related to manifesting. In the event of a dedicated launch, the spacecraft customer generally does not use a launch broker or aggregator. In both cases, however, key aspects for integration must be managed, and a launch integrator can assist or coordinate those activities for the spacecraft customer.

Whether a spacecraft customer chooses to use a launch integrator or not, it is the responsibility of the spacecraft operator to obtain flight certifications, including radio frequency licensing, National Oceanic and Atmospheric Administration (NOAA) remote sensing licensing, and laser usage approval (4) (5). The launch integrator or the launch service provider will require proof of licensure before launching the satellite. They will also require additional analyses and supporting data prior to launch. This may include safety documentation, orbital debris information, materials and venting data, and spacecraft specific models (6).

For rideshare and multi-mission launches, many satellites are subject to a "do no harm" requirement to protect the primary satellite or other satellites on a multi-mission launch. A list of "do no harm" requirements are imposed on the rideshare satellite by the launch provider, launch integrator, or primary mission owner. These requirements vary by launch provider and launch integrator, but usually include restrictions on transmitters, post separation mechanical deployments, and hazardous materials. A comprehensive list of typical "do no harm" requirements is provided in the NASA Rideshare User Guide (RUG) (7).

10.2.1 Launch Brokers and Services Providers

A launch broker for small satellites is an individual or organization which matches a spacecraft with a launch opportunity, usually as a rideshare satellite or a multi-mission manifest spacecraft. Typically, a launch broker does not provide any additional launch integration services beyond coordinating the relationship between the spacecraft manufacturer or customer and the launch service provider. Their purpose is to fill excess capacity on a launch, and they can also bolster negotiations between the launch provider and payload for scheduling, integration, safety testing, and cost (8).

Service providers can work with the satellite customer and the launch vehicle provider to ensure that the customer's spacecraft is compatible with the launch vehicle's mission by performing analyses and physical integration services. The launch services provider can assist with hardware integration for the CubeSat dispenser, separation system, or other hardware, or these may be provided by the spacecraft customer. It should be noted that there is no universally accepted definition of "launch broker" and the term can be used interchangeably with "launch aggregator" and "launch integrator."

10.3 Launch Paradigms

The SmallSat market has grown considerably over the past decade experiencing a 23% compound annual growth rate from 2009 to 2018 (9), and this growth continues unabated. From 2013 to 2017 there was an average of ~140 SmallSats (less than 200 kg) launched per year. From 2017 to 2021 this number increased to an average 332 SmallSats per year, with more than 550 SmallSats launched in 2022 (1). Of the total spacecraft launched in 2022, 200-600 kg type spacecraft accounted for 71%, while micro, nano, pico, and femto spacecraft were the next most launched spacecraft (1). The 200-600 kg type spacecraft have seen a considerable launch growth since 2020, primarily due to mega constellation Starlink by SpaceX.

This increase in small satellite demand has caused a shift in the launch vehicle market, with many companies creating or advertising launch platforms centered around small satellites. While other chapters in this report cite specific companies providing "state-of-the-art" technologies, this section will provide an overview of the three types of launch methods for SmallSats and the current state of these markets.

10.3.1 Dedicated Launches

In the context of this report, dedicated launches for SmallSats are those generally used to launch satellites with a mass less than 180 kg. However, this does not mean that the maximum mass to orbit is 180 kg or less, as some dedicated launchers have a payload maximum of 1000 kg, and many launch vehicles marketed for SmallSats can deliver masses to orbit that are higher than 180 kg. The primary orbit for this type of launch is low-Earth orbit (LEO), with very few companies currently targeting highly elliptical orbit (HEO), medium-Earth orbit (MEO), or geostationary equatorial orbit (GEO).

Dedicated launches for SmallSats have many advantages. A SmallSat on a dedicated launch controls the mission requirements in whole—what they need, when they want to launch, and where they want to go. They generally have a readiness "go/no-go" call on launch day in case something goes wrong with their satellite pre-launch. They can also request special launch accommodations, such as a nitrogen purge or late battery charge, that are generally not available to a rideshare launch (this may be as a standard service or with an additional cost as mission-unique). The downside to a dedicated launch is that they are generally more expensive than a rideshare launch.

10.3.2 Traditional Rideshare Launches

Until recently, there were only a few launchers that allowed small spacecraft to ride as primary spacecraft. The majority of small spacecraft are carried to orbit as secondary spacecraft, using the excess launch capability of larger rockets. Standard ridesharing consists of a primary mission with surplus mass, volume, and performance margins which are used by another spacecraft. Secondary spacecraft are also called auxiliary spacecraft or piggyback spacecraft. For educational small spacecraft, several initiatives have helped provide these opportunities. NASA's CubeSat launch initiative (CSLI) for example, has provided rides to a significant number of schools, non-profit organizations, and NASA centers. As of September 2023, the initiative launched 162 CubeSats, and continues to select CubeSats for launch (10). The European Space Agency (ESA) "Fly Your Satellite" program is a similar program which provides launch opportunities to university CubeSat teams from ESA Member States, Canada, and Slovenia (11).

From the secondary spacecraft designers' perspective, rideshare arrangements provide far more options for immediate launch with demonstrated launch vehicles. Since almost any large launcher can fit a small payload within its mass and volume margins, there is no shortage of options for craft that want to fly as a secondary spacecraft. On the other hand, there are downsides of hitching a ride. The launch date and trajectory are determined by the primary spacecraft, and the smaller craft must take what is available. In some cases, they need to be delivered to the launch provider and be integrated on the adapter weeks before the actual launch date. Generally, the secondary spacecraft are given permission to be deployed once the primary spacecraft successfully separates from the launch vehicle, but there are instances where the rideshare spacecraft separate prior to the primary satellite (12).

Multi-mission manifest launches are those that exclusively use launch vehicles to launch multiple SmallSats. These launches have shown the ability to hold and deploy many satellites to multiple altitudes, though these orbits tend not to be vastly different. These types of launches are growing in popularity with many launch vehicle providers offering regular launches to the same altitude at regular intervals throughout the calendar year. While challenging, the logistics of these missions are managed by various integrators throughout the market, many of which are new to industry but are forging a new path in rideshare. Multi-mission manifest launches provide the opportunity to place large numbers of satellites into orbit on a single launch.

10.4 Deployment Methods

The method by which SmallSats are deployed into orbit is a critical part of the launch process. The choice of deployment method depends on the form factor of the satellite. This section will discuss the deployment of CubeSats, which generally use CubeSat dispensers, and the deployment of free-flying SmallSats.

10.4.1 CubeSat Dispensers

The CubeSat form factor is a very common standard for spacecraft up to approximately 24 kg (12U CubeSat) but can also be extended to approximately 54 kg in a 27U configuration (13). The most updated CubeSat Design Specification document is found at http://www.cubesat.org, a website maintained and operated by California Polytechnical State University, San Luis Obispo, the creators of the CubeSat form factor.

The CubeSat form lends itself to container-based integration systems, or dispensers, which serve as an interface between the CubeSat and the launch vehicle. It's a rectangular box with a hinged door and spring mechanism. Once the door is commanded to open, the spring deploys the CubeSat. Many companies currently manufacture dispensers for the CubeSat form factor which follow one of two constraint systems: the rail-type dispenser, and the tab-type dispenser. Due to the large number of dispenser manufacturers, the different companies are not listed here. Instead, a brief overview of the two types of dispensers is provided.

A rail-type dispenser (figure 10.1) supports CubeSats that have rails which extend the length of the CubeSat on four parallel edges. The rails on the CubeSat prevent it from rotating while inside the dispenser. After the dispenser door has been commanded to open, the rails slide along guides inside the dispenser and the CubeSat is deployed. As such, it is important that any rail-based CubeSat follow the current development specifications to ensure compliance. This type of dispenser is the most widely manufactured configuration, with more than fifteen manufacturers worldwide.

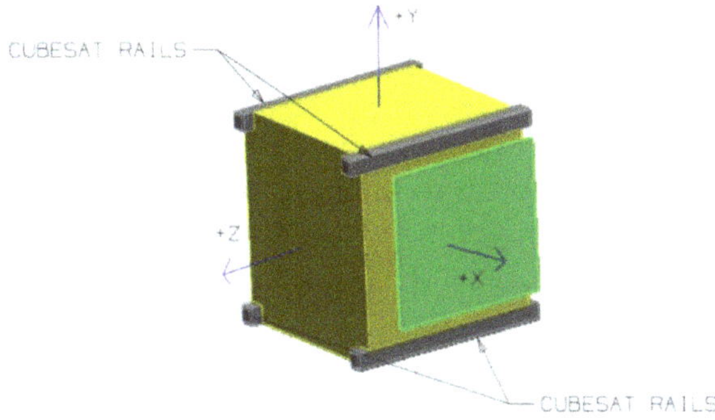

Figure 10.1: The Rail-type CubeSat. Credit: CalPoly's CubeSat Program.

A tab-type dispenser (figure 10.2) supports CubeSats with tabs which run the length of the CubeSat on two parallel edges. Typically, the dispenser grips the tabs to hold the CubeSat in place, only releasing it after the door has been commanded to open. In the past, this type of dispenser was not widely manufactured as Planetary Systems Corporation (recently acquired by Rocket Lab USA) held the patent for the design. Recently however, more developers are beginning to develop their own tab-based designs for CubeSat dispensers. Many are based on the original Planetary Systems Corporation (now Rocket Lab) standard, however some offer features such as built-in isolation to accommodate for the launch vehicle environment. In addition, there are some tab-based dispensers that do not grip the tabs. Rather, they provide a slot to accommodate the tab, which slides freely within the slot. While use of tab-type dispensers is growing, they remain a minority among dispensers purchased and used by developers.

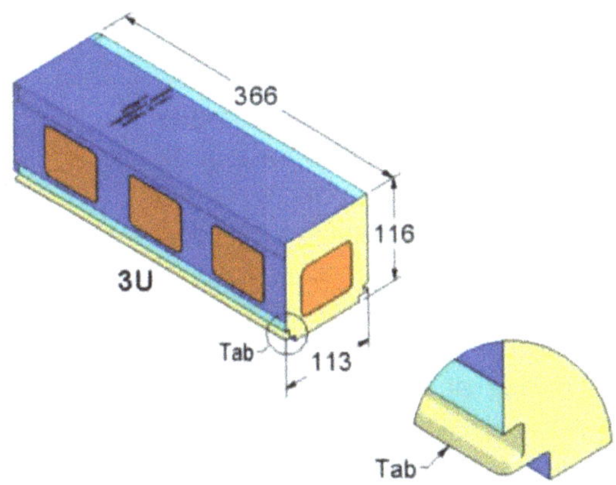

Figure 10.2: The Tab-type CubeSat. Credit: Planetary Systems Corporation.

While CubeSats can generally pick their dispenser type (rail vs. tabbed), the choice of the actual dispenser is not always a decision made by the CubeSat. In many cases, the launch vehicle provider or launch aggregator/integrator has already determined which dispensers will be installed on the launch vehicle. As each dispenser manufacturer has slightly different volumes and requirements, it is beneficial for the CubeSat to design for as wide a range of dispensers as possible to maximize launch opportunities.

Additionally, some dispenser manufacturers offer accommodations which may violate the "do no harm" requirements set forth by the launch vehicle or launch integrator, such as inhibits on deployables and transmitters. Therefore, it is beneficial for the CubeSat to evaluate "do no harm" recommendations from a variety of organizations, as these requirements can vary from flight to flight on the same launch vehicle based on the risk posture of the primary payload and/or the mission "owner" (7).

10.4.2 SmallSat Separation Systems

Small satellites which do not meet the form factor of a CubeSat, or will not be using a CubeSat dispenser for integration to the launch vehicle, require a different separation mechanism. Separation systems for SmallSats generally follow either a circular pattern or a multi-point (3 or 4 point) pattern. Depending on the launch vehicle, separation systems may already be in place and available to secondary spacecraft. It should be noted that separation systems are often some of the most complicated pieces of hardware involved with launching spacecraft. If a spacecraft is given the option to bring its own separation system to launch, great care should be taken in selection, including the development maturity and flight heritage for any separation system.

Circular separation systems use two rings held together by a clamping mechanism. One ring is attached to the launch vehicle and the other ring is attached to the spacecraft. Once the clamping mechanism is released, the two rings separate and are pushed apart by springs. Each ring then remains with the spacecraft or the launch vehicle. There are two primary types of clamping configurations, motorized light bands (MLB) and Marman clamps.

The MLB (figure 10.3) is a motorized separation system that ranges from 8 inches to 38 inches in diameter. Smaller MLB systems are used to deploy spacecraft less than 180 kg, while larger variations may be used to separate larger spacecraft or other integration hardware such as orbital maneuvering systems, which are discussed below. The MLB's separation system eliminates the need for pyrotechnic separation, and thus deployment results in lower shock with no post-separation debris.

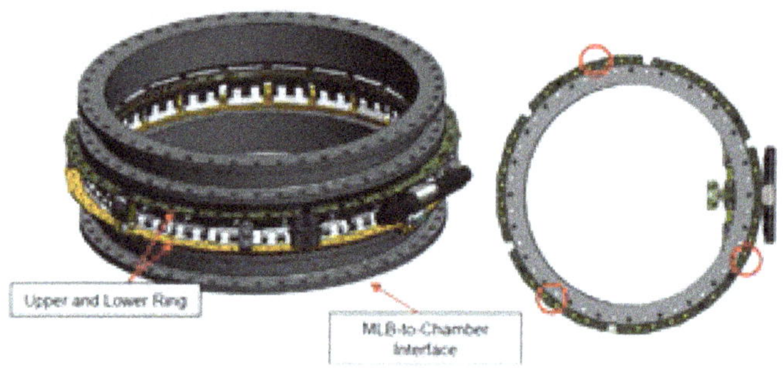

Figure 10.3: MkII Motorized Lightband. Credit: Planetary Systems Corporation.

Marman band separation systems use energy stored in a clamp band, often along with springs, to achieve separation. The Marman band is tensioned to hold the spacecraft in place. Some Marman bands use pyrotechnic devices to cut the clamping bolt, however many companies offer a low shock release mechanism which is potentially better for the spacecraft. Sierra Nevada produces a Marman band separation system known as Qwksep, which uses a series of separation springs to help deploy the spacecraft after clamp band release. RUAG Space provides several circular separation systems which use their Clamp Band Opening Device (CBOD) release mechanism to reduce shock impact on the spacecraft (14).

Several companies are now providing multi-point separation systems instead of the circular band. Using a multi-point separation system may result in mass savings over a circular separation system. However, some systems require additional simultaneous signals from the launch vehicle provider to ensure proper release. The RUAG PSM 3/8B is a low-shock separation nut developed to fit OneWeb satellites (15). It requires additional firing commands from the launch vehicle or a dedicated sequencing system. ISISPACE has also developed the M3S Micro Satellite Separation System (see figure 10.4) which is designed for satellites up to 100 kg but can be configured for higher masses (16).

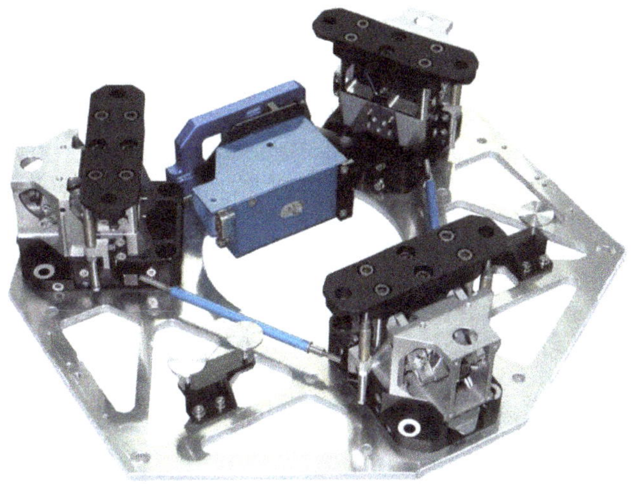

Figure 10.4: ISISPACE M3S Micro Satellite System. Credit: ISISPACE.

Given the stiffness and fundamental frequency requirements of traditional rideshare missions, many companies are also shifting to 4-point separation systems for MicroSats and SmallSats as a viable alternative to traditional MLB or clamp band systems. These systems function in a similar way to the systems above and are typically rated for microsatellites (≤100 kg), however boast less complexity than a traditional MLB or Clampband. The rapid acceptance of this launch solution is driven by the fundamental frequency requirements of traditional rideshare launches, with the hope that reduced stiffness at the interface will increase the compatibility of SmallSats and MicroSats for those types of launches. In addition to reduced complexity, many of these result in cost savings as well, which can be passed on to both the integrator and the SmallSat manufacturer. Many integrators are exploring the addition of such systems into their portfolio to accommodate launches in the near future.

Cake Topper and Plate System for Rideshares

SpaceX has developed a system that differs from the SPA system for rideshare missions to SSO (Transporter Missions) and mid-inclination orbits (Bandwagon Missions). This system of plates rather than a ring is intended to allow more payloads to be included in the circumferential space for flight on their commercial rideshare missions. In addition, for larger spacecraft or spacecraft that cannot be horizontally mounted during flight, they also offer a cake topper option. Figure 10.5 shows these two options. The blue box shows the plate option which has a specific set of rideshare loads for the missions' part of transporter or bandwagon. The green box denotes the cake topper option which also has separate environments. User guides for both configurations are available on the launch provider's website. For missions that will fly under the VADR contract, please contact LSP for additional guidance and enveloping environments.

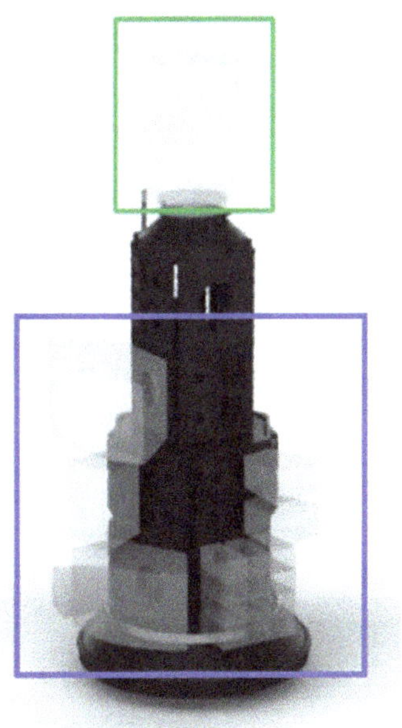

Figure 10.5: SpaceX Cake Topper (green) and Plate system (blue) configurations. Credit: SpaceX.

10.4.3 Integration Hardware

A main driver for CubeSat utility is their adhesion to a standard that can be integrated into several different launch configurations. The physical hardware that attaches both a containerized and non-containerized small spacecraft and keeps it insulated from a rocket body include deployers, adapters, dispensers, and launchers. The purpose of this hardware is to eject the spacecraft safely into orbit, and most services offer different features, interfaces, connections, and designs for small spacecraft specifications. The exact configuration and standards vary by launch vehicle, and the determination of an appropriate and reliable launch option is part of the launch qualification process (8). With this rise in CubeSat constellations, integration hardware capable of launching multiple SmallSats simultaneously and consecutively is now a standard. This section will highlight some existing examples of integration flight support hardware applicable to both SmallSats and CubeSats, but the reader is highly encouraged to identify other integration services.

Evolved Expendable Launch Vehicle (EELV) Secondary Payload Adapter (ESPA)

The ESPA ring (figure 10.6, top) is a multi-payload adapter for large primary spacecraft originally developed by Moog Space and Defense Group. Six 38 cm (15") circular ports can support six auxiliary payloads up to 257 kg each. It was used for the first time on the Atlas V STP-1 mission in 2007. The ESPA Grande (figure 10.6, lower) uses four 61 cm (24") circular ports which can carry spacecraft up to 450 kg (991 lb) (17). Although developed by Moog, several other companies now offer similar designs in different configurations.

Figure 10.6: [top] ESPA Ring [lower] ESPA Grande Ring. Credit: Moog, Inc.

Small Spacecraft Mission Service (SSMS) Dispenser

ESA has developed the Small Spacecraft Mission Service dispenser for the Vega launch vehicle (figure 10.6). This dispenser comes in a variety of different modular parts which can be configured based on the satellite launch manifest. The modularity of the dispenser provides greater flexibility for accommodating different customers (18).

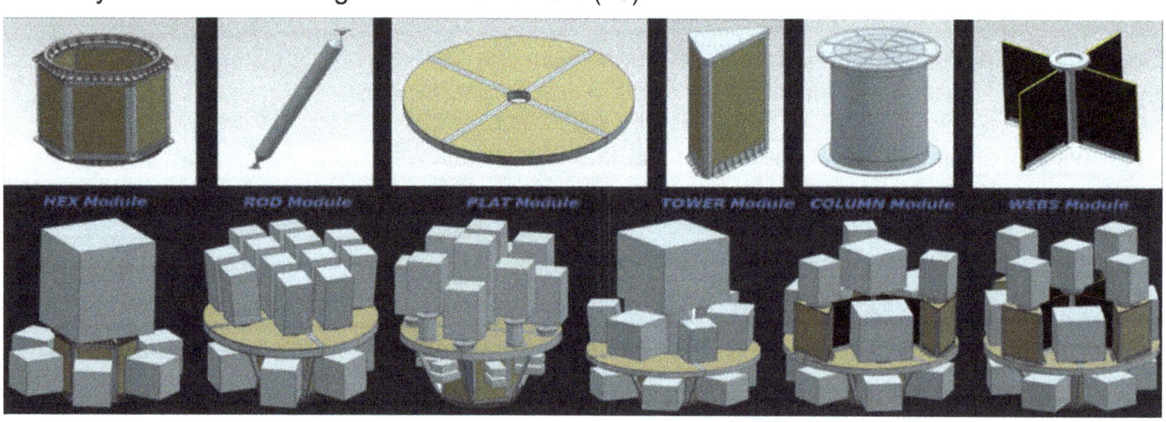

Figure 10.6: The European Space Agency Small Spacecraft Mission Service Dispenser for the Vega Launch Vehicle (19). Credit: European Space Agency.

Dual / Multi Payload Attach Fittings (DPAF / MPAF)

Many launch vehicle providers have existing accommodations for two or more payloads which are sometimes referred to as Dual Payload Attach Fittings (DPAF) or Multi Payload Attach Fittings (MPAF). As these are generally launch vehicle specific, and occasionally mission specific, they are not discussed here.

10.5 Orbital Transfer / Maneuvering Vehicles (OTV/OMV)

One of the main disadvantages of riding as a secondary spacecraft (even on a dedicated rideshare mission) is the inability to launch into the desired orbit. The primary spacecraft determines the orbital destination, so the secondary spacecraft orbit usually does not perfectly match the customer's needs. However, by using an OTV or OMV, secondary spacecraft can maneuver much closer to their desired orbits. OTVs are generally more coarsely propulsion-capable, while OMVs may offer hosted systems more in terms of power, pointing, and communications. The OTV / OMV market is nascent, with many planned systems but few with existing flight heritage. This emerging technology is an area of interest in the near term for both SmallSats and CubeSats, as it offers a significant capability to reach destinations not previously achievable with systems of this scale. The ability for small spacecraft to reach new orbits will enable a much wider range of mission designs for destinations both near and far.

Commercial delivery and deployment system launches are increasing in cadence. Below is a launch timeline of OTV/OMVs with commercial payload services as part of the vehicle's documented purpose. This cadence is expected to grow as more vehicles become space qualified and as new companies emerge. The data is derived from publicly accessible data and only contains launches by the providers described in this report. There may be more launches than indicated herein, and the success of the launch/mission is not factored.

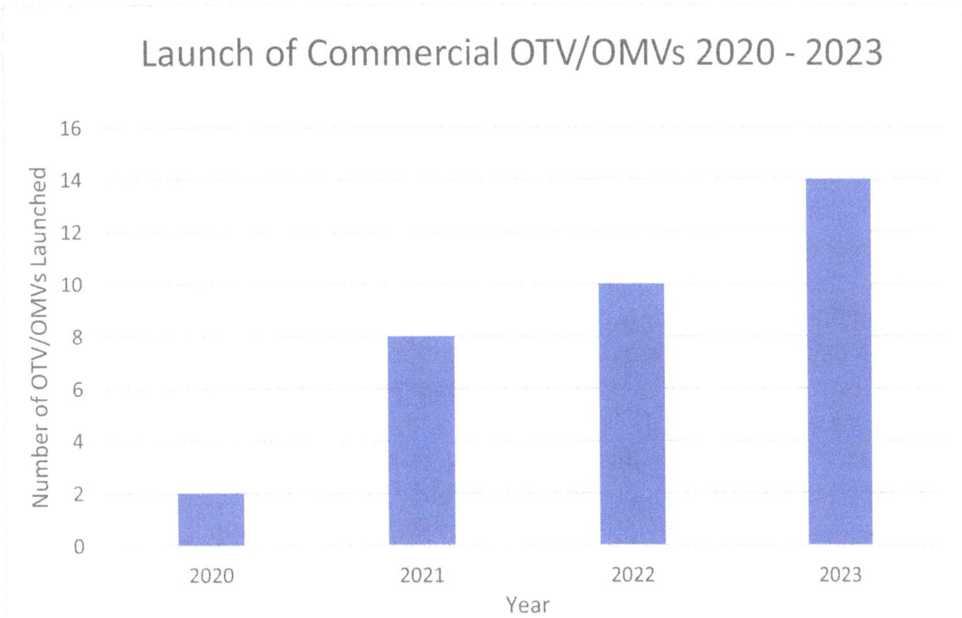

Figure 10.7 - Number of commercial OTV/OMVs launched per year since 2020. Credit: NASA

10.5.1 Commercial Development/Services

As discussed above, the ESPA ring provides structure to which SmallSats or CubeSat dispensers are mounted. The idea of adding propulsion to an ESPA ring led to many of the early commercial

OMV/OTV vehicle designs. Systems that were originally derived from the ESPA ring include the Spaceflight Inc. (now Firefly) SHERPA (19), the Orbital Sciences Corporation (now Northrup Grumman Space Systems) ESPAStar Product Line (20), and the Moog METEOR (21). The ESPAStar product line has gone through multiple programs with multiple names. The ESPAStar line has significant flight heritage from the Long Duration Propulsive ESPA (LDPE) satellites, now known as Rapid On-Orbit Space Technology Evaluation Ring (ROOSTER) United States Space Force (USSF) program. While the SHERPA system is no longer commercially available, FireFly is now offering rides on its Elytra OTV/OMV product line, which appears to build upon the ESPA ring heritage for specific compatibility with FireFly's small- and medium-lift launch vehicles. Rocket Lab is another launch provider offering an OMV/OTV platform. The Photon system is an evolution of the Electron launch vehicle Kick Stage (22).

Other systems are being developed from the ground-up by commercial entities to provide orbital transport, hosting, and deployment services. These include the Momentus Vigoride vehicles, the Epic Aerospace Chimera vehicles, the Exolaunch Reliant, the Nanoracks Outpost, the Moog SL-OMV, the TransAstra WorkerBee, the Atomos Quark, and the D-Orbit ION vehicles. Northrup Grumman, in addition to its ESPAStar series, has a proven Mission Extension Vehicle (MEV) that can host payloads, but is primarily developed to dock and extend the life of large GEO satellites. To date, the MEV has successfully extended the life of two IntelSat systems.

The current state-of-the-art OTV vehicle offerings tend to advertise 1000's of km/s of delta-V, which enables the following services:

- Altitude change / planetary transfer – changing the altitude from where the vehicle was deployed by the launch vehicle. This could be an altitude raising or lowering operation, depending on the needs of the integrated payloads. In some cases, if escape velocity is reached, this maneuver can be used to transfer to lunar orbits or beyond. A payload delivery and deployment vehicle could perform multiple altitude raises and lowers during a single mission.
- Inclination change – changing the angle of the orbit with respect to the angle at which the vehicle was initially deployed by the launch vehicle. This can be done during the same burn as an altitude change to increase efficiency, or as a standalone operation. This operation dramatically changes what is viewable on the ground swath by the orbiting spacecraft. It can also dramatically change the solar incidence of the orbit.
- Phasing – this refers to changing the position of a spacecraft within a given orbit. Assuming the same orbit of two spacecraft, the phasing of each of them would dictate at what time each of them are over a specific ground swath. This is typically achieved with a "slow-down" burn to achieve a smaller orbital period, and then a "recircularization" burn to return the spacecraft to its original orbit.
- Constellation deployment – this is the ability for a single payload delivery vehicle to drop multiple spacecraft into a constellation formation. This dramatically reduces the propulsive need of the individual small spacecraft and can enable constellations of small spacecraft to achieve much more than they could if they needed the propulsion to deploy and phase themselves.

Many OTVs and OMVs are designed with a large excess propellant volume so that after the initial contracted services, the hosting vehicle can remain in orbit and be contracted for additional future services such as deorbiting via "pushing" another spacecraft, or contracted inspection services. Some providers are developing systems specifically for these in-space services, including Astroscale, Starfish Space, and Turion Space. Technology gaps to developing all these highly capable space vehicles include RF licensing, ACDS component availability, reliable propulsion systems, and lack of standardization for payload interfacing.

The following sections contain additional details about a select number of proven and upcoming OMVs and OTVs. The OMVs and OTVS listed here are not an exhaustive list of all those being developed, but they provide an overview of current state-of-the-art technologies and their development status. Additionally, Table 10-1 contains commercial OMV/OMT vehicle performance parameters. There was no intention of mentioning certain companies and omitting others based on their technologies.

Firefly Elytra

The Elytra vehicle line is a series of orbital vehicles offered by Firefly Aerospace. The smallest vehicle is the Elytra Dawn, optimized for intra-LEO mobility. The largest vehicle is the Elytra Dark, which offers payloads transportation to lunar orbit and beyond. The Elytra line builds off Firefly's previous Space Utility Vehicle design and experience, as well as the Spaceflight SHERPA (Firefly acquired Spaceflight in 2023). The first Elytra mission will fly aboard a Firefly Alpha rocket in 2024 in support of a National Reconnaissance Office (NRO) mission.

Epic Aerospace Chimera

Epic Aerospace has a line of Chimera systems ranging from smaller LEO vehicles up to larger GEO-targeting vehicles. The LEO range of products are intend to enable altitude changes from as small as 450 km from the deployment location to as large as intensive and fast phasing maneuvers (3 hours of LTAN change in less than 90 days is advertised, (23)). The GEO systems are advertising capabilities beyond Earth, with trans-lunar injection from GTO/LEO as an option. The first LEO Chimera system was launched in January of 2023 aboard a SpaceX Transporter and is currently operational. At the time of writing, the system is in the process of deploying its integrated CubeSats.

D-Orbit ION

The D-Orbit ION system is one of the most used OTV/OMV platforms on the market. The system was first used in 2020 to deploy Planet Labs constellation satellites. The system has persisted and has been used nine times as of writing to deploy small satellites, the most of any of the described OMV/OTV platforms.

Rocket Lab Proton

The Kick-Stage-derived Proton system is a flight-proven vehicle that deploys payloads to target orbits not otherwise achievable by the Electron launch vehicle. In addition, the system can be mounted on an ESPA port of other launch vehicles as a secondary payload.

Rocket Lab has flown three Proton systems so far. The third Proton mission successfully delivered the NASA CAPSTONE spacecraft to a near-rectilinear halo orbit around the Moon via a trans-lunar injection maneuver.

Exolaunch Reliant

The launch services, mission management, and space deployment hardware company Exolaunch is developing a series of OTV/OMVs. The series has a Standard and a Pro configuration, with the Pro configuration having more available payload mass, more powerful phasing capabilities, and the ability to modify orbital planes.

Nanoracks Outpost

The company Nanoracks is building off 10+ years of experience as a provider of airlock and CubeSat deployments for the ISS with a plan for their own in-space platform called *Outpost*. The idea of the platform is to transform repurposed upper stages into a system that provides power, attitude control, data, and communication services to commercial payloads through the implementation of a Nanoracks Mission Extension Kit (MEK). A technology demonstration of a portion of *Outpost* technology flew in 2022 called Mars Demo-1. The first commercial version of *Outpost* is scheduled to fly in 2024.

Atomos Quark

Atomos has plans to offer many in-space services to small spacecraft, including deployment and orbit raising, phasing, and inclination changes. The first version of their OTV/OMV, *Quark*, is planned to launch in 2024.

Momentus Vigoride

Momentus Space has developed an in-space orbit transfer service for SmallSats, named Vigoride. The maximum payload mass on Vigoride is 750 kg to LEO, and it can be launched from an ESPA or ESPA Grande ring, from ISS airlocks, or a launch vehicle. It uses water plasma engines to change the orbit prior to releasing payloads at their final orbit (24). Like all OMVs, the Vigoride is capable of changing inclination, altitude, and orbital planes. The first flight for Vigoride occurred in May of 2022. While the inaugural flight had issues due to failed solar array deployment, two additional flights of the system occurred in 2023 and were both successful. The company has additional Vigoride systems planned to fly in 2024.

Table 10-1: Commercial Orbital Transfer / Maneuvering Vehicles						
Company	**Heritage**	**Vehicle**	**Operational Altitudes**	**Maximum Payload Capacity**	**Delta-V**	**Reference**
Moog	Upcoming	SL-OMV	400 - 700 km	6 x 12 kg (+ additional, if containerized)	Up to 200 m/s	(25)
Moog	Upcoming	METEORITE	500 – 1200 km	100 kg (or more, with caveats)	>175 m/s	(26)
Moog	Upcoming	METEOR	500 – 1200 km	750 kg	>400 m/s	(27)
Momentus	Flown Successfully	Vigoride	250 - 2000 km (GEO and LLO also available)	750 kg	up to 2000 m/s	(28)
Epic Aerospace	Flown Successfully	Chimera LEO Block 0	+/- 450 km from LEO deployment	-	200 m/s at max payload	(23)
Epic Aerospace	Upcoming	Chimera LEO Block 1	+/- 450 km from LEO deployment	-	950 m/s at max payload	(23)
Epic Aerospace	Upcoming	Chimera GEO Block 0	GTO to TLI	-	1800 m/s at max payload	(23)
Epic Aerospace	Upcoming	Chimera GEO Block 0 / Bus	GTO to TLI	-	1800 m/s at max payload	(23)
Blue Origin	Upcoming	Blue Ring	Not disclosed	-	-	(29)
D-Orbit	Flown Successfully	ION	Not disclosed	-	-	(30)
Exolaunch	Upcoming	Reliant Standard	+/- 275 km from LEO(?) deployment	200 kg	-	(31)
Exolaunch	Upcoming	Reliant Pro	+/- 275 km from LEO(?) deployment	260 kg	-	(31)
Atomos	Upcoming	Quark	LEO, with higher destinations in the near term (2024)	400 kg	-	(32)
Rocket Lab	Flown Successfully	Photon	LEO, MEO, GEO, and beyond	-	-	(22)
Firefly	Upcoming	Elytra Dawn	LEO	-	-	(33)
Firefly	Upcoming	Elytra Dusk	LEO to GEO	-	-	(33)
Firefly	Upcoming	Elytra Dark	LEO to Lunar/Planetary	-	-	(33)
Impulse Space	Flown Successfully	Mira	LEO	300 kg	500 m/s at max payload	
TransAstra	Upcoming	WorkerBee (multiple configurations)	LEO, MEO, GEO, HEO, Lunar/Planetary	200 – 2000 kg	-	(34)

Information not disclosed represented as -

10.6 International Space Station Options

The International Space Station (ISS) provides several methods for deploying CubeSats and SmallSats. The sections below discuss SmallSat deployment from the ISS as well as deployment above the ISS. The ISS also accommodates hosted payloads for experiments, but those accommodations are outside the scope of this chapter as they are for individual payloads themselves and are not satellites.

10.6.1 Deployment from ISS

The ISS also provides several options for deploying satellites. Generally, satellites are launched below the ISS to avoid potential contact with the ISS. Below are several options available for launching from the ISS.

Nanoracks ISS CubeSat Deployer (NRCSD)

Nanoracks CubeSat Deployer (NRCSD) (figure 10.9) is a self-contained CubeSat dispenser system that mechanically and electrically isolates CubeSats from the ISS, cargo resupply vehicles, and ISS crew. The NRCSD is a rectangular tube that consists of anodized aluminum plates, base plate assembly, access panels, and deployer doors. The inside walls of the NRCSD are a smooth bore design to minimize and/or preclude hang-up or jamming of CubeSat appendages during deployment, should they become released prematurely.

For deployment, the platform is moved outside via the Kibo Module's Airlock and slide table, which allows the Japanese Experimental Module Remote Manipulator System (JEMRMS) to move the dispensers to the correct orientation and provides command and control to the dispensers. Each NRCSD can hold six CubeSat units as large as a 6U (1 x 6U). The NRCSD DoubleWide can accommodate CubeSats up to 12U (2 x 6U) with Nanoracks being able to launch up to 48U per cycle. The CubeSats deploy at a 51.6° inclination, 400 – 420 km orbit 1 to 3 months after berthing at the station.

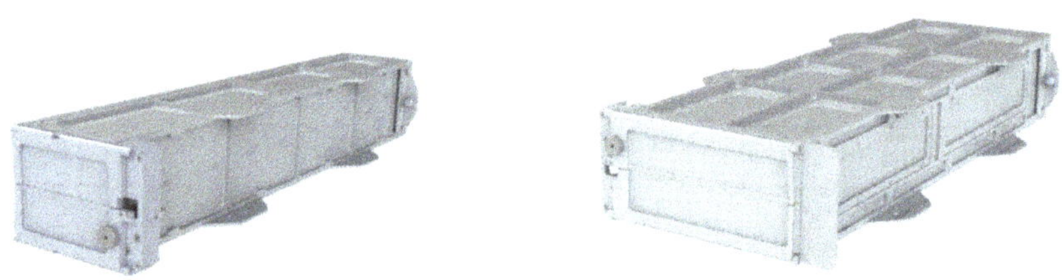

Figure 10.9: Nanoracks CubeSat Deployer. Credit: Nanoracks.

Nanoracks ISS MicroSatellite Deployment – Kaber Deployer Program

Nanoracks Kaber Microsat Deployer is a reusable system that provides command and control for satellite deployments into orbit from the Japanese Experimental Module Airlock Slide Table of the ISS. The Kaber supports satellites with a form factor of up to 24U and mass of 82 kg and uses a Nanoracks separation system with circular interface similar to the separation systems discussed above. Satellites are launched to the ISS on a pressurized launch vehicle, mounted to the Kaber deployer, and deployed outside the ISS (35).

JEM Small Satellite Orbital Deployer (J-SSOD)

The Japanese Experimental Module (JEM) Small Satellite Orbital Deployer (J-SSOD) is a Japanese Aerospace Exploration Agency (JAXA) developed CubeSat deployer used to launch CubeSats from the ISS. The J-SSOD can launch CubeSats up to the 6U form factor (2x3 configuration). The satellites, with their dispensers, are installed on the Multi-Purpose Experiment Platform prior to Kibo's robotic arm Japanese Experiment Module Remote Manipulator System (JEMRMS) transferring the Multi-Purpose Experiment Platform (MPEP) to the release location. At that point, the CubeSats are deployed (36).

Bishop Nanoracks Airlock Module

A new airlock module, Bishop, was developed for the ISS by Nanoracks, Thales Alenia Space, and Boeing, and is the first commercialized, private module for the space station (37). Bishop provides more than five times the volume of the current Japanese Experimental Module (JEM) airlock, allowing for larger satellites and payload experiments. Bishop can host satellites and payloads, as well as deploy them, based on the needs of the mission. It has been attached to the exterior of the ISS since December 21, 2020 and has been instrumental in deploying CubeSats from the ISS (38).

10.6.2 Deployment Above ISS

Regular access to the ISS is very attractive for many satellite providers. However, the lower altitude of the ISS means the in-orbit lifetime for the satellite is generally shorter. This section discusses the options that have been developed to deploy CubeSats above the ISS using a cargo resupply module.

Nanoracks Interchangeable CubeSat Launcher (Previously E-NRCSD)

The Nanoracks Interchangeable CubeSat Launcher (NICL) is a system to deploy CubeSats into orbit above the ISS by using the Northrop Grumman Cygnus ISS Cargo Resupply vehicle. The first mission to use the ENRCSD was on the OA-6 mission in March 2016; the updated E-NRSD design (NICL) was scheduled to have its first flight in March 2023, however the geopolitical situation between Ukraine and Russia has impacted the conops that would have enabled this demonstration. Specifically, the ISS program currently will not allow the Cygnus mission to boost above station to deploy CubeSats, so the NICL will be delayed until that changes.

In the past, up to 36U of CubeSats in any form factor up to 16U could be deployed above the ISS with each Cygnus mission. CubeSats are installed in the Nanoracks deployer and mounted externally to the Cygnus vehicle before launch. They remain external to the ISS for the duration of time that Cygnus is attached to the station. The deployment altitude is dependent upon the propellant margins remaining in the Cygnus but is typically 465-500 km, meeting a minimum of 45 km above the ISS altitude (39). It is hoped that this capability will return soon, allowing additional deployment options for CubeSats from the ISS.

SEOPS SlingShot

SEOPS SlingShot is a system to deploy CubeSats into orbit above the ISS using the Northrop Grumman Cygnus ISS Cargo Resupply vehicle. The first mission to use the SlingShot was in 2019. SlingShot can fly up to 72U of CubeSats per Cygnus mission; the largest CubeSat form factor it can fly is 12U. This deployment method differs from the ENRCSD in that the satellites and their dispensers are flown to the ISS as pressurized cargo on a resupply mission. Astronauts remove the satellites and install the dispensers onto the Cygnus Passive Common Berthing Mechanism (PCBM) just prior to Cygnus' departure from the station. Once Cygnus departs the ISS, it raises to an altitude of approximately 500 km and deploys the CubeSats (40). As these CubeSats are hosted in a different location and manner than the ENRCSD CubeSats, it is possible for Cygnus to carry CubeSats in both locations on a single mission. Due to the geopolitical situation in Europe, this capability is also on hold with hopes that it will return in the future.

10.7 On the Horizon

10.7.1 Integration

From a launch broker perspective, several companies have developed online booking systems for launches similar to web-based airline ticket platforms. Some companies, including SpaceX, even accept credit card payment options for launch services (29). The premise is that you click on your preferred destination and timeline and the website provides you with launch options. As the supply of launches increases, there will most likely be an increase in demand for this type of service.

10.7.2 Launch

As discussed in the launch section above, there are always several new launch vehicles in development. The number continues to grow every year, and how many become realized remains to be seen.

10.7.3 Deployment

There are several emerging capabilities for SmallSat deployment, including CubeSat dispensers, SmallSat separation systems, and orbital maneuvering and transfer vehicles. The technologies listed here are not a comprehensive list.

10.8 Summary

A wide variety of integration and deployment systems exist to provide access to space for small spacecraft. While leveraging excess LV performance will continue to be profitable into the future, dedicated launch vehicles and new integration systems for small spacecraft are becoming popular. Dedicated launch vehicles take advantage of rapid integration and mission design flexibility, enabling small spacecraft to dictate mission parameters. New integration systems will greatly increase the mission envelope of small spacecraft riding as secondary spacecraft. Advanced systems used to host secondary spacecraft in-orbit or support a rideshare to a dedicated orbit, can increase mission lifetime, expand mission capabilities, and enable orbit maneuvering for smaller spacecraft. In the future, these technologies may yield exciting advances in space capabilities. The expanding popularity of OMVs are bolstering these new opportunities.

These emerging launch capabilities associated with SmallSat structure cost efficiencies will continue to flourish in the SmallSat market, allowing more countries and new operators to procure, manufacture, and launch their own SmallSat systems. Many new launch providers are now offering spacecraft operations, testing and calibration services, and downstream services such as space-as-a-service, multispectral Earth observation, and onboard internet. These new launch providers will further increase the launch cadence of SmallSats. The previous few years have shown an increase in the number of available launch vehicles dedicated to small spacecraft. Additionally, the CubeSat Design Specification (CDS) has been revised to include the nanosatellite classification to 12U (6), which has led to the design of dispensers that can be accommodated on a variety of launch vehicles. Regardless of the evolution of the CDS, the dispenser and bus market are symbiotic and seems to be expanding.

For feedback solicitation, please email: arc-sst-soa@mail.nasa.gov. Please include a business email.

References

(1) Bryce and Space Technology. "SmallSat by the Numbers, 2023." [Online] [Accessed: September 1, 2023. Available at: https://brycetech.com/reports/report-documents/Bryce_Smallsats_2023.pdf

(2) NASA. "13 Companies to Provide Venture Class Launch Services for NASA," Jan 26, 2022. [Online] Available at: https://www.nasa.gov/press-release/13-companies-to-provide-venture-class-launch-services-for-nasa

(3) AIAA. AIAA R-099-2001. Recommended Practice Space Launch Integration. American Institute of Aeronautics and Astronautics, Reston, VA, 2001.

(4) E.M. Sims and B.M. Braun. "Navigating the Policy Compliance Roadmap for Small Satellites." s.l.: The Aerospace Corporation, November 2017.

(5) Department of Defense. DoD Instruction 3100.11: Management of Laser Illumination of Objects in Space. 2016.

(6) California Polytechnic State University, San Luis Obispo (Cal Poly) CubeSat Systems Engineer Lab. CubeSat 101 Basic Concepts and Processes for First-Time CubeSat Developers. s.l.: NASA CubeSat Launch Initiative, October 2017.

(7) NASA. Launch Vehicle Secondary Payload Adapter Rideshare Users Guide with Do No Harm (SMD SPA RUG with DNH) Rev3, 2023.

(8) E.S. Nightingale, L.M. Pratt, and A. Balakrishnan. "The CubeSat Ecosystem: Examining the Launch Niche (Paper and Presentation)". IDA Science and Technology Policy Institute. December 2015

(9) Puteaux, M and Najjar, A. "Analysis | Are smallsats entering the maturity stage?" August 6, 2019. [Online] Available at: https://spacenews.com/analysis-are-smallsats-entering-the-maturity-stage/

(10) D. Hill. "NASA Announces 12th Round of Candidates for CubeSat Space Missions," April 7, 2021. [Online] Available at: https://www.nasa.gov/feature/nasa-announces-12th-round-of-candidates-for-cubesat-space-missions

(11) European Space Agency. "New CubeSat Missions Selected for the Third Cycle of Fly Your Satellite," February 3, 2020. [Online] Available at: https://www.esa.int/Education/CubeSats_-_Fly_Your_Satellite/New_CubeSat_missions_selected_for_the_third_cycle_of_Fly_Your_Satellite

(12) Erwin, S. "Air Force cubesat successful deployed from Atlas 5 upper stage," August 8, 2019. [Online] Available at: https://spacenews.com/air-force-cubesat-successfully-deployed-from-atlas-5-upper-stage/

(13) The CubeSat Program, Cal Poly SLO. CubeSat Design Specification (CDS) Rev 14. San Luis Obispo, California, July 14, 2020.

(14) RUAG. "Payload Adapter Systems for EELV." [Online] Available at: https://www.ruag.com/sites/default/files/2016-11/payload_adapter_system_for_EELV.pdf

(15) RUAG. "Separation Nut PSM 3/8B." [Online] Available at: https://www.ruag.com/sites/default/files/media_document/2019-03/Separation%20Nut%20PSM%2038B.pdf

(16) ISISPACE. "M3S Micro Satellite Separation System." [Online] Available at: https://www.isispace.nl/wp-content/uploads/2018/09/M3S_Flyer.pdf

(17) Moog. "ESPA Overview." [Online] https://www.csaengineering.com/products-services/espa.html

(18) European Space Agency. SSMS modular parts. [Online] Available at: https://www.esa.int/ESA_Multimedia/Images/2019/06/SSMS_modular_parts#.XuOX3YQCe7o.link

(19) Andrews, J. "Spaceflight Secondary Payload System (SSPS) and SHERPA Tug - A New Business Model for Secondary and Hosted Payloads." 26th Annual AIAA/USU Conference for Small Satellites (2012).

(20) Northrop Grumman. "ESPAStar™ PRODUCT LINE Datasheet". 2022.

(21) Moog, Inc. [Online] "Meteor Space Vehicle Bus". 2023. Accessed: December 1, 2023. Available at: https://www.moog.com/products/space-vehicles/meteor.html

(22) Rocket Lab. [Online] "Photon". 2023. Accessed: December 1, 2023. Available at: https://www.rocketlabusa.com/space-systems/photon/

(23) Epic Aerospace. "Spacecraft: We are building an OTV for every mission," 2023. [Online] Available at: https://epic-aerospace-web.netlify.app/spacecraft

(24) Momentus. Vigoride User's Guide V1.0. [Online] February 23, 2020. Available at: https://momentus.space/site/wp-content/uploads/2020/02/Momentus-Vigoride-Users-Guide.pdf

(25) Moog. "SL-OMV: Small Launch Orbital Maneuvering Vehicle," datasheet. 2023. Available at: https://www.moog.com/content/dam/moog/literature/sdg/space/space-vehicles/moog-slomv-datasheet.pdf

(26) Moog. "Meteorite Space Vehicle Bus: ESPA Class Small Spacecraft Bus," [Online] Available at: https://www.moog.com/products/space-vehicles/meteorite.html

(27) Moog. "Meteor: Medium Class Spacecraft Bus," datasheet. 2023. Available at: https://www.moog.com/content/dam/moog/literature/sdg/space/space-vehicles/moog-meteor-datasheet.pdf

(28) Momentus, Inc. "Momentus at a Glance," Audited Financial Statements & Stable Road S-4A, March 8, 2021. [Online] Available at: https://ipo-edge.com/wp-content/uploads/2021/05/MOMENTUS_IPO_Edge_2021-05-04.pdf

(29) Blue Origin. "Blue Origin Unveils Multi-Mission, Multi-Orbit Space Mobility Platform," October 16, 2023. [Online] Available at: https://www.blueorigin.com/news/blue-origin-unveils-space-mobility-platform

(30) D. Werner. "D-Orbit Satellite Carrier delivers Planet SuperDoves to desired orbits," Spacenews.com October 28, 2020. [Online] Available at: https://spacenews.com/d-orbit-demonstrates-ion/

(31) ExoLaunch. "Reliant Models," [Online] Available at: https://exolaunch.com/reliant#5

(32) Atomos. "Get Your Place in Space: Services," [Online] Available at: https://atomosspace.com/capabilities

(33) Firefly Aerospace. "Orbit Elytra: On-Orbit Mobility and Services," [Online] Available at: https://fireflyspace.com/elytra/

(34) D. Werner. "TransAstra claims NASA contract for debris capture bag," Spacenews.com August 29, 2023. [Online] Available at: https://spacenews.com/transastra-claims-nasa-contract-for-debris-capture-bag/

(35) Nanoracks. Kaber Nanoracks ISS Microsatellite Deployment System. [Online] Available at: https://nanoracks.com/wp-content/uploads/NRMSD-overview.pdf.

(36) JAXA. JEM Small Satellite Orbital Deployer (J-SSOD). [Online] https://iss.jaxa.jp/en/kiboexp/jssod/.

(37) N. Natario. "Commercial Space Companies In Houston Looking Forward To SpaceX Launch." [Online] May 26, 2020. Available at: https://abc13.com/society/commercial-space-companies-looking-forward-to-spacex-launch-/6213469/

(38) A. Thompson. "The International Space Station is now home to the world's 1st commercial airlock". December 23, 2020. Accessed July 9, 2021. Available at: https://www.space.com/nanoracks-bishop-airlock-installed-space-station

(39) Nanoracks. External Cygnus Deployment. [Online] Available at: https://nanoracks.com/products/external-cygnus-deployment/

(40) NASA. SlingShot. [Online] Available at: https://www.nasa.gov/mission_pages/station/research/experiments/explorer/Facility.html?#id=7847

Chapter Contents

Chapter Glossary

(API)	Application Programming Interface
(ASGS)	ASRF SmallSat Ground Station
(ASRF)	Atmospheric Sciences Research Facility
(AWS)	Amazon Web Services
(C-STS)	Celestia Satellite Test & Simulation BV
(C2)	Command & Control
(CCSDS)	Consultative Committee for Space Data Systems
(CDMA)	Code Division Multiple Access
(CPAW)	Collection Planning and Analysis Workstation
(CRC)	Cooperative Centre
(CS)	Commercial Services
(DIU)	Department of Defense's Defense Innovation Unit
(DLR)	German Aerospace Center
(DSN)	Deep Space Network
(DSOC)	Deep Space Optical Communications
(DSS-17)	Deep Space Station-17
(DTE)	Direct-to-Earth
(DVB-S2)	Digital Video Broadcast Satellite Second Generation
(EDUs)	Engineering Development Units
(EGSE)	Electrical Ground Support Equipment
(EIRP)	Effective Isotropic Radiated Power
(ESA)	European Space Agency
(ESOC)	European Space Operations Centre
(FCC)	Federal Communications Commission
(FDMA)	Frequency Division Multiple Access
(FEP)	Front End Processors
(FFRDCs)	Federally Funded Research and Development Centers
(GBPA)	Ground-Based Phase Array
(GEO)	Geosynchronous Equatorial Orbit
(GNSS)	Global navigation satellite system
(GPS)	Global Positioning System
(GSOC)	Global Security Operations Center
(GSFC)	Goddard Space Flight Center

(HEO)	Highly Elliptical Orbit
(HF)	High frequency
(I&T)	Integration and Test
(IARU)	International Amateur Radio Union
(IDL)	Interactive Data Language
(IMEI)	International Mobile Equipment Identity
(INCOSE)	International Council on Systems Engineering
(INNOVA)	IN-orbit and Networked Optical Ground Stations Experimental Verification Advanced Testbed
(ISO)	International Organization for Standardization
(ISS)	International Space Station
(ITOS)	Integrated Test and Operations System
(ITU)	International Telecommunications Union
(JPL)	Jet Propulsion Laboratory
(KSAT)	Kongsberg Satellite Services AS
(KSC)	Kennedy Space Center
(LADEE)	Lunar Atmosphere and Dust Experiment Explorer
(LCRD)	Laser Communications Relay Demonstration
(LDT)	Lowell Discovery Telescope
(LEOP)	Launch and Early Orbit Phase
(LLCD)	Lunar Laser Communications Demonstration
(LNA)	Low-Noise Amplifier
(MCS)	Mission Control Software
(MEO)	Medium Earth Orbits
(MOC)	Mission Operations Center
(MSFC)	Marshall Space Flight Center
(MSPA)	Multiple Spacecraft Per Aperture
(NEN)	Near Earth Network
(NSN)	Near Space Network
(NICT)	National Institute of Information and Communications Technology
(NIMO)	Networks Integration Management Office
(NIST)	National Institute of Standards and Technology
(NORAD)	North American Aerospace Defense Command
(NTIA)	National Telecommunications and Information Administration

(OCTL)	Optical Communications Telescope Laboratory
(OGS)	Optical ground stations
(PNT)	Position Navigation and Timing
(PPM)	Pulse Position Modulation
(PPP)	Public-Private Partnership
(PPS)	Precise Positioning System
(R&D)	Research and Development
(RF)	Radio Frequency
(SA)	Single Access
(SDR)	Software Defined Radio
(SETH)	Science Enabling Technology for Heliophysics
(SFCG)	Space Frequency Coordination Group
(SMA)	S-band multiple access
(SN)	Space Network
(SNSPD)	Superconducting Nanowire Single Photon Detector
(SOC)	Science Operations Center
(SPD-5)	Space Policy Directive 5
(SSBV)	Satellite Services B.V.
(SSC)	Swedish Space Corporation
(SWaP)	Size, Weight, and Power
(TDMA)	Time-Division Multiple Access
(TDRS)	Tracking and Data Relay Satellites
(TDRSS)	Tracking and Data Relay Satellite System
(TLE)	Two-Line Element set
(TNC)	Terminal Node Controller
(TNO)	The Netherlands Organization
(TOGS)	Transportable Optical Ground Station
(TT&C)	Telemetry, Tracking and Control
(UHF)	Ultra-High Frequency
(USRP)	Universal Software Radio Peripheral
(VHF)	Very high frequency
(VICTS)	Variable Inclination Continuous Transverse Stub
(VMs)	Virtual Machines
(WFF)	Wallops Flight Facility

11.0 Ground Data Systems & Mission Operations

11.1 Introduction

The ground segment is a critical part of the end-to-end science data return, and it includes all the ground-based elements that are used to collect and disseminate information from the satellite to the user (figure 11.1). The primary elements of a ground system are summarized in table 11-1.

There are exciting changes in the government and commercial sector ground stations and services, and the shifting synergy between these. From its inception in 1958, whenever NASA needed to receive data from one of its Earth observing satellites or talk to its astronauts in orbit, it used equipment and services it had needed to develop and build itself. Over time, commercial enterprise acquired the proficiencies necessary to reliably and securely communicate with objects in low-Earth orbit (LEO), services NASA is now pursuing to purchase as any other near-Earth space customer.

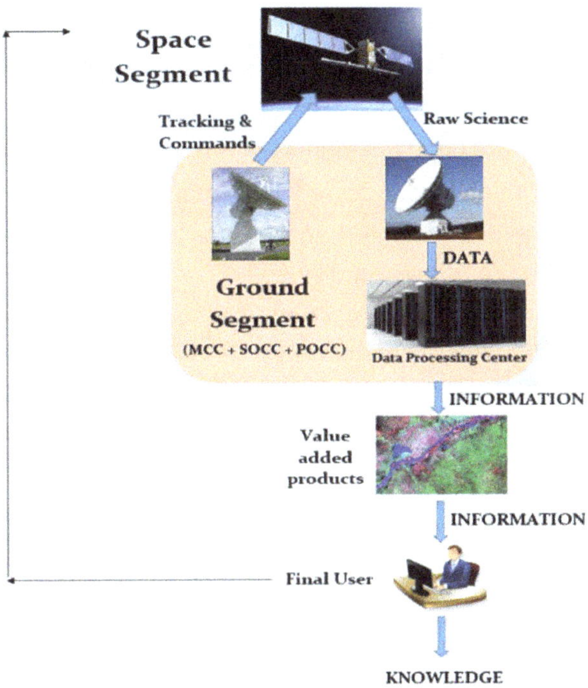

Figure 11.1: Functional relationship between the space segment, ground segment and final user for a small satellite mission. Credit: NASA.

The agency combined NASA's Near Earth Network (NEN) and NASA's Space Network (SN) into NASA's Near Space Network (NSN) in October of 2020. To support the commercialization initiative, NASA plans to have increased reliance on industry-provided communications services for missions close to Earth by 2030 (59). As of 2023, commercial providers do not service the Sun-Earth Lagrange Points or Deep Space, thus the Deep Space Network (DSN), and large NSN assets (18 m) continue to play a critical and needful role in returning science data from these regions for Heliophysics, Astrophysics and Planetary Science directorates.

Table 11-1: Primary Elements of a Ground System	
Element	**Function**
Ground Stations	Telemetry, tracking, and command interface with the spacecraft
Ground Networks	Connection between multiple ground elements
Control Centers	Management of the spacecraft operations
Remote Terminals	User interface to retrieve transmitted information for additional processing

The NSN provides Direct-to-Earth (DTE) services via a global system of commercial and NASA-owned ground stations that provide line of sight communications and tracking services to missions ranging from low-Earth orbit and extending to Sun-Earth Lagrange Points 1 & 2. These services are augmented by Space Relay services via relay satellites in geosynchronous orbit.

The ground segment design can depend on several factors which may include, but are not limited, to the following:

- Data volume to satisfy mission requirements
- Location of the ground assets relative to mission orbit parameters
- Budget limitations
- Distribution of the team
- Affiliation of who controls the spacecraft (federal vs. non-federal users)
- Regulatory requirements
- Latency requirements

The ground system is responsible for collecting and distributing the most valuable asset of the mission: the data. Using the proper ground system is key to mission success.

All small satellites use some form of a ground segment to communicate with the spacecraft, whether it be hand-held radios using an amateur frequency, or a large dish pulling down data on a non-federal or federal frequency. The commercial marketplace for Telemetry, Tracking and Commanding (TT&C) services continues to expand and has reached a maturity to enable commercialization of Direct-to-Earth (DTE) radio frequency communications services. NASA is encouraging a growing commercial market by leveraging commercial capabilities to increase efficiency and robustness of ground networks. In addition, NASA plans to enhance its communications capabilities to provide near-continuous communications support to the Artemis lunar missions through Communication Relay and Navigation services in Lunar space.

11.2 Ground Systems Architecture

A typical small satellite mission has the following elements within the ground system architecture:

- Ground Station Terminal: Transmitter and receiver or transceiver at the ground station to transmit and receive information, including related hardware such as antennas. These may be in a Radio Frequency (RF) or in an optical wavelength.
- Mission Operations Center (MOC):
    - Commands the spacecraft
    - Monitors spacecraft performance
    - Requests and retrieves dataas necessary
- Science Operations Center (SOC):
    - Generates and disseminates science data products
    - Determines science operations to be relayed to the MOC
- Ground Station Data Storage and Network:
    - Provides live connectivity to a MOC for commands and telemetry
    - Temporarily stores data to be retrieved by the MOC and/or SOC

Figure 11.2 shows a generic small satellite ground architecture that uses NASA's Near Space Network (NSN) for nominal ground passes and the NASA Space Network (SN) for low-latency messaging.

In this architecture, the MOC is responsible for all communication to and from the spacecraft, while the SOC and engineering teams can work both directly through the MOC to process commands. This is especially helpful during commissioning and troubleshooting instances where the engineering team needs direct access to the flight system. This architecture also provides a separate database generated from the MOC of telemetry and housekeeping data that is accessible to stakeholders.

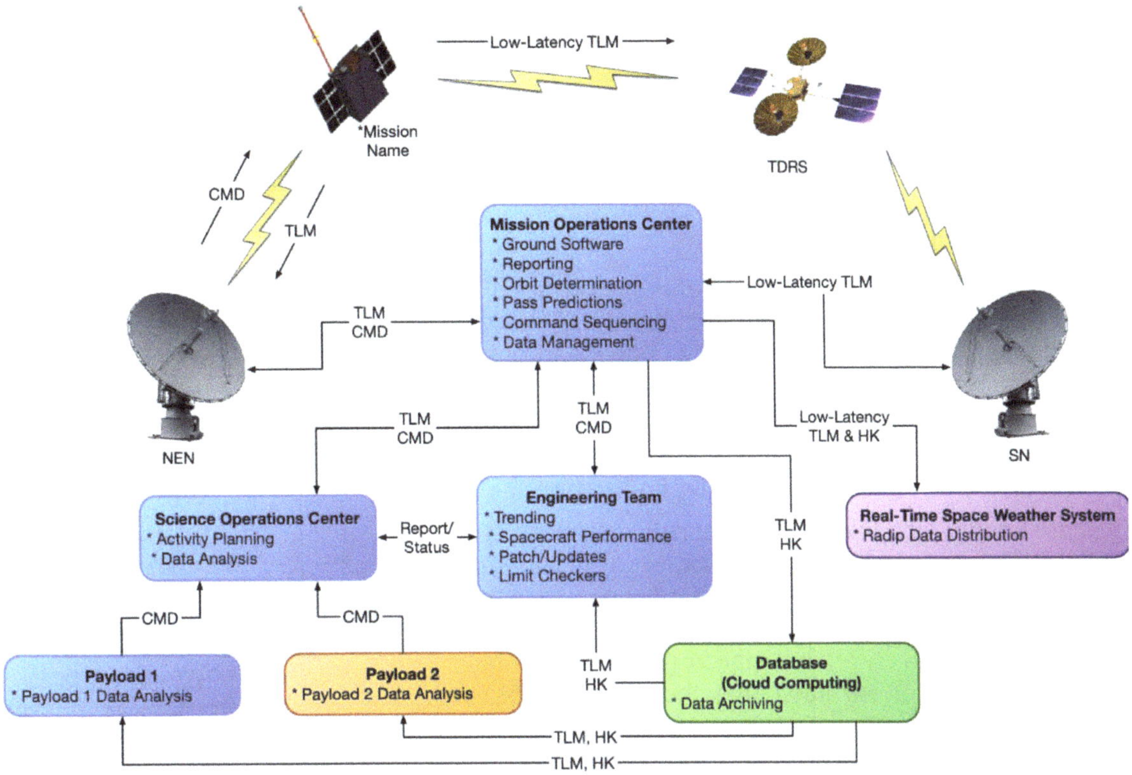

Figure 11.2: Example of a ground system architecture for a small satellite using NASA's Near Space Network. Credit: NASA.

11.2.1 Types of Communication Infrastructures

Communications services may be either Direct-to-Earth (DTE) or augmented by space relay. DTE ground stations provide direct point-to-point access with antennas at ground stations which are strategically located and equipped with telemetry, command, and tracking services. DTE antennas for NASA small satellites are high gain parabolic dish antennas used to support S, X, and Ka bands, while some universities still use parabolic or UHF Yagi antennas. DTE ground stations could also incorporate phased array antenna systems or equipment for optical communications. The DTE services are especially effective for missions needing frequent, short-duration contacts with high data throughput. They are also capable of handling longer latency durations due to orbital dynamics and station visibility.

Space relay services involve an intermediate satellite that communicates with a ground station on the Earth's surface. Relay communication satellites for low-Earth orbit spacecraft can be in Geosynchronous Equatorial Orbit (GEO), about 36,000 km from Earth, or in low-Earth orbit. Relays are essential for providing communication and tracking when direct-to-ground communications are not feasible due to physical asset visibility constraints. It is common for a low-Earth orbit spacecraft to only be in a DTE ground station's line of sight for a portion of the orbit. The addition of space-based relay assets can provide missions with full-time coverage and continuous access to communication and tracking services. They are most useful for missions that need continuous coverage, low latencies, and coverage of launch, critical events, or emergencies.

Communication with DTE ground stations can achieve much higher data rates than what is possible for space-based relays. When considering a GEO relay satellite, it can be ten times the distance from the low-Earth orbit spacecraft than the DTE ground station. With communication propagation losses being a function of the reciprocal of the distance squared, the same communications system can achieve orders of magnitude higher data rates with the DTE ground station. Achieving comparative data rates for a relay system would require a significant increase in power. The current low-Earth orbit relays have hardware limitations that permit data rates of 9.6 kbps or less, which is low relative to SmallSats being able to achieve 3 Mbps or more with DTE ground stations.

11.3 Frequency Considerations

The spacecraft transceiver and ground station need to be on a coordinated frequency to communicate. Selecting transmit and receive frequencies are a critical part of the spacecraft communications system design process. Frequencies are divided into different bands as shown in table 11-2. See a list of supported frequencies per ground station in their specific sections.

Typical bands considered for small satellites and therefore ground stations are Ultra High Frequency (UHF), S, X, and Ka. UHF was the band of choice for early small satellites, but in recent years, there has been a shift to S and X and Ka. A ground station needs to maintain antennas and receivers such that the ground receive matches the space segment's transmit frequency and vice versa. Since Transmit (Tx) and Receive (Rx) have different key drivers and requirements, many ground stations are dual or tri-band.

Table 11-2: Frequency Bands

Band	Frequency
HF	3 to 30 MHz
VHF	30 to 300 MHz
UHF	300 to 1000 MHz
L	1 to 2 GHz
S	2 to 4 GHz
C	4 to 8 GHz
X	8 to 12 GHz
Ku	12 to 18 GHz
Ka	27 to 40 GHz
V	40 to 75 GHz
W	75 to 110 GHz
mm	110 to 300 GHz

Ground Station Receive (Spacecraft Return, Telemetry)

Ground station receive frequencies are mostly S and X band from a LEO/ GEO orbit, and X and Ka band from deep space. Ka band has been implemented for transmit and is NASA's desired band for future small satellite missions. This shift has been driven by higher data return demands and frequency control. The higher frequencies permit more data to be transmitted over a given period but require more stringent pointing. UHF is appealing to some universities, due to the lower cost of hardware for both the spacecraft and ground station, good link margins, and more omni-directional pattern capability with the spacecraft but yields lower data rates and has a higher probability for interference. Higher frequencies provide wider bandwidths, and the matching antennas have narrower beamwidths or are arrayed for a higher gain, thus more stringent pointing is required.

Ground Station Forward (Spacecraft Commanding)

The key driver to successfully command a satellite from the ground, is the ability to reach it. The most critical period is after a satellite's release from the launch vehicle, at which point the satellite does not yet have full control over its attitude, thus the most important thing is a wide beamwidth for the spacecraft receiving antenna(s) in the selected frequency. For this reason, ground stations are designed with more power and Low Noise Amplifiers (LNA) to counter the low gain, ideally omni-directional single patch receive antennas in the lower frequency bands.

11.3.1 Frequency Selection: Link Budget

Calculating the RF link budget is the first step when designing a telecommunications solution. It is a calculation of the end-to-end performance of the communications link with the constraint of maintaining a required link margin. Maintaining a 3 dB link margin is adequate for data return from a satellite in low-Earth orbit at a slant range of 1,500 km. Usually commanding to a Near Earth orbit has plenty of margin because of the high power and aperture size of the ground station, and the lower required data rate on the account of the commands' low volume. When considering deep space communication, a 3 dB link margin is desired, but for distant spacecraft, such as New Horizons at 7 billion kilometers from Earth, 1 dB or less margin may be all that is practically possible. The budget calculation adds and subtracts all the power gains and losses that a communication signal will experience within the system. Factors such as uplink amplifier gain and noise, transmit antenna gain, slant angles and corresponding free space loss, satellite transceiver noise levels and power gains, receive antenna and amplifier gains and noise, cable losses, and atmospheric attenuation are considered. There is a duality to frequency effects: free space loss over the same range is less for lower frequencies; however, the wavelength is much smaller for higher frequencies, thus a same size ground aperture provides a much higher Gain over temperature (G/T). On the spacecraft end, a multi-element high-gain Ka-band antenna array for example fits in the palm of a hand. For high volume data return, which is where communications bottlenecks occur, higher frequencies are desirable – all the way up to optical wavelengths at 1550 nm (see section 11.10.1, *Free Space Optical Communications*).

11.3.2 Frequency Licensing

RF communication frequencies are intentionally protected. Within each frequency band there are government and non-government designations amongst the frequencies. Some frequencies are government use only, others are non-government only, and some are shared. Government bodies that regulate the frequency usage in the United States are the Federal Communications Commission (FCC) and the National Telecommunications and Information Administration (NTIA). Other countries may have their own national governing bodies, and all national bodies around the world must coordinate with the International Telecommunications Union (ITU), which is the governing body at the international level. The FCC is responsible for issuing communications licenses to non-government users and the NTIA handles government users. Licenses are required for both the satellite and ground station to transmit on a designated frequency or frequencies. It is becoming more common for small satellites to use multiple bands. For example, some missions have used UHF for uplink and S- band for downlink, while others have used S-band for uplink and X-band for downlink. Some of the non-government frequencies are dedicated for amateur usage.

Early university small satellites relied heavily on the use of amateur frequency bands. In recent years, there has been movement by the International Amateur Radio Union (IARU) and the FCC to significantly limit the use of amateur frequencies for small satellites. Those interested in using these frequencies are expected to first communicate their intention with the IARU and obtain a coordination letter prior to submitting an application with the FCC. It is recommended that missions with a new communication system design apply with the FCC or NTIA once an operations concept and a spacecraft design are defined, in order to verify a proper communications approach and associated hardware has been selected. Missions using a legacy communications approach can typically wait until they have been given a launch manifest. The licensing process can take several months and needs to be completed prior to launch. Some of the processing time is associated with the FCC and NTIA having to also coordinate with the ITU. Both the FCC and ITU are working to implement more streamlined small satellite licensing options. Such improvements will be necessary as constellations of small satellites become more prevalent.

BODY-333

11.4 Ground Segment Services

Ground segment services may include the below four categories. The NSN is a full-service ground station network and offers all four major service categories. Not all commercial services offer all services.

1) Mission Integration – this includes development of service agreements, interfaces, documentation, support of reviews, etc.

2) Mission Planning and Scheduling – this includes performing link and loading analyses, supporting service requests, and generating and implementing operational schedules.

3) User Mission Data Transfer – this includes primarily spacecraft forward command and return telemetry data.

4) Position, Navigation and Timing (PNT) – this includes navigation.

Position information (4) is critical for commanding the spacecraft (3). Commanding may be scripted by the mission and is actuated through ground services. Challenges are usually associated with the initial satellite-to-ground station link closure. Typically, two-line elements (TLE) or state vectors are established and shared by the launch provider after deployment. This information can be used to create an initial orbit solution for ground station antenna pointing. Low-Earth orbit missions can use North American Aerospace Defense Command (NORAD) TLE data (see https://www.space-track.org) for satellite location. However, it could take up to a week or more for NORAD to add the new object to their tracking list. This process could be delayed further if multiple spacecraft are ejected in close proximity, and it may not be clear which NORAD element set corresponds to which spacecraft. It is not uncommon to spend weeks attempting contact with different NORAD-tracked objects until the correct one is found. The position prediction accuracy based on the NORAD TLE also diverges over time and a new TLE will be needed to maintain data link. This is typically not an issue since the TLE is updated regularly, but on-board Global Positioning System (GPS) data (if equipped) can help determine the orbital parameters for the ground station to define latest orbital parameters.

Another method is to locate the satellite as it rises from the horizon. Ground station operators can point a directional antenna 5-10 degrees above the horizon to detect the satellite and synchronize with the radio. Most antenna tracking software will commence automatic tracking after the initial acquisition is successful. A half-duplex or full-duplex system could make a difference as well. Program track instead of auto-track is used for half-duplex. With a full-duplex system, the ground antenna attempts to acquire the downlink first. Predicts (NORAD or state vectors) are still used to initially acquire the spacecraft. If the predicts are off, the antenna can initiate a mechanical scan to increase the search area. Once the downlink is acquired, the ground antenna can auto-track and automatically point at the satellite for the duration of the pass. Additional passes are scheduled during spacecraft and payload commissioning. Table 11-3 describes NSN's transport and tracking capabilities.

Table 11-3: NSN Interfaces and Capabilities		
Interface/ Capability[1]	**Direct to Earth**	**Space Relay**
Terrestrial Link Data Transport Capabilities		
Data Storage [1]	Station Storage: 5-30 days Cloud-based: Mission-driven	7 days
Network Data Rate [1]	Mission-driven (up to 1.2 Gbps)	
SLE Protocols	F-CLTU, EF-CLTU (Forward) RAF, RCF, ROCF (Return)	
SLE Versions Supported [2]	CCSDS 910.4, CCSDS 911.1, CCSDS 911.2, CCSDS 911.5, CCSDS 912.1, CCSDS 912.11, CCSDS 912.3, CCSDS 913.1	
Offline-Data Transfer	CFDP, SFTP	
Security	Trusted Networks (Access Controls, Firewalls, Authentications, etc.)	
Spacecraft Navigation Tracking Capabilities		
Radiometric Tracking Services [1]	Tone Ranging 1-way or 2-way Doppler Antenna Angle Data	Spread Spectrum Ranging 1-way or 2-way Doppler Antenna Angle Data
Radiometric Measurement Accuracy [1]	Range: S-band: < 5 meters, 1σ Doppler (Range-Rate): S-band 1-way: ≤ 30 mm/s, 1σ S-band 2-way: ≤ 15 mm/s, 1σ X-band 1-way: ≤ 7 mm/s, 1σ Ka-band 1-way: ≤ 2 mm/s, 1σ Antenna Angles: S: 0.03°, X: 0.05° Ka: 0.01° (auto), 0.05° (program)	Range: ≤ 2.73 meters, 1σ Doppler (Range-Rate): 1-way ≤ 1.55 mm/s, 1σ 2-way ≤ 3.1 mm/s, 1σ Antenna Angles: ≤ 0.1°
Radar Tracking Service Bands	C-band (5.4-5.9 GHz) Single Object X-Band (10.499 GHz) Multi Object	N/A
Radar Tracking Loop Gain (dB)	C-Band: 212-245 (227 Typical) X-Band: 246 (nominal)	
Other [1]	Ground Antenna Slew Rate: Azimuth and Elevation: ≥ 10°/sec (10°/sec^2) * Train: ≥ 5°/sec (5°/sec^2) * WS1 18-m system ≥ 2°/sec (1°/sec^2)	Time Transfer Measurement: User Spacecraft Clock Calibration System: ≤ ±5 μs Return Channel Time Delay: ±25% of a bit period

[1] Services and performance (Data Rates, EIRP, G/T, etc.) are not uniform across assets.
[2] Additional capabilities above those listed could be supported as well.
[3] NASA may consider adding technologies not currently on its roadmap.
[4] 2nd and 3rd Generation TDRS only.

Another critical time in the life of a spacecraft is commissioning; either commissioning of the spacecraft bus, or commissioning of science instruments, including in-space calibration. During commissioning phases, additional time and support personnel are typically scheduled (1).

11.4.1 Ground Networks – NASA and Partners

The ground stations, MOC, SOC, and the supporting infrastructure connecting them together, make up a ground network. Ground station antenna dish diameters, LNAs, frequency feeds, station gain over temperature (G/T) requirements are carefully selected for each network and are optimized for targeted ranges. NASA's NSN ground network provides services to satellites up to 2 million km range from Earth; NASA owns and JPL maintains the DSN for missions beyond two million km, including planetary.

At NASA's Goddard Space Flight Center, the Exploration and Space Communications (ESC) projects division oversees the operations, maintenance and advancement of the Space Communications and Navigation (SCaN) program office's NSN. Operating at a high-level of reliability and proficiency, the NSN provides communications and navigation services for missions within 2 million kilometers of our planet, bringing down an average of almost 30 Terabytes of critical data daily. Through space relays and ground-based assets, The NSN provides data delivery and satellite tracking services, empowering new discoveries about the universe and our home planet. JPL is responsible for managing and maintaining the DSN.

NASA Near Space Network

"The newly established NSN is more than just an aggregation of the NEN's and SN's space-based technologies, ground stations and antennas; it's the network through which NASA and other space users will now arrange for support services for their near-Earth missions. Critically, those support services may be provisioned through government or commercial network assets in a way that is seamless to users—a cornerstone in SCaN's effort to incorporate increasing levels of commercial service while ensuring mission needs are met." (2)

The NSN provides direct-to-earth telemetry, commanding, ground-based tracking, and data and communications services to a wide range of customers. The network consists of NASA, commercial, and partner S-band, X-band, and Ka-band ground stations supporting spacecraft in low-Earth orbit, GEO, Highly Elliptical Orbit (HEO), Lunar orbit, and Lagrange point L1/L2 orbit up to one million miles from Earth. The NSN supports multiple robotic and launch vehicle missions with NASA-owned stations and through cooperative agreements with interagency, international, and commercial services. Table 11-4 shows the radio frequencies that the NSN supports via the NTIA.

Table 11-4: NSN Supported Radio Frequencies and Bandwidths

Band	Function	Frequency Band (MHz)
S Uplink	Earth to Space	2,025 – 2,110
X Uplink	Earth to Space	7,190 – 7,235 (Two NEN sites to 7,200)
S Downlink	Space to Earth	2,200 – 2,300
X Downlink	Space to Earth, Earth Exploration	8,025 – 8,400
X Downlink	Space to Earth, Space Research	8,450 – 8,500
Ka Downlink	Space to Earth	25,500 – 27,000

A comprehensive list of Forward and Return capabilities per frequency are in Table 11-5. Systems are compliant with most CCSDS recommendations. The NSN consists of geographically-dispersed ground stations operated by NASA and its commercial partners (figure 11.3).

Government

- NASA's Alaska Satellite Facility, Fairbanks — Supports: S/X Band — Assets: 11.3m, 11m, 9.1m
- NASA's Kennedy Uplink Station — Supports: S-band - Assets: 6.1m
- NASA's Ponce de Leon Station — Supports: S-band - Assets: 6.1m
- NASA's Wallops Ground Station (GS), Virginia — VHF, S/X Band — Assets: 11m/5m
- NASA's White Sands GS, New Mexico — Supports: VHF, S/Ka Band — Assets: 18.3m
- NASA's White Sand Complex, New Mexico — Supports VHF, S/Ka Band — Assets: 11m
- NASA's McMurdo Ground Station, Antarctica — Supports: S/X Band — Assets: 10m
- Fairbanks Command and Data Acquisition Station (NOAA partnership), Gilmore Creek, Alaska

Figure 11.3: NSN Global Ground Station Locations. Credit: NASA

Commercial

- KSAT Singapore — Supports: S/X Band — Assets: 9.1m
- KSAT Svalbard, Norway — Supports: S/X Band — Assets: 11.3m/11.3m/13m
- KSAT TrollSat, Antarctica — Supports: S/X Band — Assets: 7.3m/7.3m
- KSAT Punta Arenas – Supports: S/X Band — Assets: 11.5m
- KSAT Inuvik -- Supports: S/X Band — Assets: 13m
- SANSA Hartebeesthoek, South Africa — Supports: S/X Band — Assets: 12m/10m
- SSC Kiruna, Sweden — Supports: S/X Band — Assets: 13m/13m
- SSC Santiago, Chile — Supports: S Band — Assets: 9m/12m/13m
- SSC Space US North Pole, Alaska — Supports: S/X Band — Assets:

5m/7.3m/11m/13m

- SSC Space US Dongara, Australia — Supports: S/X Band — Assets: 13m
- SSC Space US South Point, Hawaii — Supports: S/X Band — Assets: 13m/13m

Table 11-5: NSN Direct to Earth Command and Telemetry Capabilities per Frequency

Interface/ Capability[1]	Direct to Earth	Space Relay
Forward (Command) Communications		
Frequency Bands (Near-Earth Use)	S-band: 2025-2110 MHz X-band: 7190-7235 MHz	S-band: 2025-2110 MHz Ku-band: 13.775 GHz Ka-band: 22.55-23.55 GHz [4]
Maximum Bandwidth	S-band: 5 MHz X-band: 10 MHz	S-band: 6 MHz Ku-band: 50 MHz Ka-band: 50 MHz [4]
Forward Max Data Rate [1,2] (prior to encoding)	S-band: 5 Mbps X-band: 5 Mbps	S-band MA: 300 Kbps S-band SA: 4.2 Mbps Ku-band: 50 Mbps Ka-band SA: 50 Mbps [4]
Antenna System EIRP (dBW) [1]	S-band: 51-81 (56 Typical) X-band: 85-86	S-band MA: 42 [4] S-band SA: 48.5 [4] Ku-band SA: 48.5 [4] Ka-band SA: 63 [4]
Modulation [2,3]	PM, FM, PCM, PCM/PM, PCM/PSK/PM, BPSK, QPSK, OQPSK, UQPSK	Spread spectrum: BPSK or UQPSK Non-spread: BPSK, QPSK, OQPSK, PCM/PM, or PCM/PSK/PM
Encoding [2,3]	Uncoded, or LDPC ½ or 7/8	Uncoded, Rate ½ Conv., Reed-Solomon, Concatenated (½ Conv. + RS), LDPC ½ or 7/8
Polarization	Circular (LHC, RHC)	Circular (LHC, RHC) (LHC only for MA services)
Return (Telemetry) Communications		
Frequency Bands (Near-Earth Use)	S-band: 2200-2290 MHz X-band: 8025-8400 MHz X-band (SRS): 8450-8500 MHz Ka-band: 25.5 – 27 GHz [4]	S-band: 2200-2290 MHz Ku-band: 15.0034 GHz Ka-band: 25.25 – 27.5 GHz [4]
Maximum Bandwidth	S-band: 5 MHz X-band: 375 MHz X-band (SRS): 10 MHz Ka-band: 1500 MHz	S-band (MAR & SAR): 6 MHz Ku/Ka-band: 225 MHz [4] Ka-band (Wide): 650 MHz [4]
Return Max Data Rate [1,2] (prior to encoding)	Rates will vary – examples: S-band: 2.2 Mbps (PACE) X-band: 220 Mbps (ICESat-2) X-band (SRS): 13.1 Mbps (IRIS) Ka-band: 3.5 Gbps (NISAR)	S-band MA: 1 Mbps S-band SA: 14.1 Mbps Ku/Ka-band: 600 Mbps [4] Ka-band (Wide): 1200 Mbps [4]
Antenna System G/T (dBW) [1]	S-band: 19.1-29.6 (21 Typical) X-band: 30.5-37.8 (32 Typical) Ka-band: 38-45 (41.3 Typical)	S-band MA: 3.2 (for LEO) S-band SA: 9.5 (for LEO) Ku-band: 24.4 (for LEO)

		Ka-band: 26.5 (for LEO)[4]
Demodulation [2,3]	PM, FM, PCM, PCM/PM, PCM/PSK/PM, BPSK, QPSK, OQPSK, AQPSK, SQPN, 8PSK	Spread spectrum: BPSK or UQPSK Non-spread: BPSK, QPSK, OQPSK, PCM/PM, or PCM/PSK/PM
Decoding [2,3]	Uncoded, Rate ½ Conv. and/or Reed-Solomon, LDPC ½ or 7/8, or Turbo Rate ½	Uncoded, Rate ½ Conv., Reed-Solomon, Concatenated (½ Conv. + RS), LDPC ½ or 7/8, Rate 7/8 TPC
Polarization	Circular (LHC, RHC)	Circular (LHC, RHC) (LHC only for MA services)

[1] Services and performance (Data Rates, EIRP, G/T, etc.) are not uniform across assets.
[2] Additional capabilities above those listed could be supported as well.
[3] NASA may consider adding technologies not currently on its roadmap.
[4] 2nd and 3rd Generation TDRS only.

While NASA's NSN is often reserved for NASA-funded missions, other ground network options exist for non-government-funded satellite operators. One common option, especially amongst amateur operators, is to take advantage of the UHF and VHF amateur network around the world.

The NSN is exploring how to provide higher data rates for CubeSat missions with techniques such as Digital Video Broadcast Satellite Second Generation (DVB-S2). Higher data rates either increase science return or reduce the number of minutes per day of required ground station contacts. Higher data rates also enable mother-daughter small satellite constellations, where the mother spacecraft handles the communication with Earth for multiple daughter spacecraft. Functions such as Multiple Satellite per Aperture (MSPA) are planned to be implemented on the Lunar Exploration Ground Sites (LEGS) mission (see the State-of-the-Art section).

The NSN facilitates Commercial Services (CS) and negotiated a bulk-buy discount for all NASA missions. This allows for contacts on the NSN Contractor/University Operated and CS apertures to be at no-cost for NASA missions. The NSN does schedule CS in accordance with NASA mission-defined priority. The Networks Integration Management Office (NIMO) at NASA GSFC is the liaison for customers that wish to use NSN services. NIMO has a variety of services and capabilities available and can coordinate support from providers throughout NASA, other US agencies, US commercial entities, and foreign governments. Some of the services that NIMO can provide include:

- Requirements Development
- Communications Design Support & Guidance
- Optical Communications Analysis
- Network Feasibility Analysis
- Spectrum Management
- RF Compatibility Testing
- Launch Support

Network Feasibility Analysis includes determining NSN station loading as a function of the mission's priority and determining the availability of planned stations for the contacts requested. Prior to the mission deployment, the NSN commits to providing the requested stations and contact time as outlined in the network feasibility analysis.

For new customer mission service requests please fill out the NSN Service Inquiry Form at:

http://go.nasa.gov/NSNServiceInquiry.

If interested in more information on using the Near Space Network (NSN), please also refer to https://esc.gsfc.nasa.gov/projects/NSN.

NASA Deep Space Network

The DSN is optimized to conduct telecommunication and tracking operations with space missions in GEO. This includes missions at lunar distances, the Sun-Earth LaGrange points, and in highly elliptical Earth orbits, as well as missions to other planets and beyond. The DSN has supported, or is currently supporting, missions to the Sun as well as every planet in the Solar System (including dwarf planet Pluto). Two missions (Voyager I and Voyager II) have reached interstellar space and still communicate with the DSN. The DSN offers services to a wide variety of mission customers, as shown in table 11-6.

Table 11-6: DSN Customers, Mission Characteristics, Frequencies, and Services

Customers	**Mission Phases**
• NASA • Other Government Agencies • International Partners	• Launch and Early Orbit Phase (LEOP) • Cruise • Orbital • In-Situ
Mission Trajectories	**Frequency Bands** – Includes Near-Earth and Deep Space Bands, Uplink and Downlink, Command, Telemetry, and Tracking Services
• Geostationary or GEO • HEO • Lunar • LaGrange • Heliocentric • Planetary	• S-Band (2 GHz) • X-Band (7, 8 GHz) • Ka-Band (26, 32 GHz)

DSN services include:

- Command Services
- Telemetry Services
- Tracking Services
- Calibration and Modeling Services
- Standard Interfaces
- Radio Science, Radio Astronomy and Very Long Baseline Interferometry Services
- Radar Science Services
- Service Management

Custom and tailored DSN services can also be arranged for missions and customers. DSN-provided data services are accessed via well-defined, standard data and control interfaces:

- The CCSDS
- The Space Frequency Coordination Group (SFCG)
- The ITU
- The International Organization for Standardization (ISO)

- De facto standards widely applied within industry
- Common interfaces specifiedby the DSN

The use of data service interface standards enable interoperability with similar services from other providers.

Figure 11.4 shows the DSN antennas and their locations. Each DSN ground station in California (United States), Madrid (Spain), and Canberra (Australia) currently as of 2021 was operating four 34 m Beam Wave Guide antennas and one 70 m antenna. By the late 2020s, this is planned to increase to include one 70 m plus four 34 m antennas at each DSN site.

Figure 11.4: DSN antennas and their locations. Credit: NASA.

The DSN supports RF testing using the following facilities:

- Development and Test Facility (DTF-21), located near NASA Jet Propulsion Laboratory (JPL)
- Compatibility Test Trailer (CTT-22), able to come to the spacecraft site

For more information on DSN, please see:

https://www.nasa.gov/directorates/space-operations/space-communications-and-navigation-scan-program/scan-services-and-scheduling/
https://deepspace.jpl.nasa.gov/about/commitments-office/
https://deepspace.jpl.nasa.gov

Swedish Space Corporation

Swedish Space Corporation (SSC) is a global provider of ground station services, including support to launch and early operations, on-orbit Telemetry, Tracking and Control (TT&C) and data downlink, and even lunar services (see https://sscspace.com/). The SSC Infinity Network is specifically designed for constellations of small satellites in low-Earth orbits. The global network provides TT&C and data download and delivery services to SmallSat operators, and customer interfaces consist of web- based portals for pass scheduling on 5-meter and smaller antennas. SSC Infinity also uses standard configurations and standardized ground system hardware, limiting the number of mission configurations to help keep costs lower for satellite operators.

Using ground services will generally require some degree of pre-coordination (or "onboarding") between the operator and provider, which is usually done before launch. This will vary between providers but may include contracting mechanisms; frequency licensing and coordination between the operator and the provider; compatibility testing; and the sharing of mission and vehicle specific information to ensure the ground stations are properly configured for the operator to use. Once the onboarding process is complete, satellite operators can schedule passes between their satellite(s) and desired ground station(s) in advance (the time window varies for each provider). The schedules for each ground station are deconflicted based on scheduling priority, and all frequency and modulation adjustments for the satellite are completed in advance of the pass by the service provider.

KSAT^LITE

The baseline of Kongsberg Satellite Services AS (KSAT)'s 3.7-meter KSAT^LITE antennas provide X-band and S-band for downlink and S-band for uplink. KSAT operates 150+ KSAT^LITE antennas at 15+ ground station sites across the globe, supporting over 1.5 million passes in 2022 alone.

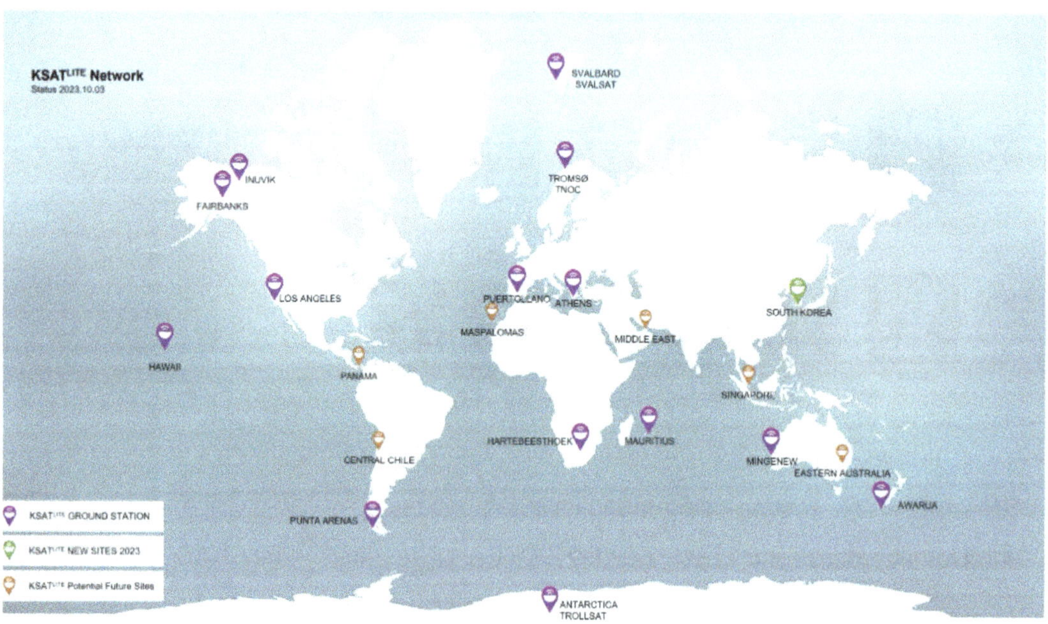

Figure 11.5: 2023 KSAT^LITE ground network map. Credit: KSAT.

located polar stations in the Arctic and Antarctic regions, providing 100% availability on passes for spacecraft in polar orbit. The network also includes mid-latitude ground stations, providing access for diverse orbits and mission profiles.

In addition, KSAT^{LITE} offers a global Ka-band network capable of supporting missions with higher data rates. To further improve latency and service quality, KSAT is installing edge compute capabilities at all their stations to support L1/L0 processing. Together with the European Space Agency (ESA) European Space Operations Centre (ESOC), KSAT^{LITE} is integrating a network of optical ground stations, and the first station of the Optical Nuclear Network was installed in Greece in January 2021. These stations will support both SmallSats and larger missions that demand a higher throughput or more secure downlink solutions. In 2023, KSAT invested into a network of three 20m, LEGS compliant, antennas to provide 100% coverage of lunar missions. See figure 11.5 for 2023 KSAT^{LITE} ground network map.

11.4.2 Ground Segment as a Service (GSaaS)

Ground Station as a Service (GSaaS) is a managed service which enables customers to communicate, downlink, & process data from their satellites/spacecrafts on as a pay-as-you go basis without needing them to build their own satellite ground stations. These services are usually scalable and use edge cloud services as an intermediate for customers data (5) (6).

AWS Ground Station

AWS Ground Station is a managed service (figure 11.6.) that lets customers build ground segment architectures in the cloud to control their satellites, process satellite data, and scale satellite operations without having to worry about building or managing their own antenna infrastructure. Customers can stream satellite data from any of the AWS antennas to the Amazon Elastic Compute Cloud (EC2) for real-time processing or to directly store data in the Amazon Simple Storage Service (S3). Additionally, customers can easily integrate their space workloads with other AWS services in near real-time using Amazon's low-latency, high-bandwidth global network. For example, customers who downlink terabytes of data daily can easily access AWS

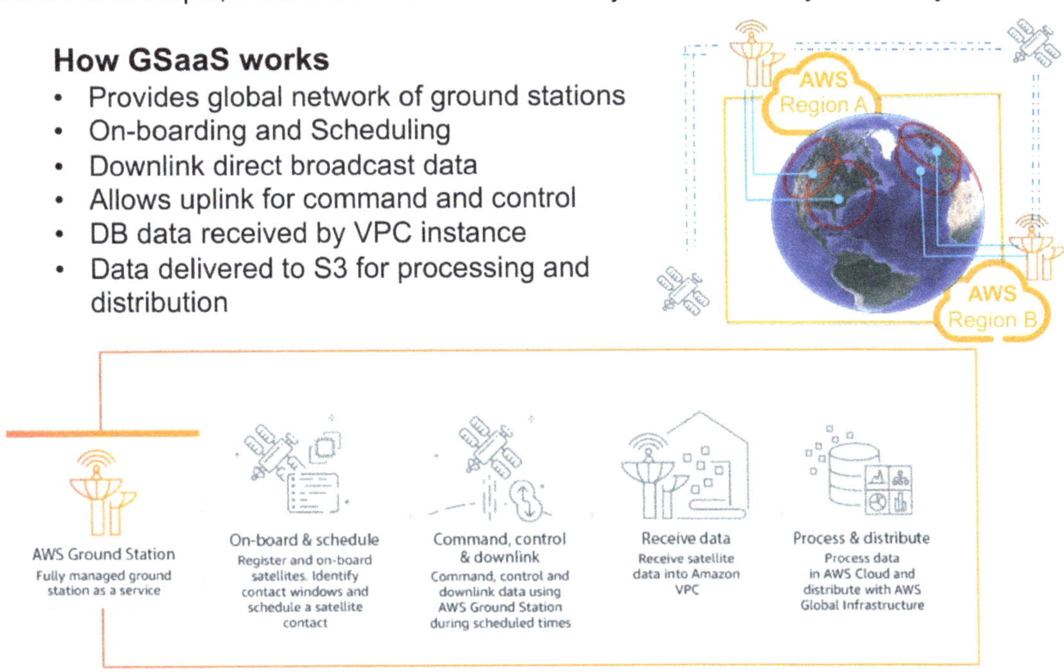

Figure 11.6. GSaaS Flow Chart. Credits: NASA/Amazon Web Services.

services such as Amazon SageMaker to quickly derive useful information. Other AWS services include Amazon VPC, Amazon Rekognition, and Amazon Kinesis Data Streams. These services allow operators to reduce data processing and analysis times for use cases like weather prediction or natural disaster imagery from hours to minutes or seconds. This also enables operators to quickly create business rules and workflows to organize, structure, and route the satellite data before it can be analyzed and incorporated into key applications such as imaging analysis and weather forecasting (7). A map of the AWS Ground Station antenna regions is shown in figure 11.7.

AWS Ground Station offers 5.4-meter apertures at each of the antenna regions. AWS Ground Station provides satellite antennas direct access to AWS services for faster, simpler, and more cost-effective storage and processing of downloaded data. Frequencies and link parameters are as follows:

- S-band uplink: 2025-2120 MHz
- S-band downlink: 2200-2300 MHz with G/T of 16 dB/K
- X-band downlink: 7750-8400 MHz with G/T of 30.5 dB/K

AWS Ground Station antennas are interconnected via Amazon's low-latency, highly reliable, scalable, and secure global network backbone. As of August 2023, the AWS Cloud spans 102 Availability Zones within 32 geographic regions around the world, with announced plans for 4 more AWS Regions in Australia, Canada, Malaysia, and Thailand.

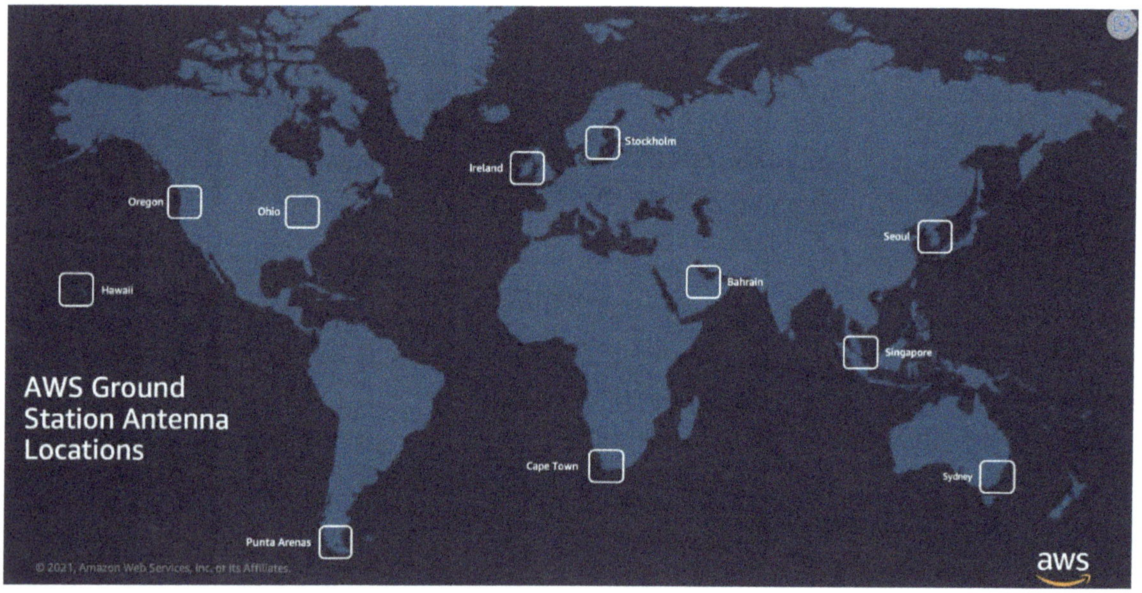

Figure 11.7. AWS Ground Station Locations (2023). Credit: Amazon Web Services

Scheduling: AWS Ground Station provides an easy-to-use graphical console that allows operators to reserve contacts and antenna time for their satellite communications. They can review, cancel, and reschedule contact reservations up to 15 minutes prior to scheduled antenna times. Access can be scheduled to AWS Ground Station antennas on a per-minute basis, so operators only pay for the scheduled time. They can access any antenna in the ground station network, and there are no long-term commitments.

Viasat

Viasat, a leader in ground antenna design, with multiple ground antennas delivered to NASA and industry since the 1960s, has recently become a ground station provider. This move leverages decades of hardware engineering, with an envisioned Real Time Network encompassing "in-house" assets and partner stations, complemented by a satellite relay network philosophy in GEO orbit for low latency (near-real time) applications. A map of Ground Assets is shown in figure 11.8. Viasat operates generous 7.3 m and 5.4 m antennas operating in 3 frequency bands (table 11-7) (8).

Viasat's Real-Time Earth (RTE) ground segment service enables communications for next-generation and legacy LEO satellites using S, X, and Ka-bands. RTE offers downlinks from low megabits per second to multiple gigabits per second empowered by cutting edge software-defined radios. The service includes high-speed connectivity for backhaul, real-time data streaming, and real-time monitoring of passes.

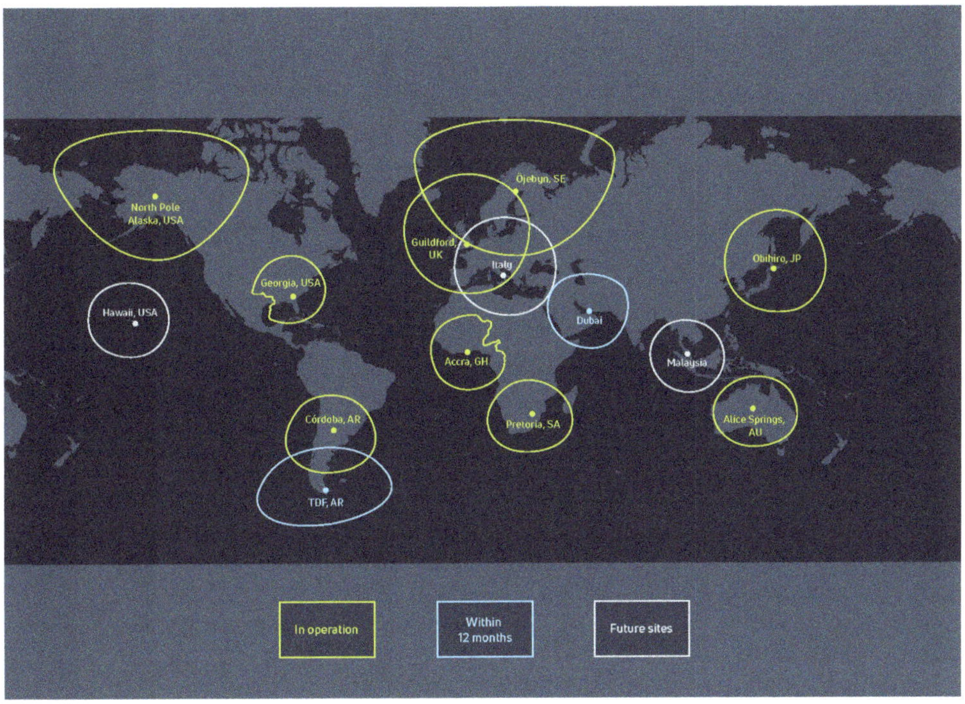

Figure 11.8: Viasat RTE Global Ground Stations. Credit: Viasat.

Table 11-7: Viasat RTE Global Ground Stations and Parameters

Site Location	Antenna	S-band Uplink	S-band Downlink	X-band Downlink	Ka-band Downlink
Pendergrass, GA	5.4 m X-Y Full motion 5 °/sec max speed	2025-2110 MHz EIRP: 53.2 dBW Selectable RHCP or LHCP	2200-2290 MHz G/T: 17 dB/K Selectable RHCP or LHCP	8025-8400 MHz G/T: 30 dB/K Simultaneous RHCP and LHCP	N/A

Guildford, UK	5.4 m X-Y Full motion 5 °/sec max speed	2025-2110 MHz EIRP: 53.2 dBW Selectable RHCP or LHCP	2200-2290 MHz G/T: 17 dB/K Selectable RHCP or LHCP	8025-8400 MHz G/T: 30 dB/K Simultaneous RHCP and LHCP	N/A
Alice Springs, AU*	7.3 m X-Y Full motion 6 °/sec max speed quantity 2	2025-2110 MHz EIRP: 65.0 dBW Selectable RHCP or LHCP	2200-2290 MHz G/T: 18 dB/K Selectable RHCP or LHCP	8025-8400 MHz G/T: 32 dB/K Simultaneous RHCP and LHCP	25500-27000 MHz G/T: 34.5 dB/K Simultaneous RHPC and LHCP
Ghana*	7.3 m X-Y Full motion 6 °/sec max speed	2025-2110 MHz EIRP: 65.0 dBW Selectable RHCP or LHCP	2200-2290 MHz G/T: 18 dB/K Selectable RHCP or LHCP	8025-8400 MHz G/T: 32 dB/K Simultaneous RHCP and LHCP	25500-27000 MHz G/T: 34.5 dB/K Simultaneous RHCP and LHCP
Cordoba, AR	5.4 m X-Y Full motion 5 °/sec max speed	2025-2110 MHz EIRP: 53.2 dBW Selectable RHCP or LHCP	2200-2290 MHz G/T: 17 dB/K Selectable RHCP or LHCP	8025-8400 MHz G/T: 30 dB/K Simultaneous RHCP and LHCP	N/A
Öjebyn, Sweden	7.3 m X-Y Full motion 6 °/sec max speed	2025-2110 MHz EIRP: 55.2 dBW Selectable RHCP or LHCP	2200-2290 MHz G/T: 18 dB/K Selectable RHCP or LHCP	8025-8400 MHz G/T: 32 dB/K Simultaneous RHCP and LHCP	25500-27000 MHz G/T: 34.5 dB/K Simultaneous RHCP and LHCP
Pretoria, South Africa (PRY)	7.3 m X-Y Full motion 6 o/sec max speed	2025-2110 MHz EIRP: 55.2 dBW Selectable RHCP or LHCP	2200-2290 MHz G/T: 18 dB/K Simultaneous RHCP and LHCP	8025-8400 MHz G/T: 32 dB/K Simultaneous RHCP and LHCP	25500-27000 MHz G/T: 34.5 dB/K Simultaneous RHCP and LHCP
Obihiro, Japan (OBO)	7.3 m X-Y Full motion 6 °/sec max speed	2025-2110 MHz EIRP: 55.2 dBW Selectable RHCP or LHCP	2200-2290 MHz G/T: 18 dB/K Simultaneous RHCP and LHCP	8025-8400 MHz G/T: 32 dB/K Simultaneous RHCP and LHCP	25500-27000 MHz G/T: 34.5 dB/K Simultaneous RHCP and LHCP

Shimoji-shima, Japan (SHI)	7.3 m X-Y Full motion 6 °/sec max speed	2025-2110 MHz EIRP: 55.2 dBW Selectable RHCP or LHCP	2200-2290 MHz G/T: 18 dB/K Simultaneous RHCP and LHCP	8025-8400 MHz G/T: 32 dB/K Simultaneous RHCP and LHCP	25500-27000 MHz G/T: 34.5 dB/K Simultaneous RHCP and LHCP
North Pole Alaska USA 7.3 m (NTP01)	7.3 m El-Az-Train El & Az: 15 °/s max speed Train 6 °/s max	2025 – 2120 MHz EIRP: 53 dBW Selectable RHCP or LHCP	2200–2300 MHz G/T: 19 dB/K Simultaneous RHCP and LHCP	8000–8500 MHz G/T: 32 dB/K Simultaneous RHCP or LHCP	N/A
North Pole Alaska USA (9.1 m (NTP02)	9.1 m El-Az-Train	2042–2052 MHz EIRP 38 dBW Selectable RHCP or LHCP	N/A	8000–8500 MHz G/T: 34 dB/K RHCP	N/A
Canada (YEV)	7.3 m X-Y Full motion 6 °/sec max speed	2025-2110 MHz EIRP: 55.2 dBW Selectable RHCP or LHCP	2200-2290 MHz G/T: 18 dB/K Simultaneous RHCP and LHCP	8025-8400 MHz G/T: 32 dB/K Simultaneous RHCP and LHCP	25500-27000 MHz G/T: 34.5 dB/K Simultaneous RHCP and LHCP

*Additional L-band uplink 1755-1850 MHz. EIRP: 63.4 dBW

Viasat's fully automated system allows users their choice of machine-to-machine or manual GUI-based scheduling over a highly resilient cloud-based platform. Instantaneous confirmation of your pass without human intervention lowers your cost and risk of human error. Data is delivered to your choice of cloud provider, operations center, or government cloud destination.

Viasat's Real-Time Earth Space Relay is designed to leverage the ViaSat-3 global constellation with a newly developed terminal that will enable on-demand and cost-effective communications for LEO spacecraft anywhere and at any time in their global orbit. Viasat won a portion of the NASA Communications Service Provider contract with its integrated RTE ground solution, which will include on-orbit demonstrations in 2025 that allow users to select between the low latency space relay and the high throughput global ground network (8).

Leaf Space

Leaf Space is a Ground-Segment-as-a-Service provider, operating a globally distributed ground stations network tailored to support SmallSats in low-Earth orbit. Called *Leaf Line*, Leaf Space's network enables TT&C and payload data transmissions to and from the satellite operators' mission control software and cloud platform of choice (see figure 11.9). It does so via a simple interface, a proprietary autonomous scheduling software, and global coverage ideal for both mid- and high-inclination orbits. Leaf Line antennas are either fully owned or managed on an exclusive basis, ensuring maximum availability, flexibility, and independence of operations. Leaf Line is powered and orchestrated by a cloud architecture designed to support multiple satellite missions and operators at the same time. The network's architecture allows seamless use of any of the antennas with a single integration procedure and no difference in operations or performances. Leaf Space works with its customers to execute both routine operations and complex tasks such as LEOP or MSPA data downlink for formation flying satellite clusters.

Leaf Line is characterized by ~3.7m antennas supporting operations in S-band uplink (2025-2110 MHz, EIRP: 50 dBW) and downlink (2200-2290 MHz, G/T: 12.8 dB/K) and X-band downlink (8025-8400 MHz, G/T: 25.4 dB/K), as well as by UHF antennas (uplink and downlink). In 2024, the Leaf Line network will be upgraded to support Ka-band downlink (25.5-27 GHz) at selected sites, with the first installations expected in Q1 2024. Leaf Space will keep expanding the network in terms of sites, capacity, and antenna performances. With an eye towards the future of the sector, Leaf Space is working to integrate optical ground systems as well as cislunar-oriented solutions to its existing network.

In addition to providing access to the Leaf Line network, Leaf Space can procure, deploy and operate dedicated ground stations (service offering called Leaf Key), enabling operators with particular frequencies, access or data requirements to leverage Leaf Space's cloud architecture and ground segment experience without compromising on antenna time. For more information on the company or on how Leaf Space can support current and future missions, please contact sales@leaf.space or visit https://leaf.space.

Figure 11.9: The Leaf Line network, September 2023. Credit: Leaf Space.

Laser Light

Laser Light Communications operates a Global Network platform, delivering a first-of-a-kind 21st century data service that will transform the way high volume communications traffic is carried. Using a hybrid infrastructure spanning terrestrial, subsea, and space domains; an end-to-end software defined architecture offering all-optical: up to 400 Gbps connectivity and provisioning within minutes (see https://www.laserlightcomms.com/).

<u>Network On Demand</u>

- Pay-as-you-go: only pay for the duration you use the service with no upfront or fixed costs
- Cost Effective: automation and end-to-end control yields significant operating cost savings

- Secure: highly targeted, dynamic, laser links are extremely difficult to intercept and can be encrypted
- High Capacity/ Performance: data delivery at up to 400 Gbps in the most direct route from origination to destination, automatically bypassing points of congestion

Data Transport as a Service

- Pay-as-you-go: only pay for volume you use when you use it with no upfront or fixed costs
- Cost Effective: automation and end-to-end control yield significant operating costs savings
- Secure: highly targeted, dynamic, laser links are extremely difficult to intercept and can be encrypted
- High Capacity/Performance: data delivery at optical speeds -- up to 400 Gbps directly from the point of origination to the point of destination, automatically bypassing points of congestion along the way

Atlas Global Network

ATLAS Space Operations, Inc. provides satellite RF communications services to the government and commercial sectors (see https://atlasground.com/). Through geographical dispersion and cloud services, ATLAS provides satellite communications services to the government and commercial sectors and offers GSaaS through a global network of 34 ground stations and 51 antennas.

All ATLAS ground stations are built upon the Freedom™ Software Platform (figure 11.10) within a cloud-based distributed operations center and facilitates dynamic demand and scalable growth. Through geographical dispersion of its ground stations and cloud services, ATLAS provides a resilient capability that delivers dependable low latency data. ATLAS bases its mission success model on our global network of operational and traditional RF parabolic ground stations that are integrated with our network management and scheduling software. These platforms work together as a mission

Figure 11.10. The Freedom™ Software Platform is a simple and scalable ground network management system. Credit: ATLAS Space.

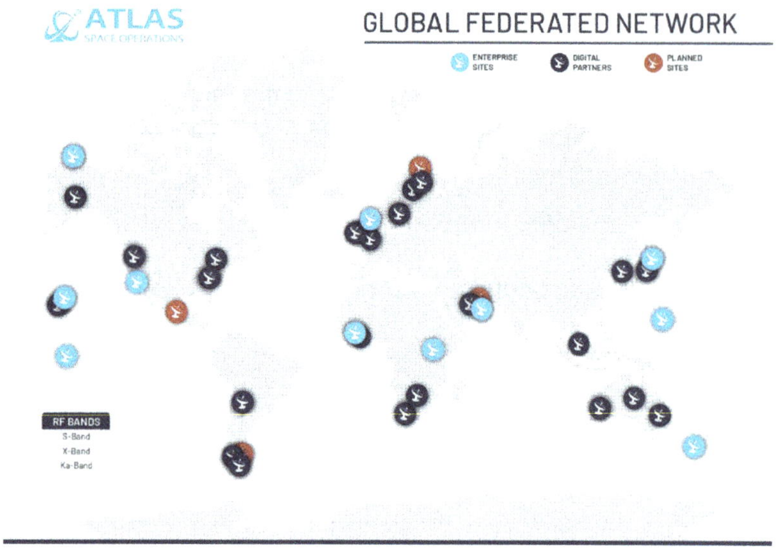

Figure 11.11: ATLAS Space Operations ground network map. Credit: ATLAS Space Operations

architecture to meet Client requirements for routine satellite to ground communications.

ATLAS' Global Antenna Network is fully integrated with the Freedom Software, providing users with a robust low latency, secure communications solution, including: automated network operations, set-and-forget scheduling, mixed modem capability, real-time metrics, and single secure VPN access. Once integrated into the ATLAS Network, a single secure VPN enables access and load balancing of network resources. Freedom Core Services advance operations beyond legacy constructs and enable users the freedom and flexibility to reliably schedule satellite passes with minimal human interaction via machine-to-machine capabilities. Entire data processing and forwarding workflows can be automated within the cloud to ensure your data is ready for use as soon as it arrives at the Mission Operations Center.

ATLAS supports deploying clusters and serverless instances onto AWS. ATLAS hardware parameters are highlighted below (ATLAS Enterprise Sites and Digital Partner Sites). In addition to S- band and X-band capabilities, ATLAS can provide VHF, UHF services in Japan and Guam. The existing and planned ATLAS antenna systems support RF connectivity for low-Earth orbit, MEO, GEO, and L1 orbits, and ATLAS is actively pursuing technology development for deep space capabilities. Atlas's global federated network is also displayed below with future sites planned for Finland (North Base), Mexico (Mexico City), UAE (Dubai), and Tierra Del Fuego (Argentina). A summary of government and commercial ground stations and antennas in table 11-8, and shown in figure 11:11 is ATLAS's Global Federated Network of ground stations.

Freedom Space Technologies, Inc. dba Freedom Space was established in 2023 as a non-traditional small business and wholly-owned subsidiary of ATLAS Space Operations. Freedom Space, headquartered in Colorado Springs, CO, is purpose built to address and support the Department of Defense (DoD) and Intelligence Community (IC) mission requirements.

Table 11-8: ATLAS government and commercial ground stations and antennas.

Location	Lat	Long	Antenna Size	Rx Freq (MHz)	G/T (dB/K)	Tx Freq (MHz)	EIRP (dBW)	Polarization
Utqiagvik (Barrow), AK, USA	71.27	-156.81	3.7m	S: 2200-2300 X: 7900-8400	12.8 25.4	S: 2025-2120 -	60.0 -	R/LHCP
Dundee, Scotland	56.40	-3.18	3.7m	S: 2200-2400 X: 7800-8400	13.6 26.5	S: 2025-2120 -	48.0 -	R/LHCP
Chitose, Japan	42.77	141.62	3.4m	S: 2200-2300 X: 7900-8500	11.5 26.4	S: 2025-2120 -	52.0 -	R/LHCP
Mojave, CA, USA	35.05	-118.15	3.0m	S: 2200-2300 X: 7900-8500	11.3 25.9	S: 2025-2110 -	54.4 -	R/LHCP

Location	Lat	Long	Antenna Size	Rx Freq (MHz)	G/T (dB/K)	Tx Freq (MHz)	EIRP (dBW)	Polarization
Dubai, United Arab Emirates	24.94	55.35	3.7m	S: 2200-2300 X: 7900-8400	12.8 25.4	S: 2025-2120 -	50.0 -	R/LHCP
Paumalu, Hawaii, USA	21.67	-158.03	7.3m	S: 2200-2300 X: 7900-8500	20.7 31.0	S: 2025-2120 -	65.0 -	R/LHCP
Harmon, Guam	13.51	144.82	3.7m	S: 2200-2300 X: 7900-8400	13.6 25.1	S: 2025-2110 -	52.4 -	R/LHCP
Sunyani, Ghana	7.34	-2.34	3.0m	S: 2200-2300	12.4	S: 2025-2110	50.1	RHCP
Mwulire, Rwanda	-1.96	30.39	9.3m	S: 2200-2300 X: 7900-8500	21.5 36.0	S: 2025-2120 -	60.0 -	R/LHCP
Tahiti, French Polynesia	-17.63	-149.60	3.7m	S: 2200-2300 X: 7900-8400	13.9 27.4	S: 2025-2110 -	47.5 -	R/LHCP
Awarua, New Zealand	-46.52	168.38	3.7m	S: 2200-2300 X: 8025-8500	13.7 27.0	S: 2025-2120 -	49.0 -	R/LHCP

Location	Lat	Long	Antenna Size	Rx Freq (MHz)	G/T (dB/K)	Tx Freq (MHz)	EIRP (dBW)	Polarization
Sodankylä, Finland*	67.36	26.63	7.3m	S: 2200-2300 X: 7900-8500	19.8 32.1	S: 2025- 2110 -	54.8 -	R/LHCP
Öjebyn, Sweden	65.33	21.42	7.3m	S: 2200-2290 X: 8025-8400 Ka: 25500-27000	18.0 32.0 35.7	S: 2025- 2110 - -	55.2 - -	R/LHCP
North Pole, Alaska, USA	64.79	-147.53	7.3m	S: 2200-2300 X: 8000-8500 Ka: 25500-27000	19.0 32.0 35.7	S: 2025- 2120 - -	53.0 - -	R/LHCP
Stockholm, Sweden*	-	-	5.4m	S: 2200-2300 X: 7750-8400	16.0 30.5	S: 2025- 2120 -	50.0 -	R/LHCP
Dublin, Ireland*	-	-	5.4m	S: 2200-2300 X: 7750-8400	16.0 30.5	S: 2025- 2120 -	50.0 -	R/LHCP
Guildford, United Kingdom	51.24	-0.61	5.4m	S: 2200-2290 X: 8025-8400	17.0 30.0	S: 2025- 2110 -	53.2 -	R/LHCP
Portland, Oregon, USA*	-	-	5.4m	S: 2200-2300 X: 7750-8400	16.0 30.5	S: 2025- 2120 -	50.0 -	R/LHCP
Obihiro, Japan	42.59	143.45	7.3m	S: 2200-2290 X: 8025-8400 Ka: 25500-27000	17.9 31.5 35.0	S: 2025- 2110 - -	55.2 - -	R/LHCP
Columbus, Ohio, USA*	-	-	5.4m	S: 2200-2300 X: 7750-8400	16.0 30.5	S: 2025- 2120 -	50.0 -	R/LHCP
Seoul, South Korea*	-	-	5.4m	S: 2200-2300 X: 7750-8400	16.0 30.5	S: 2025- 2120 -	50.0 -	R/LHCP
Pendergrass, Georgia, USA	34.17	-83.67	5.4m	S: 2200-2290 X: 8025-8400	17.0 30.0	S: 2025- 2110 -	53.2 -	R/LHCP

*Denotes antenna redundancy

Location	Lat	Long	Antenna Size	Rx Freq (MHz)	G/T (dB/K)	Tx Freq (MHz)	EIRP (dBW)	Polarization
Zallaq, Bahrain*	-	-	5.4m	S: 2200-2300 X: 7750-8400	16.0 30.5	S: 2025- 2120 -	50.0 -	R/LHCP
Kapolei, Hawaii, USA*	-	-	5.4m	S: 2200-2300 X: 7750-8400	16.0 30.5	S: 2025- 2120 -	50.0 -	R/LHCP
Accra, Ghana	5.74	-0.30	7.3m	S: 2200-2290 X: 8025-8400 Ka: 25500-27000	18.0 32.0 35.7	S: 2025- 2110 - -	65.0 - -	R/LHCP
Singapore*	-	-	5.4m	S: 2200-2300 X: 7750-8400	16.0 30.5	S: 2025- 2120 -	50.0 -	R/LHCP
Alice Springs, Australia*	-23.75	133.88	7.3m	S: 2200-2290 X: 8025-8400 Ka: 25500-27000	18.0 32.0 35.7	S: 2025- 2110 - -	65.0 - -	R/LHCP
Pretoria, South Africa	-25.88	27.70	7.3m	S: 2200-2290 X: 8025-8400 Ka: 25500-27000	18.0 32.0 35.7	S: 2025- 2110 - -	55.2 - -	R/LHCP
Mingenew, Australia*	-29.01	115.34	5.0m	S: 2200-2300 X: 8025-8500	14.0 29.5	S: 2025-2120 -	55.0 -	R/LHCP
Cordoba, Argentina	-31.52	-64.46	5.4m	S: 2200-2290 X: 8025-8400	17.0 30.0	S: 2025- 2110 -	53.2 -	R/LHCP
Cape Town, South Africa*	-	-	5.4m	S: 2200-2300 X: 7750-8400	16.0 30.5	S: 2025- 2120 -	50.0 -	R/LHCP
Dubbo, Australia*	-	-	5.4m	S: 2200-2300 X: 7750-8400	16.0 30.5	S: 2025- 2120 -	50.0 -	R/LHCP
Punta Arenas, Chile*	-	-	5.4m	S: 2200-2300 X: 7750-8400	16.0 30.5	S: 2025- 2120 -	50.0 -	R/LHCP
Ushuaia, Argentina	-54.51	-67.11	7.3m	S: 2200-2290 X: 8025-8400 Ka: 25500-27000	17.9 32.0 33.0	S: 2025- 2110 - -	56.0 - -	R/LHCP

*Denotes antenna redundancy

11.4.3 Space Relay Network - NASA

Unlike a traditional ground network that goes direct from a "client" satellite to a ground station on the ground, space relay networks consist of communication satellites that relay data from the "client" satellite down to a ground station. While some no longer consider it state-of-the-art, NASA's Tracking and Data Relay Satellite System (TDRSS), shown in figure 11.12, is one of the most well-known space relay networks. TDRSS relays data from the International Space Station (ISS) and the Hubble Space Telescope to NASA ground stations around the world.

Space relay networks can be beneficial for small satellites in low-Earth orbit because those SmallSats are only in view of a ground station for a portion of their orbit. However, depending on the orbit of the relay satellites, a low-Earth orbit SmallSat could be in view of a relay satellite for most of its orbit. This makes a relay network beneficial for a SmallSat, especially right after SmallSat deployment when a ground station is still trying to locate the satellite. The space relay can transmit satellite telemetry, tracking, and control data to the ground, enabling faster satellite identification. This proves to be even more valuable when the satellite is deployed with several others for a given rideshare opportunity. This data can also contain satellite health information to give mission teams either peace of mind while awaiting acquisition by the ground station, or information for troubleshooting prior to the commissioning phase. Another benefit is the ability to obtain real-time event notifications without the need for prior scheduling. Scientists are interested in using this technology to alert the science community when certain phenomenon are observed. Space relay networks often require special hardware or software that must be added to a satellite before launch. In general, a satellite operator will purchase a modem compatible with the relay network and fly it on their satellite to access the network. It is common for the network providers to license their proprietary chipset to developers who build the modem hardware and serve as a service broker. Because of this added hardware component, the decision to leverage a space relay network must be made before the satellite has been launched.

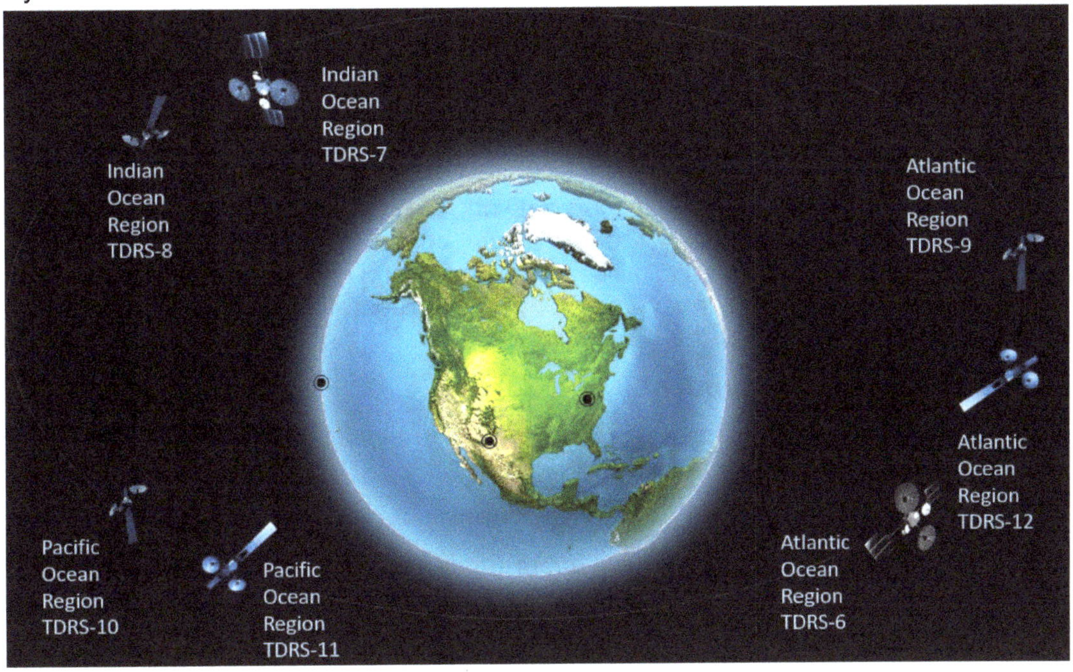

Figure 11.12: NASA's Tracking and Data Relay Satellite System. Credit: NASA.

11.4.4 Low Latency, Low rate (Short burst) Space Relay Providers

Space relay solutions are less common than traditional direct-to-Earth solutions, but there are a few options that exist for small satellites (see table 11-9). To access the space relay, a satellite operator purchases a modem from the relay manufacturer and flies that on their satellite to access the relay services. In general, space relays are ideal for obtaining satellite TT&C data (health and safety of the vehicle) rather than for large data downlinks.

Table 11-9: Service Providers for Space Relay Networks

Product	Manufacturer	TRL	Specifications
Iridium Global Network	Iridium	9	LEO relay requiring 9600 series transceivers onboard the satellite
Fast Pixel Network	Analytical Space	6	Establish a data transport network in LEO

The Iridium network is one example satellite customers can rely on for delivering low latency messages. Iridium uses a combination of Frequency Division Multiple Access (FDMA) and Time-Division Multiple Access (TDMA) for its communication waveforms. L-band (1616 – 1626.5 Mhz) is used for uplink and downlink between the user spacecraft and the Iridium spacecraft. Inter-satellite communication links between Iridium satellites is accomplished through Ka-band (23.18 – 23.28 GHz). Operators install an Iridium transceiver (9600 series) onboard their spacecraft to communicate with the Iridium network. Messages are relayed through Iridium's Short Burst Data Service, which is hosted on Iridium's cloud platform for easy user operation. For each transceiver unit, a data plan must be chosen and purchased, much like cellular phone data plans, and the plan details are linked to the unit's ID, which is referred to as International Mobile Equipment Identity (IMEI). The special feature of this system is that it has as an option for "IMEI-to-IMEI" transmission. When an Iridium IMEI is activated, five output destinations may be specified. Most vendors allow for a combination of emails addresses, fixed IP address, or another device with an IMEI ID (6).

Iridium has announced the commercial availability of its Certus 100 'midband' service providing 88 kbps connectivity via small antennas and battery-powered devices, for basic data communications and IoT applications (9).

Analytical Space is another company to watch for future services. Their recent contract puts a LEO relay in space to aggregate data from GEO satellites. Through Fast Pixel Network, Analytical Space Inc. plans to establish a data transport network in low-Earth orbit to ingests data from geospatial intelligence satellites, send the data from node to node via high speed optical intersatellite links, and deliver the data to military, intelligence and commercial customers (10 High speed MEO and GEO commercial relays are not presently operating, but several are planned. These are listed under the State-of-the-Art Ground Data and Supporting Systems section (11.9).

11.5 Ground Stations Components

The hardware for ground stations consists of the tracking antenna, its feed, and the modem that converts the RF waveform into digital packets and vice versa.

11.5.1 Ground Station Operation

A DTE ground station is comprised of a system of hardware and software working together to convert the RF signal from a satellite signal into digital data. The first key element of the system is the antenna. It is chosen based on the frequency and gain required to talk with a satellite. NASA uses parabolic reflector antennas for RF ground satellite communications, while some universities use dish or Yagi antennas.

The dish antenna uses a parabolic reflector to collect signals from the spacecraft and focus them onto a feed antenna. The feed antenna is typically a horn antenna with a circular aperture. The size of a dish is at least several wavelengths in diameter at the frequency of operation and can be increased for higher gains. The distance between the feed antenna and parabolic reflector can also be several wavelengths. For example, a Ka-band 34 m deep space antenna with a feed distance of 15 m would be approximately 3,000 wavelengths for the dish diameter and 1,500 wavelengths for the feed distance relative to a 1 cm Ka-band wavelength. The gain of a dish reflector (figure 11.13) is frequency-dependent and it is directly related to the square of its diameter. Dish antennas are available in sizes from 1 meter to 70 meters in diameter.

The antenna collects RF waves and the antenna feed converts the electromagnetic waves into conducted RF electrical signals. The feed consists of a resonant pickup that is tuned to the transmit or receive frequency, a low gain low-noise amplifier, a sharp filter, and a second LNA with more gain than the first amplifier. These elements condition the signal. The signal then traverses through a coaxial cable to a nearby location where a radio demodulates the RF signal into digital data. In the uplink direction, the radio modulates the data bits onto an RF carrier which is amplified to 10 or more watts. The amplified RF resonates in the antenna feed, and the antenna amplifies the electromagnetic waves and focuses them towards the satellite.

Figure 11.13: Ground Antenna in Fairbanks, Alaska. Credit: NASA/ Clare Skelly.

It is desirable to have significant antenna gain, but as the gain increases, the beamwidth of the antenna decreases. There is a practical compromise where the beamwidth is so small that tracking is difficult and when the antenna gets so large that it is difficult to procure or manage. A typical antenna pattern is shown in figure 11.14. There is a center lobe where most of the transmitted energy is contained. The remaining energy is stored in the sidelobes on either side of the main lobe. The diminished side lobes are intentional so that ground noise from other emitters on Earth are not collected when receiving and so that interference to terrestrial systems is not created when transmitting. The blue arrows in the figure indicate the full-width-half-max gain point at about ±6°, which should result in an antenna pointing error of less than 6° and the full-width- half-max gain of 16 dBi to be used in a link budget. If more gain is needed, then the antenna will increase in size and the beamwidth will correspondingly decrease.

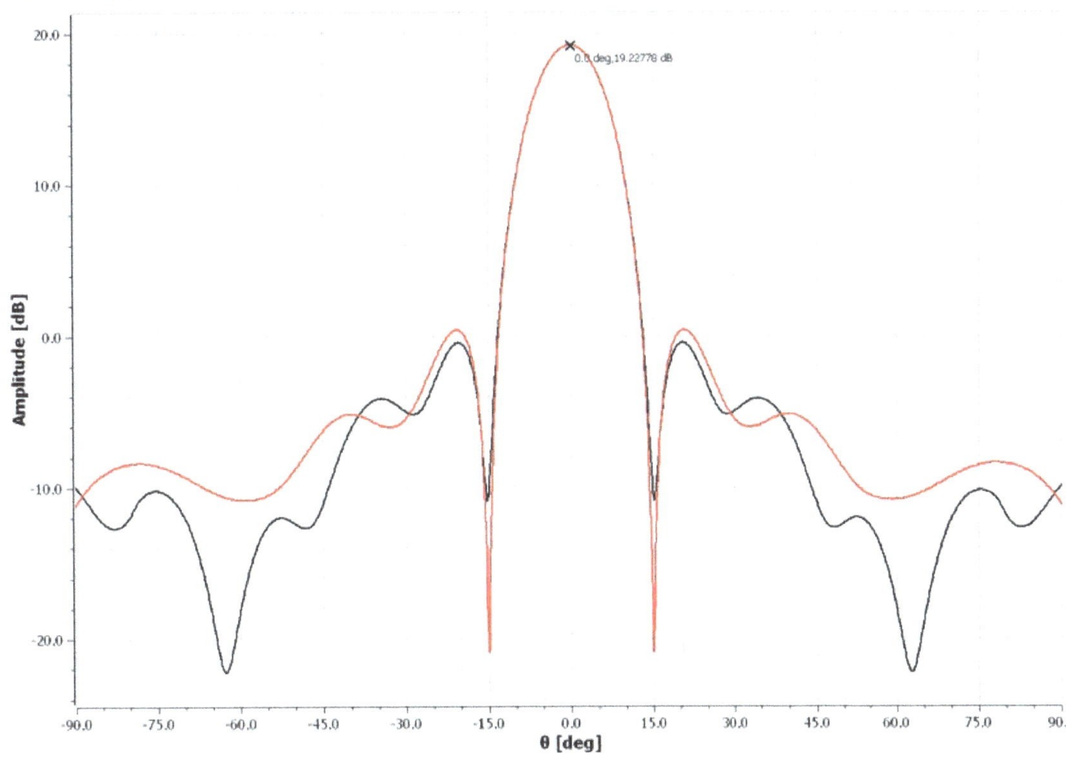

Figure 11.14: Antenna pattern from a 1.8-meter diameter parabolic dish operating at 915 Mhz with a high gain center lobe and diminished side lobes. Credit: NASA.

Directional antennas point towards the satellite as it moves over the ground station. Pointing adjustments are necessary in both the vertical (elevation) and horizontal (azimuth) directions. These movements are accomplished using motors and gears. Tracking software is used to predict the satellite's future location. The satellite position and time are processed through additional software that converts this information into commands for the motor controller. Time is an important factor and GPS time is used by the computer generating the satellite position estimate. A dedicated GPS receiver is connected to the computer for that purpose.

The cost of a DTE ground station is directly correlated with the size of the aperture, which drives the ground station foundation, pedestal, motors, and gears. The Yagi is less expensive. It sustains low wind loads and therefore can use a smaller foundation for support. In contrast, the dish antenna reflector sustains comparatively high wind loading and therefore needs a stronger concrete foundation and larger motors and gear box elements than the Yagi antenna.

11.5.2 Component Hardware for Ground Systems (GS)

This section lists examples of GS components and some supporting equipment. Table 11-10 lists example products in each category. The antenna feed consists of the RF pickup, LNA and mechanical filters located directly on the antenna. A radome is an RF transparent enclosure that protects the antenna from weather.

Table 11-10: Ground System Components

Product	Manufacturer	Type of Product
Tracking Antenna	Viasat, Safran	Antennas for small satellites in and S, X and Ka-band frequencies
Antenna Feed	See End-to-End Hardware Section 11.7.2	RF pickup, mechanical filters, LNA
Radio, Software Defined	NI Ettus Research	USRP X410, up to 7.2 GHz with RFSOC advanced FPGA and meeting wide bandwidth requirements. USRP X310: DC-6 GHz with up to 160 MHz of baseband bandwidth, multiple high-speed interfaces
Data Receiver	Safran Data Systems	Cortex CRT (low data rate) and HDR (high data rate (previously by Zodiac)
Modem, for TT&C and Payload Reception	Safran Data Systems	Satcore, plug-and-play modem for TT&C and Payload Reception
Digital Processing	Kratos	SpectralNet: Digital IF product that converts analog signals at RF frequencies up to S-band into digital IF packets.
Radome	Infinite Technologies	Antenna radomes
Ground Station Dongle	GAUSS	A USB low-power board to simulate ground station in laboratory conditions. The USB dongle integrates both a low-power UHF transceiver and a TNC, thus miniaturizing common ground station rack systems
Integrated Testing Systems (EGSE) & Ground Station TT&C Modems	Celestia Satellite Test & Simulation	Hardware and software elements all operating within a single reference platform and environment

Cortex HDR

Several NSN, SSC and NOAA stations use the Cortex HDR High Data Rate Receiver, which performs demodulation, decoding and frame synchronization on the (X-band) data stream. Each virtual channel in the AOS frame that is received by the station X-band receiving system will be written into separate files. Files are separated into small one-minute file sizes for a single VCID that allow for faster turn-around time on the data and smaller transmission cycles in case of transfer problems. File-based data is stored in a buffer (e.g., for 7 days) used for retransmissions and failure recovery when necessary. At the end of a pass, ground stations such as the NSN ones, perform an automatic secure file transfer protocol (SFTP)/secure copy protocol (SCP) push to the customer. If the customer wants to "replay" a dataset, they may use the self-service /SFTP/SCP interface on the system to pull their data to their site. Alternatively, the customer may choose to manually retrieve files and not select automatic file transfer (11).

USRP X310 and X410 Open-Source Software Defined Radio for SatCom Applications

The NI Ettus Research brand includes the Universal Software Radio Peripheral (USRP™) family of products. The USRP is one of the most popular open platforms for small satellite communications with options from high-performance to low-cost to highly deployable. One of the most popular hardware units for satellite communication applications is the USRP X310 with the UBX RF daughterboard. The USRP X310 is a high-performance software defined radio with the ability to transmit and receive modulated signals. With up to 160 MHz of instantaneous bandwidth and a frequency tuning range up to 6 GHz, the X310 with UBX has the raw hardware performance to cover many ground station satellite communication needs. The USRP family supports a wide range of software tool chains from LabVIEW to GNU Radio, with many existing IP modules for modulation and demodulation. The USRP X310 is intended for lab environments, however, it can be built in rugged weatherproof configurations. Many small satellite researchers are using the USRP as their ground station equipment for its adaptability with open-source software and its embedded FPGA pre-processing capability. The USRP X310 offers 2 channels, 10 GIGE and PCIE bus, whereas the NI Ettus USRP X410 is equipped with two dual 100 GbE interfaces capable of moving much more data (12).

Kratos OpenSpace SpectralNet

OpenSpace SpectralNet® eliminates the distance constraints of RF transport by digitizing RF signals for transport over IP networks in a way that preserves both frequency and timing characteristics, and then uniquely restores the RF signals at their destination. By eliminating the distance constraints between antennas and signal processing equipment, this technology enables operators to deploy new ground architectures with numerous advantages, such as the ability to mitigate the effects of rain fade for Ku/Ka satellites, reduce costs by centralizing operations, simplify disaster recovery and system maintenance, optimize antenna placement, and develop a migration path toward virtual ground systems. SpectralNet does all of this while protecting the operator's current investment in existing equipment (13).

Integrated Testing Systems & Ground Station TT&C Modems

Celestia Satellite Test & Simulation BV (C-STS) provides ground-based solutions in the domains of satellite simulation, testing, communication, and data processing. Established in 1985, Satellite Services B.V. (SSBV) was acquired by Celestia Technologies Group in 2016 and re-branded to Celestia Satellite Test & Simulation B.V. to continue as a competence center for Electrical Ground Support Equipment (EGSE) and TT&C solutions. Celestia STS has more than 30 years of experience in the space industry. More than 300 EGSEs and TT&C modems were delivered to space agencies, large system integrators, and specialized flight-equipment manufacturers around the world.

On-board computers, mass memory units, and transponders are tested every day with C-STS equipment. Celestia EGSE solutions have been used in more than 80% of all European Space Agency (ESA) missions. Celestia STS testing equipment is available in standard functionality or configured to meet specific customer needs. System options include:

- Telemetry and Telecommand Processing System
  - TM acquisition and simulation
  - TC generation and acquisition
  - Bit error rate tester
  - TC authentication
  - TM/TC deciphering (API/DLL/LAN)
  - Includes control and monitor software for data processing and visualization
- Wizardlink High-Rate Interface System
  - Up to 4 Wizardlink channels in parallel
  - Up to 2Gbps data rate per channel
  - Includes software for high speed ingest, processing, data archiving, and export
- LVDS High-Rate Interface System
  - Up to 4 parallel LVDS inputs and outputs
  - 8-bit parallel up to 1Gbps per channel
  - Teaming of 2 LVDS input and output channels to 16-bits
  - 16-bit parallel up to 2Gbps per channel
  - Includes software for high speed ingest, processing, data archiving, and export
- TT&C Integrated Modem and Baseband unit
  - Single or dual channel modulation and demodulation
  - Ranging measurement
  - Doppler simulation
  - Bit error rate tester
- Level Zero Processor Software for High-Speed Data Processing
  - Data directly from the local disk drive or shared network drive
  - Processing of TM data from bitstream to frame and packet level
  - Configurable frame and packet checking rules
  - Configurable frame and packet output data storage and sorting
  - Live frame and/or packet distribution via LAN
  - Real-time statistical analysis, error checking, and reporting
- Optical Digital Convertor
  - Processing of optical detector signals to simulate optical communications

Efforts are on-going to improve product capability with a focus on modular, flexible, scalable multichannel systems that take advantage of the latest technologies. In June 2021, an agreement between the Netherlands Organization (TNO), Celestia, and the Netherlands Organization for Applied Scientific Research to commercialize optical modems (14).

Infinite Technologies Radomes

A successfully designed radome provides a protective cover and has minimal effect on the electrical functionality of the antenna. Figure 11.15 provides an example of a radome supplied by Infinite Technologies. Radomes provide the antenna system with a controlled environment, shielding sensitive equipment from weather related stresses such as wind, snow, ice, salt spray, etc. A radome can increase the useful life of the antenna and decrease overall maintenance costs for the system. Consideration for a radome should be given early in the design phase of the system, as a radome will allow for lighter duty and less expensive components such as drive motors and foundations due to the elimination of wind loads on the antenna. Also, the controlled environment inside the radome provides greater system availability allowing the antenna to operate in more adverse environmental conditions with minimal signal degradation. A radome will also provide maintenance personnel protection from weather during antenna maintenance (15).

Figure 11.15: Infinite Technologies small radome. Credit: Infinite Technologies.

For a radome to be a benefit, the unique attributes of the system being protected must be taken into consideration. A well-designed radome addresses these factors and can avoid negatively affecting the performance of the antenna system. Careful selection of a radome can improve overall system performance and readiness by:

- Allowing operation in severe weather by protecting the antenna from wind, rain, snow, hail, sand, salt spray, insects, animals, UV damage, windblown debris, and wide temperature fluctuations
- Providing security for the antenna system and protecting it from observation, vandalism etc.
- Providing a controlled environment which minimizes downtime, extends component and system operating life
- Permitting the use of more economical antenna pedestals, foundations, and drive system components

11.5.3 Ground Software

Ground station visualizes and calculates the satellite location in orbit and controlling the tracking antenna. Command and control software manages command scripts to be sent to the satellite and can display and analyze telemetry. Many software options are open source and free. Other software may be purchased from companies with a long history in ground segment solutions who had previously provided hardware products to do these tasks (table 11-11).

Table 11-11: Software for Ground Systems

Product	Manufacturer	TRL	Type of Product
softFEP	AMERGINT	9	Emulation ground systems software
OpenSpace quantumFEP	Kratos	9	Software that performs data formatting and interface conversion for commands and telemetry, with full support for NSA Type 1 and AES encryption/decryption devices
Gpredict	Alexandru Csete	9	Open-source software that tracks satellites and provides orbit prediction in real-time. Radio and antenna rotator control for autonomous tracking
GNU Radio	GNU Project	9	Free software development toolkit that provides signal processing blocks to implement software-defined radios and signal processing systems
HWCNTRL	DeWitt & Associates	9	Ground station control program with an automation software package

AMERGINT softFEP

AMERGINT softFEP applications are deployed virtually on cloud architectures or hosted on dedicated servers. The applications perform control center data formatting and interface conversion for commands and telemetry, with full support for NSA Type 1 and AES encryption/decryption devices. SoftFEP applications are built on a proven library of more than 1,000 software devices. This allows each softFEP application to be tailored to the requirements specific to the ground system. Processing chains configured via Python scripts move satellite downlink data from Earth receipt for processing and uplink data to the radiating site. Deploying softFEP on multiple virtual machines (VMs) or within the cloud is inherent in the product architecture. Virtualized softFEP deployments support a wide range of ground system architectures while taking advantage of cloud-computing benefits. When applications are deployed in VMs, they can be hosted locally or run remotely in a cloud and interoperate across network connections. Customers have deployed their softFEP applications as independent network gateways, black front-end processors, red front-end processors, and data recorders, flowing data between the VMs as a satellite contact is processed (16).

Kratos quantumFEP

OpenSpace quantumFEP (qFEP) is a virtual front-end processor for Telemetry, Tracking, and Command (TT&C). The system has been specifically designed to match the requirements, schedules, and budgets of quick turn programs. qFEP connects C2 systems to RF signal processing equipment, handling command and telemetry stream formatting, encryption/ decryption devices, CCSDS processing, and network interfaces to either quantumRadio (qRADIO) or third-party ground antenna networks. Virtual machine environments are supported, and qFEP's small memory footprint allows for more efficient use of system resources. In addition, the product achieves independence by eliminating custom drivers, firmware, and hardware cards. Figure 11.16 provides an illustration of the qFEP system architecture (17).

Key features of quantumFEP are:

- Pure software implementation for signal processing functions
- Deploys on bare metal machines, in a private or public cloud
- Suitable for all types of satellite programs
- Compatibility tested with widely used ground modems
- Built-in test functions reduce Integration and Test (I&T) effort and costs
- Configurable as mission requirements change or as new missions come online
- Commercial AES Encryption/Decryption standard feature with built in AES Key Manager
- Standard TCP/IP, GEMS, REST and VITA-49 interfaces make integration seamless
- Access and control from anywhere through the web. No client software to install or maintain

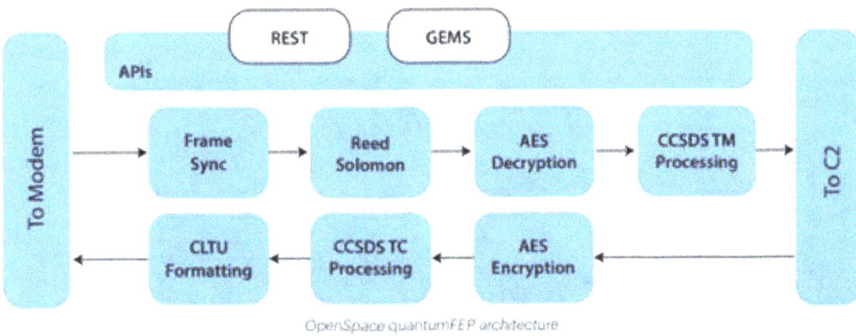

Figure 11.16: Kratos qFEP system architecture. Credit: Kratos.

Gpredict

Gpredict is a real-time satellite tracking and orbit prediction application. It can track a large number of satellites and display their position and other data in lists, tables, maps, and polar plots (radar view) as shown in figure 11.17. It can also predict the time of future passes for a satellite and provide detailed information about each pass. Gpredict is different from other satellite tracking programs in that it allows the satellites to be grouped into visualization modules. Each of these modules can be configured independently from others, allowing unlimited flexibility in the look and feel of the modules. It will also allow satellite tracking relative to different observer locations at the same time (18).

The following are key features of the software:

- Fast and accurate real-time satellite tracking using the NORAD SGP4/SDP4 algorithms
- No software limit on the number of satellites or ground stations
- Appealing visual presentation of the satellite data using maps, tables and polar plots (radar views)
- Allows satellites to be grouped into modules, each module having its own visual layout, and being customizable on its own. Of course, several modules can be used at the same time
- Radio and antenna rotator control for autonomous tracking
- Efficient and detailed predictions of future satellite passes. Prediction parameters and conditions can be fine-tuned by the user to allow both general and very specialized predictions
- Context sensitive pop-up menus allow future passes to be quickly predicted by clicking

on any satellite

- Exhaustive configuration options allowing advanced users to customize both the functionality and look & feel of the program
- Automatic updates of the Keplerian Elements from the web via HTTP, FTP, or from local files
- With a robust design and multi-platform implementation, Gpredict can be integrated into modern computer desktop environments, including Linux, BSD, Windows, and Mac OS X
- As free software licensed under the terms and conditions of the GNU General Public License, it can be freely used, learned from, modified, and re-distributed

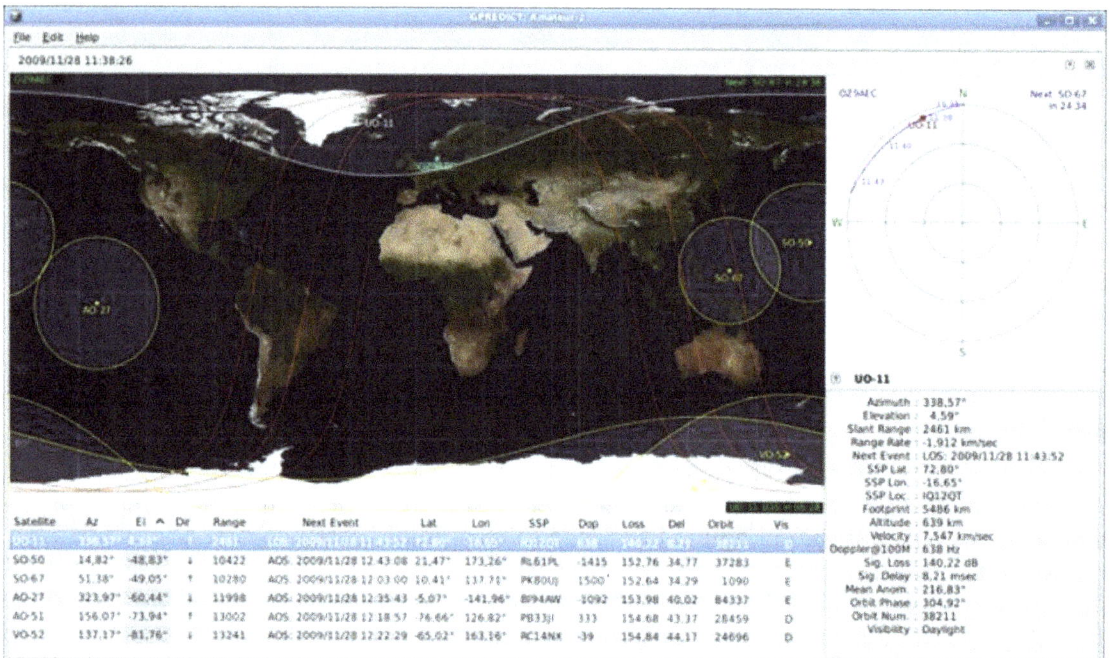

Figure 11.17: Gpredict graphical display with multiple satellites. Credit: Gpredict.

GNU Radio

GNU Radio is a free & open-source software development toolkit for developing radio systems in software as opposed to completely in hardware. It can be used with readily available low-cost external RF hardware and runs on most modern computers to create software-defined radios. It can also be used without hardware in a simulation-like environment.

GNU Radio performs all the signal processing and can be used to write applications to receive data out of digital streams or to push data into digital streams, which are then transmitted using hardware. GNU Radio has filters, channel codes, synchronization elements, equalizers, demodulators, vocoders, decoders, and many other elements (referred to as blocks) typically found in radio systems. More importantly, it includes a method of connecting these blocks and then manages how data is passed from one block to another. Blocks can be implemented in C++ or Python.

Since GNU Radio is software, it can only handle digital data. It operates with digitally sampled waveforms that can correspond to an RF signal with digital or analog modulation. Usually, complex baseband samples are the input data type for receivers and the output data type for transmitters. Analog hardware is then used to shift the signal to the desired center frequency. That requirement aside, any data type can be passed from one block to another–be it bits, bytes, vectors, bursts, or more complex data types. GNU Radio supports heterogeneous computing, where some of the blocks run on an FPGA or GPU. These acceleration techniques are particularly important for processing large bandwidths or data rates, and also in embedded platforms, where size and power consumption are usually constrained. Execution scheduling and data movement in a heterogeneous environment are active areas of development and research in the GNU Radio runtime. GNU Radio can be used on embedded systems, such as ARM SoCs running Linux, as well as on more powerful desktop or server PCs. GNU Radio is frequently used as part of the ground station, both for standard protocols such as CCSDS and for custom modems. Some commercial ground-station-as-a-service solutions that support GNU Radio modems are Azure Orbital Ground Station and AWS Ground Station. Another example is the open-source community-driven SatNOGS network. GNU Radio is also very useful for prototyping and lab testing. Additionally, some small satellites run GNU Radio on-board, usually as part of a highly flexible SDR payload. Figure 11.18 shows an example GNU Radio block diagram (19).

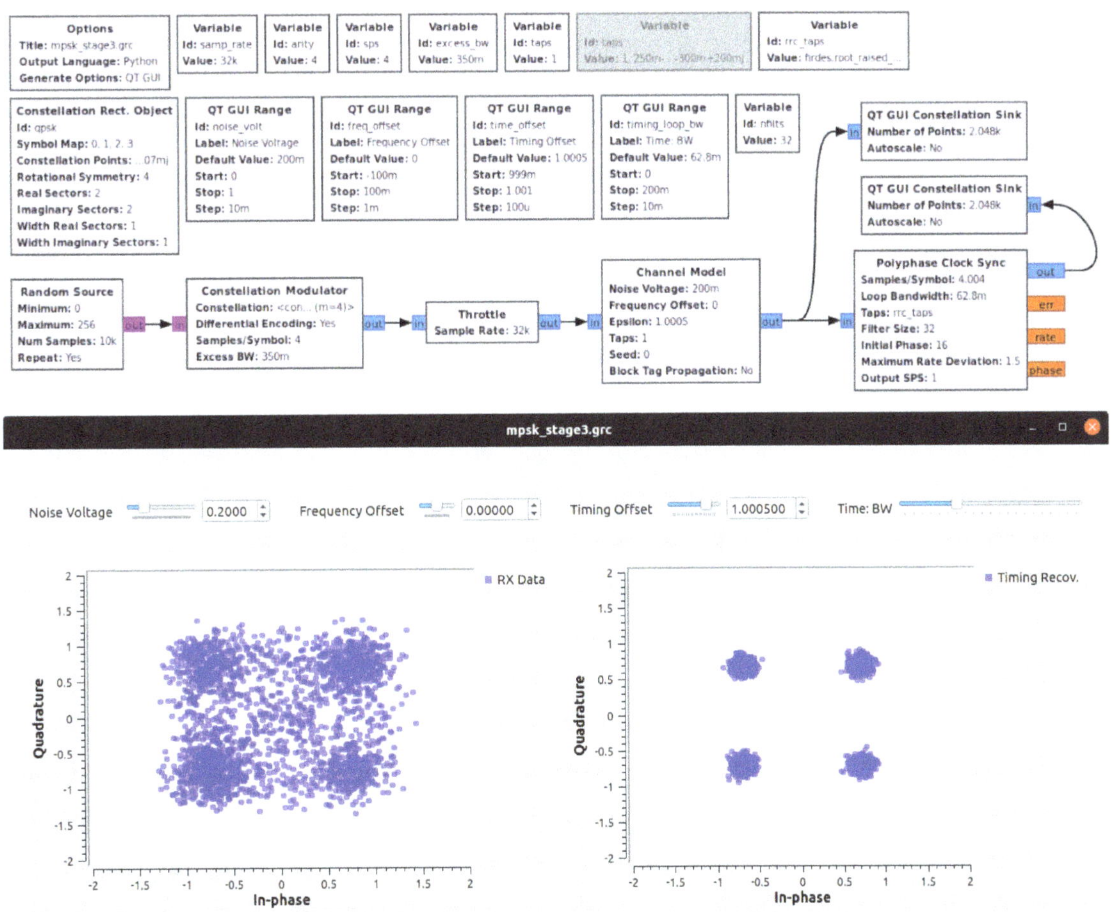

Figure 11.18: GNU Radio block diagram example for a 2-meter NBFM receiver. Credit: GNU Radio.

HWCNTRL

HWCNTRL is a satellite ground station control program that is installed in more than 30 sites throughout the world. This automation software package can support multiple antennas and instruments simultaneously. Satellite passes are generated by user request based on the ephemeris set, and users can select specific passes to be added to the schedule. Scheduled events can be single-use or reoccurring on a daily or weekly basis. A control/status screen is accessible for each instrument in the system, and the user can view and change the settings of any instrument through these screens (20).

11.6 Mission and Science Operations Centers

The MOC is where all satellite commanding is generated, ground station control is managed, and satellite telemetry is archived. It is typically a physical location where everything required to operate the satellite is located. It is often in a secure room with controlled access to protect the satellite operating equipment and prevent unauthorized satellite control. Inside the room are typically several terminals so that multiple subsystem experts can be reviewing telemetry or running their analysis programs concurrently. An example of a MOC with multiple terminals is shown in figure 11.19.

The size of the MOC is determined by the complexity of the mission. There are more experts on during critical events or to resolve an anomaly. For a SmallSat mission, the complexity is usually lower and the MOC is a much smaller room. In addition to the terminals and telemetry analysis software are other resources for managing the satellite. These may include physical models of the satellite to study when contemplating anomalous telemetry. In the case of CubeSats, due to their small size, a functioning spacecraft engineering model may be useful to test commands and reproduce anomalies.

All tasking requests for future satellite operations are managed by the mission operations team. They will generate command plans, simulate satellite response to verify those plans, and if confidence in the simulations is not sufficient, they will run the commands on engineering model hardware prior to approving them for upload. The MOC team will also manage downloads. They will decide what ground resources are available when. If the MOC does not own its ground stations, a request for contact will be submitted to the ground station managing company. The

Figure 11.19: MOC at NASA Ames Research Center. Credit: NASA

MOC submits data necessary for commanding the satellite for upload which includes commands and parameter settings for the payloads, a schedule of events for the flight computer, and ephemeris and pointing tables for the attitude control system along with its own timeline of events. When the contact is complete, the data will be sent back to the MOC by the ground station.

Prior to a launch, there will be rehearsals with everyone at their stations, and simulated telemetry with anomalous readings inserted will be used to test the team. This ensures that they are ready with the proper analysis software or integration test data available to quickly diagnose the problem and propose aplan of action. At the time of launch, the MOC will be fully populated, as this is a critical event. Telemetry will have to be interpreted and acted upon in short order.

The SOC is the focal point for all mission science data. The science team will use it to store and analyze the data. From that analysis, the science team generates satellite tasking requests that are sent to the mission operations team. External requests for additional data collection come through the science team first to assess feasibility with the instrumentation before tasking requests are made to the operations team.

The SOC is typically separate from the MOC. The payload developer will have their own operations center located at their facility with easy access to supporting resources. Before cloud data storage, the SOC was a physical place was where data servers resided to archive the mission science data. Prior to secure network solutions, dedicated computers were located inside the SOC that would run programs written specifically to analyze the science data. If the mission was secure and the data classified, then the physical SOC would be protected behind a locked door. Missions that do not produce classified data can take advantage of a virtual SOC instead of a physical location and the science data and special programs for analyzing data can reside in the cloud. The virtual SOC allows scientists to log on from anywhere and perform work without the need to come to a physical location and pass through secure doors. In the future, as cyber security techniques improve, it is likely that more and more secure missions will be comfortable with the virtual SOC solution and only the highest classification missions will maintain a secure physical SOC.

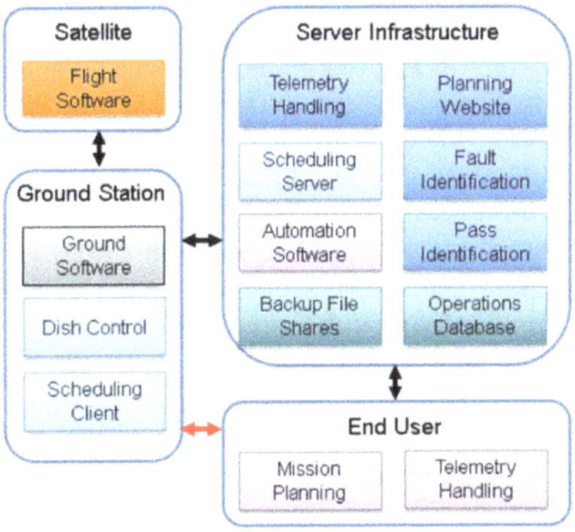

Figure 11.20a: Software functions for elements within the ground segment. Credit: NASA.

11.6.1 Software for Mission Operations

Mission operations rely on software across all the elements of the ground segment. Figure 11.20a outlines software functions for each of these elements. Satellite flight software not only manages state-of-health telemetry and payload data, but also software specific to the ground segment. Figure 11.20b provides an example of a command and telemetry data flow for a mission using DTE and relay services. Transmission can start autonomously by programming the satellite to know when it is over a ground station or within sight of a relay satellite. It can also be triggered by a command received from the ground station or relay satellite. When a communications link is established, the radio enters a higher power transmit mode and sends the data. The flight software manages the flow control of information into and out of the radio, making sure that no buffers are overflowed. It also formats the housekeeping and science data to be transmitted into a packetized file format that can be accepted by the ground station. Ground networks have specific data protocol standards developed from experience. For example, the NASA's NSN incorporates standards proposed by the CCSDS. The flight software unpacks received packets, retrieving the uploaded commands and data.

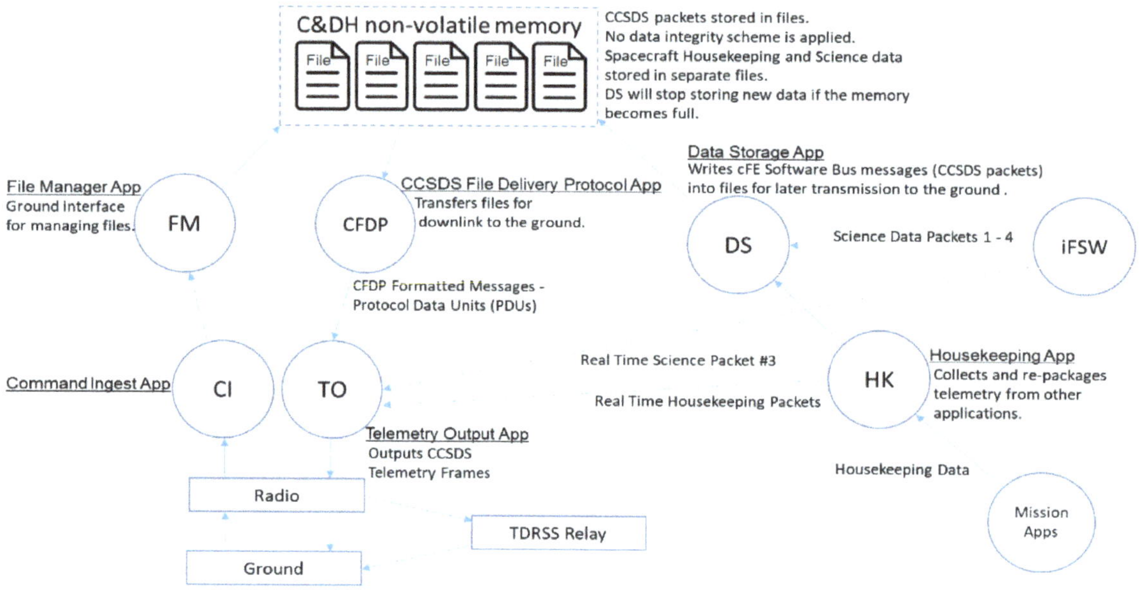

Figure 11.20b: Command and telemetry data flow for a SmallSat mission using DTE and relay services. Credit: NASA.

Software supporting the ground segment exists in the satellite, at the physical ground stations and in the MOC (server infrastructure and end-user software). The ground station uses various software for controlling the antenna, commanding, signal formatting and encoding, scheduling passes, and interfacing with the MOC. One software computes the pointing direction by using a Two-Line Element set (TLE) to define the satellite motion, an accurate model of the pointing system mount, and GPS time. It generates motor commands as a function of time. The motor controller uses these commands to actively track the satellite during a pass. During the pass, another software suite is used to monitor the link, process and encode commands for transmission, handle any signal formatting or encryption, and demodulate and decode the received transmissions. This software also manages the network connection with the MOC over which the TLE is passed, as well as data for uploading and requests for data to be downloaded. When the contact is complete, the data received from the satellite is transferred back to the MOC. The ground station may also have its own telemetry for that contact. That data is used to trend its performance. Trending the performance of each contact provides insight and notice of degradation for both the satellite and the ground station. The ground station may also use scheduling software when handling multiple missions. This software uses orbit simulation and current TLE information to determine when the contacts are expected. It will indicate when there are conflicts between contact opportunities and can assist with schedule optimization. A schedule is generated for a given period and then programmed into the ground station control system for execution. This process can be automated, but there is typically an operator on staff to monitor the system.

For the MOC, mission planning software is necessary for missions that require complex satellite behavior such as pointing at a target during science data collection. The software will include a model of the satellite dynamics and the capability of its components. The event is planned by listing a series of actions that must occur in a certain order and are spaced out by times that are approximated. The software will simulate the satellite response and then the times and actions are iteratively adjusted as needed to optimize the plan and not cause a satellite fault condition. The output of the plan is all the commands and databases that are required by the satellite. This output is submitted to the ground station ingest software for upload at a time prior to the planned event.

The SOC uses software to handle the receipt, unpacking, reconstruction and post processing of the mission science data. Using an ISS payload as an example, the science data is downlinked via TDRSS to NASA Marshall Space Flight Center (MSFC) where it is separated into different science streams and piped to the correct payload SOCs. At the SOC, but outside the company firewall, a computer is constantly running and ready to receive the data from MSFC. On that computer, the TReK software provided by NASA is running and it properly handshakes with the MSFC software assuring the data transfer. The science team periodically retrieves the data and safely brings it through the corporate firewall into the SOC. The science team writes parsing software to unpack the data which is stored in CCSDS format. They write another software to arrange the data back into the original image seen by the payload. Still more custom software will process the image to produce post-processed data products that are stored in the SOC archive and distributed to interested customers. The computer languages vary but Interactive Data Language (IDL) and Python are common choices for this type of software.

11.7 End-to-End Communications and Compatibility Testing

A SmallSat undergoes various tests through its development cycle to verify proper functionality. For the communication subsystem, end-to-end communication and compatibility testing with the selected ground network is its most critical test. Compatibility testing verifies that the ground station can properly communicate with the satellite on the uplink and downlink RF channels. Ideally, compatibility would be validated by testing the flight spacecraft with the actual ground station that will be supporting the mission. This may not be practical for larger or high-cost satellites, due to logistics associated with shipping and risk of damage. Two alternatives to shipping the satellites are typically used. One includes sending a replicate set of ground station hardware to the satellite facility for testing. A second option is to test with only the flight or an ETU radio (also common to include the flight computer) at the ground station or at a test lab configured with the ground station hardware. Drawbacks to the alternative options would include not testing the exact command path or determining whether ground sensitivity is sufficient.

For CubeSats, it is commonly feasible to bring the CubeSat to the ground station for testing. If that is not feasible, then at a minimum, the radio and flight processor (or Engineering Development Units [EDUs]) should be used. Testing at the ground station allows for the entire equipment chain to be part of the test, including the low-noise amplifier (LNA) and transmit/receive switch, if used. It is desirable to first test in a closed-loop configuration, where the satellite is connected to the ground system at the antenna port via a cable (with appropriate attenuators in line). If the satellite is fully integrated, disconnecting the flight antenna may not be feasible. In this case, a small monopole antenna located indoors near the CubeSat can be connected to the ground system. The monopole antenna connection to the ground system may vary depending on the ground antenna configuration but should include as much of the ground system electronics as practical.

Some missions elect to include an outdoor open-loop test with the CubeSat and ground antenna. This method allows for the entire ground system, including the ground antenna, to be included in the test. However, the ground antenna typically cannot point directly at the CubeSat due to mechanical limitations or to limit the received signal so the ground system RF components will not be overdriven. Off-pointing and reflections from the ground and local structures can also make it difficult to achieve a valid test.

End-to-end network testing primarily validates the ground station to MOC interface. This test verifies that the MOC can properly receive downlink data from the ground station and verifies that the ground station can receive and process uplink command data from it. Initial end-to-end testing will validate network connectivity, showing that network connections can be established and firewall rules at the ground station and MOC are in place. Once network connectivity is established, the MOC can transmit commands to the ground station for capture. The ground station can then transmit simulated or recorded data to the MOC for validation.

It is preferable to conduct initial end-to-end network testing prior to compatibility testing. In cases where the satellite can be brought to the ground station, a full end-to-end test can be conducted. Command transmissions from the MOC, through the network and ground system to the satellite can be validated. A complete end-to-end telemetry dataflow from the satellite to the control center can also be validated.

11.7.1 End-to-End Hardware for Ground Systems

A complete ground system can be provided as a kit with all the necessary components bundled together and setup to work seamlessly. These end-to-end solutions include the antenna, its controller, and the RF feed with all the necessary filtering and low noise amplification for the particular wavelength of interest. They use a software defined radio or a dedicated transceiver to convert between digital packets and RF waveforms. Software is included to process the satellite position and direct the antenna to track it. Additional software is used to archive and display the information within the digital packets. Three vendors, GAUSS, Innovative Solutions In Space (ISISPACE) and GomSpace, listed in table 11-12 provide solutions for the low-cost CubeSat and small satellite market. One vendor, Surrey Satellite Technology Limited, offers a higher end system, installation service, and personnel support. The final vendor listed, Kratos, offers a different end-to-end solution that begins with the digitized RF waveform. The Kratos Quantum software then demodulates, filters, unpacks, parses, displays, and archives the data (17).

Table 11-12: End-to-End Hardware for Ground Systems

Product	Manufacturer	TRL	Type of Product
Complete Ground Solution	GAUSS	9	Small satellite provider offering a complete ground solution. UHV, VHF, and S-band
Complete Ground Solution	ISISPACE	9	Small satellite provider offering a complete ground solution. UHV, VHF, and S-band
Complete Ground Solution	GomSpace	9	Small satellite provider offering a complete ground solution. UHV, VHF, and S-band
Surrey Ground Segment	Surrey Satellite Technology Ltd.	9	Major contractor who will install ground stations capable of S-band for U/L and D/L and X-band for D/L
Quantum	Kratos	9	Major contractor with a complete ground solution

GAUSS Ground Station Kit

The GAUSS ground station is a turnkey solution. It can be configured with UHF, VHF and S-band on the same pointing system. An example of the associated hardware is shown in figure 11.21. Hardware features of the systems offered include (21-23):

- High gain Yagi-Uda VHF and UHF antennas (>16 dBi for UHF)
- Low-noise amplifiers and band-pass filters for VHF and UHF bands
- Low-loss RF coaxial cables
- 1.5-meter parabolic dish for higher frequencies downlink (up to 6 GHz, default feed is for S-band)

- VHF: uplink and downlink up to 100 W using radio and Terminal Node Controller (TNC), software defined radio (SDR) optional
- UHF: uplink and downlink up to 70 W, using radio and TNC, SDR optional
- TX using ICOM-9100 hardware, RX recording and decoding via SDR
- Several RF and electrical fuses for lightning protection
- S-Band: downlink using SDR for recording and post-processing of I/Q RF data
- Az/El rotor for high-torque maneuvering
- Hardware components power switch on/off to minimize power consumption
- Full HD camera for instant antenna monitoring and picture logging

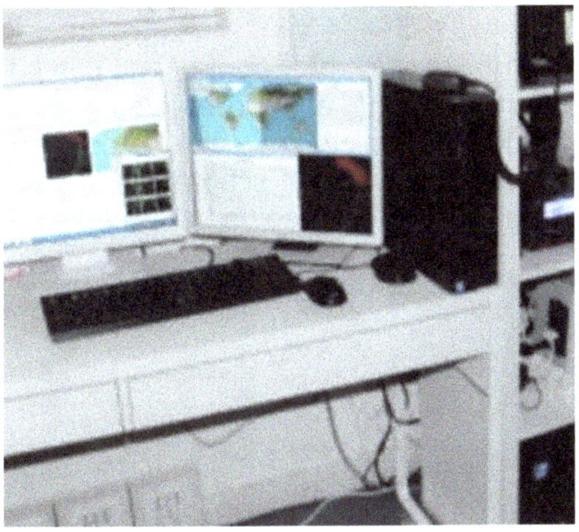

Figure 11.21: (left) GAUSS ground station hardware, transceiver and (right) tracking antenna. Credit: GAUSS Srl.

The features of the software that accompanies the system include:

- Automatic TLE download from publicly available repositories
- SGP4 propagator as suggested by USAF NORAD's Space-Track
- Rotor control (compatibility with several rotor controllers, e.g. Yaesu, RF Hamdesign)
- Assisted rotor pointing calibration and verification using Sun position
- Fully compatible with ICOM-9100 satellite radio and GAUSS USB ground dongle
- Separated Doppler shift corrections for uplink and downlink frequencies
- DUPLEX TX/RX mode
- Instant weather check and logging to operate the ground station safely
- Lightning detection for safe antennas operation
- Instant logging of all subsystems operation
- Ground map with live Earth clouds
- Compatible with several TNCs (Kantronics, Symek, Paccomm, Kenwood)
- Email report to ground station operators
- Instant email alerts for non-nominal conditions of the satellite or GS hardware components
- Session programming for weeks of unattended ground station operations

- GUI command recording for easy session programming
- One button programming to include a whole set of commands in the session
- Manual override during pass for last-minute command addition
- Control and handling of multiple satellites using configurable priorities
- Satellite TLM decoding, graphing, and archiving into a database accessible by web
- Integrated satellite payload data handling and decoding (e.g., for image file processing)
- TCP/IP connections for remote ground station & TNC operations

Innovative Solutions In-Space Ground Station Kit

The ISISPACE small satellite ground station is a low-cost, turnkey solution that is designed to communicate with satellites in low-Earth orbit that operate in either amateur frequency bands or commercial bands. The frequency bands covered are S-band, UHF, and VHF. The ground station consists of an antenna and a 19" rack which houses the transceiver, rotor control and computer which make the system very compact. Examples of these components are shown in figure 11.22. The transceiver makes use of a SDR that provides flexibility to swiftly reconfigure modulation/coding/data-rate on the run. Most commonly used modulation schemes and coding methods are already implemented, and any customization requests can also be handled (24).

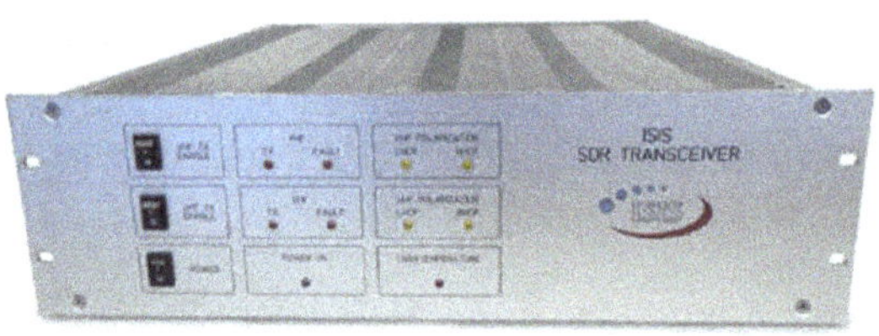

Figure 11.22: (left) ISIS ground station hardware, transceiver rack and tracking antenna (right). Credit: ISISPACE.

GomSpace Ground Station Solutions

The GomSpace end-to-end solution is unique from other vendor offerings because a generic software defined radio is replaced with their AX100 or TR-600 radios, depending on the type of radio the in-orbit satellite uses to communicate. Using the same transceiver hardware on both sides of the link simplifies the configuration and validation testing steps in the integration and test (I&T) phase of the project. While the GomSpace solution does not work with satellites that do not use the GomSpace transceivers, the benefit is lower cost and simpler ground segment equipment (25). GomSpace also offers the Hands-off Operations Platform (HOOP), a satellite operations service to permit autonomous satellite operations for single spacecraft and constellations. The HOOP is compatible with GomSpace ground stations and leading ground station network providers and has been operational since 2020 (26).

Surrey Satellite Technology Ltd. Ground Station Kit

Surrey can provide complete turnkey ground segment solutions for a range of space platforms, including all the hardware and software necessary to operate, maintain, process and archive data. Services provide by Surrey include:

- S- and X-band ground stations with full motion antenna systems from 2.4 meter to 7.3

meter in diameter, with radome options available for harsh climates

- SSTL Pilot Satellite Control Software
- Mission planning systems
- Radiometric and geometric image processing
- Data storage solutions
- Site surveys, ground segment installation and training
- Technical and maintenance support packages

In addition, Surrey can work with customers to integrate their ground segment solutions with existing ground infrastructure or with 3rd-party ground station networks (27).

Kratos Ground Station Solutions

The Kratos unique ground solution begins with the OpenSpace SpectralNet® digitizer that converts analog signals at RF frequencies up to S-band into digital IF packets. It is the start of the Kratos digital processing product line chain. Kratos quantum software products operate on a fully digitized RF waveform. For example, a ground station service company would maintain the antennas and modems and use a very good internet connection to ship a huge amount of data either into the cloud for storage and processing with the Kratos quantum software, or to the customer Mission Operations Center. Figure 11.23 provides a visualization for the system concept.

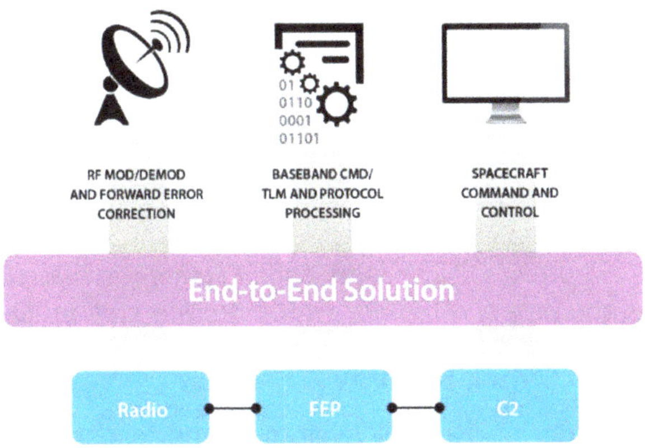

Figure 11.23: Visualization for the Kratos quantum system concept. Credit: Kratos.

All components are available separately to support an existing C2 solution or third-party ground network with existing signal processing and antenna resources. The quantum products include:

1) quantumCommand (qCMD), COTS software application for command and control (C2) of small spacecraft;
2) qFEP, connects C2 systems to RF signal processing equipment: handling command and telemetry stream formatting, encryption/decryption devices, CCSDS processing, and network interfaces to either qRADIO or third-party ground antenna networks;
3) qRADIO, the software modem for RF signal processing on premise or in the cloud;
4) quantumDRA (qDRA), a data recording and archiving application supporting CCSDS/non-CCSDS header and channel data routing with IP-based interfaces;
5) quantumRX (qRX), a fully virtualized wideband software receiver, specifically tuned to streaming earth observations in near-real time with 600 MHz bandwidth using Digital IF digitizers.
6) quantumTX (qTX), a fully virtualized wideband software transmitter, specifically tuned for earth observation and Remote Sensing satellites with over 1Gbps throughput for uplinks.

In 2021 Kratos introduced a virtual, software-defined architecture solution called the OpenSpace Platform. As an enterprise level, end-to-end system, it provides the SmallSat community the flexibility to scale on-demand as their operations grow in size and capability. By leveraging Digital IF over IP with time deterministic latency and software defined networks, the Platform allows virtualized functions such as modems, channelizers, recorders and combiners to be orchestrated in a cloud environment. The virtual architecture lends itself to upgrades and/or updates automatically, ensuring ongoing reliability and security. In addition, the ability to test software releases in real-time, allows ground equipment strings to be included in continuous integration and continuous delivery cycles. Software defined architectures are more agile, programmable, and automated, enabling the ground system to work in tandem with dynamic satellite payloads. By shifting from RF signals and analog equipment to a virtualized, IP-based infrastructure, orchestration can occur on the fly. Figure 11.24 provides an illustration of the OpenSpace architecture concept (29).

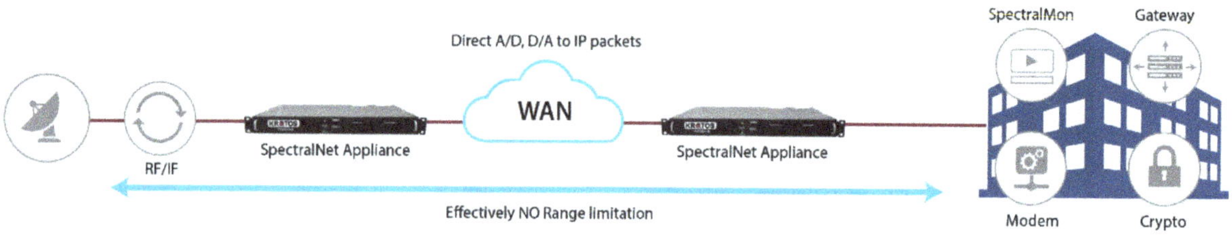

Figure 11.24: Kratos OpenSpace architecture concept keeps most of the RF ground equipment remote from the ground station. Credit: Kratos.

11.8 Cyber Security

Security of a space system needs to consider all aspects of the system, including the space platform, payloads, and all supporting functions. To a remote attacker, the most accessible portion of a spacecraft is the end-to-end command path as accessed through direct contact (RF link), subversion of any command path transports (space or ground networks), and subversion of the command authority (e.g., MOC). Another enticing option for an attacker is to cause a mission impact through manipulation of key space system dependencies, such as the Global Navigation Satellite System (GNSS), ground stations, and external service or data providers. Related concerns include degrading or denying the use of the command path through jamming or denial of service of the command path and supporting functions. Finally, vulnerabilities in any system, sub-system, or component, can be exploited by an attacker in creative ways to ultimately gain the ability to affect the overall system. Supply chain risk management is needed to address some portions of these challenges.

Effective space system security efforts begin early in the lifecycle and continue through to mission termination. Early architecture and design decisions will have the most significant impact on the overall system's security outcomes and can help avoid rework in later phases. Integrating cybersecurity capabilities into a system fits well into standard system engineering practices. Many organizations are establishing a sub-discipline for systems security engineering, to serve as a focal point within the mission team to enable system security. This may overlap with or work closely with related roles such as a cybersecurity systems engineer. The International Council on Systems Engineering (INCOSE) has established a working group to support system security engineering, and NASA is actively implementing a system security engineering capability (30).

Operators need to maintain command authority over their spacecraft, preventing unauthorized access. Use of authenticated encryption, between the point where the command sequence is generated and the spacecraft, is the best method to ensure command authority and data integrity. Encryption provides confidentiality, and authentication ensures that a trusted source initiated the command. This approach generally addresses initial concerns about an attacker attempting to gain access through use or subversion of the command path to the spacecraft.

Key spacecraft external dependencies can be manipulated to cause a mission impact. Example dependencies might include the ground station and GNSS signal. Ground station impacts could disrupt data flows with the spacecraft. Jamming and measurement spoofing of GNSS signals has become common in various regions, with observed impacts in the maritime and airspace domains. Orbiting systems have detected similar effects. Space systems should be prepared to tolerate loss or interference with GNSS signals.

For the overall space system and its component parts to function properly and securely, each part must in turn be sufficiently secure. Traditional cybersecurity efforts apply well to most software and terrestrial systems. Attention to the security of each component, as well as ensuring the interplay between components is secure, will help protect the overall system. On-board security should also be considered, particularly for multi-customer and multi-payload spacecraft. Increasingly, spacecraft developers should consider that external defenses (e.g., command authentication or encryption) may be bypassed or subverted (similar to how network firewalls may be bypassed). Without further consideration, the on-board systems are likely vulnerable to further exploitation.

Supply chain risk management is an essential related discipline. Particularly for key components and software, understanding the vendor's sourcing, manufacturing approach, and cybersecurity and assurance management will help ensure any associated risk can be identified and managed. If a vendor relies on other vendors, additional scrutiny may be appropriate. Review of the sourcing may address whether the vendor is rebranding, assembling/integrating, or internally manufacturing components, and whether there is sufficient control to deliver a trustworthy product. Reviewing the manufacturing approach, whether hardware, software, or some combination, allows the customer to determine whether the vendor uses repeatable processes that yield deterministic and trustworthy results. And understanding how the vendor addresses cybersecurity, quality, and protection topics in their components provides insight into whether or not it is appropriate to integrate the component into the overall system.

U.S. National Guidance and Regulations

In September 2020, the U.S. National Space Council issued Space Policy Directive 5 (SPD-5), Cybersecurity Principles for Space Systems. Amongst other elements, the directive calls for use of "risk-based cybersecurity-informed engineering," anticipating and adapting to evolving malicious activities, and recommending capabilities to maintain positive control of space vehicles. Another element implies that Federal agencies may issue or update guidance, rules, or regulations to adopt the principles in this directive (31).

U.S. regulatory agencies, such as the Federal Communications Commission (FCC), have considered and not yet issued rules that may require specific cybersecurity measures to be adopted. Prior proposed (and not issued) rules included requirements for encryption on the telemetry, tracking, and command communications for propulsive spacecraft.

U.S. Agency Guidance

Several U.S. government agencies and their support ecosystem have made various frameworks, standards, and other guidance available to address cybersecurity and protection concerns.

NASA Technical Standard

NASA requires its missions, including small satellites, to comply with NASA STD-1006 "Space System Protection Requirements." This standard covers protection of the "command stack," critical information, Position Navigation and Timing (PNT) sub-system resilience, and reporting detected and unexplained interference. The tailoring guidance within the standard allows for some flexibility in certain small satellite scenarios, such as non-maneuverable systems. Protecting the command stack involves use of encryption complying with FIPS 140 (level one). PNT resilience addresses the loss of or temporary interference with external PNT services, such as a GNSS (32).

National Institute of Standards and Technology (NIST)

NIST has several useful publications addressing cybersecurity and system security engineering. The NIST Cybersecurity Framework is a voluntary guidance framework that can be used by organizations to manage cybersecurity risk. The framework provides a structured and tailorable approach for organizational security capabilities and includes informative references to other NIST documents (e.g., SP 800-53), as well as other standards or guidance organizations (e.g., International Organization for Standardization, ISO). NIST's SP 800-160 Volumes I and II offera thorough approach for system security engineering practices. In particular, SP800-160 Vol I's Appendix F, Design Principles for Security, can be used as an effective foundation for systems security (33-36).

Additional Resources

Two U.S. Federally Funded Research and Development Centers (FFRDCs), the Aerospace Corporation (37) and MITRE (38), have published guidance for space system cybersecurity. These FFRDCs are expecting to make additional recommendations widely available. Aerospace's "Defending Spacecraft in the Cyber Domain" includes a brief survey of known cybersecurity initiatives and standards, challenges with legacy engineering approaches, emerging threats, and principles for "cyber-resilient spacecraft." The paper also includes a section specific to small satellites. MITRE has published a paper "Cyber Best Practices for Small Satellite" that briefly addresses cyber threats to space systems and includes a discussion on applying lessons learned from other industries to space systems.

Consultative Committee for Space Data Systems (CCSDS)

CCSDS provides a variety of guidance documents for implementing security measures in various aspects of the mission. For an introduction to the CCSDS approach and available guidance, see CCSDS 350.7-G-2 "Security Guide for Mission Planners" that provides a perspective on approaching security in space systems (39). The CCSDS 350.0-G-3 "The Application of Security to CCSDS Protocols" informative guidance document provides an introduction and discussion on various topics, including protecting the command path (40).

11.9 State-of-the-Art – Ground Data and Supporting Systems

11.9.1 Technologies

Multiple Spacecraft Per Aperture

The Annual Small Satellite Conference on the grounds of Utah State University is the premier event amongst small satellite stakeholders, and its themes reflect the trends of the times. In 2021 one of the key talks was on how to "Maximize Contact Availability of SmallSat Clusters through MSPA Technique on GSaaS," (41).

As scientists are increasingly interested in characterizing fields (going beyond single point measurements) requiring swarms of satellites, as well as the emergence of Distributed Satellite Missions (multiple satellites working in concert towards one common goal), MSPA is a critical enabler. While it is not a new concept, few ground stations have invested in such upgrade. The DSN has the capability to track multiple spacecraft per antenna (MSPA) (up to four) if they are all within the scheduled antenna's beam. The 34 m antennas at each complex can be combined into an array, with or without the co-located 70 m antenna. The combined G/T depends on several factors but is approximately increased by the sum of the antenna areas from the arrayed apertures minus approximately 0.3 dB combining loss. For instance, arraying four 34-meter antennas results in an increase of 5.72 dB.

Automation and Modeling

The MOC of the future will include a "lights out" or fully automated option. This requires software on the ground station side to run the antenna automatically. Automation software will receive a list of times that the antenna should track the satellite and it will manage that list. It will send TLEs and data to the antenna with no one present, receive downlinked telemetry, and archive it. Software automatically parses the telemetry, compares key watch items to defined limits, and alerts the team via email or phone text message. FreeFlyer by a.i. solutions combine astrodynamics/ spacecraft propagation, coverage and contact analysis (including swarms), attitude and maneuver modeling, and orbit determination (42).

Figure 11.25: Viasat's new Large-Aperture Space-to-Ground Communication Antennas ready to support Lunar, Cislunar, Deep Space and DoD Missions. Credit: Viasat.

Large Ground Antennas: to the Moon and Beyond

For years there had been a gap between NSN's largest 18m and DSN's 34/70m antennas, and such large antennas were not available from commercial ground providers. This gap has now been filled.

In 2022 Viasat introduced new 19/24m aperture antennas (figure 11.25) at their Antenna Systems campus in Duluth, GA supporting several ongoing programs. The size and architecture of these larger apertures support current programs while offering the flexibility and scalability to support future and forward planning missions. (43).

NASA Lunar Exploration Ground Sites (LEGS)

SCaN announces the LEGS with its mission is to provide direct-to-earth communication and navigation services for missions operating from 36,000 kilometers (km) in the GEO to cis Lunar and other orbits out to 2 million km. To fully support distant orbits there will be three LEGS sites equally spaced around the Earth. The Ground sites use CCSDS Modulation and coding schemes for forward and return data. Specialized/unique Mod-Cods are optional. User Local Equipment on-site is optional. The 18m assets are listed as White Sands, USA: 32.544863, -106.612504

Matjiesfontein, South Africa: -33.231224, 20.58163 (TBD) and Pacific Region TBD. MSPA is planned for up to 4 simultaneous return services per aperture (Max 3 for Ka). Use of LEGS for other than Artemis support is TBD. See table 11-13 for projected performance of LEGS assets.

Table 11-13: Projected Performance of LEGS Assets Pending Finalization			
RF Performance Criterion	**Radio Frequency Performance (Return)**		
	S-Band	**X-Band**	**Ka-Band**
G/T (minimum)	28 dB/K	39 dB/K	47.5 dB/K

11.9.2 Ground Aggregators

Table 11-14 lists those Ground Service providers who own and or operate their own brand of ground assets. Irrespective of the nature of ownership, satellite operators are reliant on the limited ground stations they have access to. Satellite operators have well defined windows for exchanging information with their satellites. To meet evolving demand from within the fast-growing segment of the space industry, multiple aggregator models have evolved from private market participants. Services from companies such as RBC Signals, Infostellar, Amazon Web Services, and Spaceit are offered through specialist ground station capacity aggregator platforms. These are digital solutions enabling ground station operators to provide their excess capacity to a global user base. Since this is very similar to the business model of Uber, these aggregator services represent the ongoing "Uber-ization" of ground station services within the space industry. The downstream service markets are observing new players with new products and services. With increasing competition, the differentiating factors are shrinking in number. When the upstream capabilities start resembling each other, the key differentiators will include the ability to communicate with the satellites on-demand (44).

RBC Signals

RBC Signals is an innovative provider of global satellite data communication products and solutions. It offers secure space communication solutions in every major frequency band using a worldwide network of company-owned and partner-owned systems. RBC Signals delivers dynamic solutions offering affordability, flexibility, and resiliency. Its diverse products and services offer complete end-to-end solutions for best-in-class multi-network solutions (see https://rbcsignals.com).

RBC has aggregated a growing network of over 80 antennas in nearly 60 locations worldwide offering unmatched capabilities. A map of these locations is shown in figure 11.26. As of 2023, RBC owns about 15% of the ground stations, and the rest are partner stations. For customers needing turnkey access to existing antennas, RBC Signals offers ground station antenna-as-a-service, with the flexibility to secure unlimited satellite passes ("core") or 'pay-by-the-pass/minute/GB'. This is made possible through a combination of their own network of highly capable systems and the unique 'sharing economy' model, wherein they leverage the unused excess capacity of dozens of partner-owned antennas worldwide. RBC Signals also offers turnkey bring-your-own-antenna hosting solutions that pair customer-owned equipment with reliable, high-end ground infrastructure almost anywhere in the world. RBC employs a distributed compute architecture where most processing occurs at a data center/cloud, with some processing on the satellite or at the terrestrial edge at the ground station. RBC Signals can host AWS and Microsoft on premise cloud infrastructure, as well as virtual servers at the ground.

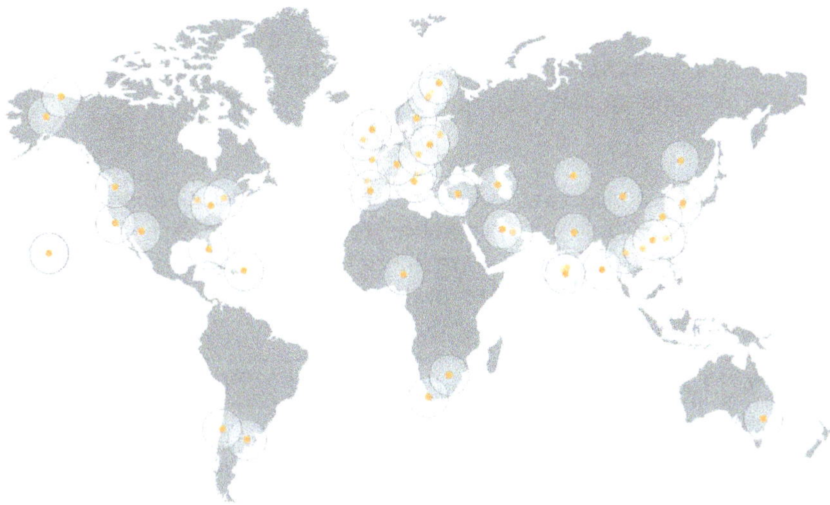

Figure 11.26: RBC Signals ground network map. Credit: RBC Signals.

Table 11-14: Service Providers for DTE Ground System Network

Product	Dish Sizes	Services	MSPA
ATLAS Global Network	Various partners with other stations	S-band, X-band, UHF (Ka-band in 2017) Built on AWS cloud infrastructure	Partner dependent
KSAT and KSAT^{LITE} by Konsberg Sat. Services	> 10 m part of NSN or 3.7 m (KSAT^{LITE})	X-band and S-band D/L and S-band U/L. VHF, UHF, Ka-band D/L. KSAT^{LITE} designed specifically for SmallSats	No
SSC Infinity by Swedish Space Corporation	13 m, 7.3 m NSN partner	Designed specifically for SmallSats; Uses standardized HW	Not found
AWS Ground Station by Amazon	5.4 m	Built on AWS cloud infrastructure	Not found
Viasat	7.3 m, 5.4 m		No
Leaf Space	3.7 m		Yes
NASA, Near Space Network	9 to 11 m, & 5 m	Global network operating in S, X, and Ka- bands that can reach LEO, GEO, HEO, and Lunar orbits; up to 2 mil km	Legacy: NoLEGS (18m): planned, Yes
NASA/ JPL, Deep Space Network	34 m, 70 m	Operating at S, X, K, Ka bands. Includes Morehead State 21m in X-band. 8 m optical receive aperture planned for 2030s	Yes
NASA UHF Ground Station	18 m	Operates in UHF (400 – 470 MHz)	No
RBC Signals Global Ground Station Network		VHF, UHF, S, C, X, Ku, and Ka-bands	Partner Dependent

11.9.3 Scheduling and Mission Operations Software

With the growing number of ground operators and aggregators, to take advantage of the plethora of assets, scheduling is emerging as the single most important enabler. As individual providers may have their own scheduling formats, for seamless operations, a common scheduler is critical, and this is true for the NASA commercialization efforts as well.

Scheduling: InfoStellar

InfoStellar offers communication services in the VHF, UHF, S, X, and Ka bands. Table 11-15 lists the frequency bands offered by InfoStellar which can be filtered by the tool on their website. Figure 11.27 shows all 34 antennas on the platform that are either live or for which integration is planned. With multiple commercial small-satellite operators in the market, the need for enhanced mission operations is much more than an industry-wide requirement (44).

Table 11-15: Select Frequency Bands by InfoStellar

Downlink		Uplink	
X Band 8 – 12GHz	S Band 2 – 4GHz	S Band 2 – 4GHz	UHF Band 300MHz – 1GHz
Ka Band 27GHz – 40GHz	None No Uplink Channel	VHF Band 30 – 300MHz	None No Uplink Channel

The following section provides an overview of mission operations and scheduling software products that can be integrated into a MOC (see table 11-16). While the specific aspects of each of these products is discussed below, they all have some common features. In general, these software applications cover functions related to mission scheduling and tasking, commanding and telemetry, and monitoring and control. Many of them also have automation features that enable "lights-out" operations or reduced manpowerrequirements.

All these products are highly customizable. They can not only adapt to multiple missions, satellites, and ground stations, but these products also allow for customized visualizations, analyses, and user interface views. Additionally, many of these products are cloud-based or have a web interface to enable easier access for an operator from almost anywhere.

Figure 11.27: All 34 antennas on the StellarStation platform that are live (25 - red) or for which integration is planned (9 - blue). Credit: InfoStellar.

Table 11-16: Mission Operations and Scheduling Software			
Product	**Manufacturer**	**TRL**	**Type of Product**
COSMOS	OpenC3	9	Open-source command and control system that can be used in all phases of testing and operations
Galaxy	The Hammers Company	9	Command and telemetry system that has been available since 2000
Orbit Logic Family of Products	Orbit Logic	9	Group of mission planning and scheduling products for both aerial and satellite imaging applications
ACE Premier Family of Products	Braxton Technologies	8+	Group of hardware and software components for end-to-end Satellite Operations Center (SOC)
Mission Control Software	Bright Ascension	8+	Monitoring and control interface with "lights-out" automation features built-in
Major Tom	Xplore	8+	Cloud-based command and telemetry system that can interface with some COTS flight software

OpenC3 COSMOS

OpenC3 COSMOS is a free, open-source, open-architecture, command, control and communications system providing commanding, scripting, and data visualization capabilities for embedded systems and systems of systems. With the release of version 5, COSMOS is now a fully containerized and microservice based architecture, with a web frontend. COSMOS is intended for use during all phases of testing (board, box, and integrated system) and during operations. OpenC3 COSMOS is made up of a set of applications that can be grouped into four categories: real-time commanding and scripting; telemetry visualization; offline analysis; and utilities. Figure 11.28 shows how data flows through the microservices and is made available to users through an API and from a web-based interface. Any embedded system that provides a communication interface can be connected to COSMOS. All real-time communications of both commands and telemetry are logged in a cloud-native data store, which can use local hardware, or cloud-based buckets for potentially infinite storage. Additionally, program specific tools can be written using the open-source OpenC3 COSMOS libraries, and these tools can interact with the commands and telemetry of all targets connected to the system as well (45).

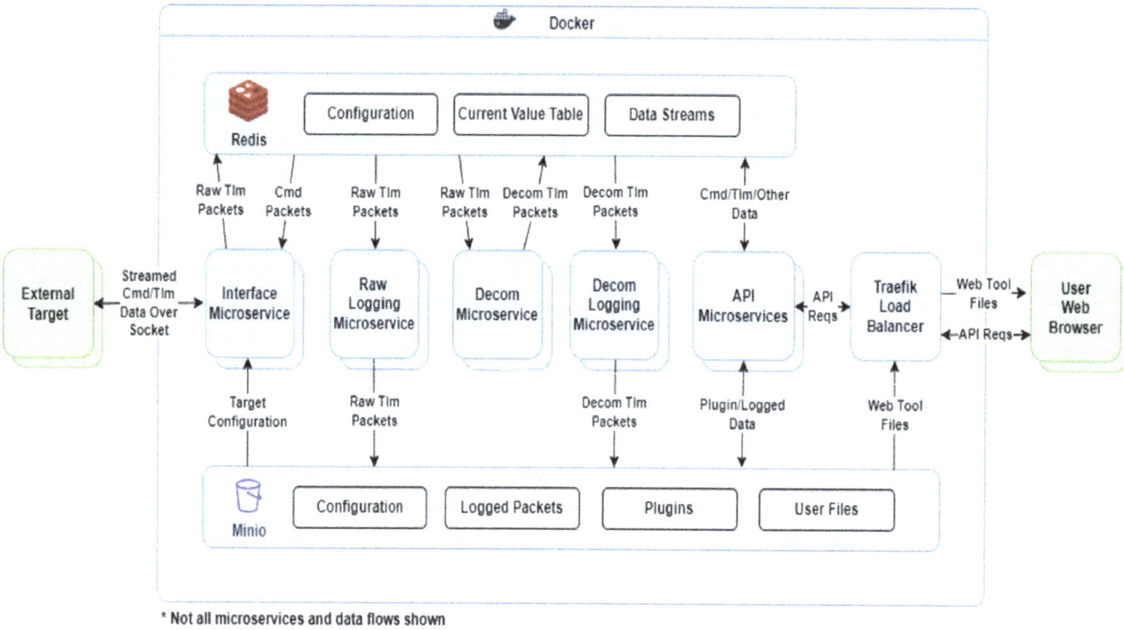

Figure 11.28: COSMOS5 architecture and context diagram. Credit: OpenC3, Inc. https://openc3.com/

Galaxy

Galaxy is a command and telemetry system that is derived directly from the Integrated Test and Operations System (ITOS) telemetry and command system developed by the Hammers Company with NASA GSFC. It has been available commercially since 2000. Galaxy can accept telemetry from, and send commands to, multiple spacecraft and ground stations simultaneously. Users can customize Galaxy for a particular mission via a database in which they provide telemetry and command specifications. Users can design telemetry displays, plots, sequential prints, configuration monitors, and spacecraft commands and table loads in simple text files stored on the computer's file system. Most displays can be viewed remotely over the web or by using remote Galaxy instances. Additionally, Galaxy is CCSDS compliant, and it can communicate over a wide variety of transports and protocols including TCP/IP networking, synchronous and asynchronous serial ports, SpaceWire, military standard (MIL-STD-1553), and the GMSEC message bus (46).

Major Tom

Xplore's Major Tom® is a scalable, cloud-based mission control software that provides satellite mission operations and planning tools for ground station scheduling, satellite tasking and telemetry monitoring. It provides satellite operators the ability to integrate and control ground segment applications and services, and further de-risks mission operations with features that include out-of-the-box ground network integrations, data analytics, real-time dashboards, and a customizable commanding API. Major Tom's gateway API architecture, definitions, and protocols can integrate wth custom ground station provider(s). This software is compatible with a heterogenous network of ground station providers and supports constellation and mission operations flexibility in satellites communication and commanding. New ground station network locations and providers can be added, and the same data model will unify accessing their functionality. Interface capabilities include control panel scheduling and pass management as well as data interface for downlink and uplink. Integrated ground station providers are ATLAS Space

Operations, Inc., Leaf Space and Microsoft Azure Orbital. Figure 11.29 provides a screenshot of the dashboard (47).

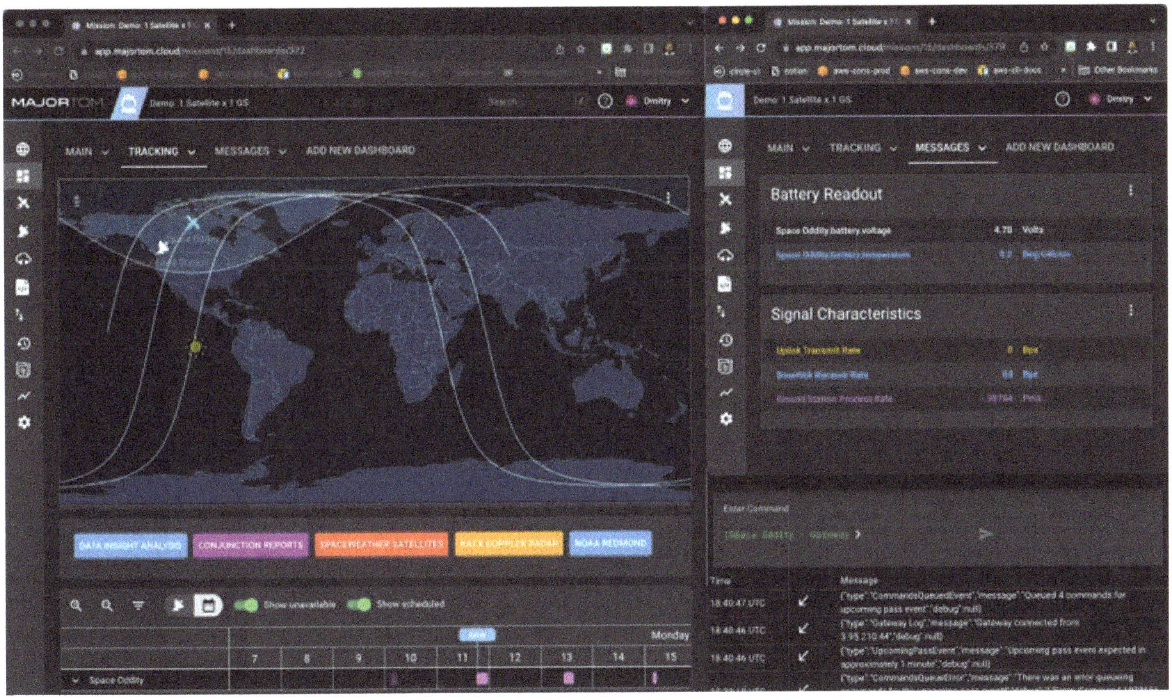

Figure 11.29: Major Tom's configurable dashboards allow operators to oversee and interact with the mission in the way they are most comfortable. Credit: Xplore.

Orbit Logic Family of Products

Orbit Logic specializes in mission planning, scheduling, and space situational awareness software. The software suite consists of multiple applications that support analysis and operations for aerial and satellite imaging and space-to-ground networking. The mobile, web, desktop, and onboard scheduling applications have a variety of features, including: configurable systems, constraints, and goals; high performance algorithms; deconflicted scheduling plans; visualizations

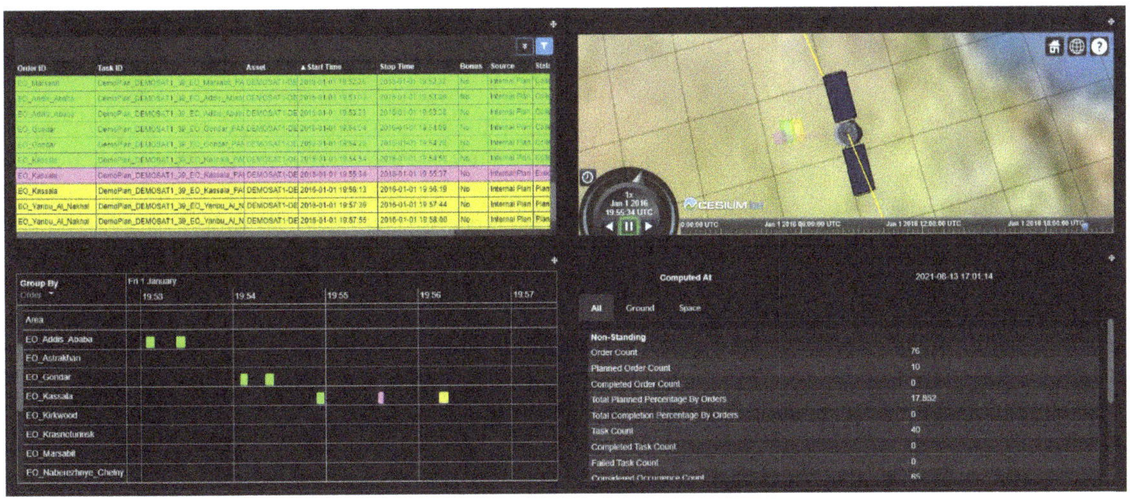

Figure 11.30: Orbit Logic CPAW dashboard. Credit: Orbit Logic.

and animations on the user interface, and flexible process flows and automation. Figure 11.30 shows Orbit Logic's Collection Planning and Analysis Workstation (CPAW) (48).

ACE CrtlPoint

The ACE CrtlPoint from Parsons (acquired Braxton Technologies January 2021) is an automated space vehicle and ground station command and control (C2) application with a plug-in architecture that provides nearly lights-out TT&C operations. ACE CrtlPoint includes the hardware and software components necessary for a satellite MOC. Key applications include: integration with antenna scheduling; ground station control and status; data forwarding for analysis; command plan execution; anomaly detection; and a turnkey TT&C system. COTS capabilities plug into standardized environments, allowing the product to be ready immediately within a range of mission architectures (49).

Bright Ascension Mission Control Software

Bright Ascension is currently developing an end-to-end suite of highly integrated software solutions for the entire mission from flight software development, testing and simulation to ground operations and end user insights and analytics. As part of the suite, Bright Ascension will offer two new products for mission operations – Groundkit, which provides reusable components and services for assembling complete ground software systems, and Ops, a suite of applications for mission operations designed to reduce the complexity and cost of operations, scalable from single satellite to multiple constellations. Ops is expected to be released in Q1 2024, with Groundkit coming out in Q2 2024.

11.10 On the Horizon

Ground data systems must continue to evolve to keep up with the furious pace of small satellite technology. Advancements in onboard processing and data storage will demand more capability in getting data to the ground. Mass production of small satellites is quickly becoming a reality and large constellations are now starting to find their way to orbit. This will require ground system technology that can communicate with multiple satellites simultaneously. Free Space Optical communications and phased array ground systems are emerging solutions to these needs. While both technologies have seen years of investment, they are now just starting to find their way into the ground networks. While it may still be years before becoming a staple for these networks, the following sections provide insight to the state of these technologies and where they are headed in the future.

11.10.1 Free Space Optical Communications

Increasing demand for data from NASA missions has led to a migration over the past few decades to increasingly higher RF bands (X, K, and Ka) and ultimately to the optical and near-infrared regime. Free Space Optical (FSO) communications are expected to increase data rates by two orders of magnitude over traditional RF links (see Communications chapter for more on FSO communications). The next generation systems will incorporate optical communications, and several early flight demonstrations and uses of optical communications in the coming decade are expected to be transformational for NASA and other space organizations. Whereas Ka-band frequencies go up to 40 GHz frequency, the optical signal reaches up to 200,000 GHz. Higher frequencies have the potential for huge increases in data rates, theoretically proportional to frequency-squared if all other factors are equal. At optical wavelengths, other factors, such as atmospheric losses, receiver sensitivity, aperture, and power, must also be considered, but nonetheless, optical communications offer the potential for orders of magnitude improvement in data throughput.

For space applications, lasers are being used as the light source. Laser systems with dynamic

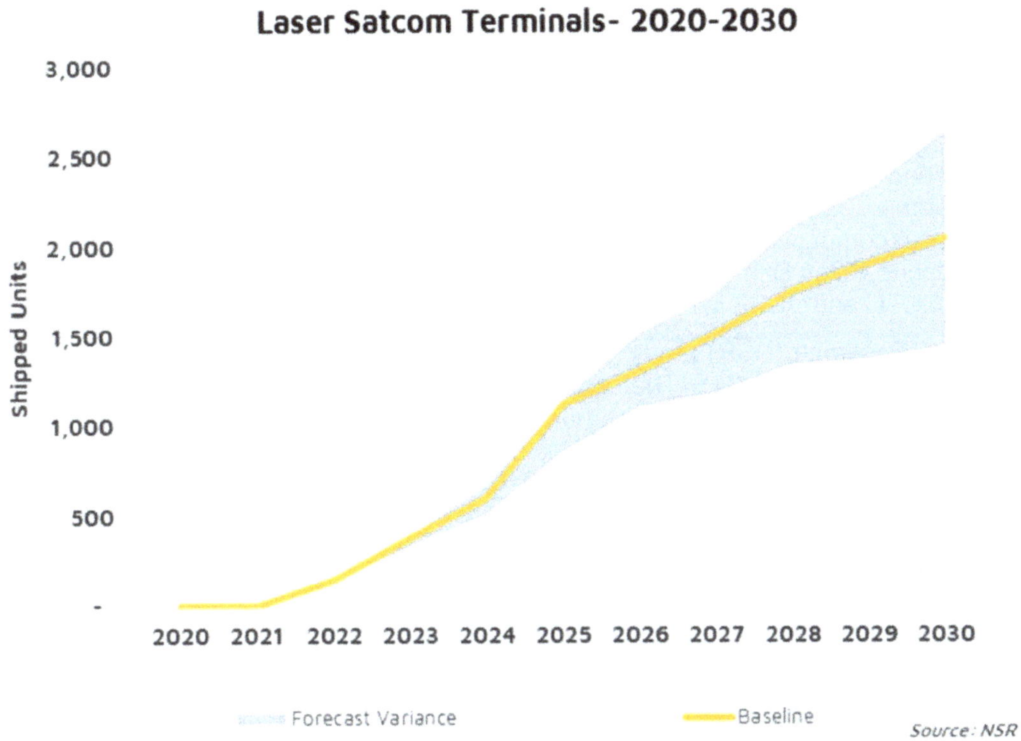

Figure 11.31: Laser terminals future forecast. Credit: Northern Sky Research.

systems such as fast-steering mirrors are used to accurately point the laser on the spacecraft to the ground terminal. Other methods using laser arrays for beam pointing are also being developed to reduce the need for complex dynamic systems. Data is transmitted in the form of hundreds of millions of short pulses of laser light every second. The light is made of photons and the optical ground terminals are setup to collect the light at the photon level. In fact, the ground terminals are designed for an environment where relatively few photons may be received from the transmitter spacecraft, especially from deep space. Direct photon detection with Pulse Position Modulation (PPM) is used instead of the common RF technique of direct carrier coherent modulation to convey information. PPM modulation uses a time interval that is divided into a number of possible pulse locations, but only a single pulse is placed in one of the possible positions, determined by the information being transmitted. To detect extremely faint optical signals with relatively few photons through the atmosphere, optical ground stations can use a superconducting nanowire single photon detector (SNSPD), which, to increase the sensitivity of the nanowires, uses a 1-Kelvin cryocooler. A real-time signal processing receiver uses time-stamped photon arrivals to synchronize, demodulate, decode, and de-interleave signals to extract information code-words. Hence, while the specific technologies employed differ in some respects from those used in radio frequency ground terminals, the higher-level functions performed by the optical communication ground terminal are similar.

Optical communication is attractive for mission designers using small, resource-constrained spacecraft, because it offers a path to relatively high data rates with relatively small, low-power spacecraft equipment. The same volume and power savings can be experienced on the ground terminal side as well. This is driven by the size of the wavelengths. Because RF wavelengths are longer, the size of their transmission beam covers a wider area, therefore, the capture antennas for RF data transmissions must be very large. Laser wavelengths are 10,000 times shorter, allowing data to be transmitted across narrower, tighter beams. This results in the ability to deliver the same amount of signal power to much smaller collecting areas. The reduction in antenna size applies for ground and space receivers, which allows for size and mass reductions on the spacecraft side.

NASA made great strides with its optical communication demonstration on the Lunar Atmosphere and Dust Experiment Explorer (LADEE) mission. The pivotal NASA Lunar Laser Communications Demonstration (LLCD) was able to achieve 622 Mbps from a lunar distance. NASA has several exciting optical communications demonstrations, including O2O, Illuma-T, T-Bird and the Laser Communications Relay Demonstration (LCRD). LCRD is supported by Optical Ground Station (OGS)-1 at OCTL, and OGS2 in Hawaii (51).

Northern Sky Research (NSR) predicts growth for optical communications, which needs to be matched by OGS (figure 11.31). "The demand now is not for just one gigabit per second, not 10 gigabits per second, but tens if not hundreds of gigabits per second. And it's growing exponentially. The only way to achieve that is by starting to use optical communications or laser communications to augment or to complement RF communications" (Barry Matsumori, CEO of BridgeComm, 2022) (52).

11.10.2 Optical Ground Stations and Future Infrastructure Requirements

OGS contain notably different equipment than RF stations, including an optics assembly, photon counter assembly (usually involving a photon counting nanowire detector and cryostat), and signal processing assembly with a time-to-digital converter. Since optical communications use a frequency higher than RF, (e.g., 1,550 nm downlink and 1,065 nm uplink wavelengths), the optical dishes can be smaller than RF antennas. To receive optical signals from a low-Earth orbit, 40 – 60 cm telescopes are sufficient. For successful deep space optical communications, calculations show that 3 m, 4 m, or even 8 m diameter ground apertures are required, depending on the distance from Earth. For these size apertures, when a 3 – 8 m OGS is not available, partnerships can be formed with large astronomy telescopes. For example, JPL-designed OGS equipment has been integrated at the Palomar Observatory (Hale 5-m telescope) for future use by the Deep Space Optical Communications (DSOC) demonstration. Note that OGS for LEO and deep space need different types of modems. It is also important for OGS to have spatial diversity. Weather, atmospheric conditions, turbulence, and aerosols in the air can degrade laser propagation. Because certain types and depth of cloud covers can cause signal loss, probability of link success increases with multiple diverse locations.

For interoperability between SmallSats and public and private optical ground stations, a common communications standard is key. The Consultative Committee for Space Data Systems (CCSDS) and Space Development Agency (SDA) provide recommendations for communications standards, including optical communications. Adhering to these standards by both SmallSats and ground stations allows for multi-mission optical ground stations.

JPL is operating the Optical Communications Telescope Laboratory (OCTL) at Table Mountain, CA, with a 1 m telescope, as shown in figure 11.32. This dish was used for the LADEE mission and offered great performance from a lunar distance (53).

Figure 11.32: JPL's OCTL showing a 1-meter optical aperture. Credit: NASA JPL.

JPL most notably operates the Deep Space Network (DSN), supporting 2-way RF communications and ranging services. Given the existing infrastructure, it is advantageous to augment a DSN RF antenna by installing optical segments at its center, making it a dual-purpose, RF-Optical hybrid antenna. The installation is being implemented in two phases.

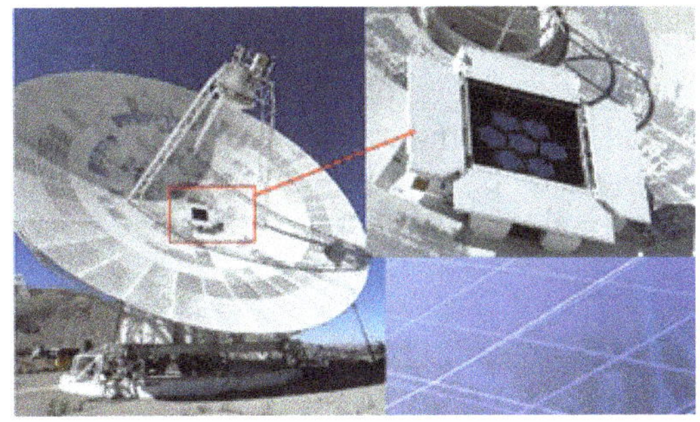

Figure 11.33: JPL's DSS-13, a 34m RF antenna, showing a 1.3-meter optical aperture in its center. Credit: NASA JPL.

In 2022 a small prototype RF-Optical system, including the mirror, cameras, and backend has been installed into DSS-13 (figure 11.33) at the Goldstone Deep Space Communications Complex. DSS-13 is the R&D 34 m BWG antenna at Goldstone. The combination of seven small (0.5m) mirrors comprises a synthesized prototype optical aperture of about 1.3 m diameter. As of 2022, the seven-segment prototype mirror system has been undergoing alignment, test, and checkout. Control was verified to maintain segment position to <1 microradian, with first light successfully received from natural light sources. 1550 nm light was measured through the 100-meter fiber at the pedestal. JPL was able to track multiple sources across the sky from 20-80 degrees elevation.

Figure 11.34: Artist overlay of built DSN RF antenna and planned 8m optical segments at its center. Credit: NASA.

A JPL designed communications detector and optical receiver been installed and tested over the air on DSS-13 as well. Other demo opportunities which could occur during the wait for Psyche (launch now delayed until next year) are being investigated.

The operational RF-Optical hybrid will ultimately include 64 mirrors each of 1m diameter, installed as a segmented 8 m optical receive aperture/mirror physically inside one of the new DSN 34 m radio frequency ground terminals (DSS-23, in California). The final phase includes completion of the full 64- segment aperture on DSS-23, as illustrated in figure 11.34, including a full year of field tests. This 8 m equivalent optical ground aperture will be operational in the early 2030s.

DSS-23 will be capable of a full set of RF services with the 34 m antenna in addition to high-rate optical communications with its 8 m optical assembly. Before the full operational readiness dates for optical communications, the 1.3 m partial optical systems will be usable at various times for best effort demonstration optical communications passes in the near-Earth or Lunar regimes, as well as for deep space missions.

The DSS-23 optical receiver is the same design that JPL is delivering to the Palomar Observatory for use with the DSOC optical communications technology demonstration on the NASA Psyche mission. This receiver is also being installed in ground terminals at White Sands and other locations for other near- and deep-space missions, as well as Artemis. One exciting implication of this 8m equivalent optical aperture is that it meets the 230 Mbps downlink data rate requirement for human exploration of Mars.

Looking at the broader optical communications landscape, over the past decade the community has been confronted with a chicken and egg problem: due to the lack of a ground segment, the FSO space segment has been slow to develop, and additional investments have not been made due to the low number of satellites flying an FSO ground segment. Further, due to the experimental nature of the first flying optical payloads, there has not been sufficient operational budget to pay for optical ground station services. Nevertheless, optical downlinks have the potential to become an essential part of data downlinks, especially in the new-space domain (54).

Europe

Figure 11.35: KSAT's low-complexity optical ground station in Nemea (2021).

The European Optical Nucleus Network was formed between ESA ESOC, Germany Aerospace Centre (DLR), Global Security Operations Center (GSOC) and KSAT. Parties agreed to have an interoperable multi-mission approach based on the Consultative Committee for Space Data Systems (CCSDS) standards. Starting locations and characteristics are summarized in table 11-17. The Nemea OGS (figure 11.35) was operational as of 2022 after initial investments in the first commercial OGS, with the other stations soon to follow. The goal is to bring all building blocks together into an automated, cost-efficient, operational, multi-mission optical ground station service. This groundbreaking OGS uses modified COTS components to reduce cost, and could also be a blueprint for future stations. KSAT co-located its first OGS with KSAT RF antennas in the mid-latitudes at Nemea, due to the temperate weather and sharing the existing infrastructure. The OGS is compliant to the Optical On-Off Keying (CCSDS 141b1) draft standard and is designed to be cost competitive to the KSAT^{LITE} service which KSAT is currently offering in the RF domain. The telescope is mounted >3 m above the ground to avoid ground layer turbulence. The dome is a one-part, completely retractable structure with UV resistant plastic fabric. It is not connected to the telescope foundation to avoid coupling of vibrations caused by wind to the telescope system.

Table 11-17: European Optical Nucleus Network OGS Key Parameters

	Nemea	Almeria	Tenerife
Downlink wavelength support range	*1529-1569nm*	*1529-1569nm*	*1529.5-1568 nm*
Telescope main aperture diameter	*50 cm*	*60 cm*	*80 cm*
Elevation angle support range	*20° to 90°*	*10° to 85°*	*10° to 87° (LEO)* *10° to 89° (GEO)*
Max. operation wind speed	*15 m/s* *(18 m/s gust)*	*15 m/s* *(18 m/s gust)*	*14m/s*
Operational min. sun-distance	*20°*	*20°*	*TBC*
Tracking modes	*Program-track* *Auto-track*	*Program-track* *Auto-track*	*Program-track* *Auto-track*
Temperature operational range	*-15°C to 40°C*	*-5°C to 35°C*	*TBC*
Acquisition Beacon Source	*1589.3nm, 5W (peak)*	*1589 – 1591nm, 5W (peak) TBC*	*1591.26 nm, 5 W* *1590.4 nm, 8 W*
Acquisition beacon divergence angle *(1/e² full-angle)*	*632 µrad*	*1000 µrad TBC*	*500 µrad*
FoV of auto-track system	*8.5x6.8 arcmin*	*15x12 arcmin TBC*	*8.6 arcmin (diameter)*
Absolute pointing error*	*< 5 arcsec*	*< 10 arcsec*	*< +- 103 arcsec (99% confidence)*
Auto-track accuracy	*1 arcsec RMS* *max. error 3 arcsec*	*1 arcsec RMS*	*0.41 arcsec RMS*
Site (mean) clear-sky probability	*Summer: 80-95%* *Winter <60%*	*Summer: >75%* *Winter: >55%*	*81%*
Site long term average seeing	*Day: 2.7 arcsec* *Night: 3.6 arcsec*	*TBD*	*Day: 2.2 arcsec* *Night: 1.8 arcsec*
Site WAN connectivity	*10 Gbps*	*10 Gbps*	*10 Gbps*
Location coordinates	*Lat. 37°50'42.5"N* *Long. 22°37'24.0"E* *278.7 m above sea level*	*Lat. 37°5'36.3''N* *Long. 2°21'31,1''W* *498 m above sea level*	*Lat. 28°17'58.7"N* *Long 16°30'38.5"W* *2382 m above sea level*

- KSAT co-located its first OGS with KSAT RF antennas in the mid-latitudes at Nemea, due to the temperate weather and sharing the existing infrastructure,
- The OGS is compliant to the Optical On-Off Keying (CCSDS 141b1) draft standard and is designed to become cost competitive to the KSAT^{LITE} service, which KSAT currently offers in the RF domain.
- The goal was to bring all building blocks together into an automated, cost-efficient, operational, multi-mission optical ground station service.
- For cost reduction modified COTS components have been selected.

- The telescope is mounted more than three meters above the ground to avoid ground layer turbulence.
- The dome is a one-part completely retractable structure with UV resistant plastic fabric. It is not connected to the telescope foundation to avoid coupling of vibrations caused by wind to the telescope system.

The First Data Link Between Commercial Optical Terminals has been validated through a temporary Sony optical terminal placed on the ISS, with a channel data-rate of 150 Mbit/s with BER varying from 1e-3 and <1e-6.

Japan

The National Institute of Information and Communications Technology (NICT) optical ground station in Japan also received transmission from the SOLISS system by Sony CSL installed on the Kibo's exposed facility on the ISS (55). In 2021 NICT also reported that the 1 m optical ground station in Koganei, Tokyo received via optical communication images taken by the satellite's camera the using the SOTA optical communications device mounted on a 50 kg class microsatellite.

Germany

The DLR German Aerospace Center is another organization active in optical communications. About 25 km west of Munich, Germany is their Optical Ground Station Oberpfaffenhofen (OGS-OP) that houses a 40 cm Cassegrain telescope (56). The German Aerospace Center has also developed a transportable optical ground station (TOGS). It has a 60 cm deployable telescope in a Ritchey- Chretien-Cassegrain configuration with a focal ratio of f/2.5. The telescope is supported by an altazimuth mount on a structure with four adjustable legs for leveling the mount and compensating for rough terrain. It has been successfully used to track the OPALS instrument on the ISS and serves as the primary ground station for the OSIRIS payload on the BiROS satellite. The German Aerospace Center OGS-OP and TOGS are shown in figure 11.36.

Figure 11.36: (left) OGS-OP and (right) TOGS. Credit: German Aerospace Center. https://creativecommons.org/licenses/by/3.0/de/legalcode.

WORK Microwave, a leading provider of satellite communication technologies and products, has emerged as a System Integrator and main supplier of the Optical Ground Stations (OGS) as a turn-key solution which can be customized upon mission request. WORK Microwave collaborated with KSAT and Astelco to establish the first commercial ground station for ESA at Nemea Greece and will install a complete SDA standard compliant OGS for KSAT in the mid of 2024. Figure 11.37 shows the schematics of the OGS with the Digital Optical Ground station (DOG) Suite manufactured by WORK Microwave for system configuration, M&C, and communication. Telescope with its control from WORK Microwave's partners. The DOG Suite is the technology platform for further developments in optical communications across LEO, MEO and GEO orbits, as well as for upcoming Lunar and Deep Space missions. More info on https://work-microwave.com/digital-optical-groundstation/.

Australia – New Zealand

The Australian Optical Ground Station Network (AOGSN) will eventually be made up of four ground stations in Western Australia, South Australia, the Australian National University (Australian Capital Territory), and New Zealand. The plan is to tie these stations together to

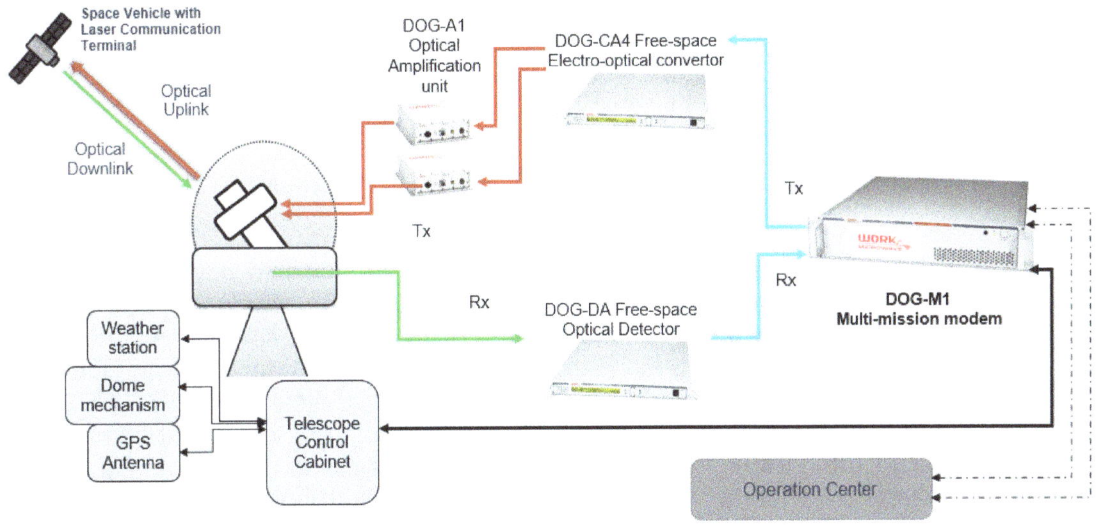

Figure 11.37: Schematics of the OGS with the Digital Optical Ground station (DOG) Suite. Credit: WORK Microwave Inc.

produce a communication network that can support optical, RF, and future quantum communications. In spring 2021, Thales Australia signed a research extension with SmartSat Cooperative Centre (CRC) for the development of advanced optical communications technologies (57).

Sascha Schediwy, head of the research group responsible for designing and building the WA Optical Ground Station (figure 11.38), believes lasers will play a crucial role in the next human missions to the Moon. "It's likely to be how we'll see high-definition footage of the first woman to walk on the Moon," Dr Schediwy said (abc.net.au).

Figure 11.38: The 70 cm Western Australian Optical Ground Station (WAOGS) is installed on a rooftop at the University of Western Australia. Credit: The International Centre for Radio Astronomy Research.

United Sates

In the U.S. the Aerospace Corporation is a player in the Optical Communication arena. Their manned OGS 40 cm telescope, located in El Segundo, CA, demonstrated 200 Mbps from 725 km. It is operating at 1064 nm wavelength, thus it is not compatible with other optical ground stations or most COTS optical space terminals.

11.10.3 Techniques to Improve Optical Comm Reliability

Laser communication is essential for future telecom networks to supplement RF communications and enable:

- Very high throughput links (> 10 Gb/s and up to Tb/s)
- Communication without frequency band limitation
- Highly secure, stealthy, non-interceptable links

It is essential for operational use cases:

- Multispectral observation of Earth from space (very bandwidth intensive)
- Securing sovereign communication
- Telecommunication constellations that rely on very broad bandwidth links

Cailabs

Cailabs has developed a unique range of beam shaping products to counter the effects of atmospheric turbulence (see https://www.cailabs.com/en/). The company has used this technology to develop turnkey laser communication solutions, from single components to the entire station. Based on Cailabs' Multi-Plane Light Conversion (MPLC) technology, TILBA-ATMO compensates for atmospheric turbulence, improving free-space optical links. TILBA-ATMO is an easy-to-integrate product that takes a perturbed beam, corrects and couples it into a standard single-mode fiber (figure 11.39). TILBA-ATMO makes it possible to use conventional telecom equipment and direct or coherent modulation formats to provide robust high-throughput links to optical ground stations. In an open-air trial with DLR, link stability with TILBA-ATMO was similar to that of the adaptive optics unit in low and medium turbulence, and more effective in strong

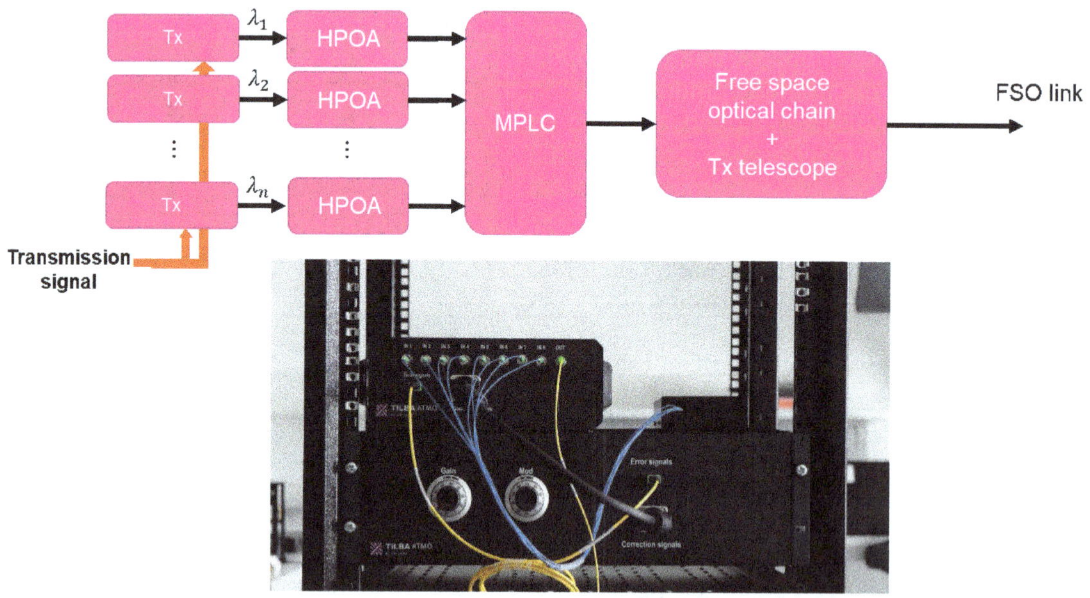

Figure 11.39: Cailabs Tilba Atmo equipment and schematic of TILBA-IBC. Credit: Cailabs.

scintillation conditions. They have also developed a second building block called TILBA-IBC, providing spatial diversity based on incoherent beam combination of spatial modes generated by the MPLC. This component can mitigate atmospheric turbulence in uplinks from ground to space, without the need to use multiple telescopes.

Cailabs is also now providing turnkey Optical Ground Stations, ready for 10+ Gbps laser communication, with built-in TILBA-ATMO and TILBA-IBC turbulence mitigation. Cailabs is involved in multiple government and commercial OGS projects featuring 60 to 80 cm telescopes, compatible with both CCSDS and SDA standards, and is remotely operable.

11.10.4 Role of Optical Relays

Optical inter-satellite links are "critical to the success of the Space Development Agency's low Earth orbit constellation" known as Transport Layer. SDA-Funded Laser Terminal Technology could connect to multiple satellites simultaneously. Each satellite in the Pentagon's "planned mesh network of communications satellites could have as many as many as four laser links so they can talk to other satellites, airplanes, ships and ground stations." BridgeComm, which recently received an SDA contract, "developed a so-called 'one-to-many' optical communications technology for point-to-multipoint transmissions" which could "help reduce the cost of building constellations by requiring fewer terminals," Michael Abad-Santos, senior vice president of business development and strategy at BridgeComm] said. BridgeComm first demonstrated "point to multipoint optical communications in 2019 in a project with Boeing, and has since continued to mature the technology," Abad-Santos said (58).

WarpSpace is a private Japanese company developing an inter-satellite communication system based on laser communication (figure 11.40). The WarpHub InterSat link relays will enable low latency data delivery from Leo and GEO orbits, and in the future it will connect to the Lunar Gateway or planetary deep space via optical communications (see https://warpspace.jp/).

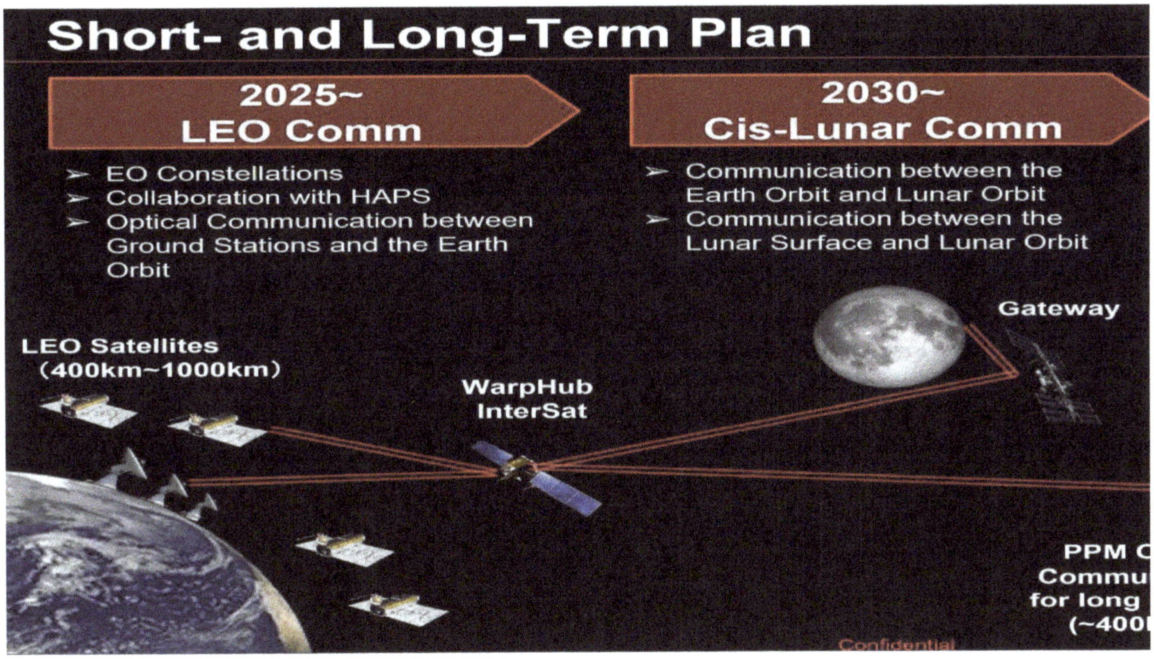

Figure 11.40: WarpSpace plans a satellite network in MEO orbit by 2025 with optical communications hardware able to be reached by any LEO satellite by 2025. Credit: WarpSpace.

11.11 Summary

The ground segment serves as the gateway to getting valuable data collected by the satellite into the hands of the user. It is a critical component of the satellite system that requires attention at the earliest stages of mission planning. Understanding what ground solution best meets the needs of the mission has a direct impact on the spacecraft design, concept of operations, launch schedule, mission operations cost, and expected data volume for processing. Much effort also goes into preparing for the interaction between the satellite and ground network. Developing software and simulations, drafting operations manuals, conducting operations rehearsals, and performing compatibility tests are all par for the course. Post launch, the ground station also plays a key role in locating and commissioning the spacecraft.

In looking forward to the future of ground systems, the clear objective is how to bring the data down more efficiently. Great strides are being made with optical communications where it is possible to have increases in data per pass that are orders of magnitude above what can be achieved with RF communications. Optical communication technology is now being infused into ground system architectures, and flight hardware is becoming miniaturized enough to fit within small satellites. The ability of these systems to quickly change beam directions and acquire multiple targets will be critical for communicating with constellations of small satellites.

While the tried and true RF ground system solution remains the workhorse for small satellites, the innovative nature of the small satellite platform will soon challenge the community to adapt to systems capable of handling hundreds of satellites and high data volumes. Efforts are ongoing to keep pace, but only time will tell whether ground systems will advance or impede the small satellite revolution.

For feedback solicitation, please e-mail: arc-sst-soa@mail.nasa.gov. Please include a business e-mail so someone may contact you further.

References

(1) NASA. "NSN Brochure." [Online] 2021. Accessed July 2022. Available at: https://explorers.larc.nasa.gov/2023APPROBE/pdf_files/Prog05e.%20NSN%20Brochure.pdf

(2) NASA. "Space Communications and Navigation," [Online] 2021 Accessed July 2022. Available at: https://www.nasa.gov/directorates/heo/scan/services/networks/near_space_network

(3) KSAT, "KSATLITE," [Online] Accessed July 2022. Available at: https://www.ksat.no/ground-network-services/ksatlite/.

(4) KSAT: Kongsberg Satellite Services: "Optical Communications: ESA-ESOC Optical Nucleus Network for Direct-to-Earth communications services up to Lunar distances." [Online] [Cited July 30, 2021]. https://www.ksat.no/ground-network-services/new-technologies/optical-comms/

(5) Uppal, Rajesh. "Satellite Ground Segment as a Service (GSaaS) driven by New Space requirements," [Online] Accessed July 2022. Available at: https://idstch.com/space/satellite-ground-segment-as-a-service-gsaas-driven-by-new-space-requirements/

(6) Nguyen, Louis. "Ground Stations as a Service (GSaaS) for Near Real-time Direct Broadcast Earth Science Satellite Data," [Online] Accessed June 2021. Available at: https://esto.nasa.gov/forums/estf2021/Presentations/June10/Nguyen_GSON_ESTF2021.pdf

(7) Amazon. AWS Ground Station Features. AWS. [Online] 2020. Accessed 2022. Available at: https://aws.amazon.com/ground-station/features/.

(8) Viasat, "Real-time Earth Rethinking ground segment as a service" [Online] Accessed July 2022. Available at: https://www.viasat.com/content/dam/us-site/antenna-systems/documents/viasat-real-time-earth-brochure.pdf

(9) R. O'Dwyer. "Iridium 88kbps service commercially available," [Online] November 11, 2021. Accessed July 2022. Available at:

https://smartmaritimenetwork.com/2021/11/11/iridium-88kbps-service-commercially-available/
(10) D. Werner. “Analytical Space wins $26.4 million to establish optical network,” [Online] February 10, 2021. Accessed: July 2022. Available at: https://spacenews.com/analytical-space-wins-26-4-million-to-establish-optical-network/
(11) Safran Data Systems. “Cortex HDR, High Data Rate Receiver for Space Science and Earth Observation,” [Online] Accessed July 2022. Available at: https://www.safran-group.com/sites/default/files/2021-05/col000016.4.0_cortex_hdr_a4_2.pdf
(12) NI Ettus Research. “Products,” [Online] Accessed July 2022. Available at: https://www.ettus.com/product-categories/usrp-x-series/
(13) Kratos. “SpectralNet,” Online] Accessed July 2022. Available at: https://www.kratosdefense.com/products/space/networks/network-devices/spectralnet
(14) Celestia STS. “TNO and Celestia STS sign agreement to commercialize Optical Modem.” June 30, 2021. [Online] Accessed July 18, 2021. Available at: https://celestia-sts.com/2021/tno- and-celestia-sts-sign-agreement-to-commercialise-optical-modem/
(15) Infinite Technologies Incorporated. “RADOMES FUNCTION” *Radomes & Composite Structures,”* [Online] Accessed July 2022. Available at: https://compositeradomes.com/radomes/what-is-a-radomes- function/
(16) Amergint Technologies. “softFEP Datasheet,” [Online] Accessed July 2022. Available at: https://www.amergint.com/wp-content/uploads/AMERGINT_Datasheet_Software_Front-End_Processor_V2-2.pdf
(17) Kratos. “quantumFEP Datasheet.” [Online] Accessed July 2022. Available at: https://www.kratosdefense.com/-/media/k/pdf/s/sa/os-014-openspace-quantumfep.pdf
(18) Csete, Alexandru. “About Gpredict,” [Online] October 2020. Accessed: July 2022. Available at: http://gpredict.oz9aec.net/
(19) GNU Radio. GNU Radio Main Page. Wikipedia. [Online] March 5, 2022. Accessed July 2022. https://wiki.gnuradio.org/index.php/Main_Page
(20) DeWitt, Henry. Satellite Ground Station Control. DeWitt & Associates. [Online] Accessed: July 2022. Available at: http://www.dewitt-assoc.com/groundStation.html
(21) GAUSS SRL “Group of Astrodynamics for the Use of Space Systems (G.A.U.S.S. Srl): . Ground Station,” [Online] 2022. Accessed July 2022. Available at: https://www.gaussteam.com/services/ground-station/
(22) GUASS SRL. GAUSS UHF Mini Ground Dongle Datasheet. [Online] Accessed July 2022. Available at: https://www.gaussteam.com/wordpress/wp-content/uploads/2018/04/GAUSSRadio_Miniground_Datasheet.pdf
(23) GAUSS Srl. “Ground Station & Software.” [Online] Accessed July 2022. Available at: https://www.gaussteam.com/wordpress/wp-content/uploads/2018/01/Flyer_GroundStation_Software_1.pdf
(24) Innovative Solutions in Space. “Ground Stations,” [Online] Accessed July 2022. Available at: https://www.isispace.nl/products/ground-stations/
(25) GomSpace. “Ground Systems,” [Online] Accessed July 2022. Available at: https://gomspace.com/shop/subsystems/ground-systems/default.aspx
(26) GomSpace. “Hands-off Operations Platform (HOOP),” [Online] Accessed October 2022. Available at: https://gomspace.com/hoop.aspx
(27) Surrey Satellite Technology Ltd. “Space Services,” [Online] Accessed July 2022. Available at: https://www.sstl.co.uk/what-we-do/space-services
(28) Kratos. “Quantum Radio,” [Online] Accessed July 2022. Available at: https://www.kratosdefense.com/products/space/satellites/ttc-devices-and-software/quantumradio
(29) Kratos. “ Be Future-Ready to Manage Tomorrow’s Space Ground Network,” [Online] Accessed July 2022. Available at: https://www.kratosdefense.com/systems-and-

platforms/space-systems/dynamic-ground/architecture

(30) INCOSE - International Council on Systems Engineering: "Systems Security Engineering: Mission and Objectives." [Online] Accessed July 12, 2021. Available at: https://www.incose.org/incose- member-resources/working-groups/analytic/systems-security-engineering

(31) National Security and Defense: "Memorandum on Space Policy Directive-5—Cybersecurity Principles for Space Systems" [Online] September 4, 2020. Accessed July 12, 2021. Available at: https://trumpwhitehouse.archives.gov/presidential-actions/memorandum-space-policy- directive-5-cybersecurity-principles-space-systems/

(32) NASA. "Space System Protection Standard" NASA-STD-1006 w/Change 1. [Online] October 29, 2019. Accessed July 12, 2021. Available at: https://standards.nasa.gov/standard/nasa/nasa-std-1006-wchange-1

(33) NASA. "NASA's Commercial Communications Services," [Online] Accessed August 01, 2022. Available at: https://www.nasa.gov/directorates/heo/scan/services/nasas_commercial_communicatio n s_services

(34) R. Ross (NIST), V. Pillitteri (NIST), R. Graubart (MITRE), D. Bodeau (MITRE), R. McQuaid (MITRE): "Developing Cyber Resilient Systems: A Systems Security Engineering Approach." SP 800-160 Vol. 2. November 2019.

(35) R. Ross (NIST), M. McEvilley (MITRE), J. Oren (Legg Mason). "Systems Security Engineering: Considerations for a Multidisciplinary Approach in the Engineering of Trustworthy Secure Systems. " NIST SP 800-160 Vol. 1, November 2016 (Updated March 21, 2018). Accessed July 12, 2021. Available at: https://csrc.nist.gov/publications/detail/sp/800-160/vol-1/final

(36) National Institute of Standards and Technology : "Cybersecurity Framework" [Online] Accessed July 12, 2021. Available at: https://www.nist.gov/cyberframework

(37) The Aerospace Corporation: "Defending Spacecraft in the Cyber Domain." [Online] [Cited July 30, 2021]. https://aerospace.org/paper/defending-spacecraft-cyber-domain

(38) S. Visner and S. C. Kordella, The MITRE Corporation: "Cyber Best Practices for Small Satellites." (November 2020). https://www.mitre.org/publications/technical-papers/cyber-best-practices-for-small-satellites

(39) Consultative Committee for Space Data Systems: "Report Concerning Space Data System Standards." CCSDS 350.7-G-2, Green Book, Issue 2. April 2019.

(40) Consultative Committee for Space Data Systems: "Report on the Application of Security to CCSDS Protocols." CCSDS 350.0-G-3, Green Book, Issue 3. March 2019.

(41) G.P Bortoletto, D. Melli, F. Stigliano, D. Martin, F. Gardosi. "Maximize contact availability of smallsat clusters through MSPA technique on GSaaS," Presented at SmallSat Conference, Logan, UT. 2021. Available at: https://digitalcommons.usu.edu/smallsat/2021/all2021/44/

(42) Shannon Terry, A.I. Solutions, Inc. "What are Space Ground Systems?" [Online] Accessed July 2022. Available at: https://ai-solutions.com/about-us/news-multimedia/what-are-space-ground-systems/

(43) R. Jewett, "Viasat Releases New Large-Aperture Space-to-Ground Communication Antennas," ViaSatellite, January 20, 2022. [Online] Accessed July 2022. Available at: https://www.satellitetoday.com/ground-systems/2022/01/20/viasat-releases-new-large-aperture-space-to-ground-communication-antennas/

(44) Sampathkumar, Arun Kumar. Uber-ization of Ground Stations. "The Small Satellite

Market's Enabling Aggregator Wave," [Online]. Accessed September 23, 2022. Available at: https://interactive.satellitetoday.com/via/august-2021/uber-ization-of-ground-stations-the-small-satellite-markets-enabling-aggregator-wave/

(45) OpenC3, Inc. "Documentation," [Online] Accessed 2022. Available at: https://openc3.com/docs/v5/

(46) The Hammers Company. Galaxy. [Online] Accessed 2020. Available at: https://hammers.com/galaxy

(47) Xplore. Major Tom. [Online] Accessed July 2022. Available at: https://www.xplore.com/services/operations-as-a-service/major-tom.html

(48) Orbit Logic. [Online] Accessed 2020. Available at: https://www.orbitlogic.com/index.html

(49) Parsons. "Automated Command And Control For Smarter Space Operations," [Online] Accessed 2022. Available at: https://www.parsons.com/products/ace-ctrlpoint/

(50) Bright Ascension. Mission Control Software. [Online] Accessed 2020. Available at: https://www.brightascension.com/products/ground-control-software/

(51) NASA. July 15, 2018. "LLCD: 2013-2014," [Online] Accessed 2022. Available at: https://www.nasa.gov/directorates/heo/scan/opticalcommunications/llcd/

(52) Waterman, S. "Space Lasers Come of Age: Optical Communications for Satellites Are Ready for Prime Time," [Online] February 22, 2022. Accessed July 2022. Available. https://interactive.satellitetoday.com/via/march-2022/space-lasers-come-of-age-optical-communications-for-satellites-are-ready-for-prime-time/_fragment.html

(53) NASA. Lunar Laser Communication Demonstration: NASA's First Space Laser Communication System Demonstration Fact Sheet," [Online] Accessed 2022. Available at: "https://www.nasa.gov/sites/default/files/llcdfactsheet.final_.web_.pdf

(54) Krynitz, Martin and Heese, Clemens and Knopp, Marcus Thomas and Schulz, Klaus-Jürgen and Henniger, Hennes. "The European Optical Nucleus Network," 16th International Conference on Space Operations (SpaceOps 2021), 03.-05. May 2021.

(55) National Institute of Information and Communications Technology (NICT). "Small Optical Link for International Space Station (SOLISS) Succeeds in Bidirectional Laser Communication Between Space and Ground Station," [Online] April 30, 2020. Accessed 2022. Available at: https://www.nict.go.jp/en/press/2020/04/30-1.html

(56) Florian Moll, Amita Shrestha, Christian Fuchs. Ground stations for aeronautical and space laser communications at German Aerospace Center. Publisher: Proc. SPIE 9647, Advanced Free-Space Optical Communication Techniques and Applications. https://core.ac.uk/reader/31021696

(57) Thales: "Beam me up! Thales & Goonhilly Earth Station collaborate on research into laser beams as data pipes." June 8, 2021. [Online]. Accessed July 23, 2021. Available at: https://www.thalesgroup.com/en/australia/press-release/beam-me-thales-goonhilly-earth-station-collaborate-research-laser-beams

(58) S. Erwin. "DoD space agency funds development of laser terminal that connects to multiple satellites at once," [Online] March 10, 2022. Accessed July 23, 2021. Available at: https://spacenews.com/dod-space-agency-funds-development-of-laser-terminal-that-connects-to-multiple-satellite-at-once/

(59) NASA. "NASA to Commercialize Near-Earth Communications Services," [Online] October 26, 2022. Accessed July 23, 2021. Available at: https://www.nasa.gov/feature/Goddard/2020/nasa-to-commercialize-near-earth-communications-services

Chapter Contents

Chapter Glossary

(18 SDS)	U.S. Space Force 18th Space Defense Squadron
(19 SDS)	U.S. Space Force 19th Space Defense Squadron
(CARA)	NASA's Conjunction Assessment Risk Analysis program
(CAESAR)	Conjunction Analysis and Evaluation Service, Alerts and Recommendations
(CCR)	Corner Cube Reflectors
(CNES)	Centre National d'Etudes Spatiales (French Space Agency)
(COTS)	Commercial-off-the-Shelf
(CUBIT)	CubeSat Identification Tag
(D/T/I)	Detection, Tracking and Identification
(EGTN)	ExoAnalytic Global Telescope Network
(ELROI)	Extremely Low Resource Optical Identifier
(EUSST)	European Union Space Surveillance and Tracking program
(FCC)	Federal Communications Commission
(GEO)	Geosynchronous Equatorial Orbit
(GPS)	Global Positioning System
(GUI)	Graphical User Interface
(HEO)	Highly Elliptical Orbit
(HUSIR)	Haystack Ultrawideband Satellite Imaging Radar
(IDs)	Identification
(ILRS)	International Laser Ranging Service
(JSpOC)	Joint Space Operations Center
(LEDs)	Light Emitting Diodes
(MEO)	Medium Earth Orbit
(NPR)	NASA Procedural Requirement
(NTE)	Nanosatellite Tracking Experiment
(NTIA)	National Telecommunications and Information Administration
(OCAP)	Orbital Conjunction Assessment Plan
(OEM)	Orbit Ephemeris Message
(PNT)	Position, Navigation, and Timing
(RF)	Radio Frequency
(SRI)	Stanford Research Institute
(SSA)	Space Situational Awareness
(SSN)	Space Surveillance Network
(SWaP)	Size, Weight, and Power
(TLE)	Two-Line Element
(TraCSS)	Traffic Coordination System for Space
(USIR)	Ultrawideband Satellite Imaging Radar

BODY-397

12.0 Identification and Tracking Systems

12.1 Introduction

In the past, most launches involved a single, large satellite launching on a dedicated launch vehicle. Small satellites as secondary payloads were sometimes 'dropped off' along the way to the primary payload's orbit or rode along to the final orbit with the primary payload. In either case, it typically was not that difficult for tracking radars to distinguish between primary and secondary payloads via size and operational parameters.

Recently, however, multi-manifest or "rideshare" launches have become more common, and providers (1-3) are launching multiple CubeSats, or bundling CubeSats and other smaller payloads with larger payloads to fill up the excess capacity of almost any given launch vehicle. For technical and cost reasons, such launches generally deploy small satellites and CubeSats into very similar orbits over a short time window. "Batch" launches with a lack of separation between satellites can prevent effective tracking and create "CubeSat confusion" (4). When CubeSats are deployed close together in space and time they can be hard to distinguish from each other by tracking radars, making it difficult to determine which orbits correlate to which spacecraft, preventing a unique orbital state for each object from being added to the catalog of on-orbit objects (5, 6). At times it can take weeks to months to sort out which object is which, while some may never be uniquely identified at all. If the orbital states cannot be determined and the spacecraft cannot be added to the catalog of on-orbit objects, neighboring satellites may be unaware of close approaches, hindering their ability to mitigate risks of collision.

Due to their standardized shape and size, CubeSats look very similar to one another, especially when they are in orbit hundreds of kilometers away. If there are unidentified objects from a launch, then the possible number of associations of object identifications (IDs) to tracked objects scales as n! (n-factorial, where n is the number of unidentified space objects from the launch). For example, if there are just two objects, say a payload and an upper stage, there are two ways in which you can associate the IDs with the tracked objects, and even that can be a challenge (7). However, if there are ten unidentified objects, there are 3,628,800 possible combinations; with 20 this rises to 2.4×10^{18} combinations. The magnitude of the problem gets big quickly.

Small satellites can improve their chances of being identified and tracked through good coordination with tracking agencies pre-launch, through community sharing of position and covariance data in clearly defined, consistent, standard formats (such as Orbit Ephemeris Message (OEM), and through careful selection of deployment direction and timing (8). Good spacecraft design choices can also improve the chances of small satellites surviving launch and early orbit (9) and can even make use of in-space commercial radio networks as a "back-up" method of communicating should primary systems fail (10). However, despite improvements in both design and coordination, many small satellites still go unidentified. This has led to the introduction of tracking aids – independent systems that help owners and trackers identify small satellites and CubeSats, in some cases even if the satellite is malfunctioning.

The information described below is not intended to be exhaustive but provides an overview of current state-of-the-art technologies and their development status for a particular small spacecraft subsystem. It should be noted that Technology Readiness Level (TRL) designations may vary with changes specific to payload, mission requirements, reliability considerations, and/or the environment in which performance was demonstrated. Readers are highly encouraged to reach out to companies for further information regarding the performance and TRL of described technology. There is no intention of mentioning certain companies and omitting others based on their technologies or relationship with NASA.

BODY-398

12.2 Identification and Tracking Ground Systems

The United States Space Force Delta 2 is responsible for performing space surveillance and providing foundational Space Situational Awareness (SSA) for the US Department of Defense as well as for other agencies and space entities. Delta 2's 18th Space Defense Squadron (18 SDS), located at Vandenburg Space Force Base in California, performs all catalog maintenance functions including detection, tracking and identification (D/T/I) of artificial objects in Earth orbit and maintaining the space catalog which is publicly available on space-track.org. As part of their activities, they provide launch support, re-entry assessment, and other SSA functions: Orbital safety activities, such as conjunction assessment (which identifies close approaches between launch and other catalogued in-orbit objects) are provided by Delta 2's 19th Space Defense Squadron (19 SDS), located at the Naval Support Facility at Dahlgren, VA. Maintaining the catalog is achieved via the US Space Surveillance Network (SSN) that is formed by a suite of sensors around the world (29). 18 SDS is currently tracking more than 45,000 objects in Earth orbit and can provide data for pieces as small as 10 cm^3. They issue two-line elements (TLEs) that are updated on a regular basis and can be used to compute predicted orbit position for spacecraft communications acquisition and other purposes. They also produce precision vectors with covariance that can be used to perform conjunction analyses. TLEs are not accurate enough to be used for conjunction assessment.

The US Space Force next generation SSA sensor, known as the Space Fence, was declared operational in March 2020 and can track objects below the previous 10 cm^3 limit. It is located on Kwajalein Island, in the Republic of the Marshall Islands and consists of a S-band radar system to track objects primarily in low-Earth orbit, although it can track objects in medium-Earth orbit (MEO) and geostationary equatorial orbit (GEO) as well. The 20th Space Control Squadron based in Huntsville, Alabama, manages the Space Fence and provides data to 18 SDS to augment the space catalogue (30). Another important sensor in the SSN is the Haystack Ultrawideband Satellite Imaging Radar (HUSIR), which is the highest-resolution, long-range sensor in the world. HUSIR simultaneously generates X- and W-band images that can provide valuable information about the size, shape an orientation of Earth orbiting objects (31). These are just two examples of sensors that make up the SSN, many having specific unique capabilities that support the SSN's various functions, including conjunction assessment.

The NASA Conjunction Assessment Risk Analysis (CARA) program acts as the intermediary between 18/19 SDS and NASA satellite missions. CARA gathers orbit ephemeris and covariance files from the NASA spacecraft operations teams and provides this data to NASA's Orbital Safety Analysts (OSAs) at Vandenberg for screening and close approach prediction. CARA provides risk assessment of these predicted close approaches to NASA missions beyond the 19 SDS support provided to non-NASA users, including operations concept development, probability of collision computation, high interest event notification, and conjunction geometry analysis among other functions. In 2012, the French Space Agency (CNES) created a conjunction risk assessment team called Conjunction Analysis and Evaluation Service, Alerts and Recommendations (CAESAR) that provides risk assessment services to their missions (34) (35) as part of the European Union Space Surveillance and Tracking (EUSST) program.

In 2020, NASA released a best practices handbook entitled "Spacecraft Conjunction Assessment and Collision Avoidance Best Practices Handbook," which is a great reference for satellite operators with respect to collision avoidance topics (32). NASA Procedural Requirements (NPR) 8079.1 codifies those best practices as requirements for NASA space flight programs, projects, and vehicles to protect the space environment and reduce the risk of collision (33).

Besides USSF Delta 2, several commercial entities are providing tracking capabilities that can be purchased by stakeholders. The COMSPOC Corporation which provides data from a network of commercial sensors. ExoAnalytic also has a global telescope network (EGTN) consisting of over 25 observatories and 275 telescopes tracking orbiting objects in GEO, highly elliptical orbit (HEO), and MEO. The EGTN can collect angles and brightness measurements. They maintain a proprietary catalog of satellites and space debris that are regularly tracked and cataloged. This includes a historical archive of over 100 million object measurements (26).

LeoLabs is another commercial entity providing spacecraft tracking and support services. They use a group of distributed Earth-based, phased-array radars to make a commercial-off-the-shelf (COTS) satellite tracking service targeted to the specific requirements of SmallSat operators in low-Earth orbit. They currently have two radar stations in New Zealand, Costa Rica, the Azores, Poker Flat, AK, Midland, TX, Western Australia, and Argentina. There are currently seven functioning radars as of 2023, with plans for more radars strategically located around the world to robustly track objects down to 2 cm in size. The predicted performance also includes a revisit time of over 10 observations per day for specific objects, and a low-Earth orbit catalog of over 250,000 pieces. Through their LeoTrack platform, they can use their radar data to perform precision tracking and curate orbit information products for satellites as small as 1U. Their system includes an open-source graphical user interface (GUI) capable of displaying all the catalog in real time, as well as fundamental orbit information about each individual object. They also offer a commercial launch and early orbit service, with SpaceX as a customer (39). Since March 2020, LeoLabs tracks all Starlink satellites within the first few hours after deployment and provides orbital data during orbit raising to SpaceX. During deployment, LeoLabs differentiates any non-Starlink objects (third-party satellites for rideshare missions) and provides location and separation information to SpaceX (43).

Catalogs provided by these commercial entities are different from the one maintained by 18 SDS in accuracy and objects included. Spacecraft owner/operators should be aware of the differences before choosing to use a particular service for a particular purpose. For conjunction assessment purposes, having multiple differing solutions can be confusing. The Department of Commerce was charged in Space Policy Directive-3 with creating a space traffic coordination system that enables commercial capabilities for conjunction assessment. In the future they may offer a conjunction assessment service that merges data form multiple sources in one solution as they work to transition the service currently provided by 19 SDS. Initial capability of the Traffic Coordination System for Space (TraCSS) is planned for September 2024. Lessons learned from NASA's recent Starling (launched July 2023) and Starlink constellations will be fed into the design of TraCSS.

NASA's Starling 1.5 mission extension, using the four-CubeSat Starling swarm of spacecraft, will build on the Starling primary mission demonstration of autonomous maneuver planning and execution. Partnering with SpaceX and their Starlink constellation, Starling 1.5 will demonstrate an advance space traffic management solution for co-located groups of spacecraft from different owner/operators. The onboard autonomous maneuvering software for the original Starling mission, consisting of Emergent Space Technologies (acquired by York Space Systems) Navigator and Autopilot products, can perform probability of collision calculations and propose risk mitigation maneuvers for conjunctions with co-located vehicles. A space traffic management hub on the ground will be able to receive maneuver responsibility claims from an operator, indicating that they intend to maneuver to reduce the conjunction risk; receive and screen proposed new trajectories; and provide operators conjunction data messages with screening results. This will enable highly automated operations including conjunction risk assessment and mitigation between satellite owner/operators.

12.3 Tracking Aids

For spacecraft that cannot be routinely tracked by the SSN, it is important to ensure trackability by another means to enable other owner/operators to know where your spacecraft is to prevent debris-producing collisions. This is especially important for SmallSats that have orbital lifetimes that exceed operational lifetimes, and the risk to orbital neighbors remains after tracking activities have stopped. For NASA spacecraft, trackability until demise is required in NPR 8079.1 and assessed as part of the required Orbital Conjunction Assessment Plan (OCAP) during design in NPR 8079.1.

Tracking aids come in several categories, each with benefits and drawbacks (11). Table 12-1 discusses the broad categories available, with representative examples discussed below. Size, weight, and cost vary for each of the examples, but all can be considered compatible with a CubeSat mission; see the references for detailed information on size, weight, and power (SWaP) and cost. Once the augmented tracking data is collected, the ephemeris data is made available to CARA/19 SDS for screening.

Several commercial companies offer services that will process the data produced by tracking aids and produce predicted orbit ephemeris data that can be used to perform conjunction assessment. These include SpaceNav, COMSPOC, and Kayhan Space, who has a capability that focuses on small spacecraft support.

Table 12-1: Types of Tracking Aids

Technology scheme	Description and reference mission	TRL	Citation
CubeSat position and ID via radio	A position, navigation, and timing (PNT) receiver is attached to a CubeSat, along with a radio to transmit the information via a LEO communications provider (or directly to the ground); example: BlackBox, Blinker.	7-9	(12) (13)
Coded light signals from light source on exterior of CubeSat	Exterior-mounted LEDs with large-aperture telescopes to receive the signal or diffused LED lasers with ground-based photon-counting cameras.	6-7	(14) (15)
Radio Frequency interrogation of an exterior Van Atta array	For example, exterior mounted radio frequency identification (RFID) tag & commensurate radar.	7-9	(16)
Laser interrogated corner cube reflectors (CCR)	One or several small CCRs can be attached to CubeSat exterior; ground-based laser and receiver telescope needed to distinguish number of CCRs.	7-9	(17)
Passive augmentations to visibility	Use of high-albedo paint or tape, improving overall conductance of the exterior of the satellite or other methods to increase visibility.	7-9	
Orbital Whereabout Locator (OWL)	A tunacan mounted plug and play device equipped with a battery that helps identify and keep track of a satellite with a GNSS based	7-9	

	tracker, monitor key parameters of a satellite and download data using an omnidirectional antenna in any orientation.		

12.4 Devices that Communicate Position and ID via Radio

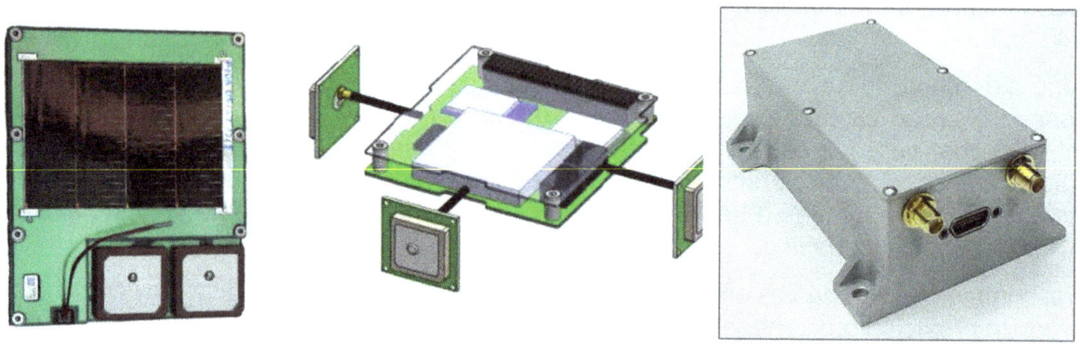

Figure 12.1: (left) Thin Patch or Stamp Black Box for side mounting. (Middle) PC104 Black Box for internal stack mounting. (Right) Standard Black Box for larger satellites. TRL 9: flown on spaceflight launch. Solar array and antennas not shown. Credit: NearSpace Launch, Inc.

The most comprehensive (but also potentially the most complex and SWaP-intensive) option involves equipping a small satellite with an independent positioning, navigation, and timing (PNT) receiver and independent radio capable of transmitting data to an independent communications provider. An example technology is the Black Box system (figure 12.1), described by NearSpace Launch, Inc., in a recent conference paper (18). This system comes in several form factors for mounting internally or externally to a small satellite or CubeSat. The patch antenna shown in the first image is approximately 10 cm by 8 cm and can weigh as little as 22 grams; larger systems such as the one shown in the third image of figure 12.1 have flown and are considered TRL 9. Another example often used for CubeSats is the NovAtel OEM7700. JAXA offers a strap-on tracking device called mini Mt. Fuji; they can provide sensor tracking or operators can track it themselves. These systems combine a low-power GPS receiver with a low-power radio capable of communicating with a low-Earth orbit communication provider (in the case of Black Box, the Global Star network) and operate independently from the spacecraft's regular command and telemetry links. Externally mounted versions often include solar cells for independent power generation. A Black Box system is currently flying on Spaceflight Sherpa-FX orbital transfer vehicle, launched on January 24, 2021, and is returning GPS fixes to the developer. The GPS fixes were analyzed and reports were presented at the October 2021 and October 2022 International Astronautical Congress (41, 42).

A similar concept under development is The Aerospace Corporation's 'Blinker' (13), in which a GPS receiver and low-power radio are externally mounted to a CubeSat. GPS positions ("tags") are recorded, stored, and then radioed when the satellite is over an Aerospace Corporation ground station (which is separate and independent from the CubeSat's mission ground station). Research and development are being conducted to automatically convert the GPS tags into ephemeris information that can be directly ingested by space situational awareness centers (in this case the 18 SDS via Space-Track.org) as an owner/operator initial ephemeris that would be propagated by 19 SDS and screened for conjunction assessment.

The advantages to such a system are that it provides complete data on a satellite's position and requires no specialized ground equipment (other than the equipment used by the communications provider). Some such systems are independently powered and can provide data even if the host satellite never powers up, though others are dependent on spacecraft power to function. These systems are the most complex of the tracking aids described, however, and despite their relatively small size, are still the most SWaP-intensive of the options examined. Systems that rely on power from the host vehicle are also useless if the host vehicle suffers a power anomaly or failure. Having an additional onboard radio that communicates with other space assets necessitates additional oversight by the Federal Communications Commission (FCC) (or National Telecommunications and Information Administration (NTIA) for US Government missions licensing and coordination).

In choosing such systems, it is important to choose one that is space rated, otherwise simulation is required to perform that testing pre-launch. The system should also provide fast convergence after launch to allow rapid computation of an orbit determination solution to provide to 18/19 SDS for conjunction assessment screening as soon as possible after separation. Other available features, like multi-frequency, multi-GNSS, and availability of pseudoranges are nice-to-have, but the benefit for conjunction assessment is marginal vs. just having a receiver. If flying a GPS system, it is important to ensure that the GPS data is available in the download stream and not just internally available to the spacecraft computer for onboard use.

12.5 Devices that use Coded Light Signals

Figure 12.2: (left) ELROI PC104 beacon unit that was installed on NMTSat.d (right) Two ELROI beacon units delivered for a launch in 2021. Credit: Los Alamos National Laboratory.

Identification systems and devices that make use of light emitting diodes (LEDs) and coded light signals have the advantage of being relatively simple and capable of identifying satellites uniquely. However, all systems flown to date have required power from the host satellite, leading to issues with detection (19) if the host satellite does not power up. Current implementations are also relatively large, though future systems are expected to be much smaller and may include independent power. Devices such as the Extremely Low Resource Optical Identifier (ELROI) beacon (figure 12.2), under development by Los Alamos National Lab (19), use exterior-mounted LEDs or diode lasers that blink in a prescribed sequence to uniquely identify satellites. The ELROI system is designed to be independently powered by a small solar cell and battery, and is packaged into a system as small as a Scrabble tile, though only larger systems – with power provided by the host satellite – have flown.

The emitters on such devices can be regular LEDs or diffused diode lasers but require specialized ground equipment – either a large-aperture telescope or a photon-counting camera –to track the object as it passes overhead. Figure 12.3 shows how the ELROI system works: a photon-counting camera attached to a telescope tracks the signal from a diode laser and decodes the ID of the host satellite from the on/off pattern of flashes.

Another similar system (36) proposes to use red, blue, and green LED lights on specific faces of the satellite, which blink in a unique pattern, and standard astronomical optical telescopes to track and identify the LED flash pattern (14). LEDSAT, a CubeSat to test this concept on-orbit, launched in August 2021 (37). A test of an exterior-mounted blue LED on a CubeSat was attempted in March 2021, but was indeterminate due to a lengthy period of bad weather at the single designated telescope site.

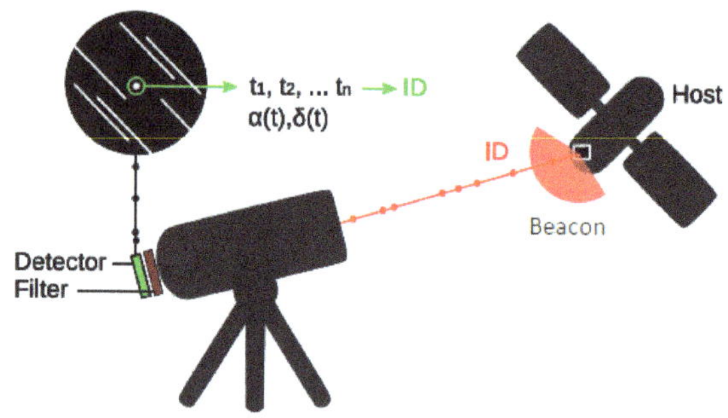

Figure 12.3: ELROI Optical Detection System. Credit: Los Alamos National Laboratory.

LED-based systems require relatively clear night-time skies for identification, and dedicated ground equipment (telescope and sensor). The light sources are too faint to allow blind searching of the sky for the satellite; orbital information from a SSA provider is also required to find and track the CubeSat, although the process of tracking the satellite via an optical telescope allows the orbital ephemeris to be updated. Therefore this tracking enhancement alone cannot be used for identifying and cataloging the spacecraft. Issues with attitude control on the host satellite can also complicate the identification process. In addition, using LEDs or other light sources on a satellite while in Earth's shadow should be done carefully to minimize interference with astronomical observations. The SatCon1 report (38) on page 6 lists several recommendations to be followed: 1) assure the light source is fainter than apparent magnitude of V ~7 (and the fainter the better), and 2) advance notice of any illumination times, including accurate orbital elements.

12.5.1 Van Atta Arrays and RF Interrogation Receivers

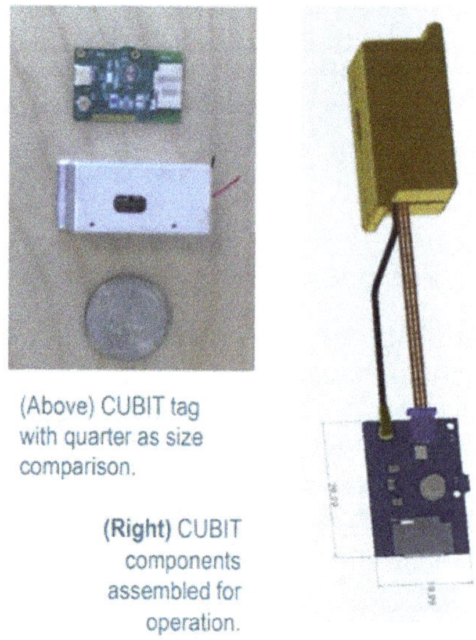

Figure 12.4: CUBIT. Credit: SRI International.

Another method for increasing the ability to track and possible identification of small satellites involves devices that respond when interrogated by a radio frequency (RF) signal of appropriate wavelength. One such system, the CubeSat Identification Tag (CUBIT) shown in figure 12.4, is similar to the RFID devices used in proximity badges (16). Built by SRI International and partnered with NASA Ames, CUBIT responds with a short burst of information when interrogated by a radio signal of the correct frequency. CUBIT is relatively small and designed to be independent of host vehicle power. The implementations that have flown contain a small battery suitable for 30 days of in-orbit life, which covers the most critical early orbit identification period. It could therefore be coupled with a coded light emitter to overcome the inability of that system to allow object identification. The device is separated into an internally mounted electronics unit attached to an exterior antenna to minimize the exterior footprint of the unit. Two units have flown and were successfully demonstrated in space onboard TechEdSat-6 in 2017 and TechEdSat-7 in 2020. A relatively large ground architecture (in CUBIT's case, a 30 m antenna and an array of antennas) are required to interrogate the system and successfully acquire the low-power response. CUBIT is patent-pending, and SRI has reached commercialization agreements with potential vendors. Future research will continue with a recently awarded AFWERX Phase 1 study.

Another example of an RF-interrogated device is a Van Atta array, a passive device which re-radiates RF energy back toward the source of that energy (20). One such device, the Nanosatellite Tracking Experiment (NTE) consists of a 64-element Van Atta array of tiny, paired antennas tuned to a Ku-band RF frequency, as shown in figure 12.5 (21). When interrogated at the proper frequency range, the incident RF field received by each antenna is fed to a corresponding antenna via a passive transmission line, where it is re-radiated. This significantly increases the radar cross-section of the object, allowing it to be more easily tracked. Unique identification is difficult, however, and requires specialized ground stations which tend to be expensive to operate. A satellite carrying a Van Atta array device will be distinguishable from one not carrying such a device, or from one carrying a device tuned to a different frequency band, but two satellites carrying the same Van Atta array will return the same signature. The RF interrogation also requires a ground source of the appropriate frequency. However, Van Atta array devices are entirely passive and extremely low SWaP, making them easy to include on small satellites and CubeSats. NTE devices have flown in space but results from those flight experiments have not been published to date.

Figure 12.5: NTE Van Atta array retro-reflector in the Ku-band, fits standard 1U panel, tuned to HAX RADAR frequency. Credit: Naval Information Warfare Center.

No capability currently exists at the SSN to process Van Atta Array data, so it would be incumbent on the spacecraft operator to produce its own orbit determination solution from the data to provide an ephemeris to 18/19 SDS for CA screening.

12.5.2 Laser-Interrogated Corner Cube Reflectors

Corner cube reflectors (CCRs), long used in the space industry, are special mirrors designed to reflect laser light back in the direction from which it arrived. They require no internal energy source. When illuminated by a laser, they provide a return signal that can be detected on the ground by a fast camera, as seen in figure 12.6. Putting a different number of CCRs on a set of CubeSats allows the ground station to differentiate between the CubeSats (i.e., a CubeSat with one CCR will produce a different return signal from another with two CCRs or three CCRs, etc.). One can use a laser and telescope system like those employed by the International Laser Ranging Service (ILRS) (23), which are high TRL and have been operating for decades. Precise orbital information is required to lase the CubeSat and receive a return signal, and the number of satellites that can be uniquely identified is limited by the number of corner cube reflectors that can be attached.

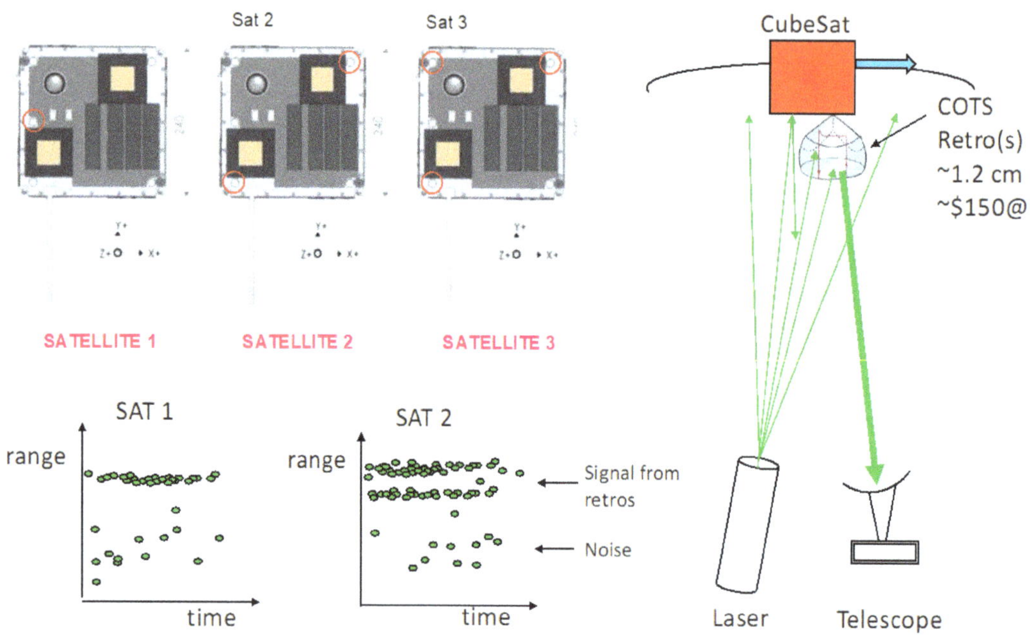

Figure 12.6: Corner Cube Reflectors. Credit: The Aerospace Corporation.

12.5.3 Passive Increase in Albedo

The simplest method of increasing trackability of satellites involves using high-albedo paint, special tape, or other simple methods to increase the optical or radar visibility of a small satellite, allowing it to be more easily detected by ground-based systems (24). White-colored thermal paint has been used for years to increase the ability of satellites to reject heat, which also helps make the satellites more visible and more trackable. Additionally, CubeSats often deploy a mission-specific configuration of wire antennas and/or cylindrical boom structures which can serve as unique identifiers using ground-based optical or radar characterization (25). Such approaches are simple, require little to no SWaP, and are readily available, but don't uniquely identify the satellite, and are limited in their effectiveness.

12.6 Future Efforts

Many in the community are aware of the "CubeSat confusion" issue, and there is a ground-swell of desire to make progress in mitigating this problem. Regulators have recognized the issue (27), and one of the consolidators, SpaceFlight, Inc., has announced their Sherpa orbital transfer vehicle will take tracking and identification technologies into space as hosted payloads aboard some of their upcoming dispenser satellite flights to increase their TRLs (28).

On the horizon, High Earth Robotics plans to create the Argus constellation – twelve optical 6U HERO-1 nanosatellites with space telescope payloads in GEO that can identify objects, take high resolution images of damaged satellites, and help identify solutions to avoid further decomposition. The constellation is intended to be resilient to interference and communications link interruption (40).

12.7 Summary

Small satellites and CubeSats are likely to continue increasing in popularity, and multi-manifest launches provide a very cost-effective way to get large numbers of satellites to space. Improving the ability to identify and track small satellites in space – especially those deployed in batches from a single launch vehicle – can help both small satellite owners and the entire space enterprise avoid the pitfalls of "CubeSat confusion." Tracking SmallSats from launch through demise is critical to avoid collisions and preserve the space environment from orbital debris. It is important that the end-to-end cost and resulting capability are evaluated when choosing a tracking option to ensure that the needed functionality is available.

For feedback solicitation, please email: arc-sst-soa@mail.nasa.gov. Please include a business email so someone may contact you further.

References

(1) Spaceflight. [Online] 2021. Accessed March 3, 2021. Available at: https://spaceflight.com

(2) Nanoracks. "Nanoracks Books CubeSat Rideshare and Habitat Building Demonstration in Single SpaceX Falcon 9 Launch". [Online] 2019. Accessed March 3, 2020. Available at: https://nanoracks.com/rideshare-habitat-building-demonstration/

(3) Kirtland Airforce Base. "Space Test Program". [Online] 2007. Accessed April 8, 2020. Available at: https://www.kirtland.af.mil/About-Us/Fact-Sheets/Display/Article/826059/space-test-program/,

(4) M.A. Skinner. "CubeSat Confusion: how to avoid 'dead on arrival'." International Astronomical Congress 71st meeting, October 12-14, 2020.

(5) I. Klotz. "SpaceX Rideshare Mission Spawns 12 Orphan CubeSats." Aerospace Daily and Defense Report, August 12, 2019.

(6) L. Grush. "Why the Air Force Still Cannot Identify more than a Dozen Satellites from One December Launch: The case of the unknown satellites." [Online] 2019. Available at: https://www.theverge.com/2019/4/2/18277344/space-situational-awareness-air-force-tracking-sso-a-spaceflight-CubeSats

(7) J. Bloch et al. "The ALEXIS Mission Recovery." 17th annual American Astronomical Society (AAS) guidance and control conference, Keystone, CO, 1994.

(8) B. Braun and S. Herrin. "The More, the Messier: ORS-3 lessons for multi-payload mission deployments." 1-10. 10.1109/AERO.2016.7500582. 2016.

(9) C.C. Venturini. "Improving Mission Success of CubeSats." [Online] Aerospace Report number TOR-2017-01689. June 12, 2017. Accessed May 15, 2019. Available at: https://www.nasa.gov/sites/default/files/atoms/files/improving_mission_success_of_CubeSats_-_tor-2017-01689.pdf

(10) H.D. Voss et al.: "Globalstar Communica0on Link for CubeSats: TSAT, GEARRS1, and GEARRS2." 29th AIAA/USU Conference on Small Satellites, Logan UT, August 9, 2015.

(11) M.A. Skinner. "Making Small Satellites Visible: Nanosat tracking and identification techniques and technologies." International Astronomical Congress 70th meeting, Washington DC, October 21-25, 2019.

(12) H.D. Voss, J.F. Dailey, M.B. Orvis. "Black Box' Beacon for Mission Success, Insurance, and Debris Mitigation." 32nd Annual AIAA/USU Conference on Small Satellites, Logan UT. 2018.

(13) A.J. Abraham. "*GPS Transponders for Space Traffic Management*." Center for Space Policy & Strategy, Aerospace Corporation, April 2018.

(14) J. Cutler, P. Seitzer, C.H. Lee *et al.* "Improved Orbit Determination of LEO CubeSats: Project LEDsat." AMOS Technologies Conference, Maui, Hawai`i. September 2017

(15) D.M. Palmer R.M. Holmes. "*ELROI: A License Plate for Your Satellite*." Journal of Spacecraft and Rockets, Vol. 55, No. 4, pp. 1014-1023, 2018.

(16) S. Phan. "'*Spaceflight Industries' SSO-A Flight Launches with SRI International's CUBIT Technology Onboard, Developed to Track and Identify Low Earth Orbit Satellites*." [Online] Accessed October 30, 2019. Available at: https://www.sri.com/blog/spaceflight-industries-sso-flight-launches-sri-internationals-cubit-technology-onboard

(17) G. Kirchner, L. Grunwaldt, R. Neubert, F. Koidl, M. Barschke, Z. Yoon, H. Fiedler, C. Hollenstein. "Laser Ranging to Nano- Satellites in LEO Orbits: Plans, Issues, Simulations." 18th International Workshop on Laser Ranging, Fujiyoshida, Japan. 2013.

(18) H.D. Voss et. al.: "'Black Box' RF Sat-Link for Space Debris, Mission Success and Risk Mitigation." First International Orbital Debris Conference, Sugar Land, TX. 2019.

(19) D.M. Palmer et. al. "*Progress Towards the ELROI Satellite License Plate.*" 34th Annual Conference on Small Satellites, Logan, UT. 2020.

(20) D. Bird. "Design and Manufacture of a Low-Profile Radar Retro-Reflector." RTO SCI Symposium on "Sensors and Sensor Denial by Camouflage, Concealment and Deception", Brussels, Belgium, April 19-20, 2004, and published in RTO-MP-SCI-145.

(21) D. Lane et. al. "Nanosatellite Tracking using Passive Retro-Reflectors." SSC20-WKIV-04, 34th Annual Conference on Small Satellites, Logan, UT, 2020.

~~(22) N/A~~

(23) NASA. "ILRS Overview," [Online] May 9, 2016. Accessed July 2022. Avaiblae at: https://ilrs.gsfc.nasa.gov/about/Overview.html

(24) D. Hall. "Optical CubeSat Discrimination." AMOS Technical Conference, September 2008.

(25) C. Chong and M. Shozo. "*Metrics for feature-aided track association*." 9th International Conference on Information Fusion, IEEE, 2006.

(26) ExoAnalytics Solutions. "Space Domain Awareness (SDA)." [Online] 2020. Accessed March 18, 2020. Available at: https://exoanalytic.com/space-domain-awareness

(27) US Federal Communications Commission, "Orbital Debris in the New Space Age, Report and Order and Further Notice of Proposed Rulemaking*.*" FCC-20-54, paragraph 62, April 24, 2020.

(28) J. Sorensen. "SpaceFlight Inc. Unveils Next-gen Orbital Transfer Vehicle to Fly Aboard Next SpaceX Rideshare Mission." [Online] July 15, 2020. Accessed: 2020. Available at: https://spaceflight.com/spaceflight-inc-unveils-next-gen-orbital-transfer-vehicle-to-fly-aboard-next-spacex-rideshare-mission/

(29) Launch Conjunction Assessment Handbook. LAUNCHCONJUNCTION ASSESSMENT HANDBOOKVERSION 1.0, DECEMBER 2018. 18 SPCS Process for Launch Conjunction Assessment, JFSCC, 18th Space Control Squadron Combined Space Operations Center (CSpOC)Vandenberg Air Force Base, California, www.s p a c e-track.org

(30) S. Erwin. "Space Fence surveillance radar site declared operational". [Online] Accessed September 29, 2020. Available at: https://spacenews.com/space-fence-surveillance-radar-site-declared-operational/

(31) Massachusetts Institute of Technology, Lincoln Labs. "Haystack Ultrawideband Satellite Imaging Radar." [Online] 2022. Accessed July 2022. Available at: https://www.ll.mit.edu/r-d/projects/haystack-ultrawideband-satellite-imaging-radar

(32) NASA. "NASA Releases Best Practices Handbook to Help Improve Space Safety." [Online] December 15, 2020. Accessed 2021. Available at: https://www.nasa.gov/press-release/nasa-releases-best-practices-handbook-to-help-improve-space-safety

(33) NASA. NID 7120.132 Collision Avoidance for Space Environment Protection. November 19, 2021.

(34) D. Murakami, S. Nag, M. Lifson, P. Kopardekar. "Space Traffic Management with a NASA UAS Traffic Management (UTM) Inspired Architecture." AAIAA SciTech Forum, San Diego, California, 7-11 January 2019.

(35) NASA. "Conjunction Assessment Risk Analysis." [Online] 2022. Accessed 2022. Available at: https://cms.nasa.gov/conjunction-assessment

(36) P. Marzioli, A. Gianfermo, L. Frezza, D. Amadio, N. Picci, F. Curiano ,M.G. Pancalli, E. Vestito, J. Schachter, M. Szczerba, D. Gu, A. Lin, J. Cutler, S. Pirrotta, F. Santoni, P. Seitzer, F. Piergentili. "Usage of Light Emitting Diodes (LEDs) for improved satellite tracking." Acta Astronautica 179 (2021) 228–237.

(37) European Space Agency. "LEDSAT passes Flight Acceptance Review and now has a ticket to orbit." [Online] June 15, 2021. Accessed 2021. Available at: https://www.esa.int/Education/CubeSats_-_Fly_Your_Satellite/LEDSAT_passes_Flight_Acceptance_Review_and_now_has_a_ticket_to_orbit

(38) C. Walker et al. "Impact of Satellite Constellations on Optical Astronomy and Recommendations Toward Mitigations." NSF's NOIRLab (Communications, Education &

(39) Engagement division).

(40) M. Boggett. "LeoLabs Partners with SpaceX to Track Starlink Satellite Deployments". [Online] 2020. Accessed June 16, 2021. Available at: https://medium.com/@leolabs_space/leolabs-partners-with-spacex-to-track-starlink-satellite-deployments-4e579052546f

(41) M. Skinner, M. Coletti, M.C. Voss, T. Svitek, J.C. Lee, K. Auman, H. Patel, E.J. Moyer. "Mitigating CubeSat Confusion: results of in-flight technical demonstrations of candidate tracking and identification technologies." 72nd International Astronautical Congress (IAC), Dubai, United Arab Emirates, 25-29 October 2021.

(42) M. Skinner, K. Auman, E.J. Moyer, M.C. Voss, J.C. Lee, R. Olcott, H.E. Weiher, A.M. Goodyear, D.A. Hinkley. "Mitigating CubeSat Confusion: further results of in-flight technical demonstrations of candidate tracking and identification technologies," 73rd International Astronautical Congress (IAC), Paris, France, 18-22 September 2022.

(43) LeoLabs. "LeoLabs Partners with SpaceX to Track Starlink Satellite Deployments," October 27, 2020. [Online] Available at: https://leolabs-space.medium.com/leolabs-partners-with-spacex-to-track-starlink-satellite-deployments-4e579052546f

Chapter Contents

BODY-410

Chapter Glossary

(ADCS)	Attitude Determination and Control System
(AEOLDOS)	Aerodynamic End-of-Life Deorbit system for CubeSats
(AFRL)	Air Force Research Laboratory
(ARC)	Ames Research Center
(CRD2)	Commercial Removal of Debris Demonstration
(D3)	Drag Deorbit Device
(DOM)	De-orbit Mechanism
(EOL)	End-Of-Life
(FURL)	Flexible Unfurlable and Refurlable Lightweight
(GCD)	Game Changing Development
(GTO)	Geosynchronous Transfer Orbit
(HSC)	High Strain Composite
(IADC)	Inter-Agency Space Debris Coordination Committee
(ISS)	International Space Station
(JAXA)	Japan Exploration Space Agency
(MSFC)	Marshall Space Flight Center
(RODEO)	Roll-Out DeOrbiting Device
(SBIR)	Small Business Innovation Research
(SSO)	Sun-synchronous orbit
(STMD)	Space Technology Mission Directorate
(TRL)	Technology Readiness Levels
(UTIAS-SFL)	University of Toronto Institute for Aerospace Studies Space Flight Laboratory
(VESPA)	Vega Secondary Payload Adapter

13.0 Deorbit Systems

13.1 Introduction

Space debris, also known as orbital debris or space pollution, are derelict artificial objects left in space on purpose and accidentally that include larger nonfunctional spacecraft and rocket bodies, and smaller disintegrated mission-related objects such as lens caps, ejected bolts, or even paint flakes. Additionally, larger space debris are commonly broken up into even smaller fragments due to collisions, erosion, or expelled particles from the spacecraft or rocket bodies. This presents a major problem in the space environment as spacecraft can be damaged or destroyed by space debris collisions due to the very high velocities of the debris objects, and thus producing even more space debris.

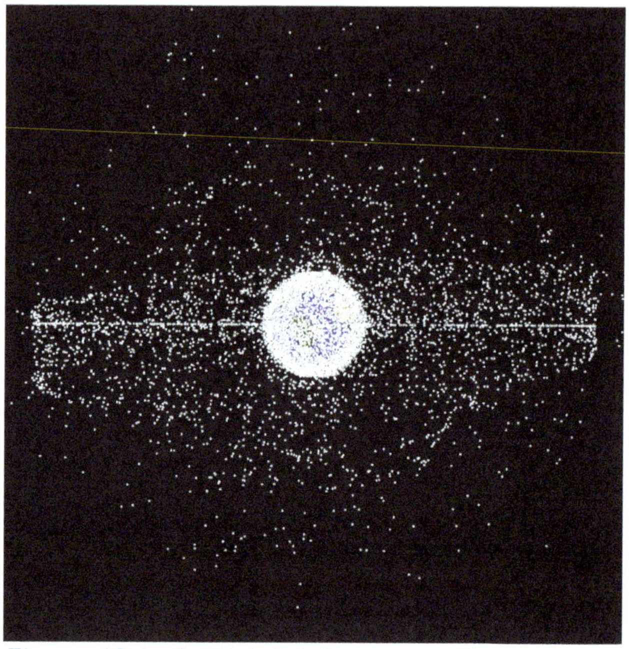

Figure 13.1: Orbital debris around Earth. Credit: NASA.

While space debris is present throughout space, there is a large accumulation around Earth particularly in low-Earth orbit (LEO) where most space operations take place. This is also attributed to the increased launch cadence of small spacecraft and the recent surge in constellations in the past decade. Improved access to space has made LEO accessible and less expensive for more countries, organizations, and institutions to launch a small spacecraft mission which only adds to the associated space debris risks and threats. Estimates of the accumulation of orbital debris suggest approximately 100,000,000 objects with a diameter 1 – 10 cm, and over 36,500 pieces with diameters >10 cm, are in orbit between geostationary, equatorial, and LEO altitudes (1). Figure 13.1 shows a representation of the orbital debris around Earth. Additionally, the orbital lifetime of space debris can be extremely long since atmospheric drag is only really helpful at <250 km (2).

Due to the inherent problem of space debris, there are ongoing policy measures to establish the importance of mitigating and removing space debris. The general guideline is that spacecraft in LEO must deorbit, also known as decay, or be placed in graveyard orbit within a maximum of 25 years after the completion of their mission (3).This standard spacecraft lifetime regulation has been recently updated that directs NASA and other national agencies to reassess current mitigation policies, especially regarding the potential advantages and cost implications resulting from limits on the space debris orbit lifetime (4). These regulations have incorporated spacecraft decay capabilities into mission design.

The rate of spacecraft decay in LEO depends on several factors, including the initial orbit insertion, the ballistic coefficient of the spacecraft, and solar weather conditions, which all play a fundamental role in the ability to comply with decommissioning regulations. Small spacecraft designers have examined various strategies for complying with decommissioning regulations to accelerate spacecraft decay post mission: spacecraft are launched in a lower orbit for a natural decay within a few years or equipped with deorbit systems to encourage altitude decay and ultimately reenter and burn up in the Earth's atmosphere. Natural decay in <5 years can be achieved for most smallsats at altitudes <400 km, however several smallsat missions must be in orbits beyond 400 km making them susceptible to use deorbit methods.

Spacecraft deorbit methods are either passive or active. Passive deorbit methods have gained maturity since the last iteration of this report, and there are more devices with high Technology Readiness Levels (TRL ≥ 8) that are guaranteed to satisfy current lifetime requirements. Traditionally, passive systems were the main option for deorbiting due to their increased simplicity, however recently active methods are gaining traction. Common active deorbiting requires attitude control and, in some cases, surplus propellant post-mission, such as a steered drag sail that relies on a functioning attitude control system, or on actuators for pointing the sail. Propulsion devices for deorbiting techniques are considered risky due to potential failure or malfunction of either the spacecraft, up until its final stage of decommission, or the propulsive technology itself. Adequate attitude control and navigation capabilities after the mission for a controlled reentry are never a guarantee. Some of the new active deorbiting solutions include a separate spacecraft that can attach to the defunct satellite to bring it down to lower orbits where the satellites can complete the deorbit using their own drag decay.

The influx of small spacecraft in LEO has also developed space situational awareness and space traffic management data. For information on this, please see the *Identification and Tracking Systems* chapter.

13.1.1 Chapter Organization

This chapter is organized as follows:

- Orbital Debris Regulations (13.2)
- Passive Deorbit Systems (13.3)
- Active Deorbit Systems (13.4)

Orbital Debris Regulations provide the reader with a comprehensive understanding of the current policy regulations for deorbit mitigations, when they were initiated, and the organizations that implement space orbiting debris regulations. The Passive and Active Deorbit System sections contain technology description, summary table of devices; and previous, current and planned missions. This chapter provides a comprehensive guide to existing commercial technologies and technology demonstrations for both methods, and the authors have attempted to highlight technology gaps within existing deorbiting capabilities and current development status on each deorbit method.

The information described below is not intended to be exhaustive but provides an overview of current state-of-the-art technologies and their development status as discussed in open literature. It should be noted that TRL designations may vary with changes specific to payload, mission requirements, reliability considerations, and/or the environment in which performance was demonstrated. Readers are highly encouraged to reach out to companies for further information regarding the performance and TRL of described technology. There is no intention of mentioning certain companies and omitting others based on their technologies or relationship with NASA.

Definitions

- *Disposal* refers to removal of spacecraft from orbital environment.
- *Deorbit* refers to lowering spacecraft's orbital altitude, also referred to as *Decay.*
- *Decay* refers to a gradual decrease of the distance between two orbiting bodies.
- *Atmospheric Drag* refers to molecular collisions with the spacecraft body.
- *Drag Area* refers to the spacecraft surface area experiencing atmospheric resistance.
- *Orbital Lifetime* refers to total time spacecraft is in orbit.

13.2 Orbital Debris Regulations

Space debris has been a concern for several decades, but with visible sightings of reentry fragments of spacecraft and rocket bodies, the urgency to address space debris has grown. NASA's Orbital Debris Program Office was created in 1979, the Air Force Space Debris Research Program was initiated in the 1980s. NASA was among the first organizations to implement plans for mitigation and remediation of space debris in the early 1990s, and in 1993 the Inter-Agency Space Debris Coordination Committee (IADC) was founded internationally.

NASA collaborated with the Department of Defense in 1997 to develop the U.S. Government Orbital Debris Mitigation Standard Practices (ODMSP) (5). The agency's most updated orbital debris guidelines can be found in NASA NPR 8715.6B "NASA Procedural Requirements for Limiting Orbital Debris and Evaluating the Meteoroid and Orbital Debris Environments" (6) and NASA Standard 8719.14C "Process for Limiting Orbital Debris" (3). These technical documents describe the processes and requirements to limit orbital debris for all NASA spacecraft missions. The guidelines, among other considerations, include a limit on the risk of potential human casualties caused by reentering debris, which shall not be greater than 1 in 10000 (5). Of the three spacecraft disposal methods identified – direct retrieval, atmospheric re-entry, and maneuvering into a storage orbit – atmospheric reentry was deemed as the most feasible for the majority of spacecraft missions. Therefore, a maximum 25-year post-mission orbital lifetime (no longer than 30 years after launch or a move into a graveyard orbit for safe storage) was established for all US spacecraft. The rationale for this specific orbital lifetime was based on the least amount of propellant required to maneuver to a lower orbit as predicted by various orbital debris models (7).

The IADC is an entity formed by national and multi-national space agencies, including NASA, ESA, JAXA and several others, and is widely recognized by the international community as the technical authority on space debris. In 2002, the IADC established the Space Debris Mitigation Guidelines to address orbital debris. Their findings and procedures are submitted to the United Nations (UN), as space debris has been one of the main interests of the UN Committee on the Peaceful Uses of Outer Space (COPUOS). In 2007, space debris mitigations guidelines based on the IADC procedures were accepted by the COPUOS and endorsed by the UN (5). The IADC adopted the 25-year orbital lifetime guideline for space objects in LEO.

The U.S. Federal Communications Commission (FCC) regulates all radio communication across the U.S. and all U.S. spacecraft must be licensed for space communications. Since the early 2000s, the FCC has deliberated over how best to mitigate orbital debris from FCC-authorized space activities, and formally adopted debris mitigation regulations (2) in 2004. These FCC regulations include orbital debris mitigation plans as part of license applications, and require applicants to disclose "the design and operational strategies that they will use, if any, to mitigate orbital debris," and to "identify particular methods by which a proposed satellite system will mitigate orbital debris" (2). The FCC adopted the ODMSP 25-year lifetime guideline as well, and commented that the 25-year "rule" should be tightened, as this no longer adequately addresses current orbital debris issues arising from the launch of large constellations and the expected increase in future LEO space activity. On September 29th, 2022, the FCC adopted a new rule for all FCC-licensed satellites within the LEO region (<2000 km) to reduce the lifetime requirement to 5 years after launch (8). As of 2023, there are discussions at the agency and federal level to determine the final policies.

Since this updated "5-year lifetime rule" by the FCC, there has been increased focus on space debris removal activities. In April 2023, the FCC created a new Space Bureau responsible for the regulation of satellites and space debris (9). The World Economic Forum (WEF) released the 'Space Industry Debris Mitigation Recommendations' document in June 2023 to standardize a series of recommended behaviors for satellite operations. One of these listed recommendations is to target five years or less after end-of-life for spacecraft removal. The document was signed by several companies including Airbus, The Aerospace Corporation, SES, and Planet. The WEF collaborated with ESA, the MIT Media Lab, and other stakeholders to establish a Space Sustainability Rating (SSR) system to provide a measurable score that can characterize spacecraft mission compliance with the international space debris remediation guidelines (10)(11).

The Federal Aviation Administration (FAA) announced on September 20th, 2023, a proposal to create a new rule to limit the growth of debris from commercial launch vehicles in order to reduce collision risk and limit space debris in populated LEO environments (12). The new regulation will give commercial launch operators specific options for orbital debris countermeasures, requiring disposal of their rocket upper stages by performing a controlled reentry within 30 days after mission completion, moving to a less congested or graveyard orbit (within 30 days), placing them in an Earth escape trajectory (within 30 days), retrieving them with active debris removal within five years after launch, or performing an uncontrolled atmospheric disposal within 25 years.

13.2.1 Considerations for Orbital Lifetime Requirements in LEO

Small spacecraft launched at or around the 400 km altitude naturally decay in under five years, however at orbital altitudes beyond 500 km, there is no guarantee the spacecraft will deorbit within that timeframe and some may have trouble deorbiting in under 25 years. This is due to potential low atmospheric density conditions and the effects on various ballistic coefficients, as seen in figure 13.2. This graph displays various cases of SmallSats with distinct masses, drag areas, and initial orbits, under the atmospheric density conditions during the 11-year solar cycle maximum and minimum.

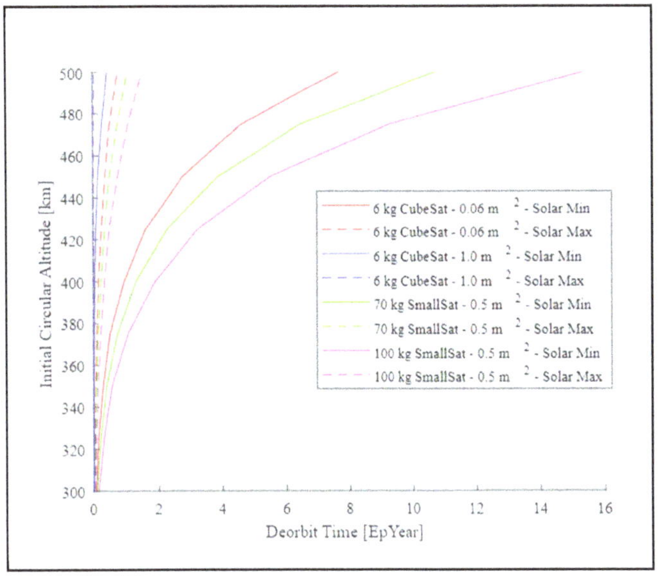

Figure 13.2: Initial orbit altitudes yield different lifetimes depending on the ballistic coefficient of the spacecraft. Three representative area-to-mass ratios are shown. Note that the propagation stops at 16 years, but the initial altitudes yield even longer times. Credit: NASA.

The varying solar weather conditions can affect the deorbit performance for a given altitude and can have a significant impact on orbital lifetimes. The atmospheric drag force that satellites experience is increased during solar maximum, resulting in a faster decay. In this situation, the Sun emits extra energy in the atmosphere and creates higher density layers in LEO altitudes that produce a stronger drag force on the satellites (13). It is common for some missions to plan their launch periods around the solar cycle, and if the stricter 5-year orbital lifetime requirement becomes widely accepted, more companies may want to consider this, as the deorbit time can be reduced by more than 10 years as seen in figure 13.2.

Another important factor that affects orbit propagation in LEO is the spacecraft's Ballistic Coefficient (BC). The BC is defined in this chapter as the mass to area ratio multiplied by the inverse of the drag coefficient, that is assumed to equal 2.2. By this definition, a spacecraft with a lower ballistic coefficient will decay faster due to the smaller mass to area ratio. As shown in figure 13.2, a 6U spacecraft with an area of 0.06 m^2 and an assumed mass of 6 kg has a ballistic coefficient of 45, which is significantly lower than a 100 kg spacecraft of 0.5 m^2 with BC of 90.

Since timing the launch for a particular solar weather scenario may not be feasible, another strategy for satellite operators to comply with orbital lifetime requirements is to decrease their spacecraft ballistic coefficient or mass to area ratio. Deorbit technologies such as drag devices can effectively increase the spacecraft's drag area and may become even more important for spacecraft operations in LEO.

13.3 State-of-the-Art – Passive Systems

Passive deorbit methods require no further active control after deployment. Recent developments have increased the number of available options with flight heritage. This chapter will emphasize recent developments rather than past missions. In addition, the chapter aims to discuss devices used exclusively for deorbit purposes, excluding technologies such as solar sails that are used for other propulsive applications.

13.3.1 High TRL Drag Sails

Drag devices are the most common deorbit device for satellites orbiting in LEO. They are advantageous due to simplicity and small stowed volumes. For certain area-to-mass ratios in altitudes equal to or lower than 800 km, drag devices can be deployed to increase the drag area for faster deorbiting in compliance with the new 5-year requirement. Recently, this technology has been implemented in several small spacecraft missions, and several companies and institutions are developing prototypes that are increasingly more mature, providing solutions to the space debris problem for missions that do not have resources for an active system. Table 13-1 displays current state-of-the-art technology for passive deorbit systems.

Table 13-1: Drag Sails									
Product/Mission	**Manufacturer**	**Mission host and launch mass (kg)**	**Device mass (kg)**	**Initial orbit (alt and inc.)**	**Launch Year**	**Deployment Year**	**Drag area (m^2)**	**TRL**	**Ref.**
NanoSail-D2	NASA MSFC/ARC	FASTSAT (4.2 kg)	N/A	650 km 72 deg	2010	2011	10	7-9	(3)
Drag-Net	MMA Design	ORS-3 Deployed a Minotaur Upper Stage (100 kg)	2.8	N/A	2016	2016	14	7-9	(14)
Drag-Net	MMA Design	General Atomics GAzelle Satellite	2.8	N/A	2022	TBC	14	7-9	(15)
Icarus-1	Cranfield Aerospace Solutions	SSTL TechDemoSat-1 (157 kg)	3.5	635 km	2014	2019	6.7	7-9	(16)
Icarus-3	Cranfield Aerospace Solutions	Carbonite-1 (80 kg)	2.3	650 km 98 deg	2015	Future (in-orbit)	2	7-9	(16)
DOM	Cranfield Aerospace Solutions	ESEO (45 kg)	0.5	572 km × 588 km 97.77 deg	2018	Future (in-orbit)	0.5	7-9	(16)
Terminator Tape	Tethers Unlimited, Inc.	Prox-1 (71 kg)	0.808	717 km 24 deg	2019	2019	10.5	7-9	(17)
DragSail	Surrey Space Centre	InflateSail (3.2 kg)	N/A	505 km 97.44 deg	2017	2017	10	7-9	(18)
Exo-Brake	NASA	TechEdSat 5 (3.4 kg)	N/A	405 km 51.5 deg	2014	2015	0.35	7-9	(19)
Exo-Brake	NASA	TechEdSat 7 (3 kg)	N/A	485 x 513 km 60.7 deg	2021	2021	1.2	8-9	(20)
Exo-Brake	NASA	TechEdSat 13 (4 kg)	N/A	505 km 45 deg	2022	2022	0.083	8-9	(20)
Exo-Brake	NASA	TechEdSat 15 (4.5 kg)	N/A	215 x 270 km 137 deg	2022	2022	0.087	8-9	(20)
removeDebris	Surrey Space Centre	removeDebris (100 kg)	N/A	405 km 51.5 deg	2018	2019	16	7-9	(21)

BODY-417

CanX-7	UTIAS-SFL	3U CubeSat (3.6 kg)	0.8 (4 modules of 0.200)	688 km 98 deg	2016	2017	4	7-9	(22)
NABEO-1	HPS	1U CubeSat (attached to Rocket Lab Kick Stage)	0.85	500 km	2018	2018	2.5	8-9	(23)
ADEO-2	HPS	1U CubeSat (attached to the D-orbit ION carrier)	3.4	N/A	2021	2022	3.6	9	(24)
ADEO-Cube series	HPS	1-50 kg	0.5	LEO	N/A	N/A	2	7	(25)
ADEO-N series	HPS	20-250 kg	0.8	LEO	N/A	N/A	5±2	9	(26)
ADEO-M series	HPS	100-700 kg	4	LEO	N/A	N/A	15±5	6	(27)
ADEO-L series	HPS	500-1500 kg	9.5	LEO	N/A	N/A	20±100	7	(25)
ARTICA (ALPHA)	NPC Spacemind	1U CubeSat	0.285 (0.3U)	5865 Km, 70.16 deg	2020	2020	2.2	7-9	(28)
ARTICA (FUTURA SM 3)	NPC Spacemind	6U CubeSat	0.285 (0.3U)	N/A	2023	N/A	2.2	7-9	(29)
ARTICA (DANTESAT)	NPC Spacemind	3U CubeSat	0.285 (0.3U)	415 km	2022	2022	2.2	7-9	(29)
ARTICA (URSA MAIOR)	NPC Spacemind	3U CubeSat	0.285 (0.3U)	450 km, 97.1 deg	2017	2019	2.2	7-9	(28)
ARTICA (1KUNS)	NPC Spacemind	1U CubeSat	0.285 (0.3U)	N/A	2018	2019	2.2	7-9	(28)
LightSail - 2	The Planetary Society	3U CubeSat	N/A	720 km	2019	2022	32	9	(30)
ACS3	NASA	12U CubeSat	1 (6U)	1000 km SSO	2024	2024 (expected)	81	8	(31)
Gama ALPHA	Gama	6U CubeSat	N/A	550 km	2023	N/A	73.3	8-9	(32)

BODY-418

Several small spacecraft missions have built and launched passive deorbit technologies in the past using a drag sail or boom. The NanoSail-D2 mission, which was deployed in 2011 from the minisatellite FASTSat–HSV into a 650 km altitude and 72° inclined orbit, demonstrated the deorbit capability of a low mass, high surface area sail. The 3U spacecraft, developed at NASA Marshall Space Flight Center (MSFC), reentered Earth's atmosphere in September 2011.

CanX-7, still in orbit at an initial 800 km Sun-synchronous orbit (SSO), deployed a drag sail in May 2017. The sail was developed and tested at University of Toronto Institute for Aerospace Studies Space Flight Laboratory (UTIAS-SFL) (figure 13.3).

Figure 13.3: CanX-7 deployed drag sail during testing. Credit: Cotten et al. (2017).

The CanX-7 deorbit technology consists of a thin film sail that is divided into four individual modules that each provide 1 m^2 of drag area. These sail sections are deployed mechanically with spring booms, which help to preserve the geometry. Each module also has electronics for individual telemetry and command. This feature allows different sections to be controlled separately to mitigate risk of a single failure, and to allow custom adaptability to various spacecraft geometries and ballistic coefficient requirements for other missions. For the 2017 deployment, all four segments functioned successfully. The deorbit performance was measured after a month. The deorbit profile showed that the effects of the sail segments accounted for an altitude decay rate at the time of measurement of 20 km/year, which results in a significant increase from the previous 0.5 km/ year. These rates are expected to increase as the atmospheric density increases exponentially with lower altitudes (22).

The Technology Educational Satellite, also known as TechEdSat-n (TES-n), program at NASA Ames Research Center (ARC) has contributed significantly to the development of drag devices. It consists of a series of nanosatellite technology demonstrations in collaboration with several universities including San Jose State University and the University of Idaho. One of the main goals of the program is to test and improve deorbiting techniques and develop a unique targeting capability with their own drag device design known as the Exo-Brake. The Exo-Brake deorbit system is an atmospheric braking system that distinguishes itself from other drag devices since it is more akin to a parachute instead of a solar sail due to its primary tension-based elements. This becomes fundamental for accurate deorbit targeting since the device must retain its shape without collapsing during those critical reentry moments occurring at the atmosphere interface altitude of 100 km, known as the Von Karman line (33). The Exo-Brake has been used as both a passive and a controlled active deorbit system, therefore it is included in both sections.

The Exo-Brake development is funded by the Entry Systems Modeling project within the NASA Space Technology Mission Directorate's (STMD) Game Changing Development (GCD) program. The Exo-Brake was first implemented as a passive deorbit device on the TechEdSat missions TES-3, TES-4, and TES-5. Recent CubeSats have also used it for controlled mission deorbiting. Two of the four TechEdSat spacecraft using a passive Exo-Brake were TES-5 and TES-7, while TES-13 and TES-15 also used variations of the TES-7 design. TES-5 was deployed from the ISS in March 2017 and demonstrated this deorbiting capability after 144 days in orbit with the Exo-Brake deploying at 400 km. TES-7, a 2U CubeSat that launched January 2021, onboard Virgin

Orbit's LauncherOne rocket, was placed into orbit at 500 km (34) and decayed May 2022. TES-13 was launched January 2022 with other CubeSats on the third successful Virgin LauncherOne flight and carried an Exo-Brake onboard to demonstrate autonomous navigation and reentry over specific Earth locations. TES-15 was launched October 2022 aboard a Firefly Aerospace Alpha Launcher. Its primary objective was to test an Exo-Brake designed to sustain much higher temperatures than in previous missions. The satellite also included a simple ablator in the nosecap that is expected to last deeper into the atmosphere before burning up. After this experiment, TES-15 should be able to validate higher heating rates and the flight dynamics ability to target an Earth entry point (20). The satellite reentered on October 7, 2022, and the team is analyzing the data to study the performance of this latest flight.

The Surrey Space Centre based in the United Kingdom has developed the DragSail technology, which was implemented in a family of missions. The Inflatesail 3U CubeSat first demonstrated this technology. The European Commission QB50 program and the DEPLOYTECH partnership that included German Aerospace Centre (DLR) and NASA Marshall Space Flight Center, among others, funded it. This mission was launched in 2017 and included a mast/drag-sail technology that successfully deorbited the satellite in just 72 days. This achievement was the first time a spacecraft has deorbited using European inflatable and drag-sail methods (18).

Figure 13.4: TechEdSat-10 deployment from the ISS in July 2020. Credit: NASA.

The RemoveDebris mission was developed under the European Commission FP7 program by a consortium of several institutions such as Airbus and the Surrey Space Centre. The mission consisted of a 100 kg small spacecraft that was deployed from the ISS in 2018. One of the experiments it carried was a passive drag augmentation device consisting of a sail. The sail was deployed in March 2019, however, trajectory data showed it only partially deployed since no significant altitude change was measured. The lessons learned from this incident were implemented in another version for the Space Flight Industries' SSO-A mission that incorporated two of these sails. In that case, the assembly did not include an inflatable boom (21).

As part of the ESA CleanSat program, Cranfield Aerospace Solutions in the United Kingdom has also developed a variety of drag augmentation systems. The first demonstrated technology was the Icarus-1, which flew in the TechDemoSat-1 mission from SSTL, launched in 2014. Another version also flew in the Carbonite-1 spacecraft, launched in 2015. The concept is similar to other drag devices in which the drag increases by deploying a membrane sustained by rigid booms. The Icarus technology consists of a thin aluminum structure located around the satellite side panel that contains four stowed Kapton trapezoidal sails and booms. The mass of the system is 3.5 kg for about 5 m^2 of sail area for the Icarus-1, and 2.3 kg for 2 m^2 for the Icarus-3 (figure 13.4). Both sails deployed successfully and are expected to deorbit both spacecraft in less than 10 years. The second technology developed by Cranfield Aerospace Solutions is a de-orbit mechanism (DOM) device which consists of a version of the drag sail presented in a smaller cuboid outline. The mechanical system varies from Icarus since the sails are triangular and the booms work as tape springs themselves. This system flew in the European Student Earth Orbiter on a 45 kg satellite that carried several student payloads. Among them, the Cranfield University DOM module will deorbit the spacecraft after decommissioning. The sail has an area of 0.5 m2 with a mass of 0.5 kg (16).

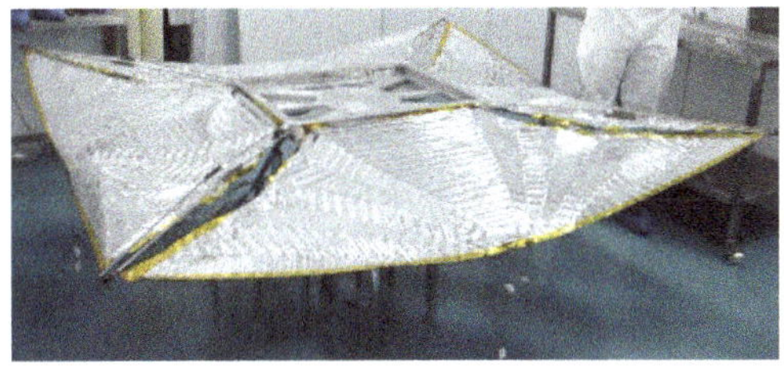

Figure 13.5: Icarus-3 drag sail implemented in the Carbonite-1 mission. Credit: Cranfield Aerospace Solutions.

MMA Design LLC, a company from Colorado, has patented the dragNET de-orbit system. The 2.8 kg module (figure 13.6) deorbited the ORS-3 Minotaur Upper Stage in 2.1 years after launch in November 2013. DragNet features four stowed thin membranes that deploy through a single heater-powered actuator. The sail has an area of 14 m^2 that can effectively deorbit a 180 kg spacecraft at an altitude of 850 km in less than 10 years (5). In October 2022, the dragNET deorbit system was launched as part of the General Atomics GAzelle satellite, as seen in figure 13.6 (15).

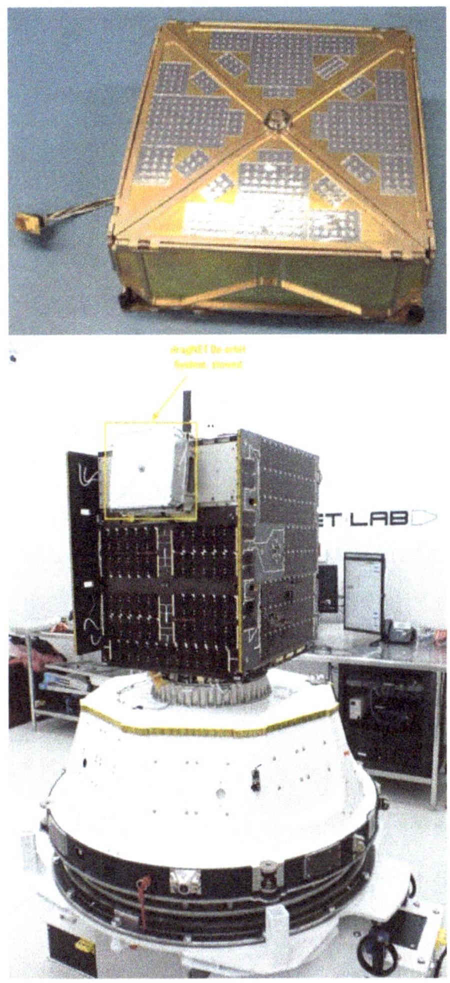

Figure 13.6: {top} The dragNET module. {bottom} dragNET module attached to the GAzelle satellite prior to its launch in late 2022. Credits: MMA design.

Redwire Space holds an exclusive license for the Flexible Unfurlable and Refurlable Lightweight (FURL) solar sail developed and tested by the Air Force Research Laboratory (AFRL). FURL extends and retracts with four booms stored around a common hub. Small satellites can employ solar sails to control attitude, change planes or remain in their proper orbits and then retract the sail once it reaches its destination. This technology could be applied to deorbit applications as well.

Purdue University has developed a drag device with a pyramid geometry that can deorbit a satellite placed in a geosynchronous transfer orbit (GTO). The Aerodynamic Deorbit Experiment (ADE), developed jointly with CalPoly and Georgia Tech, will consist of a 1U CubeSat technology demonstration deployed from a Centaur upper stage in a future Atlas V rocket from United Launch Alliance. Once deployed, the device will occupy an area of about 1 m^2 to decrease the ballistic coefficient of the spacecraft and reduce the perigee altitude during each pass. Consequently, the expected lifetime of the ADE mission will be 50 – 250 days instead of the estimated seven years (35). The technology has been licensed to Vestigo Aerospace which is commercializing the drag device with their Spinnaker series of drag sails and has been awarded funding from NASA's Phase II Small Business Innovation Research (SBIR) Program. An initial flight test was attempted in September 2021 aboard the first Firefly Aerospace Alpha rocket. The Spinnaker3 concept sail consisted of an 18 m^2 sail and was supposed to deorbit the upper stage of the launch vehicle, however the launch ended with an explosion shortly after liftoff (37). Vestigo is developing two main products, a sail targeted for small satellites that has a surface area of 1.77 m^2 and a larger 18 m^2 sail for objects weighing up to 1000 kg (38). In 2023, Vestigo was awarded a NASA Phase II-S SBIR contract to contribute to the development of a technology demonstration mission to qualify the Spinnaker 2, a 8 m^2 sail for small satellites, and the Spinnaker 3, more targeted to orbital transfer vehicles and upper stages (39).

In June 2022, China launched a Long March 2D rocket that carried a 25 m^2 drag sail attached to the payload adapter on the rocket upper stage. The 300 kg object could deorbit within two years due to this technology (40).

Figure 13.7: The ADEO-2 system deployed in LEO in December 2022, picture captured by the D-orbit's ION spacecraft carrier. Credits: ©HPS GmbH, Germany (www.hps-gmbh.com).

The Italian company NPC Spacemind has developed and launched a series of CubeSat missions that demonstrated their ARTICA deorbit system, which consists of a deployable 2.1 m^2 drag sail. The total size of the deorbiting system is 0.3 U, which makes it suitable for CubeSats as small as 1U (28). In November 2022 and in January 2023, the DanteSat 3U CubeSat, and the Future-SM3 6U CubeSat, Futura-SM3, were respectively launched and successfully operated with an ARTICA system onboard. These two new missions extend the ARTICA flight heritage after the earlier UrsaMaior, 1-Kuns, and Alpha missions, launched in 2017, 2018 and 2020 respectively (28).

The Planetary Society's LightSail-2 was a 3U CubeSat mission with a solar sail launched in June 2019 and deployed from the Prox-1 satellite once in orbit. The mission demonstrated that solar sail technology can be used in LEO by modifying its orbit altitude along the course of the mission. The 32 m^2 sail was able to extend the mission lifetime by reducing orbital decay and on some occasions, it was also able to overcome drag entirely. In late November 2022, the mission successfully reentered the atmosphere according to orbital predictions (30). The sail was intended to extend the mission lifetime of spacecraft in LEO, however the technology can be potentially use for deorbiting purposes as well.

The Drag Augmentation Deorbiting System (ADEO) is a drag sail developed by the German company High Performance Space Structure Systems (HPS). The sail is scalable, and HPS has launched already a set of missions increasing various configurations to TRL to 9. The ADEO-N series is tailored for small satellite missions of 20-250 kg, while the ADEO-M and ADEO-L target larger sizes, 100-700 kg and 500-1500 kg respectively. The ADEO-N series corresponds to sail sizes of 5±2 m^2, while ADEO-M covers areas within 15 ± 5 m^2. There are other smaller versions as well for picosatellites (ADEO-P) and CubeSats (ADEO-C) in particular, and the option to configure the sail size according to customer needs. Various missions have tested the ADEO-N product family already. The NABEO-1 was launched in 2018, attached to the center of a Rocket Lab Electron rocket Kick Stage. The sail was deployed as soon as 90 minutes after launch. There was an issue trying to measure if the drag sail was deployed initially, but optical ground observations confirmed the successful deployment and performance due to the expected change in semi-major axis (24). In late December 2022, the ADEO-2 sail was deployed from the D-orbit spacecraft carrier ION-2. The successful deployment was captured by the onboard camera from the ION carrier as depicted in figure 13.7.

In early 2023, JAXA selected Axelspace Corporation to develop the In-orbit Demonstration of Membrane Surface Deployable Deorbit Mechanism for Small Satellites (D-SAIL), together with Sakase Adtech Corporation, in Japan. D-SAIL consists of a deployable membrane mechanism to increase drag. The technology was part of the RAISE-3 satellite mission and it was launched in October 2022. However, the launch vehicle was not able to reach orbit. This new initiative results in a new opportunity to test the technology as part of the Innovative Satellite Technology Demonstration-4 mission (41).

In January 2023, the French company Gama launched its first spacecraft mission, a 6U CubeSat name ALPHA. This first technology demonstration mission aims to test the deployment and control of their 73.3 m^2 solar sail. The final phase of the mission will use the sail to rapidly deorbit the satellite (32).

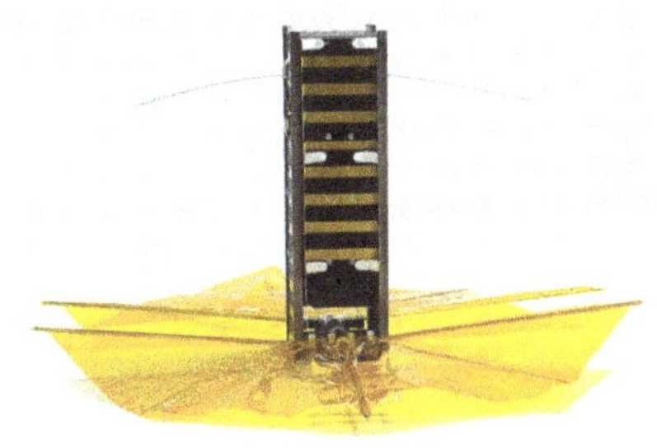

Figure 13.8: SBUDNIC CubeSat with drag sail made from Kapton polyimide film. Credit: Brown Univ.

SBUDNIC, a CubeSat designed and built by Brown University students with support from D-Orbit shown in figure 13.8, AMSAT-Italy, La Sapienza-University of Rome, and NASA Rhode Island Space Grant, demonstrated a practical, low-cost method to cut down on space debris. Rather than taking debris out of orbit after it becomes a problem, this $30 drag device can be added onto satellites to radically reduce how long they're in space. SBIDNIC was launched on a SpaceX rocket May 2022 as part of the Transporter 5 ridesharing mission. The plastic drag sail made from Kapton polyimide was deployed at about 520 kilometers, well above the orbit of the International Space Station, which helped push the satellite back down to Earth quicker than anticipated-- about five years ahead of schedule-- reentering Earth's atmosphere on Aug. 8, 2023, burning up high above Turkey after 445 days in orbit, according to its last tracked location from U.S. Space Command (42).

Figure 13.9: The ACS3 sail fully deployed during its pre-integration fit test. Credits: NASA Langley.

The Advanced Composite Solar Sail System (ACS3) is a mission developed at NASA Langley and NASA Ames that consists of a spacecraft that will deploy an 81 m^2 solar sail in a 1000 km sun-synchronous orbit (see figure 13.9). The main objective of the mission is to demonstrate that the solar wind can impulse the spacecraft to change the semimajor-axis and obtain a different orbit altitude. The sail will be composed of a combination of composite materials with distinct properties, and it will be deployed with lightweight booms from a 12U CubeSat bus, developed by NanoAvionics. The spacecraft will be launched aboard an Electron launch vehicle from Rocket LAB Launch Complex in New Zealand in 2024. Although the main objective of the mission is to show the propulsive capabilities of the solar sail, the device can be used for deorbiting purposes, and it may be used at the end of the ACS3 spacecraft lifetime for decommissioning (31).

13.3.2 Deployable Booms

Deployable booms, while not strictly a deorbit device themselves, compose a vital part of many deorbit systems. They are structural components that can be stowed during launch, then deployed once in space to provide the support structure required for various drag sail designs. More specific information regarding deployable booms can be found in the *Structures, Materials, and Mechanisms* chapter.

Built by Redwire Space, the ROC-FALL device consists of a rectangular sail supported by a High Strain Composite (HSC) boom that is co-wrapped on a spool and restrained with a strap for stowage. The ROC-FALL system is scalable both in width and length to accommodate a variety of spacecraft sizes, and the heritage system sail measures 3.8 x 0.45 m in deployed area and rolls to a 0.04 x 0.45 m tube + supporting mechanism. The ROC-FALL is tip-rolled and passively deployed from the spacecraft. Redwire Space offers a variety of deployable boom technologies with a wide range of applications on small spacecrafts including open lattice mast, rollable tubes, and telescopic booms that can be applied on small spacecraft.

The University of Florida has developed the Drag Deorbit Device (D3) 2U CubeSat which provides attitude stabilization and modulation of the satellite drag area at the same time, making the overall solution an alternative to regular ADCS units. Four 3.7 m long tape spring booms form the D3, which can deorbit a 15 kg satellite from an altitude of 700 km. A final design has already been tested and simulated, including thermal vacuum and fatigue testing (43)(44). Figure 13.10 shows two images of the final design. The mission was selected by NASA through the CubeSat Launch Initiative, and on September 6, 2022, D3 was successfully placed in orbit (45).

Figure 13.10: D3 CAD design (left), boom inside thermal vacuum chamber (center), and prototype design (right). Credit: Omar et al., 2019, and Martin et al., 2019.

Composite Technology Development, Inc. has developed the Roll-Out DeOrbiting device (RODEO) that consists of a lightweight film attached to a simple, ultra-lightweight, roll-out composite boom structure (figure 13.11). This is a self-deploying system where the stored strain energy of the packaged boom provides the necessary deployment force. It was successfully deployed on suborbital RocketSat-8 (138 kg) on August 13, 2013 (46).

BODY-424

13.3.3 Electromagnetic Tethers

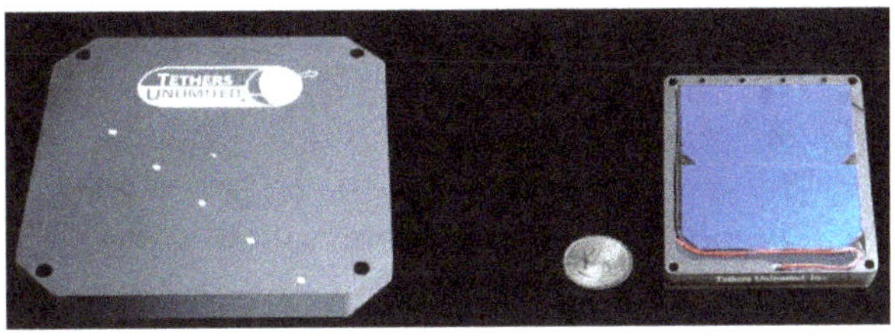

Figure 13.12: Image of the NSTT (left) and the CSTT modules. Credit: Tethers Unlimited.

In addition to drag sails, an electromagnetic tether has proven to be an effective deorbit method. This technology uses a conductive tether to generate an electromagnetic force as the tether system moves relative to Earth's magnetic field. Tethers Unlimited (now Amergint Technologies) developed terminator tape that uses a burn-wire release mechanism to actuate the ejection of the terminator's cover, deploying a 70 m long conductive tape at the conclusion of the small spacecraft mission. There are currently two main modules. The first, NSTT for NanoSats has a mass of 0.808 kg. The second, CSTT, is made for CubeSats and has a mass of just 0.083 kg. Figure 13.12 shows an image of both systems respectively. The 70 m long NSTT has been implemented in the 71 kg Prox-1 satellite, launched in mid-2019 by AFRL (17).

DragRacer, an experiment jointly developed by Tethers Unlimited, Millennium Space Systems, RocketLab, and TriSept Corp., consisted of a satellite (Alchemy) with the terminator tape, and another satellite (Augury) without it, to characterize the tape performance (47). Alchemy reentered in July 2021 while Augury is still in orbit.

13.4 State-of-the-Art – Active Systems

Several companies have been increasingly offering active spacecraft-based deorbit systems. Space startups such as Astroscale, ClearSpace, and D-orbit have long-term plans and have already started initial technology demonstration missions. These systems consist of separate, dedicated spacecraft that attach to decommissioned satellites to place them into decaying or graveyard orbits. In December 2019, Iridium stated that they would like to pay for an active deorbit system to remove 30 of their defunct satellites (48). In addition, for NASA missions, the NASA STD-8719.14C document stipulates that all spacecraft using controlled reentry processes, the designed trajectory must guarantee that no remaining debris that could impact with a kinetic energy greater than 15 Joules is nearer than 370 km from foreign landmasses, or within 50 km from any territory of the United States and the permanent ice pack of Antarctica (3).

This section covers some of the main stakeholders in the industry that are working towards the implementation of active space debris removal, as well as some other promising technologies that can potentially be used for actively deorbiting spacecraft in the future.

13.4.1 TechEdSat Series Exo-Brake

The Exo-Brake introduced earlier in the passive systems also has active control capability. The TES-6 mission was the first to implement this technology with a 3.5U CubeSat with a mass of 3.51 kg that deployed its Exo-Brake from the rear of the satellite. It targeted a reentry over Wallops Flight Facility by modulating the drag device to adjust the ballistic coefficient as orbital determination about the satellite state became available over time. The Iridium gateway enabled the command of the brake, which proved to significantly affect the reentry time and consequently, the location of the Wallops target area. The spacecraft overshot the intended target range slightly as shown in the second image, since it could not achieve a lower 4 – 5 kg m^2 ballistic coefficient configuration, which would have yielded suitable results if placed at 300 km (see figure 13.13).

However, the mission successfully demonstrated the reentry experiment and the command/control capability by overflying Wallops right before reentering. This technology was going to be demonstrated again in the TES-8 mission, although a power system failure occurred before the targeting process. It should be noted that the Exo-Brake was successfully deployed on TES-8, an improved version of the previous TES-5 and TES-6 devices. The TES-8 ballistic coefficient range was wider (6 – 18 kg m-2) and enabled better control authority for targeting. TES-10 and upcoming TES-11 are also incorporating this design (33). TES-10 (figure 13.13) marked the second targeted deorbit flight test and successfully overflew NASA Wallops Flight Facility much like TES-6 (49). TES-15 reentered seven days after deployment, and the team is evaluating the data to determine the performance of a new version of the Exo-Brake.

13.4.2 RemoveDebris Consortium Partners

The RemoveDebris mission carried two 2U CubeSats that were ejected from the mothership to simulate space debris and demonstrate active deorbit capabilities. The first CubeSat, known as DebrisSat-1, deployed at a very low velocity from the main spacecraft and subsequently inflated a balloon that provided a larger target area. A 5 m diameter net was ejected from the main spacecraft just 144 seconds after deployment, capturing the CubeSat at a distance of ~11 m from the mothercraft. The object, once enveloped in the net, re-entered the atmosphere in March 2019 (21). The RemoveDebris mission also carried another active debris technology consisting of a harpoon. In this scenario, a target platform attached to a boom was deployed from the main spacecraft. The mothership then released the harpoon at 19 m/s to hit the platform in the center. Once that occurred, the 1.5 m boom that connected the two objects snapped on one end. However, a tether secured the target in place, avoiding the creation of new debris. This resulted in the first demonstration of a harpoon technology in space. The harpoon target assembly had a dry mass of 4.3 kg (21).

13.4.3 Astroscale

Astroscale aims to provide services to address the end-of-life (EOL) scenario of newly launched satellites, and to proactively remove existing space debris. They collaborate with a variety of governmental and international organizations around the world (such as the US government, ESA, the European Union, or the United Nations) to position themselves as leaders of a more sustainable low-Earth orbit environment.

As part of the EOL campaign, the ELSA-d mission, which launched on March 22, 2021, consists of two spacecraft, with one acting as a 'servicer' and the other as a 'client' (50). They have launch masses of ~175 kg and ~17 kg respectively. The concept of operations is to perform rendezvous maneuvers by releasing the client from the servicer repeatedly to demonstrate the capability of finding and docking with existing debris. The technology demonstrations include search and inspection of the targets, as well as rendezvous of both tumbling and non-tumbling cases (50). In January 2022, the servicer spacecraft successfully released the client counterpart and initiated autonomous relative navigation over the course of multiple orbits as part of the mission plan (51). The ELSA-M spacecraft will leverage the lessons learned and technology demonstrated in this precursor mission to support a range of future satellite operators that may carry a compatible magnetic capture mechanism such as the Astroscale Docking Plate. The ELSA-M in-orbit demonstrator is planned to be launched by the end of 2024 (52). It is important to note that several science missions undertake extensive efforts to make their spacecraft magnetically neutral, which may be a concern for this method and its application in some cases.

Regarding their active debris removal campaign, Astroscale is also working with national space agencies to incorporate solutions to remove critical debris such as rocket upper stages or defunct satellites. This campaign started with a partnership with the Japanese Space Agency (JAXA) in February 2020. This collaboration will result in the implementation of the Commercial Removal of Debris Demonstration project (CRD2) which consists of the removal of a large space debris object performed in two mission phases. Astroscale will be involved in both phases. The first phase consists of a satellite that identifies and acquires data from a JAXA rocket upper stage. The Active

Debris Removal by Astroscale-Japan (ADRAS-J) satellite which will complete this first phase is scheduled to launch aboard a Rocket Lab Electron rocket in 2023 (53)(54). The ADRAS-J spacecraft has a wet mass of 150 kg and it can maneuver with its 12 green monopropellant thrusters. The spacecraft payload includes a custom rendezvous system that includes several sensors and cameras. In late September 2023, the company announced the spacecraft is completely prepared for its rendezvous mission and is ready for launch (54)(56).

In August 2022, Astroscale was also selected to participate in Phase II of the CRD2 project. The company will be responsible for the Front-Loading Technology Study which will focus on the ground test of hardware and software for close proximity operations and the capture mechanism design. This study is a requirement for satellite providers in the CRD2 Phase 2 mission (55).

Astroscale announced in May 2021, a $3.5 million funding award from OneWeb, the global communications network, to further develop their technology with the goal of commercial services starting in 2024. The next iteration consists of the ELSA-M satellite which will be capable of deorbiting multiple satellites per mission. OneWeb has also committed to including a docking plate on their satellites that would facilitate future deorbit missions (57). In September 2022, Astroscale secured funding from the UK Space Agency to keep developing the latest mission phase of the Cleaning Outer Space Mission through Innovative Capture (COSMIC). This mission will be an evolution of the Astroscale ELSA-M platform with a goal of removing two defunct British satellites by 2026 (53).

In July 2023, Astroscale announced a partnership with Astro Digital US Inc. to incorporate their Generation 2 Docking Plate into Astro Digital's modular satellite bus. The goal of this collaboration is to provide means for end-of-life servicing preparation. Having these devices on board will allow other servicing spacecraft such as ELSA-M to securely dock and achieve relocation or removal after mission completion (58).

13.4.4 ClearSpace

ClearSpace has plans include service contracts for active debris removal. One of their proposed missions, ClearSpace One, will find, target, and capture a non-cooperative, tumbling 100 kg Vega Secondary Payload Adapter (VESPA) upper stage. The chaser spacecraft will be launched into a 500 km orbit for commissioning and initial testing before raising its altitude to the VESPA's 660 km orbit, where it will attempt rendezvous and capture. ClearSpace One will use a group of robotic arms to grab the upper stage, and then both spacecraft will be deorbited together to a lower orbit for final disintegration in the atmosphere. The mission is planned to launch in 2025 to help establish a market for in-orbit servicing and debris removal (59).

ClearSpace developed a feasibility study to remove at least two UK defunct satellites and was successfully completed in March 2022. A new contract was awarded by the UK Space Agency to perform a second phase of the project, which will finish with the preliminary design review in 2023 of the Clearing of the LEO Environment with Active Removal (CLEAR) mission. This mission plans to remove two UK objects that have been in orbit for more than 10 years in an altitude of over 700 km, with a deorbit time longer than a hundred years (60).

In September 2023, the object which was intended to be the target of the ClearSpace mission, the Vespa payload adapter, was hit by several space debris pieces, too small to be tracked. Vespa was intended to be removed after the ClearSpace scheduled launch in 2026. ESA is analyzing the impacts on the ClearSpace mission, which is going to continue its development according to the initial plan as of September 2023 (61).

13.4.5 Momentus Space

Momentus operates space transportation systems that can propel or deorbit other spacecraft. Their Vigoride platform can carry satellites with masses up to 250 kg, has a wet mass of 215 kg, and can provide up to 1.6 km s-1 for 50 kg payload through a water plasma propulsion system (26). Although the main objective of this system is to provide enhanced propulsive capability to their customers, the platform is suitable for active deorbiting. Momentus launched its first Vigoride transfer vehicle (Vigoride-3) on May 25, 2022, successfully deployed three satellite payloads to their respective orbits as of September 2022 (63). As of 2023, Vigoride-5 and -6 have launched successfully. Their latest Vigoride-7 is slated for launch in 2024 (64).

13.4.6 D-Orbit

D-Orbit provides transportation services onboard their ION CubeSat carrier platform that can provide precision deployment and is able to host satellites from 1 to 12U. The first mission Origin released 12 SuperDove satellites for the Earth-observation company Planet, deploying the first in September 2020 with the last SuperDove deployed about a month later (65). The most recent Pulse mission finished deploying 20 satellites on May 11, 2021 (66). Future versions of this technology may consider other applications such as retrieving orbiting spacecraft to deorbit them. In June 2022, D-Orbit secured a contract with ESA to improve the performance and reduce the cost of its ION transfer vehicle. Over six flights, D-Orbit has already deployed over 80 satellites successfully into their orbits (67).

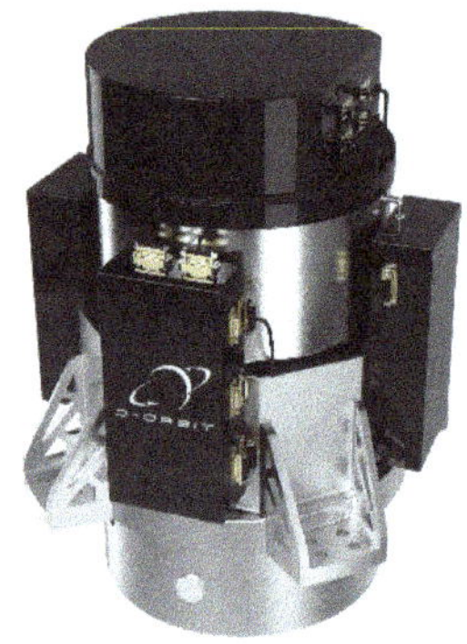

Figure 13.14: D-Orbit D3 module. Credit: D-orbit.

In addition, D-Orbit provides an external solid motor booster specifically for deorbiting purposes. This independent module, known as D-Orbit Decommissioning Device (D3) shown in figure 13.14, is a proprietary solution that is optimized for end-of-life maneuvers (44). However, it is important to note that, as compared to some other technologies in this active systems section, this technology would need to be added prior to launch.

13.4.7 Voyage Space (Altius Space Machines)

In 2019, the satellite constellation company OneWeb signed a partnership with Altius Space Machines (acquired by Voyager Space in 2019) to include a grappling fixture on all their future launched satellites in an effort to make space more sustainable. On January 14, 2021, it was

Bounding Volume	150mm x150mm x 65mm
Total Mass	250g
Mounting Interface	3x M5x0.8 threaded inserts on an 84.5mm bolt-hole circle
Compatible Gripping Methods	Magnetic Capture Adhesive Capture - Electrostatic - Gecko - Hot-Melt - Chemical Mechanical Capture - Pinch-Grasp - Snare Penetrating Capture (Harpoon)

Figure 13.15: DogTag prototype. Credit: Altius Space Machines.

announced that the first batch of DogTags were launched into space on OneWeb satellites (68). The Altius DogTag consists of a universal interface for small satellites that is inexpensive and lightweight. The fixture design enables various grappling techniques to enable servicing or decommissioning. It uses magnetic capabilities as its primary capture mechanism but is also compatible with other techniques to accommodate other potential customers and act as a standard interface (69). More specifically, it is compatible with magnetic attraction, adhesives, mechanical, and harpooning captures. Figure 13.15 includes an image of the flight DogTags and a table of its main features. In February 2022, an ArianeSpace Soyuz launch vehicle carried 34 OneWeb satellites into orbit with corresponding Altius DogTags to mitigate future space debris. In total, over 300 DogTags have already been launched to space (70).

13.5 Summary

Space debris regulations are becoming more stringent. Consequently, several deorbit technologies have matured significantly over the course of the last few years. Traditionally passive systems have been more common, have flown on various missions, and have increased to TRL 9 after successful technology demonstrations. Drag sails are the main technology for passive systems, and several companies have already commercialized and sold these products. Other systems such as electromagnetic tethers, deployable booms, or the NASA TechEdSat series Exo-Brake have also already been prototyped and demonstrated in space, now with navigation capabilities and increased reliability. The investment in active systems has also grown significantly. Several companies are offering transfer vehicles to remove debris or deorbit spacecraft at the end of their mission, and compatible systems for spacecraft rendezvous and removal are being developed in parallel as well. As an example, the RemoveDebris mission has successfully tested two different active methods: a net and a harpoon, for future implementation in active debris removal operations. Companies such as Astroscale or ClearSpace are developing missions to remove defunct satellites and are launching precursor technology demonstration spacecraft in the initial stages of their roadmaps. In conclusion, the various deorbit technologies have seen a significant TRL increase since the last iteration of this report and the robustness of the technologies is expected to grow even further as demand for deorbiting services increases with additional launches and new regulations.

For feedback solicitation, please email: arc-sst-soa@mail.nasa.gov. Please include a business email so someone may contact you further.

References

(1) European Space Agency. "Space Debris by the Numbers," December 6, 2023. [Online] Available at: https://www.esa.int/Space_Safety/Space_Debris/Space_debris_by_the_numbers
(2) Federal Communications Commission. "Mitigation of Orbital Debris: Second Report and Order," June 21, 2004. Available at: https://docs.fcc.gov/public/attachments/FCC-04-130A1.pdf
(3) NASA. "Process for Limiting Orbital Debris," NASA-STD-8719.14C. November 5, 2021.
(4) National Orbital Debris Implementation Plan, of the Orbital Debris Interagency Working Group, National Science and Security Council, July 2022.
(5) Liou, J., NASA. Orbital Debris Briefing, Executive Office of the President/Office of Science and Technology Policy (EOP/OSTP) Briefing, Washington D.C., December 8, 2017. Available at: https://ntrs.nasa.gov/api/citations/20170011662/downloads/20170011662.pdf
(6) NASA. "NASA Procedural Requirements for Limiting Orbital Debris and Evaluating the Meteoroid and Orbital Debris Environments," NPR 8715.6B. February 16, 2017-2024. [Online] Available at: https://nodis3.gsfc.nasa.gov/displayDir.cfm?t=NPR&c=8715&s=6B

BODY-429

(7) National Academies of Sciences, Engineering, and Medicine. 2011. Limiting Future Collision Risk to Spacecraft: An Assessment of NASA's Meteoroid and Orbital Debris Programs. Washington, DC: The National Academies Press. https://doi.org/10.17226/13244.

(8) Federal Communications Commission. "Space Innovation; Mitigation of Orbital Debris in the New Space Age," FCC-Fact Sheet. September 8, 2022. Available at: https://docs.fcc.gov/public/attachments/DOC-387024A1.pdf

(9) Khlystov, N. and Schrogl, K. "Space Industry Debris Mitigation Recommendations," Future of Space Network - Sustainable Space Initiative, Centre for the Fourth Industrial Revolution, World Economic Forum document, June 2023.

(10) World Economic Forum. "Space Sustainability Rating," [Online] Available at: https://www.weforum.org/projects/space-sustainability-rating/

(11) Foust, J. "U.N. opens "window of opportunity" to improve space governance" spacenews.com. June 14, 2023.]Online] Available at: https://spacenews.com/u-n-opens-window-of-opportunity-to-improve-space-governance/

(12) Federal Aviation Administration. "FAA Proposed Rule Would Reduce the Growth of Debris from Commercial Space Vehicles," September 20, 2023. [Online] Available at: https://www.faa.gov/newsroom/faa-proposed-rule-would-reduce-growth-debris-commercial-space-vehicles

(13) National Oceanic and Atmospheric Administration. "Satellite Drag," [Online] Available at: https://www.swpc.noaa.gov/impacts/satellite-drag

(14) MMA Design. "dragNET™ De-Orbit System," MMA Design: Products. 2020 [Online] Available at: https://mmadesignllc.com/product/dragnet-de-orbit-system/

(15) MMA Design. "Another MMA dragNET De-orbit Systems Launches with GAzelle Satellite," [Online] Available at: https://mmadesignllc.com/2022/10/dragnet-deorbit-launches-with-gazelle-satellite/

(16) Palla, C., Kingston, H, and Hobbs, S. Damstadt: "Development of Commercial Drag-Augmentation for Small Satellites." ESA Space debris Office, Ed. T. Flohrer & F. Schimtz, 2017.

(17) Stankey, H. C. and Hoyt, R. P. "In-Flight Performance of the Terminator Tape End-of-Life Deorbit Module," 35th Annual Small Satellite Conference, August 2021, Logan Utah.

(18) Underwood, C. et al. "InflateSail de-orbit flight demonstration results and follow-on drag-sail applications," Acta Astronautica, 2019, Vol. 162, pages 344-358. ISSN: 0094-5765

(19) Murbach, M. et al. "TechedSat 5/PhoneSat 5 (T5/P5)," 30th Annual Small Satellite Conference, August 2016, Logan Utah.

(20) NASA. "What are NASA's Technology Educational Satellites?" September 20, 2022. [Online] Available at: https://www.nasa.gov/ames/techedsat

(21) Aglietti, G. et al. "The active space debris removal mission RemoveDebris. Part 2: in orbit operations," Acta Astronautica, Elsevier, 2020, Vol. 168, pp. 310-322.

(22) Cotten, B., Bennett, I. and Zee, R. E. "On-Orbit Results from the CanX-7 Drag Sail Deorbit Mission," 10th Annual Small Satellite Conference, August 2006, Logan Utah.

(23) Sinn, T., Garcia Hemme, H., Gehly, S., and Le May, S. "The NABEO-1 CubeSat dragsail launched onboard Rocket Lab's "It's Business Time: Design, Testing, Launch and De-Orbit Mission," 33rd Annual Small Satellite Conference Presentation, August 2019, Logan Utah.

(24) Laube, P. et al. "ADEO-N – Deployable Passive De-Orbit Sail Subsystem Enabling Space Debris Mitigation for CubeSats, SmallSats and Constellations," DLR German Aerospace Center, D-Orbit S.p.A, European Space Agency.

(25) Stelzl, D., Seefeldt, P., Killian. M. et al. "The ADEO Space Sail Products," International Symposium on Space Sailing, New York, NY, June 2023.

(26) High Performance Space Structure Systems GmbH. "Deobit Module ADEO-N 1.1," datasheet. June 2023. Available at: https://www.hps-gmbh.com/wp-content/uploads/2023/06/ADEO-N1.1_Data-Sheet_1.pdf

(27) High Performance Space Structure Systems GmbH. "Deobit Module ADEO-M," datasheet. June 2023. Available at: https://www.hps-gmbh.com/wp-content/uploads/2023/07/ADEO-M_Data-Sheet_1.pdf

(28) NPC Spacemind. "Artica," [Online] Available at: https://indico.esa.int/event/234/contributions/3929/attachments/3014/3634/Clean_Space_ARTICA_Present.pdf

(29) Cowing, K. "Spacemind Successfully Launches Three Italian Nanosatellites Into Orbit From The International Space Station," cPress Release, Spacemind. February 2, 2023. [Online] Available at: https://spaceref.com/space-commerce/spacemind-successfully-launches-three-italian-nanosatellites-into-orbit-from-the-international-space-station/

(30) Davis, J. "LightSail 2 completes mission with atmospheric reentry," The Planetary Society. November 17, 2022. [Online] Available at: https://www.planetary.org/articles/lightsail-2-completes-mission

(31) Wilkie, K. "The NASA Advanced Composite Solar Sail System (ACS3) Flight Demonstration: A Technology Pathfinder for Practical Smallsat Solar Sailing". 35th Annual Small Satellite Conference, August 2021, Logan, Utah.

(32) "Gama's revolutionary propulsion Gama Alpha Solar Sail mission launches," satenews.com. January 4, 2023. [Online] Available at: https://news.satnews.com/2023/01/04/gamas-revolutionary-propulsion-gama-alpha-solar-sail-mission-launches

(33) Murbach, M., et al. "The Exo-Brake as an Inexpensive Means of Achieving Sample Return from Low Earth Orbit – Recent Flight Tests," Washington, D.C.: 70th International Astronautical Congress (IAC), 2019.

(34) NASA. "TechEDSat-7's Flight Mission Begins," Press Release. January 27, 2021. [Online] Available at: https://www.nasa.gov/image-feature/ames/techedsat-7-s-flight-mission-begins

(35) Purdue University: Missions: "ADE (Aerodynamic Deorbit Experiment)." 2017. [Online] Available at: https://engineering.purdue.edu/CubeSat/missions/ade

(36) Vestigo Aerospace. "Vestigo Aerospace Awarded NASA Phase II SBIR." 2021. [Online] https://www.altius-space.com/blog/voyager-subsidiary-altius-space-machines-inc-announces-successful-first-orbital-launch-of-dogtags-aboard-onewebs-satellites/

(37) Howell, E. "'Drag sail' to deorbit satellites receives $750K in seed funding," space.com. September 23, 2022. [Online] Available at: https://www.space.com/satellite-deorbiting-drag-sail-spinnaker-funding

(38) Martin, L. "How drag sails could help solve our space junk problem," astronomy.com. [Online] Available at: https://astronomy.com/news/2022/04/how-drag-sails-could-help-our-space-junk-problem

(39) Vestigo Aerospace. "Vestigo Aerospace Awarded Phase II-Sequential SBIR," [Online] Available at: https://vestigoaerospace.com/news/SBIR-Phase-II-S/

(40) Jones, A. "China deploys deorbiting 'drag sail' to aid fight against space junk," space.com. [Online] Available at: https://www.space.com/china-deploys-drag-sail-space-junk

(41) Axelspace. "Axelspace's D-SAIL selected for JAXA's Innovative Satellite Technology Demonstration-4," Press News. February 28, 2023. [Online] Available at: https://www.axelspace.com/news/20230227/

(42) "Satellite built by Brown students and launched by SpaceX shows a low-cost way to reduce space junk," News from Brown. March 15, 2023. [Online] Available at: https://www.brown.edu/news/2023-03-15/sbudnic

(43) Omar, S. et al. "CubeSat Mission to Demonstrate Aerodynamically Controlled Re-Entry using the Drag De-Orbit Device (D3)," 32nd Annual Small Satellite Conference, August 2018, Logan Utah.

(44) Martin, T., Omar, S., and Bevilacqua, R. "Controlled Spacecraft Re-Entry of a Drag De-Orbit Device (D3)," Journal of Undergraduate Research, 2019, Vol. 21.

(45) Satellite Evolution. “De-orbit Drag Device (D3) successfully placed in orbit by NanoRacks,” September 14, 2022. [Online] Available at: https://www.satelliteevolution.com/post/de-orbit-drag-device-d3-successfully-placed-in-orbit-by-nanoracks
(46) Turse, D. et al. “Flight Testing of a Low Cost De-Orbiting Device for Small Satellites,” Presented at 42nd Aerospace Mechanisms Symposium, NASA Goddard Space Flight Center, May 14-16, 2014.
(47) Henry, C. “Tethers Unlimited says early results of deorbit hardware test promising,” Space News. January 23, 2020. [Online] Available at: https://spacenews.com/tethers-unlimited-says-early-results-of-deorbit-hardware-test-promising/
(48) Henry, C. “Iridium would pay to deorbit its 30 defunct satellites — for the right price,” Spacenews. December 30, 2019. [Online] Available at: https://spacenews.com/iridium-would-pay-to-deorbit-its-30-defunct-satellites-for-the-right-price/
(49) Tavares, F. “TechEDsat-10 Deploys from the Space Station,” NASA.gov. August 5, 2020. [Online] Available at: https://www.nasa.gov/image-feature/ames/techedsat-10-deploys
(50) Astrolscale. “Astroscale Celebrates Successful Launch of ELSA-d,” Press Release. March 23, 2021, Spacenews. [Online] Available at: https://astroscale.com/astroscale-celebrates-successful-launch-of-elsa-d/
(51) Forshaw et al. “Operational Progress Update on the ELSA-D Debris Removal Mission,” 73rd Astronautical Congress, 2022.
(52) Astroscale, “ELSA-M,” [Online] Available at: https://astroscale.com/elsa-m/
(53) Astroscale. “Astroscale Forges Ahead with UK Active Debris Removal Mission with Support from UK Space Agency,” Press Release. September 26, 2022. [Online] Available at: https://astroscale.com/astroscale-forges-ahead-with-uk-active-debris-removal-mission-with-support-from-uk-space-agency/
(54) Foust, J. “Astroscale inspector satellite ready for launch,” Spacenews.com. September 28, 2023. [Online] Available at: https://spacenews.com/astroscale-inspector-satellite-ready-for-launch/
(55) Astroscale. “Astroscale Selected as Contract Partner for Front-Loading Technology Study in Phase II of JAXA’s Commercial Removal of Debris Demonstration Project,” Press Release, August 22, 2022. [Online] Available at: https://astroscale.com/astroscale-selected-as-contract-partner-for-front-loading-technology-study-in-phase-ii-of-jaxas-commercial-removal-of-debris-demonstration-project/
(56) Jewett, R. “Astroscale Hopes ADRAS-J Mission Will Lay the Groundwork for Commercial Debris Removal,” ViaSatellite.com. September 28, 2023. [Online] Available at: https://www.satellitetoday.com/sustainability/2023/09/28/astroscale-hopes-adras-j-mission-will-lay-the-groundwork-for-commercial-debris-removal/
(57) Astroscale. "Astroscale UK Signs £2.5 Million Agreement to Develop Space Debris Removal Technology Innovations with OneWeb," Press Release. May 24, 2021 [Online] Available at: https://astroscale.com/astroscale-uk-signs-2-5-million-agreement-to-develop-space-debris-removal-technology-innovations-with-oneweb/
(58) Astrosclae. “Astro Digital Embraces Responsible Satellite Operations with Astroscale’s Simple and Innovative Docking Plate Integration,” Press Release. July 31, 2023. [Online] Available at: https://astroscale.com/astro-digital-embraces-responsible-satellite-operations-with-astroscales-simple-and-innovative-docking-plate-integration/
(59) Mathewson, S. “ESA partners with startup to launch first debris removal mission in 2025,” space.com. May 16, 2021. [Online] Available at: https://www.space.com/esa-startup-clearspace-debris-removal-2025
(60) ClearSpace. “ClearSpace secures a major UK contract to help clean up space,” Press Release. September 26, 2022. [Online] Available at: https://clearspace.today/clearspace-secures-a-major-uk-contract-to-help-clean-up-space/

(61) Foust, J. "Target of European debris removal mission hit by other debris," Spacenes.com. August 22, 2023. [Online] Available at: https://spacenews.com/target-of-european-debris-removal-mission-hit-by-other-debris/

(62) Momentus. Vigoride User Guide – Version1.0 February 23, 2020.

(63) Momentus. "Momentus First Demonstration Mission Status Update #5," Press Release. September 8, 2022. [Online] Available at: https://investors.momentus.space/news-releases/news-release-details/momentus-first-demonstration-mission-status-update-5

(64) Momentus. "Momentus Deploys All Payloads from Vigoride-6 Mission," Press Release. July 28, 2023. [Online] Available at: https://investors.momentus.space/news-releases/news-release-details/momentus-deploys-all-payloads-vigoride-6-mission

(65) "D-Orbit announces successful ORIGIN mission," Press Release. SpaceRef.com October 29, 2020. [Online] Available at: https://spaceref.com/press-release/d-orbit-announces-successful-origin-mission/

(66) D-Orbit. "Mission: Pulse," [Online] Available at: https://www.dorbit.space/pulse-january-2021

(67) Werner, D. "D-Orbit charts ambitious course for space logistics business," Spacenews, June 15, 2022. [Online] Available at: https://spacenews.com/d-orbit-ambitions/

(68) Altius Space Machines. "Voyager Subsidiary, Altius Space Machines, Inc. Announces Successful First Orbital Launch of DogTags Aboard OneWeb's Satellites." January 14, 2021. [Online] Available at: https://www.altius-space.com/blog/voyager-subsidiary-altius-space-machines-inc-announces-successful-first-orbital-launch-of-dogtags-aboard-onewebs-satellites/

(69) Maclay, T., Goff, J., Sheehan, J.P., and Han, E. "The development of commercially viable ADR services: Introduction of a small-satellite grappling interface," 2020, Journal of Space Safety Engineering, Volume 7, Issue 3, pages 364 – 368.

(70) Altius Space Machines. Voyager and Altius Space Machines Celebrate First Successful Launch of 2022," Press Release. [Online] Available at: https://voyagerspace.com/insight/voyager-and-altius-space-machines-celebrate-first-successful-launch-of-2022/

Summary

This report provides an overview and assessment of state-of-the-art small spacecraft technologies publicly available as of September 2023. Technology maturation and miniaturization continues to expand small spacecraft capabilities, giving rise to more complex SmallSat mission designs. These improved capabilities have broadened the common SmallSat platform with larger CubeSats and smaller SmallSats; the traditional CubeSat platforms of 1U and 3U volume now include up to 16U form factors, and SmallSats once designed as <400 kg are now <100 kg with similar capability for less cost. The larger surface area of more capable SmallSat platforms can be more equipped with solar panels and subsystem arrangement options. The SmallSat industry is thinking outside the box to maximize usage of the full spacecraft volume and design increasingly complex future SmallSat missions.

While still fairly dominated by the traditional CubeSat form factor, this report is starting to reflect increased interest in the more capable SmallSat platforms. The surge in SmallSat launch opportunities and increased availability of services such as rideshares, hosted payloads, dedicated launchers, and orbital transportation is modernizing the SmallSat paradigm. Hosted payload services are increasingly available for larger SmallSats and other commercial satellites. Several SmallSat missions are actively working on rideshares (or dedicated rides) to destinations in years 2024-2026,and there is an increased interest in orbital maneuvering vehicles (OMV) that provide some autonomy from predetermined rideshares. Dedicated launches provide rapid integration and greater mission design flexibility, allowing spacecraft designers to better dictate mission parameters. A wide variety of integration and deployment systems are now available for constellations of small spacecraft, with SmallSat constellations recently launched by Starlink and OneWeb.

The pace of SmallSat technology advancement overall is rapidly accelerating and varies per subsystem. There has been significant subsystem growth in enhanced ground station support, improved technical efficiency, emerging sensor technology, and in rideshare opportunities. Recent flight missions have demonstrated innovative SmallSat technologies; the successful flights of Starling, CAPSTONE, PTD-3 and CLICK A spacecraft have each significantly contributed to SmallSat technology development. Starling successfully demonstrated intersatellite communication; CAPSTONE completed its six-month primary mission of testing the stability of the near-rectilinear halo orbit for Lunar Gateway; PTD-3 achieved a downlink of 200 gigabits per second via optical communication; and CLICK A tested the optical communication hardware that will be implemented on the second CLICK B/C mission, slated to launch later in 2024. DiskSat, expected to launch in 2024, with its revolutionary circular configuration and larger surface area will challenge the way SmallSat's are perceived. LiDAR sensor technology development is ongoing with applications for improved altimetry and relative navigation for rendezvous, docking, and formation flying. There has been particular consideration to deployment mechanisms for small spacecraft subsystems such as antennas booms, gravity gradients, stabilization, sensors, sails, and solar panels, and these technologies are gaining space heritage through operations. ACS3 is an ongoing NASA mission slated for launch in 2024 that will use a new composite boom solar sail in low-Earth orbit (LEO) for propellant-less propulsion. There is a spike in position, navigation, and timing technology progression in inertial sensors and atomic clocks, and magnetic navigation for near-Earth environments.

NASA's new Indefinite Delivery/Indefinite Quantity (IDIQ) mechanism–the Venture Class Acquisition of Dedicated and Rideshare (VADR) launch services–was developed to accommodate very low complexity CubeSats (up to more complex Class D missions) and provide FAA licensed launch services to deliver payloads to a variety of orbits. The 2023 NASA solicitation for Suborbital/Hosted Orbital Flight and Payload Integration Services included opportunities for hosted payloads on commercial orbital platforms (1). IDIQ contracts for these services will replace existing Flight Opportunities IDIQ contracts when those expire, and are expected to be in place with commercial providers in early 2024. While the deadlines for the latest opportunities recently passed in Q4 2023, readers are strongly encouraged to subscribe to the Flight Opportunities newsletter in reference 1.

There are ongoing policy measures being developed to mitigate and remove space debris. In 2022, the FCC adopted a new "5 year" rule to reduce the lifetime requirement for all FCC-licensed satellites in LEO to 5 years after launch. These new regulations have incorporated spacecraft decay capabilities into mission design. As of 2023, there are discussions at the agency and federal level to determine the final policies. To comply with new orbital lifetime requirements, satellite operators are employing strategies such as decreasing the spacecraft ballistic coefficient or mass to area ratio. Deorbit technologies such as drag devices that can effectively increase the spacecraft's drag area may become even more important for future spacecraft operations in LEO.

NASA is working with several American companies to deliver science and technology to the lunar surface through the Commercial Lunar Payload Services (CLPS) initiative. Under the Artemis program, these commercial deliveries present SmallSat designers with opportunities to perform science experiments, test technologies and demonstrate capabilities to help NASA explore the Moon and prepare for human missions. NASA has initially selected 14 companies to deliver payloads for NASA, including payload integration and operations and launch services to the surface of the Moon. The NASA CLPS program will begin delivering science payloads to the Moon in 2024. CLPS contracts are indefinite delivery, indefinite quantity contracts with a cumulative maximum contract value of $2.6 billion through 2028. Companies of varying sizes can work with selected vendors and are encouraged to fly commercial payloads in addition to the NASA payloads (2).

This report will be updated annually as emerging technologies mature and become state of the art. Any current technologies that were inadvertently overlooked in this version may be included in subsequent editions. Updates to technologies listed in this report could be also modified in subsequent revisions. This report is also available online at: https://www.nasa.gov/smallsat-institute/sst-soa. Technology inputs, updates, or corrections can be made by reaching out to the editor of this report at arc-sst-soa@mail.nasa.gov.

References

(1) NASA Flight Opportunities [Online] Jan 26, 2022. Accessed November 2022. Available at: https://www.nasa.gov/wp-content/uploads/2023/09/flight-opportunities-newsletter-august-23.pdf

(2) NASA. "Commercial Lunar Payload Services." [Online] July 21, 2022. Accessed November 2022. Available at: https://www.nasa.gov/clps

NASA Procedural Requirements

NPR 7123.1D
Effective Date: July 05, 2023
Expiration Date: July 05, 2028

COMPLIANCE IS MANDATORY FOR NASA EMPLOYEES

Printable Format (PDF)

Subject: NASA Systems Engineering Processes and Requirements w/Change 1

Responsible Office: Office of the Chief Engineer

Appendix E. Technology Readiness Levels

<table>
<tr><th>TRL</th><th>Definition</th><th>Hardware Description</th><th>Software Description</th><th>Success criteria</th></tr>
<tr><td rowspan="2">1</td><td>Basic principles observed and reported.</td><td>Scientific knowledge generated underpinning hardware technology concepts/applications.</td><td>Scientific knowledge generated underpinning basic properties of software architecture and mathematical formulation.</td><td>Peer reviewed documentation of research underlying the proposed concept/application.</td></tr>
<tr><td colspan="4">Examples:
a. Initial Paper published providing representative examples of phenomenon as well as supporting equations for a concept.
b. Conference presentations on concepts and basic observations presented within the scientific community.</td></tr>
<tr><td>2</td><td>Technology concept and/or application formulated.</td><td>Invention begins, practical application is identified but is speculative, no experimental proof or detailed analysis is available to support the conjecture.</td><td>Practical application is identified but is speculative; no experimental proof or detailed analysis is available to support the conjecture. Basic properties of algorithms, representations, and concepts defined. Basic principles coded.</td><td>Documented description of the application/concept that addresses feasibility and benefit.</td></tr>
</table>

This document does not bind the public, except as authorized by law or as incorporated into a contract. This document is uncontrolled when printed. Check the NASA Online Directives Information System (NODIS) Library to verify that this is the correct version before use: https://nodis3.gsfc.nasa.gov.

			Experiments performed with synthetic data.	
	Example: Carbon nanotube composites were created for lightweight, high-strength structural materials for space structures.			

TRL	Definition	Hardware Description	Software Description	Success criteria
3	Analytical and experimental proof-of-concept of critical function and/or characteristics.	Research and development are initiated, including analytical and laboratory studies to validate predictions regarding the technology.	Development of limited functionality to validate critical properties and predictions using non-integrated software components.	Documented analytical/experimental results validating predictions of key parameters.

Examples:

a. High efficiency Gallium Arsenide solar panels for space application is conceived for use over a wide temperature range. The concept critically relies on improved welding technology for the cell assembly. Samples of solar cell assemblies are manufactured and submitted to a preliminary thermal environment test at ambient pressure for demonstrating the concept viability.

b. A fiber optic laser gyroscope is envisioned using optical fibers for the light propagation and Sagnac Effect. The overall concept is modeled including the laser source, the optical fiber loop, and the phase shift measurement. The laser injection in the optical fiber and the detection principles are supported by dedicated experiments.

c. In Situ Resource Utilization: Demonstrated the application of a cryofreezer for CO2 acquisition and microwave processor for water extraction from soils.

TRL	Definition	Hardware Description	Software Description	Success criteria
	Component and/or breadboard validation in a laboratory environment.	A low fidelity system/component breadboard is built and operated to demonstrate basic functionality in a laboratory environment.	Key, functionality critical software components are integrated and functionally validated to establish interoperability and begin architecture development. Relevant environments defined and performance in the environment predicted.	Documented test performance demonstrating agreement with analytical predictions. Documented definition of potentially relevant environment.

This document does not bind the public, except as authorized by law or as incorporated into a contract. This document is uncontrolled when printed. Check the NASA Online Directives Information System (NODIS) Library to verify that this is the correct version before use: https://nodis3.gsfc.nasa.gov.

4			predicted.	

Examples:

a. Fiber optic laser gyroscope: A breadboard model is built including the proposed laser diode, optical fiber and detection system. The angular velocity measurement performance is demonstrated in the laboratory for one axis rotation.

b. Bi-liquid chemical propulsion engine: A breadboard of the engine is built and thrust performance is demonstrated at ambient pressure. Calculations are done to estimate the theoretical performance in the expected environment (e.g., pressure, temperature).

c. A new fuzzy logic approach to avionics is validated in a lab environment by testing the algorithms in a partially computer-based, partially bench-top component (with fiber optic gyros) demonstration in a controls lab using simulated vehicle inputs.

d. Variable Specific Impulse Magnetosphere Rocket (VASIMR): 100 kW magnetoplasma engine operated 10 hours cumulative (up to 3 minutes continuous) in a laboratory vacuum chamber.

TRL	Definition	Hardware Description	Software Description	Success criteria
5	Component and/or brassboard validated in a relevant environment.	A medium-fidelity component and/or brassboard, with realistic support elements, is built and operated for validation in a relevant environment so as to demonstrate overall performance in critical areas.	End-to-end software elements implemented and interfaced with existing systems/simulations conforming to target environment. End-to-end software system tested in relevant environment, meeting predicted performance. Operational environment performance predicted. Implementations.	Documented test performance demonstrating agreement with analytical predictions. Documented definition of scaling requirements. Performance predictions are made for subsequent development phases.

Examples:

a. A 6.0-meter deployable space telescope comprised of multiple petals is proposed for near infrared astronomy operating at 30K. Optical performance of individual petals in a cold environment is a critical function and is driven by material selection. A series of 1m mirrors (corresponding to a single petal) were fabricated from different materials and tested at 30K to evaluate performance and to select the final material for the telescope. Performance was extrapolated to the full-sized mirror.

b. For a launch vehicle, TRL 5 is the level demonstrating the availability of the technology at subscale level (e.g., the fuel management is a critical function for a re-ignitable upper stage). The demonstration of the management of the propellant is achieved on the ground at a subscale level.

This document does not bind the public, except as authorized by law or as incorporated into a contract. This document is uncontrolled when printed. Check the NASA Online Directives Information System (NODIS) Library to verify that this is the correct version before use: https://nodis3.gsfc.nasa.gov.

	c. ISS Additive Manufacturing Facility: Characterization tests compare parts and material properties of polymer specimens printed on ISS to copies printed on the ground.

TRL	Definition	Hardware Description	Software Description	Success criteria
6	System/sub-system model or prototype demonstration in a relevant environment.	A high-fidelity prototype of the system/subsystems that adequately addresses all critical scaling issues is built and tested in a relevant environment to demonstrate performance under critical environmental conditions.	Prototype implementations of the software demonstrated on full-scale, realistic problems. Partially integrated with existing hardware/software systems. Limited documentation available. Engineering feasibility fully demonstrated.	Documented test performance demonstrating agreement with analytical predictions.
	Examples: a. A remote sensing camera includes a large 3-meter telescope, a detection assembly, a cooling cabin for the detector cooling, and an electronics control unit. All elements have been demonstrated at TRL 6 except for the mirror assembly and its optical performance in orbit, which is driven by the distance between the primary and secondary mirrors needing to be stable within a fraction of a micrometer. The corresponding critical part includes the two mirrors and their supporting structure. A full-scale prototype consisting of the two mirrors and the supporting structure is built and tested in the relevant environment (e.g., including thermo-elastic distortions and launch vibrations) for demonstrating the required stability can effectively be met with the proposed design. b. Vacuum Pressure Integrated Suit Test (VPIST): Demonstrated the integrated performance of the Orion suit loop when integrated with human-suited test subjects in a vacuum chamber.			

TRL	Definition	Hardware Description	Software Description	Success criteria
	System prototype demonstration in an operational environment.	A high-fidelity prototype or engineering unit that adequately addresses all critical scaling issues is built and functions in the actual operational environment and platform (ground, airborne, or space).	Prototype software exists having all key functionality available for demonstration and test. Well integrated with operational hardware/software systems demonstrating operational feasibility. Most	Documented test performance demonstrating agreement with analytical predictions.

This document does not bind the public, except as authorized by law or as incorporated into a contract. This document is uncontrolled when printed. Check the NASA Online Directives Information System (NODIS) Library to verify that this is the correct version before use: https://nodis3.gsfc.nasa.gov.

TRL	Definition	Hardware Description	Software Description	Success criteria
7			software bugs removed. Limited documentation available.	
	Examples: a. Mars Pathfinder Rover flight and operation on Mars as a technology demonstration for future micro-rovers based on that system design. b. First flight test of a new launch vehicle, which is a performance demonstration in the operational environment. Design changes could follow as a result of the flight test. c. In-space demonstration missions for technology (e.g., autonomous robotics and deep space atomic clock). Successful flight demonstration could result in use of the technology in a future operational mission d. Robotic External Leak Locator (RELL): Originally flown as a technology demonstrator, the test article was subsequently put to use to help operators locate the likely spot where ammonia was leaking from the International Space Station (ISS) External Active Thermal Control System Loop B.			

TRL	Definition	Hardware Description	Software Description	Success criteria
8	Actual system completed and "flight qualified" through test and demonstration.	The final product in its final configuration is successfully demonstrated through test and analysis for its intended operational environment and platform (ground, airborne, or space). If necessary*, life testing has been completed.	All software has been thoroughly debugged and fully integrated with all operational hardware and software systems. All user documentation, training documentation, and maintenance documentation completed. All functionality successfully demonstrated in simulated operational scenarios. Verification and Validation completed.	Documented test performance verifying analytical predictions.
	Note: *"If necessary" refers to the need to life test either for worn out mechanisms, for temperature stability over time, and for performance over time in extreme environments. An evaluation on a case-by-case basis should be made to determine the system/systems that warrant life testing and the tests begun early in the technology development process to enable completion by TRL 8. It is preferable to have the technology life test initiated and			

completed at the earliest possible stage in development. Some components may require life testing on or after TRL 5.

Examples:

a. The level is reached when the final product is qualified for the operational environment through test and analysis. Examples are when Cassini and Galileo were qualified, but not yet flown.

b. Interim Cryo Propulsion Stage (ICPS): A Delta Cryogenic Second Stage modified to meet Space Launch System requirements for Exploration Mission-1 (EM-1). Qualified and accepted by NASA for flight on EM-1.

TRL	Definition	Hardware Description	Software Description	Success criteria
9	Actual system flight proven through successful mission operations.	The final product is successfully operated in an actual mission.	All software has been thoroughly debugged and fully integrated with all operational hardware and software systems. All documentation has been completed. Sustaining software support is in place. System has been successfully operated in the operational environment.	Documented mission operational results.

Examples:

a. Flown spacecraft (e.g., Cassini, Hubble Space telescope).

b. Technologies flown in an operational environment.

c. Nanoracks CubeSat Deployer: Commercially developed and operated small satellite deployer on-board the ISS.

Note: In cases of conflict between NASA directives concerning TRL definitions, NPR 7123.1 will take precedence.

DISTRIBUTION:
NODIS

This document does not bind the public, except as authorized by law or as incorporated into a contract. This document is uncontrolled when printed. Check the NASA Online Directives Information System (NODIS) Library to verify that this is the correct version before use: https://nodis3.gsfc.nasa.gov.

This document does not bind the public, except as authorized by law or as incorporated into a contract. This document is uncontrolled when printed. Check the NASA Online Directives Information System (NODIS) Library to verify that this is the correct version before use: https://nodis3.gsfc.nasa.gov.